2

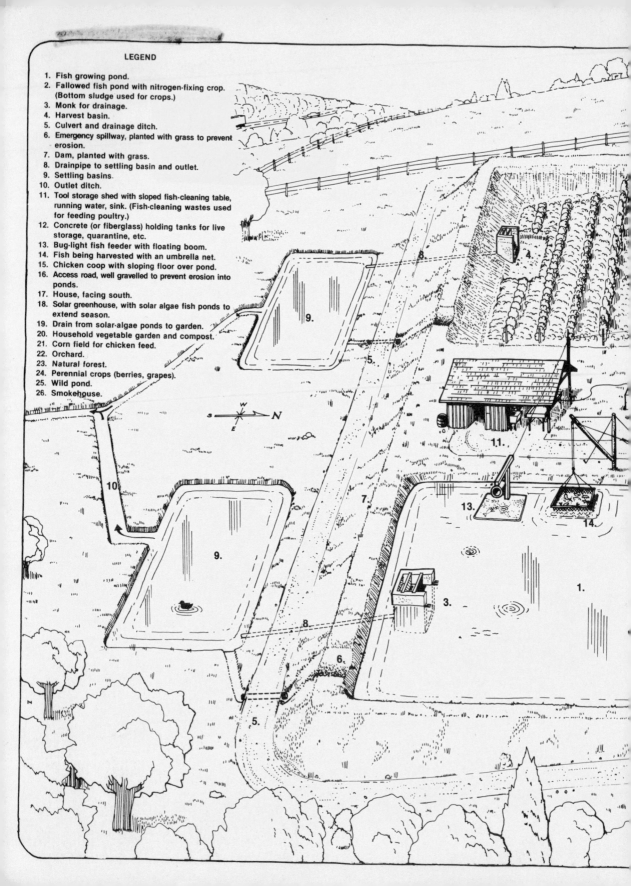

LEGEND

1. Fish growing pond.
2. Fallowed fish pond with nitrogen-fixing crop. (Bottom sludge used for crops.)
3. Monk for drainage.
4. Harvest basin.
5. Culvert and drainage ditch.
6. Emergency spillway, planted with grass to prevent erosion.
7. Dam, planted with grass.
8. Drainpipe to settling basin and outlet.
9. Settling basins.
10. Outlet ditch.
11. Tool storage shed with sloped fish-cleaning table, running water, sink. (Fish-cleaning wastes used for feeding poultry.)
12. Concrete (or fiberglass) holding tanks for live storage, quarantine, etc.
13. Bug-light fish feeder with floating boom.
14. Fish being harvested with an umbrella net.
15. Chicken coop with sloping floor over pond.
16. Access road, well gravelled to prevent erosion into ponds.
17. House, facing south.
18. Solar greenhouse, with solar algae fish ponds to extend season.
19. Drain from solar-algae ponds to garden.
20. Household vegetable garden and compost.
21. Corn field for chicken feed.
22. Orchard.
23. Natural forest.
24. Perennial crops (berries, grapes).
25. Wild pond.
26. Smokehouse.

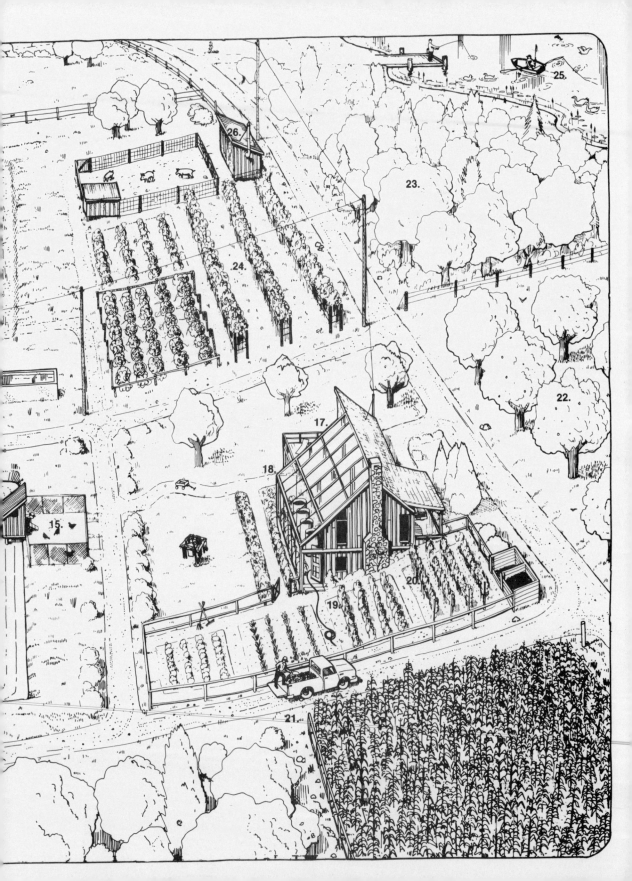

THE FRESHWATER
AQUACULTURE BOOK

THE FRESHWATER AQUACULTURE BOOK

A handbook for small scale fish culture in North America

By William McLarney

A Cloudburst Press Book

Hartley & Marks, Publishers

A Cloudburst Press Book

Published in the U.S.A. by
Hartley & Marks, Inc.
P.O. Box 147
Point Roberts, Washington 98281

Published in Canada by
Hartley & Marks, Ltd.
3663 West Broadway
Vancouver, B.C. V6R 2B8

Library of Congress Cataloging in Publication Data

McLarney, William O.
   The freshwater aquaculture book.

   "A Cloudburst Press book."
   Bibliography: p.
   Includes index.
   1. Fish-culture—North America—Handbooks, manuals, etc.
I. Title.
SH159.M438   1984   639.3'11   84-80961
ISBN 0-88179-018-4

Typeset by The Typeworks
Manufactured in Singapore

Contents

APPENDICES:

Permissions

I would like to acknowledge the following individuals, organizations, and publishers:

Academic Press, N.Y., for permission to reprint material from *Fish Nutrition*, edited by J. E. Halver, copyright © 1972 by Academic Press.

American Institute of Biological Sciences, Arlington, Va., for permission to reprint material from "The Amphibian Facility of the Unviersity of Michigan," by G. W. Nace, in *BioScience*, Vol. 30, No. 11, p. 773, copyright © 1980 by American Institute of Biological Sciences.

The Food and Agriculture Organization (FAO) of the United Nations, Rome, Italy, for permission to reprint material from "Economic Aspects of Fish Feeding in the Near East," by S. Tal and B. Hepher, in *Proceedings of the FAO World Symposium on Warmwater Pond Fish Culture*, FAO Fisheries Report 44(4), edited by T. V. R. Pillay, copyright © 1967 by the Food and Agriculture Organization of the United Nations.

Ohio State University Press, Columbus, Ohio, for permission to reprint the following illustrations from *The Fishes of Ohio*, by Milton B. Trautman, revised edition, copyright © 1957, 1981 by the Ohio State University Press. Page numbers indicate the locations of the illustrations in *The Freshwater Aquaculture Book:* Northern Largemouth Blackbass (p. 486), Northern Bluegill Sunfish (p. 488), Yellow Bullhead (p. 493), Channel Catfish (p. 495), Bigmouth Buffalofish (p. 497), Common Carp (p. 497), Rainbow Trout (p. 498), Yellow Perch (p. 501).

T.F.H. Publications, Inc., Neptune, N.J., for permission to reprint material from *Encyclopedia of Live Foods*, by Charles O. Masters, copyright © 1975 by T.F.H. Publications, Inc.

The Wildlife Society, Bethesda, Md., for permission to reprint material from "Symposium on Farm Fish Ponds and Management," by H. S. Swingle, in *The Journal of Wildlife Management* 16(3), copyright © July, 1952 by The Wildlife Society.

John Wiley & Sons, Inc., N.Y., for permission to reprint material from *Aquaculture: The Farming and Husbandry of Freshwater and Marine Organisms*, by J. E. Bardach, J. H. Ryther, and W. O. McLarney, copyright © 1972 by Wiley Interscience.

Preface

Aquaculture, as the husbandry of water organisms is called, dates back at least 2,500 years in China and 800 years in Europe, but in North America it is a new and poorly understood discipline. Its counterpart, etymologically and functionally, is agriculture (if in agriculture we include both animal and plant husbandry). Over the millenia, agriculture has become as diverse as the organisms and human cultures involved; if one were to attempt to write a book entitled "Agriculture," one would be considered grossly presumptuous. Aquaculture is potentially as diverse as agriculture, yet in 1972 the author of this book was party to the publication of a book called, simply, *Aquaculture*.

Aquaculture attempted to be a world survey of existing aquaculture practice. When in 1975 the opportunity arose to revise and update *Aquaculture*, John Bardach, John Ryther and I declined. The field had already expanded beyond the scope of any one volume. We agreed that the time had come for *Aquaculture* to be supplemented by various in-depth books dealing with particular species groups, geographic regions, and technical approaches to aquaculture. This is one of those books.

The Fresh Water Aquaculture Book differs from its predecessor in several important ways. Three are obvious from the title; the organisms and methods discussed are confined to those available and useful to residents of the United States and Canada. Within that geographic limitation, the book is addressed to the "small-scale" producer. By "small-scale" the author intends to connote everything from the smallest system which can provide a significant amount of food for an individual, up through farm systems which can be managed by a family or small group without recourse to excessively costly technology, large amounts of hired labor, or great expanses of land. Finally, the book is limited to fresh water only on account of the author's lack of experience in the salt water medium; it is hoped that someone will provide a companion volume for coastal residents.

Also, the present volume is intended as a "how-to" book, something that was scarcely possible in a work with the scope of *Aquaculture*. Small-scale aquaculture, in particular, is a young and experimental field, so the reader must excuse some of the gaps in the information presented. Sometimes tentative conclusions and frankly experimental methods are presented where a "nuts-and-bolts" approach is simply not possible. It is hoped that the author has made it clear which ideas are "tried and true" and which are speculative, and that the reader will approach the text, not as a bible, but critically.

Along with the "how-to" comes a generous dose of the author's opinions, which may make the text more interesting, or perhaps even entertaining in spots. The author acknowledges that, no matter how firmly he may hold his opinions, many of his senior colleagues are in greater or lesser disagreement on some matters. From this point of view, the tide seems to be turning in the direction indicated by this book, but the reader should keep an open mind.

The purposes of the book are two-fold. The first is simply to inform the reader on the possibilities of small-scale fresh water aquaculture and aid him or her to select the organisms and methods which will put food on the table or provide a satisfactory profit.

The second is less modest: the author must admit to the hope that this book (as well as his other publications and the aquaculture research at New Alchemy) will contribute to revolutionizing North American aquaculture. What is sought is not the sort of revolutionary technical advances which occasionally occur in any scientific field, but a change in the role of aquaculture in society and the biosphere. It is the author's contention that the majority of current aquaculture practice and research in North America is nutritionally, ecologically, economically and socially ineffectual—when it is not downright destructive. If you, the reader, succeed in harvesting aquatic crops according to the ideas presented here, you will have contributed to changing that state of affairs, and the book will have served both its purposes.

William O. McLarney
Hatchville, Massachusetts

Acknowledgements

All of the staff and volunteers of the New Alchemy Institute from 1977 through 1982 helped, directly or indirectly, to make this book possible. In a small institute such as ours, if a staff member goes part time to undertake a personal project, everybody has to pick up some of the slack. That this was done graciously has made my task easier.

Among those who helped directly, first thanks go to Meredith Olson, who did the majority of the library research and made many editorial comments, although I did not become more cheerful in the process, as she would have preferred. If at some point in the text the reader perceives that the quality of prose has dropped, it is probably evidence of the time of Meredith's departure to Minnesota, where she is carrying on the spirit of the book in the field. Peter Burgoon performed some of the same tasks toward the end of the process.

My aquaculture colleagues, Jeff Parkin, John Todd and Ron Zweig have helped more than they know. Others who made specific inputs include Gary Hirshberg (wind for water pumping and aeration), Win Munro (economics), Dave Engstrom (water chemistry) and Susan Ervin (agriculture of fish feeds and fish cookery). Susan also had to endure, at close range, the thumping and swearing which commonly accompanies my literary efforts. Thanks also to Earle Barnhart for his sporadic transfusions of unlikely literature.

Jim Lanier was most helpful in making available the materials of the National Aquaculture Information Service. John Vondruska of the National Marine Fisheries Service went out of his way to provide information on economics.

Helen Jacobson went a step beyond the usual illustrator; she endeavored to understand what she was drawing and to learn a bit about aquaculture, which made working with her all the more pleasant. Cynthia Knapp did her usual superb job of typing. (Now if she would only learn to eat fish!) Jane Palmer did a comparable job typing the appendices.

On Units and Conversions

Some clarification may be needed with regard to my use of English and metric units of measure. In almost all instances I have included both figures for the reader's convenience. Exceptions have been made in certain instances:

Where absurdly small figures would result, only units which have practical value have been used. Nothing would be gained by offering the equivalent of 2 millimeters in English units.

In examples of equations and calculations only English units are used, since the introduction of extra parentheses would render the text incomprehensible.

In a few instances I felt that convention rendered equivalents useless. For example, everywhere in North America water flow rates are measured in cubic feet per second (cfs). To convert cfs to any of its possible metric equivalents would be pedantry and aid no one.

Wherever both sets of units are used, I have converted from the original, be it English or metric, to the other. This may cause what appear to be absurdities. For example, one can buy a 15 inch pipe, but it would be odd to ask for a 38.1 cm pipe. A stocking rate of 300 to 750 fingerlings per acre is a comfortable approximation, 741 to 1853 fingerlings per hectare is not. However, less imprecision results from converting in this way than if both sets of figures were rounded.

PART I

Introduction and Fundamentals of Fresh Water Pond Biology

CHAPTER I-1

Introduction

The need for aquaculture

The need for aquaculture arises from the same root as that for agriculture. It is commonly accepted that it would be impossible to supply human demands for meat solely on the basis of hunting wild game, or to provide all our fruits and vegetables by foraging in the natural environment. Yet most of the world still obtains its fish in this manner, through traditional "capture" fisheries based on natural stocks.

The maximum sustainable yield of capture fisheries is a disputed quantity, but the most frequently quoted estimate is one hundred million metric tons per year. The current world harvest of about seventy million metric tons per year is not nearly enough to go around, much less keep pace with the demands imposed by a still-increasing human population. The obvious solution is aquaculture.

History of Chinese aquaculture

Not surprisingly, aquaculture first developed in China, where population pressure on food supplies long ago reached a level that we are only beginning to glimpse in North America. While the purpose of the earliest Chinese aquaculture may have been socio-religious or gourmet-oriented, the Chinese people, faced with the need to achieve the maximum sustainable yield from their land, wisely came to the conclusion that some land was most productive under water. (In some districts of South China, sixty percent of the land surface is covered with fish ponds.)

From the many aquatic habitats available, the Chinese chose the small pond as providing the best combination of productivity and manageability. Though aquaculture has subsequently been practiced in virtually every other type of aquatic environment, the truth of this rarely acknowledged insight remains. Ponds are naturally present in many localities and may be constructed, by hand labor if necessary, in most other places. As compared to larger lakes and the oceans, a higher percentage of pond water is relatively shallow, hence more productive and easier to harvest. Streams represent a theoretically more productive environment, but use of flowing water often raises competitive situations with regard to water rights, fishing, pollution, etc. Further, culture facilities in streams, and in large lakes and the oceans, are much more subject to damage from storms and other natural phenomena.

Fish can also be cultured in tanks or pools smaller than what is ordinarily termed a pond. However, at the high population densities of fish needed to achieve worthwhile production in very small systems, pollution and water quality problems inevitably arise. To overcome these problems requires constant vigilance together with a certain amount of technology. It is possible to grow aquatic crops in very small enclosures, but the amount of management input per unit of food harvested, whether expressed in terms of labor, financial cost or energy expenditures, will almost certainly be higher than in ponds.

Having settled on the pond as the optimal situation for fish culture, the early Chinese culturists developed another insight which Western aquaculture seems content to ignore: A pond is an ecosystem. Whether or not supplemental feed for fish is provided in a pond ecosystem, maximum production will be achieved by enhancing the fertility of the pond and by stocking a variety of fishes. In this way full use will be made of all the resulting food, as well as the physical habitats in the pond.

The Chinese appear to have recognized the value of pond fertility early on, and to have understood that fish could be grown more cheaply through fertilization with "wastes" than by direct feeding with materials which could be eaten by other live-stock or by people.

Full utilization of the food and habitat niches in the pond apparently did not come until after 618 A.D. According to Shao-Wen Ling of the University of Miami, up until the founding of the Tang Dynasty in that year, the only fish cultured in China was

Close-up of a Chinese polyculture pond. Tree by house at right is a mulberry, leaves of which are fed directly to herbivorous fishes or to silkworms which may in turn be fed to fish. Structures at right are privies placed over the water as a source of fertilizer. (Photo by Robert Sardinsky.)

the common carp (*Cyprinus carpio*). It happened that the family name of the Tang Emperors was Lee, which has the same sound as the Chinese word for the common carp. For a time it was considered sacrilegious to culture or eat common carp, and aquaculturists had to seek other fish to raise. It was found that if one stocked several kinds of fish—say one which fed on the bottom, one which fed in mid-water, and one which could eat green leaves provided by the farmer—greater yields would be achieved than if one stocked the same number of common carp only.

This was the beginning of polyculture. (See Chapter IV–1.) Through trial and error and observation, Chinese fish culturists eventually developed much more complex polycultures, in some cases including as many as a dozen species. These culture systems are often unique to a particular locality, having taken advantage of local ecological conditions, availability of stocks of fish, feed or fertilizer materials.

Chinese polyculture has only recently begun to be analyzed by Western scientific methods, but the aptness of the concept has been convincingly demonstrated. What does this mean to the North American aquaculturist? It does not mean that we should reject our native fishes or culture methods developed in North America out of hand, and seek to set up Chinese polyculture ponds here. Keep in mind that part of the secret of Chinese fish culture is in making the best use of a given local ecology and materials. However, consider this: The current average production of traditional pond polyculture in China is said to be over 4,000 lb/acre/year (4,412 kg/ha/yr). (Much higher yields are achieved in southern China and in southeast Asia, where the growing season is year-round.) This is accomplished primarily through the use of fertilizers, with no processed feeds whatsoever, with virtually no technology, and using ancient methods developed without benefit of scientific

Aerial view of fish culture ponds in southeastern China. The ponds shown vary from 1 to 10 acres (0.4 to 4 ha) in area. (Photo by Robert Sardinsky.)

research. The products of Chinese pond polyculture have traditionally been available widely and cheaply; they are an important factor in the nutrition of the Chinese people, as well as in Chinese high cuisine. Chinese aquaculture may also be regarded as ecologically beneficial, as it provides a facility for recycling organic "wastes."

From China, aquaculture spread via Korea to Japan, perhaps as early as 200 A.D. The Japanese began by copying Chinese methods, but during the Tokugawa era (1603–1868) they initiated farming of the sea (mariculture). To this day Japan leads the world in mariculture, but that is a matter which lies outside the scope of this book.

History of Japanese aquaculture

While traditional Chinese aquaculture is based entirely on fishes of the carp family (Cyprinidae), in Japan several other types of fresh water fish are cultivated, in addition to marine and anadromous fish, shellfish and plants. However, the most significant differences between Chinese and Japanese approaches to aquaculture are technological and economic. In this century, the Japanese have been quicker than the Chinese to adopt scientific methods and modern technologies and their aquaculture reflects this tendency. While in productivity Japanese aquaculture compares favorably with that of China, most of its products are not staple food sources. Japanese aquaculture is a thriving business; most of its products are sold to middle and upper class consumers. The poor person in China eats cultured fresh water fish, but the poor Japanese eats salt water fish taken by capture fisheries.

There is little information dating the origins of aquaculture in southeast Asia, but they are certainly to some extent related to the arrival of Chinese immigrants. The immigrants brought Chinese fishes to southeast Asia, but regional aquaculture also involves native fishes, of which there is a greater diversity than in China. More important, the tropical climate permits higher production rates than in China; annual harvests of up to 7,125 lb/acre (8,000 kg/ha) are reported from polyculture ponds.

History of Southeast Asian aquaculture

Southeast Asia has made several original contributions to aquaculture. Among them are the following:

• The cultivation of fish in estuaries and other brackish waters may have begun with the initiation of milkfish (*Chanos chanos*) culture in Indonesia around 1400 A.D., though ancient brackish water fish culture systems are also known from Hawaii and the Mediterranean. (In this book, estuarine aquaculture will be considered a branch of mariculture, and will not be discussed in the text.)

• The first important culture of crustaceans occurred in coastal southeast Asia. More recently, a number of forms of fresh water crustacean culture have been developed in and out of the region.

• The culture of fish in cages appears to have originated independently in Cambodia and Indonesia.

• The integration of fish culture and paddy rice farming was first achieved in southeast Asia or adjacent portions of China.

• Perhaps the most important contribution of southeast Asia to world aquaculture is the development of completely integrated systems which produce fish, other livestock, vegetables, grains and fruits with almost complete recycling of wastes. This concept, which should prove adaptable in most regions of the world, is discussed in detail in Chapter IV–2.

Aerial view of a milkfish farm in the Philippines, showing water supply canal (center) and sluice gates (white structures) to supply water to rearing ponds which may vary in size from 2.5 to 125 acres (1 to 50 ha). (FAO photo by P. Boonserm, courtesy Maurice Moore, Aquaculture Magazine.)

Japanese compared with Chinese and Southeast Asian aquaculture

Before going on to discuss Western aquaculture, it may be well to go into a bit more detail in contrasting Chinese/Southeast Asian and Japanese approaches, as they may be used to illustrate some relationships which are involved wherever aquaculture is contemplated.

In common with most Oriental cultures, both the Chinese and Japanese peoples place a high value on fish as food. Both have historically had a dense human population, so that there has been ample incentive to exploit sources of fish (and other foods) which yield a high return per unit of land or human effort.

Japan, an island nation with a long and convoluted coastline, early began to exploit the natural abundance of the sea by fishing. By the time aquaculture reached Japan, its fisheries were already well developed, and Japan continues to be among the world leaders in conventional fisheries.

China has an extensive coastline, but the Chinese coastline is much less significant than Japan's when compared to the nation's inland land mass and population. Before the development of modern storage and transportation methods, it would have been impossible for the majority of Chinese to obtain marine animals in edible condition. So China has never aspired to be a great maritime fishing nation, and Chinese coastal fisheries remain relatively undeveloped.

The natural productivity of inland waters is in some cases high, but their extent is minute when compared to the expanse of the oceans. So, China augmented the natural rate of production by constructing artificial bodies of water and managing them intensively. Today it is estimated that over half the fish consumed in China are products of fresh water pond culture. Japan had less need of inland fisheries or aquaculture as a basic protein source, and so sought to develop fresh water and marine aquaculture primarily to augment the supply of favored fishery products and to stimulate economic development. (Bear in mind that these are tendencies, not rigid patterns. Aquaculture may be more important as a source of food in China than in Japan, but Chinese fish farms must nonetheless justify themselves economically. And some forms of Japanese aquaculture are significant producers of moderately priced protein foods.)

In China the aquaculturist must seek to maximize the production of human food per unit of water, while keeping the price of the resulting product below a certain maximum. This fact influences the culture strategy at every step, beginning with the choice of species to be cultured. The criteria of high productivity and low cost virtually force the Chinese aquaculturist to choose fish occupying the lower trophic levels of the food chain. (See Chapter I–2 for a discussion of food chains and trophic levels in ponds.) Such fish will make maximum use of the natural resources of the pond and will respond well to fertilization with manures and the like. If supplemental feeding is resorted to, the feeds are usually cheap. They may even be materials which would be more costly to dispose of otherwise.

Of course it is possible to grow fish from higher trophic levels. But in a natural, or fertilized but unfed pond, the proportion of such fish in the total population is necessarily low. If growing large quantities of such fishes is desired, supplemental feeding with protein-rich materials must be undertaken. These materials will ordinarily be expensive, either because they are in demand as foods for humans or other livestock, or due to the expense of producing or harvesting them.

Nevertheless, in a society such as Japan, where capture fisheries provide an abundant and fairly inexpensive source of protein, there may be adequate incentive to produce certain inherently expensive food animals. If a relatively small crop of such a fish, produced using costly feeds, turns out to be more profitable than a larger crop of low-cost protein, the farmer is economically justified in choosing the less productive fish. However, one must understand the nature of the trade-off. The use of cultured grains in fish feed or the conversion of low cost fish meal to produce expensive fish results in a net loss in the amount of protein available for human consumption. In energetic terms, the caloric input/output ratio of an aquaculture system based on predators fed a specially processed feed will not compare favorably to that of a Chinese type system. This is especially true if one takes into account not only the conversion rate of feed and the culturist's labor but also the fuels, fertilizers and labor expended in preparing and transporting the feed. Finally, it is worth noting that while Chinese aquaculture and, especially, Southeast Asian integrated fish-livestock-agriculture systems, contribute to the solution of waste disposal problems, energy-intensive aquaculture is frequently a source of pollution.

The above is not to be taken to mean that culturing predatory fishes is never justifiable, or that in North America one might not opt for a luxury product, as opposed to a less expensive one. On the contrary, under certain ecological or economic circumstances, such practices may be preferred. However, it is my belief that, in spite of the glib acknowledgement of Oriental origins with which it is customary to preface treatises on aquaculture, the fundamental difference between traditional Chinese aquaculture, (determined by nutritional and ecological contingencies), and the more recent, economically driven Japanese aquaculture is not well understood. An understanding of this difference would certainly aid aquaculturists and policy makers everywhere.

History of European aquaculture

Aquaculture in Europe arose much later, apparently independently of developments in Asia. The first references we find to something which may be aquaculture are from Greek and Roman times, but the first indisputable references date from the Middle Ages, when various types of fish were kept in "stews" or castle moats in central and eastern Europe. This may have originated simply as a method of holding wild fish until they were wanted for the table, but sooner or later someone began feeding the captive fish and European aquaculture was born.

The early Chinese aquaculturists added manures and plant wastes to their ponds, believing them to be foods, and only later realized that these substances acted primarily as fertilizers. Early European fish culturists seemingly had a better idea what a fish might eat, and tended to feed grains and the like. Nevertheless, European fish ponds are generally less productive than their Asian counterparts.

As in China, the first fish to be cultured in Europe was the common carp. It is still the principal crop in most European countries, but other cyprinids, pikes, perches, catfishes and eels are also grown. Both supplemental feeding and fertilization are practiced. Polyculture is much less developed than in China and Southeast Asia. To oversimplify, the methods used are less technological than those used in Japan, but less ecologically derived than those of China.

A number of milestones in aquaculture were reached in Europe. One of the most far-reaching was the development in 19th century Silesia (part of present-day Czechoslovakia) of the Dubisch method for controlled spawning of common carp on grass mats. The Dubisch method and its variants spread to Asia and is now used wherever cyprinids are cultured. Development of methods for controlling spawning paved the way for the first selective breeding of food fish. (Selective breeding of ornamental fishes had already begun.) Today dozens of regional races of carp exist in Europe and elsewhere, and a number of other fishes are being selectively bred.

It is probable that the use of waterfowl in fish ponds (see Chapter IV–2) arose in eastern Europe, perhaps in Hungary. Another European contribution is the incorporation of fish culture in the process of modern sewage treatment. In Germany and Poland this is now a considerable source of food, revenue, and improved sewage treatment.

Beginnings of aquatic research in Europe

Until very recently, fish culture in Asia has developed almost totally on the basis of intuition or trial and error. Scientific research in aquaculture can be said to have begun in Europe, and the first textbooks on the subject were published there. (Oriental writings on aquaculture date back as far as *The Chinese Fish Culture Classic*, written by Fan Lee in the 5th century B.C., but they tend to be philosophical and anecdotal.) Perhaps the first aquaculture text was *La Pisciculture Industrielle* (Commercial Fish Culture), by C. Ravenet-Wattell (1914). It has been supplanted by various other works, including the acknowledged European classics *Lehrbuch der Teichwirtschaft* (Handbook of Pond Husbandry), by Wilhelm Schaeperclaus (1933), *Traite de Pisciculture* (Fish Culture Text), by Marcel Huet (1952), and C.F. Hickling's *Fish Culture* (1962), which included the first serious Western effort to describe Chinese fish culture.

Among the more important scientific developments in European aquaculture are the first technical studies of fish pathology and the first investigations of aquatic fertility, both beginning in the early twentieth century in Germany. An important recent development is the Polish attempt to describe the trophic relationships in polyculture ponds.

Of perhaps more fundamental significance than European aquaculture research per se was the beginning of limnological research in Germany in the early years of this century. Thanks to the early European limnologists (and their American successors) we have far greater knowledge of aquatic food webs, the importance of bottom soil fertility to aquatic productivity, the causes and effects of variations in concentration of dissolved gases, temperature, pH and other environmental factors, and many other natural variables which affect every aquaculture system.

While the European contributions to the culture of carp and other so-called "pond fish" are considerable, an even more influential European development was the beginning of cold water fish culture. (See Chapter I–2 for definition of warm and cold water environments.) The cold water fauna and flora are much less diverse than their warm water counterparts and the natural productivity of cold waters is considerably less. Further, cold water food chains are short; often only the creatures at the very top of the chain reach sizes convenient for human utilization.

European origins of cold water fish culture

In spite of these disadvantages the cold water habitat contains several candidates for cultivation; some are discussed in Appendix II. At present, though, "cold water aquaculture" may be taken as virtually synonymous with "salmonid (trout and salmon) culture." Salmonid culture did not arise from any need for these fish as food, nor out of any inherent superiority over other cold water fishes as objects of cultivation. Rather, it arose in response to the European sport fishermen's concern over the decline of natural trout populations. The first trout hatchery was established in Germany by Stephen Ludwig Jacobi in 1741. About a century later methods for artificial fertilization of trout eggs became well known, and the stocking of cultured trout in natural waters for the pleasure of anglers has continued to the present day in Europe, North America and elsewhere.

Though the efficacy, esthetics and economics of this form of stocking has been much questioned, it set the stage for the commercial culture of trout as food fish, beginning in 1853 in the United States and perhaps earlier in Europe. Most of the techniques used in commercial trout culture, and particularly hatchery methods, were developed by the various European and North American agencies concerned with perpetuating trout as a sport fish.

The principal limitations on salmonid culture are the high protein requirement of salmonids, the low growth rate characteristic of cold waters, and the delicacy of these fish with respect to environmental conditions, handling, and disease. But due to the high quality of trout as food, their esthetic appeal, and the fact that as yet we know no other fish which can be as productively grown in cold waters, trout culture is practiced virtually wherever there is cold water, including at high altitudes in the tropics. Current world commercial production of trout for food (almost entirely based on rainbow trout, *Salmo gairdneri*) exceeds 63,947 tons (58,000 metric tons). This places trout second only to the Chinese carps among cultured fresh water fishes of the temperate zones. Of the 18 major producing countries, 15 are European, led by Italy, Denmark and France. The exceptions are Japan (no. 4), the United States (no. 5) and Canada (no. 12).

Early developments in N. America

On this continent, aquaculture first arose as did trout culture in Europe—as a response to the depletion of game fish stocks. In the middle part of the last century, American biologists expanded techniques developed in Europe for trout to include just about every important fresh water fish. Unfortunately, this development was thoughtlessly treated as a panacea. In the U.S., with a great deal of enthusiasm, a substantial amount of government money, and a notable lack of wisdom, the Fish Commission began distributing young native and imported fishes willy-nilly throughout the land. Sometimes species were stocked in environments unfit for their survival. Just as often very small, young fish were stocked in places where they immediately provided a feast for predators ready to take advantage of the government's generosity.

A few of these introductions were successful. At least one—the European brown trout (*Salmo trutta*)—was judged a great success by sport fishermen. But that success was overshadowed by the blunder of stocking another European fish, the common carp. Carp are a valued food fish in Europe and Asia, and some American fishery officials went so far as to suggest that populations of "undesirable" native fishes like bass and pike be eradicated to make room for the carp. But it developed that most Americans disliked carp, and despite their tremendous potential as a food fish, their effect on American waters and fisheries has been largely detrimental.

By the early part of this century the stocking program so enthusiastically begun was abandoned, having failed to increase the supply of food or sport fish. Until around the time of World War II, North American aquaculture was largely restricted to federal, state and provincial trout hatcheries, with a few smaller specialized operations here and there.

The most enduring contribution of the American fish culture "boom" of the late nineteenth and early twentieth centuries was the perfection of the hatchery methods originated in Europe. It was the pioneer American fish culturist Seth Green who demonstrated that trout hatchery methods could be adapted to most fresh water fishes. Later it was shown that survival in the first weeks of life could be greatly increased in the protected hatchery environment.

A final stumbling block to application of the hatchery method was removed in 1934 when the Brazilian fish culturist Hermann von Ihering invented the process of inducing spawning by injection of pituitary hormones. Interestingly enough, this technique has found its widest application in the Orient, particularly in India and China. Pituitary injection has become the only Western technological innovation to be widely adopted in Chinese fish culture. From 1954 to 1964 this method of induced spawning was successfully adapted to all the Chinese carps of major importance. The result has been a considerable expansion in fish culture, which previously was largely dependent on eggs and fry collected from nature.

A new phase in North American aquaculture was begun in the 1930's with the work of George W. Bennett and David Thomson of the Illinois Natural History Survey, and Homer S. Swingle and E.V. Smith of Alabama Polytechnic Institute (now Auburn University). Bennett and Thomson were concerned with how the natural populations of ponds and reservoirs might be altered to provide larger crops of quality fish for fishermen. Swingle and Smith painstakingly evaluated a variety of species and species combinations experimentally stocked in artificial ponds. Their aim was to determine their potential to supply sport, subsistence, and commercial fisheries. The diversity of field studies and trial-and-error experiments carried out by the Illinois and Alabama groups is staggering; they form the basis for almost all of North American warm water fish culture today.

Pond culture beginnings in N. America

The presidency of Franklin D. Roosevelt coincided not only with the work of Bennett, Thomson, Swingle, and Smith but with:

• The "Dust Bowl," which dramatized the dangers of soil erosion and suggested the use of ponds and reservoirs in combating that problem.

• The Great Depression, which made American families conscious of the wisdom of producing at least some of their own food.

• The beginning of World War II, which threatened to cut off sources of imported foods and reduced the pool of available farm labor.

Some of the people in the Roosevelt administration were aware of the importance of fish ponds in Asia, and suggested that something similar could be done in the U.S. by adapting the research being done in Illinois and Alabama. The result was the "Farm Pond" program, through which the Soil Conservation Service, the Fish and Wildlife Service, and other agencies encouraged and assisted farmers to construct small ponds and stock them with fish. Among the many uses projected for these ponds was the production of food fish for home use. For reasons which are discussed in Chapter II-1, the great majority of farm ponds were and are of no importance as food sources, but at least two million of them exist. These may yet assume the importance seen for them thirty or forty years ago.

Swingle and his colleagues were also pioneers in experimenting with commercial fish culture systems. Their research was the basis of the United States' leading aquaculture industry, channel catfish (*Ictalurus punctatus*) farming which, in less than thirty years, has grown until in 1975 it accounted for 41,896 tons (38,000 metric tons) valued at almost 42 million dollars. During roughly the same period, commercial trout culture in the United States grew from virtual insignificance to an industry which annually produces 14,333 tons (13,000 metric tons) of fish valued at perhaps 45 million dollars.

The evolution of fresh water aquaculture in Canada has been similar but less complex. Only in extreme southern Ontario and a few small areas near the United States border is there any appreciable amount of warm water habitat, and warm water aquaculture has been very limited. The history of Canadian fresh water aquaculture, then, is a history of salmonid culture.

Evolution of Canadian fresh water aquaculture

Salmonid hatchery methods were in use in Canada as early as 1857, but the major pioneering effort occurred in 1866 when Samuel Wilmot established a private hatchery for Atlantic salmon (*Salmo salar*) on the banks of Wilmot Creek near Toronto. Wilmot and his colleagues were, like their contemporaries in the United States, proponents of the use of aquaculture as a source of animals to be stocked to augment natural production. However, commercial culture of trout as a food fish was an almost inevitable offshoot of Wilmot's work and it became Canada's only real aquaculture industry. Currently, Canada produces something less than 500 metric tons (551 English tons) of trout, valued at slightly over one million dollars annually.

Total production of all fresh water cultured foods other than channel catfish and rainbow trout in the United States probably amounts to less than 2,000 tons, worth no more than 1.5 million dollars; the comparable figures for Canada are virtually nil. No data are available on private production of aquatic crops for home consumption in the two countries, but it is surely very low.

"Third World" aquaculture

Since World War II, aquaculture has been exported from North America and Europe to Latin America, Africa and the Near East. With the exception of a single country, aquaculture has yet to achieve a fraction of the nutritional or economic impact predicted for it in these regions. A major reason for this state of affairs is the inappropriateness of the models. In Latin America, particularly, there has been a distressing tendency to push advanced North American technologies, rather than to seek methods—probably those based on tropical oriental systems—which would be appropriate to local ecologies and economies.

The exception is Israel. Here a judicious blend of European and Asian influences has been used to create an indigenous aquaculture which places Israel among the leading aquaculture producers on a per capita basis. (See Table I-1-2.) Of particular interest is the very effective use of manures; this technique has advanced to the point that production of carps on manures almost equals that on commercial feeds. (Israeli aquaculture research is well documented in the English language journal *Bamidgeh, Bulletin of Fish Culture in Israel*.)

Comparing N. American with Chinese and Japanese aquaculture

Returning to our comparison of Chinese and Japanese aquaculture strategies, where does North American aquaculture fit? First, we must bear in mind that aquaculture in North America is greatly underdeveloped and in that respect it will not stand comparison with that of either China or Japan. It is true that at least some of the nineteenth century North American fish culturists foresaw the need for aquaculture as a staple protein source, and that during the Roosevelt era the possibility was raised of an aquaculture loosely modelled on that of China. But in the first instance there was no immediate need for aquaculture products, and in the second, unforeseeable economic changes aborted the movement. The fact is that, with the possible exception of the Great Depression, the people of the United States and Canada have always enjoyed a more than adequate supply of protein. Local exceptions to this state of affairs have not been considered as problems of resource scarcity or allocation. Instead, the traditional solution has been to try to create additional employment so that the required protein foods can be purchased, usually from sources outside the problem area. In Japan, the abundance of reasonably priced fishery products dictated that aquaculture would be developed primarily as a source of luxury foods; in the United States, the combined availability of fishery products, meat, poultry, and dairy products had the same effect.

Catfish farming

Catfish farming in Arkansas (currently the number 2 producing state) provides us with a case history. Food fish culture there was begun in the 1950's in response to federal restrictions on the amount of land that could be planted to agricultural crops. (There already was a flourishing bait fish and goldfish [*Carassius auratus*] industry.) It was legal and at least theoretically profitable to grow fish on land which could not legally be conventionally farmed.

Interestingly, the biologists responsible for the early food fish culture research and development in Arkansas appear to have been inspired by Chinese fish culture. Or perhaps they arrived at a similar approach through ecological theory. In any event, their first choice as fish for culture were the buffalofishes (*Ictiobus* spp.), good quality food fish which feed on plankton, benthos and detritus. A feeding program was planned based primarily on fertilization and use of agricultural wastes. From an ecological point of view, the choice of buffalofish was excellent, but ultimately economics prevailed. While it is certainly possible to produce more buffalofish than catfish per unit area, and to do so more cheaply, catfish sell for twice the price. Under the rather special economic conditions prevailing in mid-twentieth century North America, the less productive and more expensive culture of the higher-priced fish turned out to be much more profitable. Table I–1–1 dramatically illustrates the results of this realization.

Table I–1–1. Area devoted to culture
of buffalofish and catfish in Arkansas 1958–1975

Year	Area of water primarily devoted to buffalofish culture in acres (hectares)	Area of water primarily devoted to catfish culture in acres (hectares)
1958	3,404 (1,378)	0 (0) (some catfish grown as a supplemental crop with buffalofish)
1960	3,542 (1,434)	247 (100)
1963	734 (297)	1,057 (428)
1966	247 (100)	14,326 (5,800)
1975	0 (0) (some buffalofish grown in polyculture with catfish)	6,840 (2,770)

Catfish are fed commercial processed feeds, made with fish meal as a protein base and incorporating soy meal, synthetic vitamins and other costly, high quality ingredients. Some supplementation of this diet is practiced (live minnows, fertilization, bug lights, etc.) but, due to the very high protein requirements of channel catfish, the industry is largely tied to commercial feeds.

In many cases the technologies used in catfish farming are "high" or "hard," and decisions on their use are made on an expediency basis. In addition to feed processing, among the technologies employed are mechanical harvesters, chemical fertilizers, herbicides, algicides and fungicides, semi-automatic fish processing equipment, electrical aerators and even airplanes for broadcasting feed.

All these technologies together with a considerable back-up program of scientific research and technical extension (free or very low cost technical advice from universities or government agencies) in the major producing areas, brought the average production rate of U.S. catfish culture to just slightly over 2,000 lb/acre (2,200 kg/ha). This was achieved only after most of the smaller, less efficient producers were driven out of the business.

In such a system, economies of scale favor large operations, and the tendency is toward fewer and larger producers. In 1975 in Arkansas the 6,840 acres (2,770 ha) devoted to intensive catfish farming were owned by less than 100 persons, most of whom were also involved in raising field crops and/or bait fish on a large scale. Individual ponds are usually 10 acres (4 ha) or more in size, and it has been

Aerial view of catfish farm in Arkansas. Individual ponds are usually more than 10 acres (4 ha) in area and may be as large as 40 acres (16 ha). (Photo courtesy Maurice Moore, Aquaculture Magazine.*)*

suggested that in order to realize a profit a farmer should have a minimum of 300 acres (121.5 ha) under water, and $500,000 in capital investment.

The price of cultured catfish is moderately high, and farmers sell primarily to long-distance fish haulers or processors. The distribution of catfish is similar to that for most agricultural produce in North America, though largely confined to the South and Midwest. As with agriculture, widespread distribution has led to the industry's centralizing in a few geographically favored areas. Within those areas catfish culture has made a major contribution to employment and economic development. But on a regional and national basis the nutritional effect of catfish culture is nil.

The broad range ecological effects of catfish farming have already been mentioned. Within the producing areas, catfish farming is ecologically no different than any other feedlot operation. The combination of a heavy concentration of stock and continual importation of nutrients often results in downstream pollution problems. Within the industry there is also considerable controversy over the cost and appropriateness of meeting federal water quality guidelines.

Clearly, catfish farming is an "aquabusiness," following the Japanese, rather than the Chinese model, but falling short of the high production rates of much Japanese aquaculture.

Trout farming The second most widely cultured food fish in North America, the rainbow trout, is similar to the channel catfish in its need for a high-protein diet. Indeed, the trout's requirements are even more stringent than those of catfish, and trout farming, like catfish farming, is essentially a feedlot operation, based on processed feeds. Since

Harvesting catfish from large ponds requires long nets, heavy equipment and lots of labor. Here a pond has been seined to concentrate the fish in a few pockets. (Note second seine in background, and tractor with mechanical seine hauler on dike.) Then fish are manually dipped out and into a basket for weighing and transfer to a processor's truck. (Note fish trying to escape seine behind workers.) Aerator in center background helps maintain water quality during this period of stress and crowding. (Photo courtesy Maurice Moore, Aquaculture Magazine.*)*

trout are often grown in running water, the per area yield may be many times that of a catfish farm. But production costs are comparable, as are the technologies employed, the price and distribution of the crop, the research and extension back-up, and the economic, nutritional and ecological impact of trout farming.

Ecologically, it is difficult to fault the logic of trout culture as one can catfish culture. The limitations of the cold water environment for aquaculture have already been discussed. For the present, at least, we have no alternative to salmonids if cold waters are to be used for food production. (It is worth noting that China has never exploited her cold water resource, while Japan is a major producer of salmonids.)

Trout culture also differs from catfish culture in the greater diversity of operation size and technology employed. Siting of trout farms is dictated, not so much by regional climate and geography as by the local availability of suitable spring water. While the small trout grower cannot compete with the giant operators for the supermarket trade, the scattered distribution of suitable sites, together with the greater per area production of trout, gives the small producer the chance of capturing lucrative local or specialty markets. Nevertheless, the trend in trout farming is definitely toward "aquabusiness."

Small-scale commercial trout culture illustrates one of three possible approaches to small-scale aquaculture: the production of luxury foods on a smaller scale. The choice of technology in this type of aquaculture, as in aquabusiness, will ordinarily be based on expediency, unless for reasons of principle or profit one chooses to supply the "organic" foods market. In practice, small-scale culture of luxury foods implies the identification and servicing of specialty markets. Some examples are offered in the text; see particularly Chapter VI–4.

A second approach to small-scale commercial aquaculture is the decentralized production of moderately priced items to service local or regional markets. The major example in this hemisphere is the crayfish culture of southern Louisiana. Crayfish rank a distant third to catfish and trout among North American aquaculture products. Annual production is 7.5 million pounds (3.4 million kg), almost all of it from French-speaking sections of Louisiana. Methods for crayfish culture are described briefly in Chapter II–10 and in more detail in various other publications. (See Appendix IV–1.) What concerns us here is that Louisiana crayfish farming is the only sizable North American aquaculture system which is more analogous to Chinese than to Japanese practice.

Crayfish farms are generally small, and operated without recourse to costly technologies. No commercial feeds or fertilizers are used; the crayfish derive their nourishment entirely from plants grown in the culture ponds. Many crayfish ponds are used, in rotation, as rice fields.

Crayfish production is not particularly high, averaging 301 lb/acre/year (336 kg/ha/year) and prices are substantially lower than for catfish or trout. But, as operating costs are also low and 85% of the crop is consumed locally, crayfish farming is a profitable business.

From an ecological point of view, crayfish farming is benign. It requires minimal inputs of energy and is not a significant source of pollution. Nearly half of the area under crayfish cultivation is natural swampland. Thus, crayfish culture may act as a force for conservation of the natural environment, since there are few other economic uses of swampland which do not require severe modification of the natural environment.

Attractive though Louisiana crayfish culture may be as a model, it must be admitted that the conditions which gave rise to it are rather special. Crayfish are a traditional ethnic food in French Louisiana, and the inception of crayfish culture came at a time when demand was beginning to exceed the supply available from commercial fisheries. In most parts of North America, anyone attempting to market a low-cost aquaculture crop in competition with supermarket fish and meats would probably have a more difficult time.

N. American subsistence aquaculture

In the short run, it is easier to look optimistically at the possibilities of domestic forms of "subsistence" aquaculture. Here one is not pressed to achieve any particular minimum of production in order to have a salable crop. A hundred pounds of fish in a year does not represent a significant source of cash income, but it could be a very significant factor in a family's diet. As shown in Chapter II–1, to produce fish on such a scale in a farm situation will require not new aquaculture techniques but a better understanding of what is available in ponds, and how to harvest it.

Not everyone has access to a pond; there is also a need for what Joseph Senft of Rodale Resources has dubbed "fish gardening." The precedent for such tiny aquaculture systems does not exist in the Orient or elsewhere, and those of us interested in aquaculture in the city, in greenhouses and in other space-limited environments find ourselves up against one of the knottiest research problems in the field. Researchers at The New Alchemy Institute, including John Todd, Ron Zweig and

Aquaculture on a more modest scale can be productive. This small spring-fed pond in Minnesota produced nearly 200 pounds (80 kg) of rainbow trout in less than six months. (Photo courtesy Meredith Olson, WCCA, Waverly, Minnesota.)

myself, were among the first to explore the concept of the "Back Yard Fish Farm." As a glance at Chapter IV–4 will show, we are not yet able to offer a "recipe."

If my discussion of micro-aquaculture seems to over-emphasize New Alchemy's work, it is because I want merely to give a view of the potentials of aquaculture on the smallest scale and the obstacles to its implementation, and to invite experimentally minded readers to join us in the effort. Here, I would like to acknowledge the past and continuing contributions to this area of research by The Amity Foundation, The Institute for Community Self Reliance, The Farallones Institute, Life Support Systems Inc., Rodale Resources, Solar Aquafarms and the Walden Foundation. To describe the work of each of these groups in detail would require not a chapter, but another book. Hopefully, within a few years one of us will be able to write the "compleat cookbook" of back yard aquaculture.

In my opinion, for aquaculture to assume anything approaching its potential importance in North America, the type of small-scale approaches and appropriate, ecologically sound technologies just discussed must be widely implemented. Table I–1–2, which shows the total and per capita fresh water aquaculture production of 18 selected countries[1] may illuminate the potential. Note that the countries which

"Soft" vs. "hard" technologies

[1] The countries listed have been selected for illustrative purposes, and because data, from the mid 70's, were available. A number of important aquaculture producers are not listed. Very few countries keep careful records of aquaculture production, and the reader should not be surprised to find figures differing from those cited here for some of the countries. No data from African or Latin American countries are included because of the relative underdevelopment of aquaculture there. The total production of Africa and Latin America is estimated to amount to 2.7% and 0.7% of the world crop, respectively. (See E. E. Brown's *World Fish Farming Cultivation and Economics* in Appendix IV–1.)

Table I–1–2   Per capita fresh water aquaculture production in 18 selected countries.

Country	Principal animals cultured	Production in metric tons (English tons)	% Exported	Population	Per capita consumption of cultured fresh water fish–kg (lb)
Australia	trout	less than 500	0	12,700,000	0.04 (0.08) (or less) (or less)
Canada	trout	500 (551)	Insignificant	21,568,000	0.02 (0.05)
China (main-land)	carps (also catfish, eel, tilapia)	3 million (3.3 million)	0	882,800,000	3.64 (8.02)
Denmark	trout	16,000 (17,641)	94	4,951,000	0.19 (0.43)
France	trout (also carp and crayfish)	24,000 (26,461)	3	52,900,000	0.44 (0.97)
Hungary	carps (also cat-fish, pike-perch)	19,197 (21,550)	0	10,500,000	2.00 (4.40)
Israel	carps, tilapia, mullet	13,000 (14,326)	0	3,124,000	4.16 (9.15)
India	carps, mullet	490,000*(540,243)	0	613,200,000	0.79 (1.76)
Indonesia	milkfish (also carps, tilapia, gourami, catfish, mullet)	140,000*(154,436)	0	136,000,000	1.03 (2.26)
Italy	trout (also eels, bullhead, carp)	16,730 (18,445)	23	55,000,000	0.24 (0.53)
Japan	trout, eel, carp, ayu (also loach, mullet, salmon)	68,747 (75,796)	4	111,100,000	0.62 (1.36)
Philip-pines	milkfish (also tilapia and mullet)	124,000*(136,714)	0	44,400,000	3.79 (6.14)
Taiwan	milkfish, tilapia, carps, eel (also mullet and clams)	90,105*(99,322)	0	15,130,0000	6.56 (14.44)
Thailand	tilapia, carps, catfish	80,000*(88,203)	0	42,100,000	1.90 (4.18)
United Kingdom	trout	2,000 (2,200)	0	56,400,000	0.04 (0.08)
United States	trout, catfish (also crayfish and buffalo)	53,500 (58,986)	0.1	212,500,000	0.25 (0.55)
U.S.S.R.	carps (also trout, catfish, sturgeon, whitefish and salmon)	210,000 (231,533)	0	255,000,000	0.82 (1.81)
West Germany	trout, carp (also tench, eel, pike)	9,562 (10,542)	0.1	60,651,000	0.16 (0.34)

* includes some brackish water production

have done the most to feed their people with cultured fresh water fish are, for the most part, tropical Asian countries where access to high technologies is limited. Exceptions are Israel and Hungary, both nations which have emphasized organic fertilization of fish ponds and polyculture. In the industrialized nations despite the accessibility of technology, per capita aquaculture production is quite low. Even Japan, which ranks fourth in the world (after China, India and the U.S.S.R.) in total aquaculture production, is not an abundant supplier of cultured fish for home consumption. (Even if one includes the considerable mariculture crop of Japan, total aquaculture production comes to 147,291 metric tons [162,393 English tons] or 1.33 kg per capita [2.92 lb] annually, making no allowance for exports.)

The opposite approach has prevailed in North America; the bulk of research and development effort in aquaculture is still directed toward producing luxury products on a scale and with a technology which virtually demands corporate involvement. Even with corporate backing, large-scale aquaculture has not been very successful. In a 1975 article, John Ryther of Woods Hole Oceanographic Institution stated that technologically unsophisticated, traditional, southeast Asian mariculture practices "are surprisingly reliable and productive, and have made a significant contribution, economically and nutritionally.... Such, unfortunately cannot be said for the embryonic mariculture industry of the United States, the rather dismal track record of which threatens the field with extinction before it has fairly gotten under way."

The same in general may be said of fresh water aquaculture. Traditional, low-to-intermediate technology, small-scale techniques have been effective for centuries, and their opposites, aimed mainly at producing high-priced "luxury" foods have, with notable exceptions, been failures. Robert Huke of the Department of Geography at Dartmouth College estimates that no more than 25% of the aquaculture business ventures in the U.S. operate at a profit. Some of the failures are ill-advised speculative efforts, but others are designed as tax write-offs for large corporations.

I need not take space here to present the case for decentralization of production or appropriate technology; this has eloquently been done by Peter van Dresser and the late E. F. Schumacher. (See Appendix IV–4–1 for references.) And I have taken up the cudgel for my economic and political view of aquaculture in Rich Merrill's *Radical Agriculture.*

Though the stance implied by Ryther and Huke seems compelling to me, one need not agree with my interpretation to take an interest in small-scale aquaculture. Certainly some centralization of food production is inevitable in our present industrial society. But even if one assumes that this state of affairs will last forever, one need not forego the pleasures and benefits of gardening. So it is with aquaculture.

What is unique about aquaculture, and what must be corrected if small-scale aquaculture is to evolve into anything approaching its full potential, is the lack of how-to-do-it material for the small private grower. Aquaculture in North America evolved during a time when agribusiness was in its ascendancy. In agriculture and terrestrial livestock husbandry there is at least a tradition of gardening, homesteading and small farming, with its own literature and folklore, which serves to counter the agribusiness propaganda. This is not so in aquaculture. The position of the beginner in small-scale aquaculture is like that of someone who, on deciding to garden for the first time, goes to the library and finds only books on truck farming.

If the potential of small-scale aquaculture is to be realized, we will need support, not only from small institutions like New Alchemy, but from those larger institutions which are in a position to support or carry out needed research. To date, these institutions have scarcely even shown that they are aware of small-scale aquaculture as a possibility. Let me cite a couple of examples, lest anyone underestimate the disproportion which exists:

The culture of lobsters, originally with the goal of supplementing the fishery catch and later in the hope of a commercial crop, has been the subject of research in at least nine countries since 1891. There is no estimating the sum of money, much of it tax money, which has been spent on this research. Today construction of a single hatchery may cost several million dollars. Yet to date no one has produced a single lobster which could be sold at a profit, nor has there been any demonstrable benefit to the fishery. Even if the obstacles to profitable lobster culture are eventually removed, the energy demands of lobster culture will be very high and lobsters will remain a luxury product.

Near the other end of the scale is the work of Homer Buck. (See Chapters IV–1 and 2.) By the end of his first polyculture trials in 1975, his research had been subsidized by only $1,500. Yet he had already defined a system which could economically be adapted, on a small scale, by virtually any North American livestock grower to produce fish at low cost while helping to solve waste disposal problems.

Or, as I pointed out to a recent critic of New Alchemy's aquaculture research, who wanted to see the "pay-off," in cost/benefit terms, "even if it took us twenty more years, at present rates of subsidy, to unravel the mysteries of the Back Yard Fish Farm, it would still be one of the more cost-efficient research programs in the history of aquaculture."

*Future of
N. American
aquaculture*

The absurd state of affairs in which the big money backs probable losers, while possible winners wither on the vine, is not likely to be corrected by those who back aquabusiness. They simply may become disillusioned with aquaculture. There is no logical reason to suppose that multi-national corporations will take an interest in homestead level production of anything. But someone must realize that doubling the harvest of North American farm ponds, or perfecting a system allowing urban dwellers to raise 50 to 100 pounds of fish annually in their back yards would be a substantial contribution to personal and national well-being. It would certainly enhance the quality of life as much as doubling commercial catfish production or achieving a breakthrough in lobster culture, even though the effect on the GNP (Gross National Product) might not be measurable.

There are some indications that in the United States appropriate government agencies are beginning to get this message. The Department of Agriculture has been designated the lead agency for aquaculture research, and there is talk within that agency of exploring small-scale methods, appropriate technologies, strategies based on the food chain concept, etc. Canadian freshwater aquaculture is less well developed and it would be premature to predict trends there.

As important as government participation in aquaculture is the role of the colleges and universities. Not only are there relatively few places to study aquaculture (see Appendix IV–4) but university aquaculture research is largely limited to studies oriented toward aquabusiness or sport fisheries. Increased concern for the status and potential of small-scale aquaculture is needed, not only in disciplines like fishery biology and agriculture, which have traditionally dealt with aquaculture, but also in such fields as nutrition, economics, and even Oriental studies.

If we are to speak of an aquaculture for small groups and individuals, it will ultimately be up to us, as small groups and individuals, to create it. Fortunately, despite the gaps in our knowledge, there is much that we can do right now. Some of what we can do is contained in this book. The implementation of this information and the testing of these ideas will be an important step toward a more diverse and important future for aquaculture in North America.

CHAPTER I-2

Ecology of Ponds

Given identical environmental conditions, two ponds of the same size will contain approximately the same amount of "biomass" (total weight of all living material), but it may occur in very different forms. A Chinese fish culture pond is stocked and managed so that the greatest possible proportion of that biomass will be in the form of fish to be used for human food. A natural, unmanaged pond may contain the same biomass, but much more of it will be in the form of very small fish, insects, weeds, etc. The American "farm pond," while inspired by the Chinese fish pond, is much closer biologically to a "wild" pond, hence less productive of human food.

In Chinese aquaculture as many as twelve species of fish, but more usually five or six, are stocked. This multiple-species stocking is called polyculture. Most American farm ponds are stocked with only two fish species, and American commercial aquaculture is based on true monocultures (single species culture). In the Chinese system, each species occupies a different "niche" in the pond ecosystem by virtue of its unique habitat and/or food requirements. (Figure I–2–1 is a diagrammatic sketch of the classical Chinese pond stocking scheme, illustrating the niches occupied by the several fish species.) Stocking ratios will vary from pond to pond in response to the local climate, soil type, availability of food, etc. Fertilization of the water is usually practiced; supplemental feeding (usually with materials which are of little or no use as human food) plays a relatively minor role.

These finely tuned artificial ecosystems produce as much as 6,000 lb/acre (6,618 kg/ha) of harvestable fish annually. American catfish farmers with concentrated feeds and other technological aids are seldom able to produce more than 2,500 lb/acre (2,757 kg/ha). In the American "farm pond" *standing crop* (the total biomass of food fish present at a given time) is usually from 100 to 300 lb/acre (110–330 kg/ha), and only a fraction of that is harvested annually.

To understand this discrepancy in production of desirable or "harvestable" fish, we must refer to two of the basic descriptive tools of the ecologist—the "food chain" and the "pyramid of numbers." The "food chain" outlines who-eats-whom. Cats eat

Biomass

Multiple species stocking

"Food chain" and "pyramid of numbers"

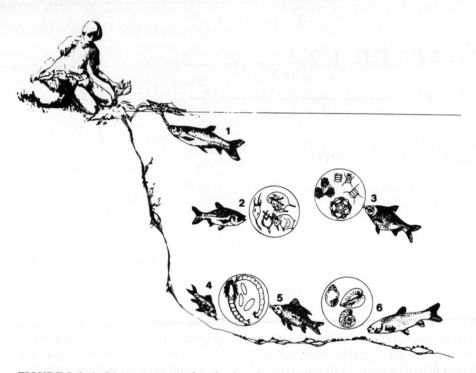

FIGURE I–2–1. *Diagrammatic sketch of a Chinese polyculture pond, showing habitat and feeding niches of the principal fish species. 1. Grass carp* (Ctenopharyngodon idellus) *feeds on greens provided by the farmer. 2. Big head carp* (Aristichthys nobilis) *and 3. silver carp* (Hypophthalmichthys molitrix) *are midwater filter feeders; the former prefers zooplankton, the latter phytoplankton. 4. Mud carp* (Cirrhinus molitorellus) *and 5. Common carp* (Cyprinus carpio) *feed on benthic animals and detritus, including grass carp feces. 6. Black carp* (Mylopharyngodon piceus) *feeds on mollusks.*

mice which eat grain which "eats" sunlight and minerals from the soil; taken together they form a chain. The "pyramid of numbers" is a graphic demonstration of the biomass of the various links in the food chain. (Two examples of such pyramids are illustrated in Figure I–2–2.) If we construct such a pyramid for any ecosystem, we will find what we logically would expect: The lower you go on the food chain, the greater the biomass. There have to be more mice than cats, more grain than mice, etc.

Of course this is oversimplified. Food chains are more accurately seen as "food webs," since most animals consume a variety of foods. And the same species may be predator or prey in relation to another, depending on its age or condition of health.

Keeping in mind that we are oversimplifying, let us look at the pyramid of numbers as it might appear in a Chinese fish culture pond and an American farm pond (Figure I–2–2). We will assume that the ponds we are comparing are of equal size, fertility, etc., and therefore that "primary production" (the total biomass of those organisms which derive nourishment directly from sunlight and minerals in soil and water) is equal.

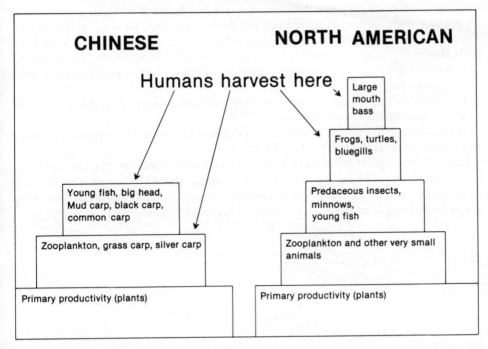

FIGURE I–2–2. *"Pyramid of numbers" showing distribution and human utilization of organisms at different trophic levels in a North American farm pond and a Chinese polyculture pond.*

Note that the food chain in the American farm pond is longer and human food is taken from the very top, by harvesting predatory fish. In the Chinese pond, the top two "trophic levels," as they are called, are entirely absent, and people feed nearer the bottom of the pyramid, on fish which are as large as the American bass and bluegill, but which consume plants, plankton, or very small animals. The presence at the lower trophic levels of animals large enough to be worth harvesting means that a greater percentage of the total pond biomass is "harvestable."

The main reason, then, for the lower productivity (in terms of human food) of the American pond is the lack of large fish which are low on the food chain. This lack is not entirely the fault of the biologists who thought up the farm pond program. By a quirk of evolutionary fate, North America, alone among the continents, received very few large herbivorous (vegetarian) or plankton-feeding fish. Nevertheless, we could improve our fish ponds by making use of more of our native fishes.

American pond productivity

There are other reasons for the discrepancy in productivity, including the following:

• The American pond stocking system (to be described below) has been oversimplified and reduced to too pat a formula. Every body of water is unique and should be managed with its individual characteristics in mind.

Reasons for low productivity in N. America

• American ponds are not designed with attention to methods of fish harvesting. The ponds we build and the fish we normally stock are among the more difficult to harvest efficiently.

• Americans are imbued with a sport fishing mentality. We stock our ponds exclusively with predatory fish, because they are fun to catch on hook and line, and we manage them as though recreation, not food production, were the primary goal. Yet we continue to complain about the poor harvests in our farm ponds. Associated with this attitude is our insistence on large fish for the table. A quarter pound (0.11 kg) bluegill may not constitute a trophy, but a few of them make as fine a meal as one large fish.

• Chinese fish ponds usually are fertilized, a practice which many Americans still consider unesthetic and unhygienic.

• The Chinese have three thousand years of experience in fish farming; we should not expect to rival them after just 30 years.

Later, I will suggest some ways in which North Americans can start eating lower on the aquatic food chain. But first, for those readers who already have farm ponds or who like that sort of ecosystem for esthetic or recreational reasons, let's examine the conventional farm pond and discuss some ways to get more food out of it.

PART II
Fishes

CHAPTER II-1

The Traditional Farm Pond—
The Bass-Bluegill Community

Definition and uses of farm ponds

"Farm pond" could be taken to mean any pond located on a farm. But North American fishery biologists generally use the term to include only small (less than 5 acres or 2 ha), artificially constructed, warm water ponds which are relatively little managed and fished chiefly by hook and line. As a working definition, consider a "warm water" pond to be any pond in which the water temperature at 3 feet (0.91 m) ever exceeds 72°F (22.2°C). Often fish production is not the primary purpose of a farm pond. Dendy (1963) lists the following additional uses:

> Livestock water
> Irrigation
> Household and barn water
> Fire protection
> Spray water for orchards
> Production of ice
> Attraction of wild fowl
> Swimming, boating, skating
> Esthetic values
> Erosion control

Pond stocking principles

Still, the majority of farm pond owners would prefer at least some fish production for recreational, nutritional or commercial purposes, and at least as early as the publication of L. L. Dyche's book "Ponds, Pond Fish, and Pond Fish Culture" in 1914, fishery biologists responded to this desire. Though stocking formulae used in North American farm ponds are far less complex and sophisticated than those used in the cultivation of food fish in China (see Chapter I-1), the same two basic principles obtain:

(1) A body of water is a three-dimensional growing space. To treat it like a field, by planting only one kind of crop, is likely to result in wasting the majority of that space.

(2) Any fertile pond will produce a number of different fish food organisms. However, most fish are not omnivorous but are rather selective in their diet. Stocking a single species wastes not only space but food.

Two species are traditionally stocked in American farm ponds—the largemouth bass (*Micropterus salmoides*) and one of the sunfishes, usually the bluegill (*Lepomis macrochirus*). This combination only partly conforms to the first principle. Bluegills probably make better use of the open water of the pond than bass, which tend to seek cover, but there is considerable overlap in habitat. With regard to the second principle, one can predict, by comparing the bass' cavernous mouth with the bluegill's tiny one, that bass feed on larger prey.

The bluegill has tremendous reproductive potential which, unchecked, can lead to overpopulation and stunting, with the end result of a pond crowded with dwarfed fish. In theory, the bass alleviate this problem by feeding on young bluegills. This aspect of the system does not always work out well, because farm ponds are usually harvested exclusively by sport fishing. Sport fishermen tend to favor the larger and more exciting bass, and sport harvesting will decrease the predators and encourage the predominance of small bluegills in the population. In some areas, this problem is partly avoided by substituting the redear sunfish (*Lepomis microlophus*) for the bluegill. While this species' maximum size is less than that of the bluegill, it has a far lower reproductive potential and its populations tend to consist of fewer but larger fish. A more recent development is the introduction of hybrid sunfishes which have extremely low reproductive potential and grow more rapidly than bluegills or redears, but these fish have just begun to be generally available. (See detailed discussion below.) Redears and hybrids notwithstanding, the commonest problem in farm ponds continues to be overpopulation and stunting by sunfish, and fishery biologists in North America are kept busy poisoning and restocking ponds.

Reproductive potential

Farm pond writings often point to "improper stocking" as the main cause of failure of farm ponds to provide satisfactory fishing. This would seem to imply that there is such a thing as "proper stocking." Of course there is, but the issue is not nearly so cut and dried as one might be led to believe. H.S. Swingle, who might be considered the foremost authority on farm pond management, has admitted "There has probably been more disagreement over the most desirable methods of stocking ponds than any other phase of pond management." I will shortly discuss the roles of both stocking and harvesting in managing farm ponds. But first I must mention an inherent disadvantage of the traditional farm pond that may send some readers on into the following chapters, which deal mostly with more intensive fish culture systems:

Improper stocking

Even with good management, the productivity of the bass-bluegill community is limited by the fact that both species are highly carnivorous. (See discussion of "food chains" and the "pyramid of numbers" in Chapter I–2.) The best standing crop of bass and bluegills that can be expected without fertilization or supplemental feeding is less than 400 lb/acre (441 kg/ha). Even with fertilization or feeding, harvests do not approach those obtainable with some of the other fish culture systems described here.

Despite its limitations, the bass-bluegill community has its advantages:

Advantages of bass-bluegill community

• Both species are familiar and accepted as food by North Americans. Bluegills in particular are excellent food fish.

• The recreational benefits of a well-managed bass-bluegill pond can be great.

• It is often possible to obtain fish for pond stocking free of charge. Until recently, the U.S. federal government provided fish for this purpose. This practice has been discontinued, but some states still distribute fish for stocking in private ponds. Or you may be able to obtain fish from a neighbor with an overpopulated pond.

• There is a wealth of literature and experience in the pond culture of large-mouth bass and bluegills, and fishery biologists are usually experienced in handling problems associated with these species.

We may assume, then, that farm ponds will continue to be built and stocked with

Harvest of half pound (0.2 kg) bluegills grown on natural foods in a 5 acre (2 ha) farm pond in Minnesota. The pond was stocked with 2 to 3 ounce (0.05 to 0.075 kg) bluegills taken from overpopulated lakes in the spring and harvested four months later. (Photo courtesy Meredith Olson, WCCA, Waverly, Minnesota.)

the bass-bluegill combination. Samples of recommended stocking ratios for bass, bluegills and redear sunfish are given in Table II–1–1. In recent years, the channel catfish (*Ictalurus punctatus*) has also become a popular farm pond fish and it, too, is included in Table II–1–1. (For detailed information on channel catfish culture, see Chapter II–3.) The stocking rates in the table are taken from pamphlets published by the listed states. If you are interested in specific recommendations for your own state, write to your conservation or fish and game department. In the U.S. most states provide free booklets on farm pond management.

As important as the numerical rates of stocking are both the time of stocking and the sizes of fish stocked. Initial stocking is usually done in the fall or winter, depending on latitude. If the pond is not to fail almost from the start, bass of suitable size must be present to crop the first bluegill hatch.

Channel catfish from an Illinois farm pond. (Photo courtesy Leo Pachner, Farm Pond Harvest Magazine.*)*

The practice of stocking a few adult bass in the spring following the first stocking is gaining in popularity as insurance of adequate bluegill cropping. (See listing for Pennsylvania in Table II–1–1.)

Size of fish and time of stocking

Timing of course varies greatly with climate. The fact that Swingle (1951) considered that all four of the schedules listed in Table II–1–2 would result in a satisfactory size and numerical distribution of bass shows the complexity of the situation.

Table II–1–2. Age and size of fish and time of stocking for largemouth bass-bluegill ponds in Alabama

Species	Age or Size	Time of Stocking
Bluegill	fingerling	fall
Largemouth bass	fingerling	fall
Bluegill	fingerling	fall or winter
Largemouth bass	1 inch (2.54 cm)	following spring
Bluegill	fingerling	fall or winter
Largemouth bass	0.5 lb (0.23 kg)	late winter
Bluegill	adult	winter
Largemouth bass	0.5 lb (0.23 kg)	winter

The ideal in farm pond stocking is the maintenance of a "balanced" population— that is, one which yields satisfactory numbers and sizes of food and sport fish, year after year, with conventional fishing methods and little additional management.

Balanced population

Table II-1-1.   Recommended stocking rates for farm ponds.

State	Number of fish per acre (hectare)			
	Largemouth bass	Bluegills	Redear sunfish	Channel catfish
Alabama				
(Unfertilized ponds)	50 (124)	500 (1,235)	–	–
(Fertilized ponds)	100 (247)	1,000 (2,470)	–	–
(Fertilized ponds)	100 (247)	–	–	500 (1,235)
Illinois	100 (247)	100 to 400 (247–988) (depending on soil fertility)	100 to 400 (247–988) (depending on soil fertility)	50 to 100 (124–247) may be added to any of these systems
	100 (247)	70 to 280 (173–692) (depending on soil fertility)	30 to 120 (74–296) (depending on soil fertility)	
Missouri	100 (247) or less	500 or less (1,235)	–	100 or less (247) (optional)
Ohio	200 (494)	1,000 (2,470)	–	Optional, rates not given
Pennsylvania	100 (247) (plus 10–20 mature fish the spring following first stocking)	1,000 (2,470)	–	500 (1,235) (plus 1,000 fathead minnows) (2,470)
	200 (494)			

Various criteria have been proposed as an aid to determining the relative balance of pond fish populations. The three most widely used by biologists are the F/C ratio, the Y/C ratio and the A_T value. These may be defined in simple terms as follows:

The F/C ratio is the ratio of the total weight of forage fishes (F), in this case bluegills, to that of piscivorous fishes (C), or largemouth bass. For bass-bluegill farm ponds it has been determined that the F/C ratio should be in the range of 3.0–6.0, but a satisfactory F/C ratio alone does not guarantee balance.

F/C ratio

In the Y/C ratio, C is defined in the same way, while Y refers to the total weight of bluegills small enough to be eaten by the average adult bass. Optimum Y/C ratios are 1.0–3.0. Ratios lower than 1.0 are rarely encountered; higher ratios will ordinarily not right themselves naturally, but require remedial measures (discussed below).

Y/C ratio

The A_T value is the percentage, by weight, of harvestable size fish in a population. Just what constitutes "harvestable" size is partly a matter of taste, but is conventionally taken to mean a minimum of 6 inches (15.4 cm) for bluegills and 10 inches (25.6 cm) for largemouth bass. Given these assumptions, optimum A_T values are 60–85%.

A_T value

Arriving at the F/C, Y/C and A_T ratios requires fairly sophisticated mathematical methods of population estimation—often beyond the means of the home fish grower—which will not be described here. (These methods are discussed in the standard texts on fishery biology. See Appendix IV–1.) Swingle (1956) offers a simpler method of gauging balance. His method requires two seines—a minnow seine 15 feet (4.6 m) long and 3 feet (0.9 m) deep with 6–8 meshes to the inch (2.3–3.1 meshes to the cm), and a 50 foot (15.2 m) long, 6 foot (1.8 m) deep one with ½ inch (1.3 cm) mesh. Samples of fish are taken at several points around the pond, using first the minnow seine, then the larger one. The seines are set and pulled as shown in Figure II–1–1. (See Chapter VI–1 for a description of seining technique.)

Simple method to gauge balance

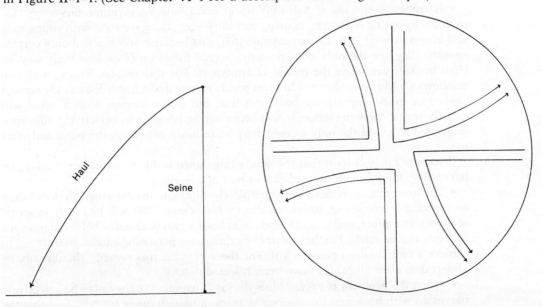

FIGURE II–1–1. *Method of setting and hauling seines for estimating degree of balance in farm pond fish populations.*

If the pond is balanced, most of the bluegills taken will be 6 inches (15.4 cm) long or more. The average bass taken will weigh 1–2 lb (0.45–0.9 kg) though both smaller and larger fish may be caught. In unbalanced ponds, either bluegills will be many, but mostly less than 6 inches (15.4 cm) long and the bass few and mostly large, or there will be a relatively few, large bluegills (¼ lb or 0.11 kg or more) and relatively many bass weighing less than 1 lb (0.45 kg) and in poor, or skinny condition.

Angling to gauge balance

Similar estimations of population balance may be made by angling, but the seining method has several advantages. It provides more data in a short time, compensates for temporarily poor fishing conditions due to weather or excess available feed, is less dependent on fishing skill, and does not bias toward one species or the other as do most anglers.

If a pond is found to be imbalanced, the cause of the problem must be determined before remedial measures can be taken. Usually the fault lies with inadequate harvesting or improper stocking, but other factors may also be involved. The first question to ask is whether the pond is suited for bass-bluegill culture. Some farm ponds, particularly those characterized by cool water or short growing seasons, might better be managed for other kinds of fish. (See Chapter II–6, and Appendix II.) If you are reasonably certain the pond is suitable for the species of fish you have stocked, you should find out if the following problems exist:

Problems in imbalanced ponds

• If aquatic plants are growing in excessive amounts, they may provide so much cover for young bluegills that bass are unable to crop them adequately. The plants also interfere with fishing. (See Chapter VIII–2.)

• Very turbid or cloudy water will inhibit both reproduction and feeding of bass and bluegills. It will also limit fishing success. (See Chapter X–1.)

• Sudden and extreme changes in water level can interfere with spawning and natural food production.

• Extraneous species of fish may upset pond balance in numerous ways (for example by eating bass eggs, causing turbidity, competing for food with young bass and bluegills, etc.). You cannot assume that, just because you stocked only certain species, they are the only ones present; "weed" fishes do often find their way in. (Bait buckets are often the means of transport. For this reason, fishing with live minnows should be prohibited in farm ponds.) Some undesirable fishes may remain unseen or evade capture by hook and line, but a few sweeps with a seine will usually reveal their presence. It will often not be possible to selectively eliminate unwanted fish, and the only answer may be to drain or poison the pond and start over.

If you are fairly certain that the pond's imbalance is due to improper stocking or harvesting, the following remedies can be considered:

• The most drastic remedy is to completely eliminate the existing fish population by draining or poisoning, following this by restocking. This will be costly in terms of money and effort, and usually leads to at least a two year delay before significant harvests can be made. Further, neither draining nor poisoning can be guaranteed to achieve a one hundred percent kill. For these reasons this remedy should only be undertaken after all others have been rejected.

• When the problem is excess bluegills (as it usually is), they may be selectively harvested with hook and line, seines or traps, although many farmers find this too time consuming. Ordinarily a 75–90% reduction in population is desired. Another suggested rule of thumb is that each 50% reduction of bluegill numbers will result in a 1 inch (2.5 cm) gain in average length, with the biomass remaining unchanged.

Bear in mind that harvested excess fish can be eaten, used as livestock feed, in compost, or to stock other ponds. But as long as unbalanced conditions persist, *no* bass should be harvested.

• A quicker way of eliminating excess bluegills is by selective poisoning. Poisoning of ponds should ordinarily be carried out only by qualified biologists, so details are not given here. However several publications dealing with this procedure are referenced in Appendix IV–1.

• Perhaps the simplest remedy is to stock additional adult bass to crop young bluegills. This approach may be used as often as every year instead of intensive bluegill harvesting but it is, of course, expensive.

Many pond owners are curious about the possibilities of increasing fish production or improving sport fishing by adding other species than the four mentioned above to their ponds. The theory of polyculture (see Chapters I–2 and IV–1) would predict that many niches in a normal pond are left unfilled in the traditionally stocked farm pond. The largemouth bass-bluegill combination and the variations just described were developed and tested in the Midwest and South. In those particular regions it is unlikely that other combinations will as effectively produce good recreational fishing and respectable harvests with a minimum of labor. However, there are other fishes which may have a place in the farm pond picture; some of them will be discussed here.

Other fish for the farm pond

The most promising substitute or supplement for the largemouth bass is the chain pickerel (*Esox niger*). In some ponds the pickerel might provide more effective control for small fish, since it is extremely temperature-tolerant. It is also easier to harvest than bass. Unlike bass, pickerel will feed on the hottest day of summer or the coldest day of winter. The pickerel, which prefers weedy habitats, is a good sport and fine textured food fish. It is rarely stocked in farm ponds, perhaps because its distribution is largely limited to the Atlantic coastal plain, where farm ponds are not common.

Chain pickerel

The black crappie (*Pomoxis nigromaculatus*) and white crappie (*Pomoxis annularis*) are popular panfishes which occupy a niche somewhere between the largemouth bass and bluegill. They will eat more insects than bass, but also eat considerable numbers of small fish. For reasons which are not understood, crappie usually produce poorly in small ponds. The various states list minimum sizes of 3 to 12 acres (1.3–5 ha) for ponds to be stocked with crappie. Factors favoring crappie are a diverse population of fishes, including both crappie predators and smaller fish for crappie feed, a fluctuating water level and a stream entering the pond.

Crappie

Crappie are just as capable of overpopulating a pond as bluegills and are much less interesting as sport fish than either bass or bluegills. They should be stocked only when heavy continuous harvesting is planned. Their main use is in turbid ponds which provide only marginal habitats for bass and bluegill. (See Chapter X–1 for a discussion of turbidity problems.) Even with heavy harvesting, good management of crappie will require periodic raising and lowering of the water level. If the water is kept low during the summer and high during the cold months, crappie will reproduce at a lower rate while their growth rate will increase. However, such manipulation is not feasible for most pond owners and is unfavorable for sunfishes.

The best species of sunfish for stocking the great majority of farm ponds are the bluegill and redear sunfish. For reasons related to the natural food supply, it may occasionally be advantageous to stock some other species of sunfish. (See Appendix

Sunfish

II.) Ordinarily, though, they are less desirable since the other *Lepomis* species do not exhibit the large size and rapid growth of which bluegills and redear sunfish are capable.

Hybrid sunfish

A recent development in farm pond stocking is hybrid sunfish. Already in the 1930's, biologists were experimenting with the many possible *Lepomis* hybrids, but only in the last few years has the knowledge gained from this research been applied. Like many hybrid animals and plants, sunfish hybrids often grow more rapidly than either parent. Hybrids are also often sterile or exhibit sex ratios in which one sex or the other is highly dominant. Either condition, of course, alleviates the normal sunfish tendency to overpopulate ponds.

To date the most suitable of the dozens of possible sunfish hybrids for pond culture appears to be a cross between male green sunfish (*Lepomis cyanellus*) and female bluegill. Such hybrids, known as "hybrid bluegills," have recently become commercially available under a variety of trade names. A "hybrid redear" (male red-ear crossed with female green sunfish), similar in behavior to the readear sunfish, is also available.

Choice of a suitable hybrid for a given situation is not as simple as it might appear, due to the wide genetic variability of the *Lepomis* species. For instance, according to Francis Bezdek of Aquatic Management Inc., Lisbon, Ohio, a pioneer in hybrid sunfish culture, the cross between male green sunfish and female bluegill occurs naturally in Ohio, but the same two fish will not cross in Illinois. Further, if a particular cross is made in two different localities, there is no guarantee that the resulting hybrids will grow or behave similarly.

Given this rather confused situation, and the present high price of hybrid stock, many pond owners will be inclined to stick with the pure strains. However, research in sunfish hybridization continues, and eventually we may see the development of superior strains adapted to particular localities. The wise sunfish farmer will watch the aquaculture trade publications (see Appendix IV–1) for news and advertisements.

Forage fishes

Another group of fishes which may play a part in farm pond culture are various forage fishes. The golden shiner (*Notemigonus crysoleucas*), for example, and the fathead minnow (*Pimephales promelas*), while of no value as human food, may still be used to augment bass production. They are discussed in Chapter III–5 and Appendix II–2.

Pond temperature problems

A problem which has been observed in New York state and undoubtedly exists elsewhere is that some farm ponds which are too cold or have too short a growing season for good largemouth bass and bluegill production, still are too warm for trout. So far no one has undertaken the task of developing an assemblage of fishes analogous to the largemouth bass-bluegill community for such ponds. Likely candidates include the smallmouth bass (*Micropterus dolomieui*), the chain pickerel, the white perch (*Morone americana*), the yellow perch (*Perca flavescens*), the rock bass (*Ambloplites rupestris*) and the pumpkinseed sunfish (*Lepomis gibbosus*).

The reader familiar with the North American fishes will probably by now have noticed the exclusion of a number of desirable fishes from this chapter. Their absence is in conformity with the standard recommendations worked out in Illinois and Alabama in the 30's and 40's and since then adopted by most fishery and agricultural agencies. Reasons for rejection of species generally considered acceptable food and/or sport fish are chiefly the following:

• Trout, while desirable pond fish, are part of an ecosystem very different from the classical "farm pond," and are discussed separately. (See Chapter II–6.)

• A number of species mature at too large a size to maintain reasonable numbers in small ponds; these include pike (*Esox lucius*), muskellunge (*Esox masquinongy*), walleye (*Stizostedion vitreum*) and paddlefish (*Polyodon spathula*).

• Certain bottom feeders, for example common carp (*Cyprinus carpio*) and goldfish (*Carassius auratus*) may create turbid conditions in ponds. These will be detrimental to other fishes, particularly such visual feeders as bass and bluegill.

• Like the bluegill, many of the smaller panfishes show the trait of overpopulating ponds, and may be even more difficult to manage. Amongst them are the yellow perch, white perch, and most sunfishes.

• A few otherwise desirable species seem not be have sufficient reproductive potential to make a significant contribution to farm pond production (e.g., the larger catfishes).

No matter which species are stocked, as long as they reproduce naturally in the pond, their proper harvesting is of the utmost importance. Next to "improper stocking" (already discussed) the single most frequently reported cause of failure of farm ponds is inadequate harvesting. This problem, in my opinion, is closely bound up with the "sport fishing mentality." The fisherman, whose main goal is recreation and who considers food merely a by-product, wants to fish when he feels like it; he does not want to be required to fish, nor will he fish for bluegills when the bass are hitting. Farmers whose interest in food fish production equals or surpasses their interest in recreational fishing (which I assume includes most readers of this book) will have already made the first step toward solving the harvesting problem. If you approach your pond like a gardener, and harvest when the crop is ready, chances are you will be able to maintain a balanced population. Suggestions for improving the rate of success of hook and line fishing together with non-traditional harvesting methods for farm ponds are discussed in Chapter VI–1.

The goal of any type of farm pond fishing should be to achieve and maintain a balanced population as described above. This generally means harvesting no bass at all for the first two or three years after stocking. After that, bass should account for about 20 percent of the weight of the harvest. Another rule of thumb is not to take more than 25 bass per acre (10/ha) annually. Large bluegills should be kept whenever they are caught, as they are relatively immune to natural predation and have a far greater reproductive potential than younger fish. Since channel catfish cannot maintain their population naturally in most farm ponds, they may be harvested as desired.

The total permissible annual harvest of bass and bluegills will range from 50 to 200 pounds per acre (55–221 kg/ha) without fertilization, supplemental feeding or additional stocking. The determining factors in how much you should harvest are soil fertility and latitude; where the growing season is longer ponds can be harvested more heavily. Based on sport fishing figures in midwestern public waters, the average harvest to hook-and-line fishing is about 4 pounds (1.8 kg) per angler day, with highly skilled anglers doing three times as well, or more. It has been suggested that in order to maximize fish production, farm ponds should be fished 500 hours per acre per year. Clearly it will pay you to keep records so that you will be able to adjust your harvesting efforts.

It would definitely enhance the nutritional importance of farm ponds if North Americans would learn to eat smaller fish. Our insistence on a certain minimum size for a food fish is rather irrational; fish which are "too small" to eat here would

Harvesting

Harvesting ratios

Harvesting large vs. small fish

be greeted with enthusiasm in Asia or Latin America. This may be partly due to laziness; it is definitely less work to clean one large fish than the same weight of small ones. But I believe this insistence on large fish is also connected to the sport fishing mentality. It is a traditional sport fishery management practice to prohibit fishermen from keeping fish under a certain minimum length. This length is usually defined by the average length of a species when it reaches maturity. The theory is that a greater number of fish thus survive long enough to reproduce at least once. This practice is valid when applied to wild species which have a low reproductive potential, to heavily fished populations, or to bass in the farm pond. But, as we have seen, one of the commonest problems in farm pond management is overpopulation by bluegills. So in a farm pond, observing the custom of "tossing back the little ones" is usually counterproductive. If my personal experience will serve as any added incentive to the harvesting of small fish, let me add that I frequently dine on 5 to 6 inch (12.8–15.4 cm) bluegills, perch, bullheads, etc., and for frying purposes I find them as good as, if not superior to, larger fish.

It is one thing to take small fish as they come, but it would be another to aim to produce them. Theoretically, one should be able to produce more 5 inch (12.8 cm) bluegills than 6 inch (15.4 cm) ones in a given pond. However, to date, no one has investigated management policies aimed at such small fish. A pond which produces mainly 5 inch (12.8 cm) bluegills is probably already imbalanced, and will produce 4 inch (10.2 cm) bluegills in the near future, and so on. By weight, the carrying capacity of a pond for a given species is generally the same, regardless of the size of the individual fish. All in all, you are better advised to manage bass-bluegill populations according to conventional notions of optimal size.

Feeding and fertilizing

Farm pond production may be increased by direct feeding or fertilization; appropriate methods are discussed in Part III. But if you are thinking of committing considerable time and/or money to feeding your fish or fertilizing your pond, and particularly if commercial culture is your goal, you may want to consider intensive culture of species and species combinations which are more productive than the farm pond ecosystem. (See Chapters II–2–5, 8 and 10–12.)

Increased production of food fish would doubtless result if farm ponds were designed and constructed with mass harvesting in mind. Construction details for conventional farm ponds and other fish culture facilities are given in Chapter VII–1, but there is one aspect of farm pond construction which merits discussion here, and that is pond size.

Most publications on farm ponds specify a minimum size of ¼–½ acre (0.1–0.2 ha); maximum size is usually limited by the amount of land and money the farmer has available for construction. Almost every summer Saturday at the New Alchemy Institute, I face one or more visitors who want to know what they can do with a pond or a pond site of ½ acre (0.2 ha) or less. In searching for an alternative to the "conventional wisdom" on the subject, I ran across a study of 33 stock watering ponds, ranging in size from 0.17 to 0.25 acre (0.07–0.10 ha) in northern Alabama which were experimentally stocked with largemouth bass, bluegill and redear sunfish. The owners of the ponds were given suggestions "that would aid in the management of a pond to obtain the maximum production of fish." Population balance was checked one and two years after stocking. It was found that 14 of 30 "usable" ponds were imbalanced after one year; 22 of 29 usable ponds contained an imbalanced population the second year. Causes reported were inadequate application of commercial fertilizer by the owners, overflow and entry of green sunfish and

Alabama stock ponds study

yellow bullheads (*Ictalurus natalis*), and reduction of water level, especially at spawning time. It was concluded that "it is not feasible to stock ponds that are smaller than 0.25 acre (0.1 ha) with a largemouth bass, bluegill and redear sunfish population." It was further explained that "the majority of owners of ponds less than 0.25 acre (0.1 ha) would not manage them properly."

I don't think it is necessary to discuss at length the differences between a person who builds a small pond for stock watering and is then offered free fish and the person who builds a similar pond with fish culture in mind from the start. I do want to draw attention to the seven ponds in the study which contained balanced fish populations after two years. Obviously it *is* possible to manage farm ponds of less than 0.25 acre (0.1 ha) for fish production. Without denying that such ponds are harder to manage than larger ones, I must admit that I have not found a minimum size for a productive farm pond.

An offshoot of farm pond management to be discussed is monoculture of the princi-pal species. Toward the northern limits of the warm-water environment, the growth rates of sunfish may be so slow as to make them unsuitable for stocking in farm ponds. Such ponds may still support respectable bass sport fishing, particularly if forage fish, (usually golden shiners), are stocked with the bass. Table II–1–3 gives examples of bass monoculture stocking rates for northern waters.

Monoculture

Bass monoculture

Table II–1–3. Monoculture stocking rates for bass in northern waters.

State or Province	Bass Species	Size	No./Acre (no./ha)	No. Golden Shiners/Acre
New York	Largemouth	fingerlings	100 (247)	400 adults (1,088)
Ontario	Largemouth or Smallmouth	fingerlings (1–3 inches) (2.5–7.7 cm)	100 (247)	—
		yearlings (5–7 inches) (12.8–17.9 cm)	25 (73)	—
		adults	10 (25)	—

Shiners stocked with bass are usually eliminated by the third year and must be restocked. Such ponds produce minimal harvests of 20–30 lb/acre/year (22–33 kg/ha/year) in the North and only slightly higher in the South. Consequently most readers will want to seek out other possibilities.

There is also an interest in some quarters in commercial monoculture of bass. The principal limitations on commercial bass culture are the same as for channel catfish and trout—the ecological inappropriateness and economic cost of rearing a fish which occupies the top of the pyramid of numbers. (See Chapter I–5.) And, in my view, the largemouth bass lacks the qualities which have compensated for these disadvantages in catfish and trout, namely, high food quality and established commercial demand. But some pond owners realize handsome profits by rearing bass to adult size for stocking in farm ponds, sportsmen's club lakes, children's fishing derbies, etc. A combination of bluegills for home use and bass for sale might prove appropriate for some small scale fish culturists. Readers interested in the details of intensive bass culture, including efforts to develop a commercial bass feed, are referred to Robbins and McCrimmon, 1977. (See Appendix IV–1.)

As for the sunfishes, bluegills, hybrid sunfish, and perhaps other varieties seem likely to be among the important commercially cultured food fishes of the future,

Sunfish monoculture

and should be particularly useful to the small-scale grower seeking to serve localized markets. Information useful to the prospective monoculture grower of sunfishes for home use or commercial purposes will be found in publications listed in Appendix IV–1.

Farm ponds— present and future

In closing, here are some interesting facts concerning the present and potential importance of farm ponds in North America. The number of farm ponds in the United States increased from about 20,000 in 1934 to well over two million by 1965. Comparable numbers for Canada are harder to arrive at, but an educated guess would be perhaps 60,000 warm or "cool" water farm ponds (excluding trout ponds, which represent the predominant habitat in Canada). Table II–1–4 lists the most recent available figures for 33 states and two Canadian provinces. While the greatest "boom" in pond construction appears to be over, all indications are that the number of farm ponds continues to increase annually.

Table II–1–4. Numbers of farm ponds
in the United States and Canada by state and province.

State or Province	Estimated No. of Ponds	Date of Estimate
Alabama	26,000	1968
Alaska	0	(author's estimate—no warm water habitat)
Colorado	22,500	1963
Connecticut	4,500	1976
Delaware	50	1965
Florida	4,000	1963
Georgia	40,000	1965
Idaho	91,000	1963
Illinois	63,000	1967
Indiana	7,500	1952
Iowa	22,000	1963
Kansas	11,000	1963
Kentucky	79,000	1963
Louisiana	19,000	1963
Maryland	4,000	1963
Massachusetts	2,000	1963
Michigan	2,500	1963
Minnesota	3,500	1963
Mississippi	84,000	1967
Montana	24,000	1963
Nebraska	15,000	1963
Nevada	386	1963
New Hampshire	1,500	1963
New York	16,000	1963
North Carolina	41,000	1963
Oklahoma	250,000	1963
Oregon	3,500	1963
South Carolina	25,000	1965
South Dakota	63,000	1963
Tennessee	55,000	1965
Texas	341,000	1963
Utah	5,000	1963
Vermont	240	1976
Ontario	10,500	1963
Prince Edward Island	0	1976

The average production of these ponds can only be guessed at, but taking into account that many of them are unstocked or unharvested and that others are badly managed, it is probably quite low. But if we assume there are still no more than two million ponds in the U.S. (Illinois estimates that 1,200 new ponds are constructed each year), that these ponds average ½ acre (0.2 ha) in surface area (a low estimate), and that all of them were managed so as to yield an average of 100 lb/acre (110 kg/ha)—obtainable without feeding or fertilization—then U.S. farm ponds *could* provide an annual harvest of 100 million lb (45.5 million kg) of fish, or 43% more than the 1976 production of catfish farming, the leading North American aquaculture industry. Another way of looking at that 100 million pounds is that it would represent 3.7% of all the food fish landed in the United States, and at a tiny fraction of the cost of conventional commercial fishing or aquaculture. *What are we waiting for?*

CHAPTER II-2

Bullheads

There are a number of fishes other than those discussed in Chapter II–1 which may be useful to the small farmer. Most of them require more intensive management than the fishes of the traditional farm pond community. A notable exception is the group of small catfishes known as bullheads, which may actually give respectable yields when managed less intensively than the largemouth bass (*Micropterus salmoides*)—bluegill (*Lepomis macrochirus*) combination. There are three widely distributed species of bullhead: the black bullhead (*Ictalurus melas*), the yellow bullhead (*Ictalurus natalis*), and the brown bullhead (*Ictalurus nebulosus*). A subspecies of the brown bullhead (*I. nebulosus marmoratus*) of markedly different appearance and known as the speckled bullhead, is found in the southern states and the lower Ohio valley. Although the brown bullhead has been used in most of the few aquaculture experiments involving bullheads, to my knowledge no one has attempted to compare the suitability of the different bullheads for this purpose since Bennett's work in Illinois in the 1940's. He noted that:

- Black bullheads were best confined to deeper ponds since they were the most likely to roil the water.
- Yellow bullheads were less prone to stunting than black bullheads.
- Speckled bullheads did not successfully reproduce in reservoir lakes. (This is contrary to their behavior in ponds and to the usual experience with the northern brown bullhead, *I. nebulosus nebulosus*, in lakes.)

I have not had the opportunity to work with the black or speckled bullheads, but in my experience the yellow bullhead is to be preferred to the northern brown. When handled, and particularly when confined in cages or aquaria, my brown bullheads have tended to develop a host of known and unknown diseases; mortalities of 90% or more have been the usual. I have had no such problems with the yellow bullhead.

The state of the art of bullhead cultivation has not advanced since publication of two papers by Swingle in 1954 and 1957. As the larger channel catfish (*Ictalurus punctatus*) came to dominate U.S. warm water fish culture, bullhead research came to a halt. This chapter, then, will consist mainly of a summary of Swingle's work together with a few suggestions, based on my own experience, for low intensity cultivation of bullheads for home use.

Swingle, in his early experiments with speckled bullheads, succeeded in producing as high as 745 pounds per acre (822 kg/ha) using very primitive supplementary feeds (soybean cake—a pressed meal—or peanut meal) in ponds where some natural food was available. However, mortalities were often very high, stunting was a problem, and production of commercial size fish (0.2–0.5 lb or .09–0.23 kg) was seasonal.

Mortalities were determined to be due in part to vitamin deficiencies in the diet. Swingle reduced losses by fortifying the diet with fish meal and distiller's dried solubles, but also by a variety of chemical prophylactic treatments, described below.

To overcome the problem of stunting which was caused by over-reproduction, the rather surprising tactic of stocking more densely was used. Swingle found that, while stunting occurred when bullheads were stocked at 1,000 or 2,000 per acre (2,470–4,940/ha), when one to four inch (2.5–10.2 cm) fish were stocked at 3,000 to 6,000 per acre (7,410–14,820/ha), reproductive behavior was suppressed. In sustained operation, consequently, it would be necessary to maintain separate reproduction and growing ponds. If additional ponds were available, stocking could be staggered to provide more than one harvest annually.

In Alabama in the 1950's, according to Swingle, restaurants preferred to buy bullheads weighing 0.2 to 0.5 lb (0.09–0.23kg) live weight, rather than larger fish. This is contrary to the experience of the channel catfish industry. One wonders if today the same restaurants would buy farm-raised bullheads at all. Despite their excellent table qualities, in most parts of the U.S. and Canada bullheads have little or no commercial value at any size.

Management procedures for commercial production of brown bullheads as recommended by Swingle in 1957 are quoted below. These procedures are based on experience in Auburn, Alabama. They would have to be adjusted to local conditions, particularly with respect to temperature, for use in other regions.

1. Select for brood fish the fastest growing bullheads each year when ponds are drained. [At the end of the growing season.]
2. Disinfect the brood fish with 10 p.p.m. potassium permanganate for one hour, followed by 15 p.p.m. formalin plus 1 p.p.m. acriflavine for from 4 to 12 hours. [Disinfection involves placing the fish in a holding tank containing the disinfectant solutions.]
3. Stock the brood fish in spawning ponds between January and July using 50 fish per acre [120/ha] weighing from 0.5 to 1.5 lb. [0.2–0.7 kg.] each.
4. Fertilize the spawning pond once monthly, using 100 lb [45 kg] 8–8–2 per acre.
5. Feed the brood fish 1 lb/acre [1.1 kg/ha] per day using the Auburn No. 1 fish feed [containing 35% soybean oil meal, 35% peanut oil meal, 15% fish meal, and 15% distillers dried solubles] until after fish have spawned.
6. The first schools of young cats will be found in June to July after the water temperature has reached 80°F [26.7°C]. Additional spawning will occur until September if feeding is adequate.

Early experiments in speckled bullhead culture

Supplementary feeds

Prevention of stunting

Brown bullhead management

7. Remove schools of fry as the small fish reach a length of 0.7–1.0 inch [1.8–2.5 cm]; disinfect, and store them in holding ponds at rates up to 60,000 per acre [148,200/ha]).

8. Increase daily feeding in spawning pond to 5 lb [2.3 kg] Auburn No. 1 fish feed to allow extra feed if the fry and fingerlings remain in the spawning pond.

9. Fertilize the holding ponds once monthly with 100 lb 8–8–2 per acre [110 kg/ha] per application until October.

10. Beginning in October, feed the fingerling cats in the holding pond at 1% of their estimated weight per day, six days weekly. Cease feeding while the water temperatures remain below 60°F [15.6°C] during November to March. [To estimate the weight of the fish in the pond, remove a sample by seining or trapping, and determine their average weight. Multiply by the number of fish stocked to determine the estimated pounds of fish in the pond.]

11. Before stocking, remove fingerlings from the holding ponds and disinfect with 15 p.p.m. formalin plus 1 p.p.m. acriflavine for four or more hours. Then stock in the commercial production ponds at the rate of approximately 3,000 per acre [7,410/ha] at any time of year that ponds are available. Use only 1 to 4 inch [2.5–10.2 cm] fingerlings for stocking during the period March to October, and the larger fingerlings for stocking only after October to prevent reproduction in the commercial ponds.

12. Fertilize the commercial production ponds once monthly for four months using 100 lb 8–8–2 per acre [110 kg/ha] per application if stocking occurs between February and October.

13. Do not feed when water temperatures are below 60°F [15.6°C]. While temperatures are higher, feed [Auburn No. 1] fish feed at rates between 3 and 5% of the body weight of fish per day, six days per week. Do not feed at rates above 25 lb/acre [27.6 kg/ha] per day.

14. For fish stocked in July, the following schedule for fertilization and feeding was successful:
   a. Fertilize once monthly in July, August, September and October using 100 lb [45 kg] 8–8–2 per acre [110 kg/ha] per application.
   b. Feed as follows:
   5 lb/acre/day [5.5kg/ha] in September
   10 lb/acre/day [11.0 kg/ha] in October
   10 lb/acre/day [11.0 kg/ha] March 15 to April 1
   20 lb/acre/day [22.1 kg/ha] in April
   25 lb/acre/day [27.6 kg/ha] in May and June

15. Cease feeding about one week before draining the pond. This procedure should yield approximately 900 lb fish per acre [993 kg/ha].

Bullhead subsistence culture

Were bullheads again to be commercially produced in N. America, it would benefit both the small fish farmer and everyone's eating habits. At present, bullheads can be recommended commercially only in those few areas where there is a traditional market. However, I can recommend bullheads, particularly the yellow bullhead, to readers interested primarily in food for their own tables. I cannot say whether Swingle's fairly elaborate stocking and harvesting procedures would be a satisfactory solution to the problem of overpopulation in a non-commercial system. This problem must be faced, for bullheads, particularly in the absence of predators, can quickly fill a pond with 1 to 4 inch (2.5–10.2 cm) stunted adults. Though one could harvest these fish, it has been my experience that the flesh of bullheads from such stunted populations is of inferior quality. A possibility still to be explored is a polyculture of channel catfish or blue catfish (*Ictalurus furcatus*) with bullheads. The larger catfish would control bullhead reproduction and provide a fish of greater commercial value with bullheads as the main food crop.

Provided they are harvested heavily, bullheads can sometimes be stocked in the farm pond with bass and bluegills, though there is a chance that they will eventually be eliminated by the bass. Bear in mind, also, that in some ponds roiling of the water by bullheads has been a serious problem. Both of these problems can be eliminated by growing bullheads in cages. (See Chapter IV–3.) Experiments at The New Alchemy Institute indicate that the yellow bullhead is well suited for this form of cultivation.

Bullheads may also be valuable in stocking small bodies of water which will not support significant populations of other food fishes. As long as a pond is deep enough not to freeze to the bottom during the winter, bullheads will find food and conditions suitable for breeding. A few years ago, John Todd and I disposed of less than 30 brown bullheads we had been using as experimental animals by tossing them into a small (perhaps 1 acre or 0.4 ha) pond in a suburban neighborhood. Until the severe winter of 1976–1977, the pond was a valuable community food and recreational resource, fished chiefly by local children. We have no way of knowing

The author (right) and Jeffrey Parkin harvesting half pound (0.2 kg) yellow bullheads grown in cages in a natural pond at the New Alchemy Institute, Hatchville, Massachusetts. (Photo by Hilde Maingay, New Alchemy Institute.)

precisely how many fish were taken, but the yield must have been in the hundreds of pounds without any management whatsoever.

It is doubtful if ordinary hook and line fishing can keep pace with the reproductive rate of bullheads, but unlike the centrarchids, they may readily be taken in traps. Figure II–2–1 illustrates a type of trap which, when baited with raw liver, chicken entrails, or fish scraps is extremely effective. It is usually necessary to experiment a bit to determine just where in a given pond bullheads are concentrated, but once they are located it is possible to bring up as many as 200 fish in such a trap. Traps should be set on the bottom and left for a period of 15 minutes to an hour, depending on water temperatures. The colder the water the less rapidly the fish will respond.

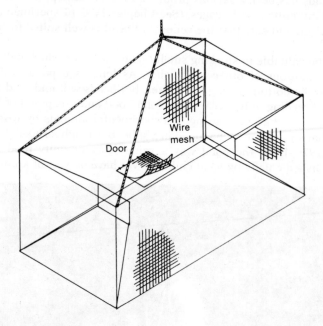

FIGURE II–2–1. *Bullhead trap.*

Small harvests may be made by hook and line. Few fishes are as cooperative about taking a worm-baited hook as bullheads, particularly in the late afternoon and evening; they are a perfect children's fish.

Wherever bullheads are stocked, fishing should be concentrated in the spring and fall. During northern winters they lie dormant, buried in pond bottom sediments. During the summer they may be captured readily enough, but in very warm water the texture of bullhead flesh is sometimes found to be excessively soft. This may be why they are seldom eaten in the deep South; I have no personal experience with bullheads there. If bullheads are taken from clean, cool water their sweet, red meat is, in my opinion, unsurpassed among small fresh water fishes, and they are less bony than most. These qualities make them one of the most suitable fishes for the homesteader or small farmer who wants good fish to eat, but cannot afford to make fish culture a full-time job.

CHAPTER II-3

Channel Catfish

When people talk about the boom in catfish farming in the United States, they are referring not to the bullheads, but to various large North American catfishes, principally the channel catfish (*Ictalurus punctatus*), and to a much lesser extent the blue catfish (*Ictalurus furcatus*) and the white catfish (*Ictalurus catus*). These fish, though closely related to the bullheads and superficially similar in appearance, are quite different animals. In addition to being larger and of far greater importance commercially, they are more finicky about their diet and environment. Their flesh is white, not red like the bullheads', but of similarly high quality.

The history of catfish culture in the lower Mississippi valley is central to an understanding of the history and current position of North American aquaculture as a whole, and so is discussed in the introduction. (See Chapter I-1.) What I would like to discuss here is the overemphasis by the media on agribusiness-scale catfish farming. It is true that giant catfish farms, some of them owned by multinational corporations, are a major force in the catfish industry. It is *not* true, as I have seen it reported in print, that you cannot farm catfish profitably without at least 300 acres (121.5 ha) under water and half a million dollars capitalization. The rapid growth of catfish "aquabusiness" has obscured the fact that some catfish farmers were making money before the arrival of processing plants, fast food chains and giant corporations. Up to that time, a higher percentage of small than large operations were successful. The business "secrets" of small growers are intelligent diversification (bait fish, soy beans, cotton, timber, etc.), good management, and a willingness to identify and secure local or ethnic markets rather than competing for mass markets. Among the commercial options best suited to the small grower are fee fishing (see Chapter IX-1), sale of live fish to other fee fishing pond operators, or to "live haulers" who carry the fish to northern markets, specialty restaurants, at-the-farm sales of live or fresh whole, dressed, or smoked fish, and sale of fingerlings to other growers.

Large and small catfish farmers generally raise their fish in fairly large (10 acre [0.4 ha] or more) ponds and use commercial feed concentrates. Their methods are well covered in print and there is no need to duplicate that information here. The reader interested in catfish farming may refer to publications listed in Appendix IV–1. There are, however, a number of non-conventional methods and ideas which, while not covered in the catfish how-to-do-it books, may be useful to the catfish farmer, particularly the small grower. Some of these are discussed in this book. (See Chapters III–3 and 5, IV–1 and 3, VI–4.)

Subsistence catfish farming

Channel, blue and white catfish, though high on the food chain, can also play a role in subsistence fish farming. Some farmers monoculture channel cats in fertilized ponds for table use. (See Chapter III–7.) The key to success in this sort of operation is regular harvesting. You should aim for an annual harvest of about 150 pounds of ½ pound (0.2 kg) fish per acre (165 kg/ha).

Of perhaps greater interest is the role of catfish in polyculture. In Chapter II–1 I mentioned that channel catfish are beginning to see use in conventional "farm ponds." Harvesting of such fish, usually by hook and line, can begin within a year of stocking if pan-sized fish are desired. Catfish will ordinarily not reproduce in farm ponds; if they do so, predation by bass generally eliminates most of the young. As a result it will usually be necessary to restock about every three years. Over a period of time this may prove uneconomic since, while the first catfish stocked can be the same size as the bass, subsequent stockings will have to be made with much costlier pan-size fish. Grass carp (*Ctenopharyngodon idellus*), brown bullheads (*Ictalurus nebulosus*), blue catfish and blacktail redhorse (*Moxostoma poecilurum*) may all have beneficial effects on channel catfish production. In Chapter II–1 it was pointed out that largemouth bass (*Micropterus salmoides*) can have a negative effect. White catfish, common carp (*Cyprinus carpio*), Chinese carps, crappies and sunfishes are also sometimes grown with channel cats, but so far no one has ventured to analyze their interactions.

CHAPTER II-4

Buffalofish

In the 1950's, when fish farming first became prevalent in the south central United States, the most commonly raised fishes were not catfishes, but a group of large and commercially important members of the sucker family (*Catostomidae*) known as buffalofishes (*Ictiobus* spp.). There are three species of buffalofish, of which the bigmouth buffalo (*Ictiobus cyprinellus*) has proven most suitable for culture; the following account refers primarily to that species. Research at the U.S. Fish Farming Experimental Station in Stuttgart, Arkansas, has shown that hybrids obtained by crossing female black buffalo (*Ictiobus niger*) with male bigmouth buffalo grow more rapidly than pure bigmouth buffalo. However, this discovery coincided with the decline of buffalofish culture, so the hybrid has been little exploited.

History of N. American buffalofish culture

From an ecologist's point of view, the buffalo is perhaps the best suited for culture of all the fishes of the lower Mississippi valley; certainly it is a more logical choice than the channel catfish (*Ictalurus punctatus*). The bigmouth buffalo may reach a weight of 80 pounds (36.4 kg).They are unique among large North American fishes in being low on the food chain, feeding on plankton, benthos and detritus.

Ecological and economic factors

But the ecologists were quickly overruled by the economists, who accurately perceived that, given the rather special conditions of the mid-twentieth century American economy, a farmer could make more money growing a high priced fish on a high priced diet (commercial fish feed) than by growing a low to medium priced fish at very low expense. In Arkansas in 1958 there were 3,032 acres (1,227.5 ha) of water devoted to buffalofish culture. Some catfish were being grown in ponds, but always as a secondary crop to buffalo. By 1963, more water was devoted to catfish than to buffalo, and today, while catfish culture is a thriving industry, there is no water devoted primarily to the culture of buffalofish.

In the long run it seems likely that the ecologists' choice will prevail; that as the economy changes there will be greater incentive to produce fish which can be grown at little expense and with reduced energy inputs. For the present, though, the channel catfish is a better choice for the commercial fish farmer. Not only is catfish culture more lucrative, it is presently possible to raise more pounds of catfish per acre. While this appears to be a violation of ecological principles it is merely a result of the greater amount of study which has been put into catfish culture together with the cumulative experience of catfish farmers over the years.

There are already signs of a new interest in buffalofish which may presage the development of a true American pond polyculture. (See Chapter IV–1.) Catfish farmers are beginning to stock buffalo in small numbers in their ponds. No additional feed is allotted to these buffalo, which derive their nourishment from excess catfish feed, catfish feces, and the rich plankton and benthos characteristic of the highly fertile waters of a densely stocked catfish pond. In this way they not only contribute a small additional crop at no expense, but may actually enhance the production of catfish by cleaning the pond and improving the water quality.

Fish-rice rotation   The role originally visualized for buffalofish was in fish-rice rotation. This adaptation of Oriental techniques never found much acceptance in the United States for a variety of reasons, mostly economic. There is reason to believe that fish-rice farming, which is applicable to a number of species in addition to buffalofish, may yet catch on here. Methods are described in Chapter IV–2.

CHAPTER II-5

Carp

The word "carp" is properly applied to any of a considerable number of large European and Asian fishes of the family Cyprinidae. In North America "carp" almost always refers to the common carp (*Cyprinus carpio*), a fish first imported from Europe in 1877 and subsequently established in 47 of the 50 states of the United States as well as the more southerly regions of Canada. The carp was brought to this hemisphere by the U.S. Fish Commission, with the intent of culturing it as a food fish. For this purpose the carp has much to recommend it, as aquaculturists throughout the world have long recognized. According to Shimon Tal, Israel's Director of the Inland Fisheries, "No other fish has yet been found that can be as easily managed for high yields." The early Chinese fish farmers recognized its value and made it an integral part of their sophisticated pond polyculture. By now, the common carp has been introduced practically everywhere fresh water aquaculture is practiced, and plays a major role in all of the Asian countries and much of Europe.

But in North America carp aquaculture has never become as important as its early proponents foresaw. This is certainly not due to any lack of adaptability to its environment, but to socioeconomic factors. The early American apostles of the carp "seem not to have considered the differences between American and European conditions. In North America human population density was low compared to Europe, land was plentiful, and farming of mammalian and avian stock, together with freshwater and saltwater fisheries, provided for a supply of protein far more abundant than that of any European country. There was simply no incentive for intensive culture of carp" (Bardach, Ryther and McLarney, 1972) or any other fish. Carp distributed by the Fish Commission to private pond owners were neglected when it was learned that there was no market for the fish. By one means or another, these fish found their way into natural waterways. Here they quickly reverted to wild type fish, which are not nearly as good food fish as the selected strains developed over many years in Europe and Asia. To make matters worse, the behavior and appearance of the carp were not appealing to sport fishermen, and it was blamed for the decline of sport fisheries, for which it was at most marginally responsible. Today, although even the wild carp can be made quite palatable, especially if smoked, most North Americans will tell you that carp are virtually inedible.

Selected and wild strains

In view of the natural suitability of the carp for intensive, low-cost culture, it would seem that the time is now ripe to polish up its image as a food fish and to begin developing carp culture systems tailored to the North American environment. Both of these ends will be best served if we begin, not with wild carp, but with a domesticated strain developed with the food fish grower in mind.

Selected strains of carp

The common carp is one of the few truly "domesticated" fish; no other fish has such a long history of successful selective breeding. Among the qualities for which carp have been selected are rapid growth, good rate of food conversion, disease resistance, tolerance of low temperatures, quality of flesh, reduced numbers of intermuscular bones, absence or scarcity of scales, desired colors, and high height-length ratio. The last of these characteristics refers to the profile of the animal's body. In Europe a "fat" fish with a high profile is preferred to a long, thin one.

Israeli carp

Most of the selected strains developed in Russia, Japan, Germany, and other countries are not available in the United States, but carp of a variety known as the "mirror" or "Israeli" carp have been imported and stock is offered for sale. The Israeli carp, which has the advantage of consuming more plant materials than most carp varieties, has only a few very large scales scattered about the body, a factor which simplifies cleaning. There is also a variety known as the leather carp, with no scales at all, but since Jewish law forbids the consumption of fish without scales, it is not grown in Israel nor in other places where Jewish consumers form a significant part of the market.

Although the Israeli carp was selected with Israel's environment in mind, it should prove suitable for most North American growers and as a basis for the eventual development of American strains. This latter task is not one for the small farmer; in fact I do not recommend that you breed your own stock at all. Although there is no particular difficulty in breeding the common carp, experience has shown that only a large operator can maintain and segregate stocks large enough to avoid inbreeding. In time, an organization such as the Carp Breeders' Union of Israel, which carries out large scale selective breeding and supplies improved carp stock to farmers, may also be instituted here. For the present, the best approach will be to make yearly purchases of the most suitable available imported strains.

Polyculture

Probably the most appropriate use of the common carp in North American aquaculture is as a major component in a polyculture. The carp adapts more easily to varied environments than most fish, and will usually make good use of whatever food is available. It is one of the most efficient converters of commercial fish feeds, yet in a pond which is at all fertile, it will fare better without such feed than most of the commonly raised food fish. Carp feces may be recycled by other fishes and vice versa. Carp are ordinarily neither piscivorous nor aggressive toward other fish.

Despite these facts, there is at present no sizable North American pond polyculture. The following comments apply to carp monoculture, which may serve as the keystone for the eventual development of an indigenous North American pond polyculture. The basis of polyculture and some of the forms a North American pond polyculture might take are discussed in Chapter IV-1.

The first two questions a carp farmer needs to ask are: How many fish shall I stock? What, and how much, shall I feed? In most forms of aquaculture, but particularly in carp culture, the answers to both questions will depend on the fertility of the pond. Few fish benefit more from fertilization than the common carp. The history of pond fertilization as practiced in the United States is discussed in Chapter III-2. Specific fertilization methods applied in carp culture in Europe and Asia are given in Chapter III-7.

Feeding and stocking carp

The appropriate rate of stocking will also depend on the size of fish stocked. The larger the fish, the better the rate of survival and the more rapid and uniform their growth. Of course the larger fish are also more expensive. The question of stocking rates is too complex to allow for a formula, but I will cite two stocking rates which may be useful to the culturist in establishing his own optimum rate. In Indonesia, where feeding is of secondary importance to fertilization, six-week old carp, one to two inches long (2.5 to 5.0 cm), are stocked at 6,000 to 12,000/acre (16,820 to 33,640/ha). For temperate climates these figures would have to be adjusted downward. In Japan's colder climate, where heavy feeding is the rule, carp of this size and age are stocked at 100,000 to 400,000/acre (247,000–988,000/ha) and progressively thinned; by four months of age only 4,000 to 12,000/acre (9,877 to 33,640/ha) remain. Beginners would do best to err on the low side in stocking carp ponds.

As for feeding, the carp has often been called the aquatic equivalent of the hog because of its high rate of growth and adaptability to different foods. Comments on feeds and feeding strategies designed especially for carp may be found in Chapter III-7.

In my opinion, the common carp will come to play an important role in commercial aquaculture in North America. For the present, however, there is little market for it and prices are low. A quality smoked product or some specialty food, such as *gefilte* fish, made from cultured carp might prove profitable. At this time, though, I can only wholeheartedly recommend the common carp as a "subsistence fish" for homesteaders, or for those interested in tinkering with the development of polyculture systems.

Commercial potential

Young mirror carp are sometimes cultured and sold for use as bait "minnows" (see Chapter IV-2.); perhaps a combination of commercial and subsistence carp farming could be arrived at on this basis. However, you should check with fish and game authorities before preceeding; in many places use of carp as bait is illegal.

CHAPTER II-6

Trout

Trout

Institutions involved in aquaculture receive more inquiries about trout than any other type of fish. Trout are popular for good reason; they are excellent table fish, and esthetically perhaps the most satisfying of all food fish to grow. As a result they command a good price on the market, and enjoy the highest status among fresh water sport fish. The world commercial production of cultured trout is 127 million pounds (57.7 million kg) and an unknown amount is grown in small ponds for home consumption.

Species of trout for culture

Fourteen species belonging to the genera *Salmo* and *Salvelinus* are commonly referred to as "trout" in North America, but rainbow trout (*Salmo gairdneri*) and brook trout (*Salvelinus fontinalis*) have proven most productive for cultivation. Rainbow trout will prove best for the majority of growers, but brook trout may be indicated where the water is extremely cold or where there is a known local preference for brook trout. Some trout culturists who grow fish for their own use prefer the brook trout simply for it great beauty and the somewhat higher quality of its flesh. A mixture of the two species may provide more interesting fishing.

Another commonly available species, the brown trout (*Salmo trutta*) is very important in the hatchery business (see Chapter IX-3) and in sport fishing, but it is not preferred for food culture. Commercial growers find that it does not tolerate crowding as well as rainbow or brook trout. Pond owners who raise trout for home use do not have crowding problems, but find that brown trout are more difficult to catch on hook and line than the others. Not only does this result in slow fishing, but the comparatively high percentage of brown trout which survive the first year or two of fishing prey heavily on subsequent stockings.

Water temperature for trout culture

Certain environmental conditions must be met before you can consider raising trout. The most critical of these is a suitable water temperature. A body of water used for raising trout should be between 50 and 68°F (10–20°C) as much of the time as possible; 64°F (17.8°C) is considered ideal. Lower temperatures are permissible some of the time (indeed they are unavoidable in standing water in the northern United States and Canada), but growth is very slow at such temperatures. The same growth-inhibiting effect will be observed between 68 and 76°F (20–24.4°C), while sustained temperatures above 74–77°F (23.3–25°C) may be lethal. (These temperatures should be adjusted slightly downward for brook trout.) Marginal waters —those which in occasional summers are too warm for trout—sometimes produce exceptional crops. If you are in doubt as to whether to stock trout or warm water fish, try trout first. Any warm water fishes surviving a failure will interfere with subsequent trout culture, but the reverse does not hold true.

In the United States, waters thermally suited to trout are likely to be found in the areas indicated on the map below (Figure II–6–1). In Canada temperatures suitable for trout will be found almost anywhere. However, isolated locations for trout culture may also be found outside what is normally considered "trout country." Trout are grown commercially as far south as northern Alabama wherever springs or other localized sources of suitable water occur, even though surface waters in the surrounding area may be too warm or too polluted for trout.

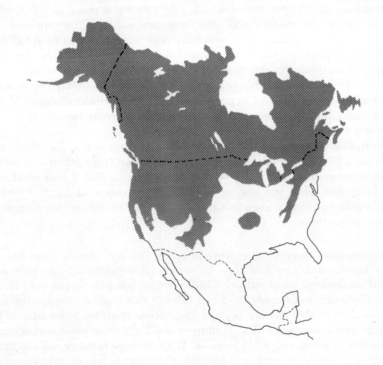

FIGURE II–6–1 *Area of North America where water temperatures suitable for trout culture are most likely to be encountered.*

Experiments in which trout were successfully grown to marketable size in cages during the winter in waters as far south as southern Alabama and Georgia may lead to further geographic expansion of trout culture. According to Collins (1972): "Inspection of surface water temperatures of lakes indicates that southern Arkansas, northeast Texas and central parts of Mississippi, Alabama and Georgia should provide about a 150–day growing season during which time 3 oz (85 g) trout may grow to 12 oz (340 g)." The same waters may be used to grow catfish during the warm months.

Water depth, pH, and D.O. for trout culture

If trout are to be maintained year-round in standing or slow moving water, twenty-five percent or more of the water will need to be at least ten feet deep to prevent "winterkill" in the North, and to provide a cold water refuge during the hot months in the South. The pH should be between 6.5 and 9 for best results and the dissolved oxygen concentration should always be at least 7 ppm.

Feeding

Of the fish recommended for culture in these pages, trout are the most exclusively carnivorous, and trout culture might seem at variance with our principle of working as close to the bottom of the food chain as feasible. But in ecological and economic terms the cultivation of trout is often the wisest choice that can be made. Typically, food chains in cold waters are shortened because some of the links between primary production and large predators are omitted. The only sizable North American cold water fishes which may occupy a place in the food chain between those of the insects and the trout are some of the suckers (family Catostomidae), and they meet with generally poor acceptance as food fish and are more difficult to culture than trout. Warm water fishes—including the common carp, (*Cyprinus carpio*), the catfishes, buffalofishes and sunfishes—may sometimes survive in environments suitable for trout, but they grow slowly there. Rainbow trout are metabolically adapted to grow most rapidly at temperatures of 55 to 68°F (12.8–20°C), and, if such waters are to be used in food production at all, there is no better known way than trout culture.

The practical implication of the feeding habits of trout is that the trout culturist must provide a diet quite high in animal protein. When trout are grown on a large scale, this usually implies the use of commercial feeds. But if you want to grow trout for your own table, you may be able to get by with fresh foods, by means of pond fertilization, or maybe even in a totally "natural" situation. (See Chapter III–7 for specific suggestions.)

Trout culture in ponds

If you contemplate commercial culture of trout, you will almost certainly need to provide a substantial, constant flow of water. (Exceptions to this rule are two specialized methods of trout culture—cage culture and prairie pot hole farming—which are discussed in Chapter IV–3 and later in this chapter, respectively.) But a standing pond or pool of any size will produce some trout for home use, as long as water conditions are suitable. The minimum size for a trout pond will depend only on the amount of trout you want to raise. With average fertility, no supplemental feeding, no perceptible current, and a growing season of eight months (months with water temperatures between 50 and 68°F or 12.8–20°C) you can figure on harvesting 50 pounds/acre (55 kg/ha) annually. If any of these conditions can be substantially improved upon, you may do better.

The basic design for a standing water trout pond is the same as that already described for a warm water fish pond, except that the trout pond should be deeper, and may be as small as 0.1 acre (0.04 ha).

Under certain special conditions, it may be possible to use a pond shallower than those generally recommended. There are a few climates where ponds do not freeze over in the winter, yet do not become too warm for trout in the summer. Or, if you wish to make a complete harvest every fall, a shallow pond may be used, and the problem of "winterkill" ignored.

The inclusion of devices to facilitate controlled drainage (Chapter VIII–1) is not as important as in warm water fish ponds, since it is usual in trout ponds to achieve harvests of virtually 100% without drainage.

You can do your first stocking in the spring, late summer, or early fall, depending chiefly on the availability of stock. If there are already fish in the pond before the introduction of trout, they should be eliminated because it is almost certain that they will not be beneficial. (See Chapter VIII–4 for a discussion of poisons and other methods of eliminating unwanted fish.)

Stocking rates

Stocking rates depend on the size of the fish to be stocked: "Advanced fry" (1–2 inches or 2.6–5.1 cm long) may be stocked at 500–1,000/acre (1,235–2,470/ha) under normal conditions. "Fingerlings" (2–5 inches or 5.1–13.1 cm long) may be stocked at 300–750/acre (741–1,853/ha). The larger the fish stocked, the higher the rate of expected survival. There is considerable leeway in the suggested rates to allow for variations in pond fertility, temperature, etc., and to provide the size fish you prefer. A trout pond will produce the same weight of fish at any reasonable stocking rate. It is simply a matter of deciding whether you want a few large fish or a lot of smaller ones. Never stock many more trout than you think you will use; the higher the population density, the greater the danger of disease and other problems.

You will want to do your second stocking, at the same rate as the first, in the late summer or fall, after the first crop of fish have been thinned out. From then on, annual or biennial late stocking is the normal procedure. Because trout rarely reproduce in ponds, and the operation of a trout hatchery is a complicated and specialized business, the vast majority of trout growers are best advised to continue to purchase stock. This fact should not discourage you—after all, most farmers do not produce their own seed.

The stocking rates mentioned earlier presume that there will be no supplemental feeding. Stocking rates may be augmented or growth rates increased if additional food is provided. This usually implies the use of commercial pelleted feeds, but there are a variety of natural foods which may be used, and fertilization may also be effective in some instances. Specific strategies for feeding trout are discussed in Chapter III–7.

Table quality

A word of caution to those who may come to trout culture from their experience as anglers: commercially cultured trout are fed almost entirely on prepared dry feeds and, from a gourmet point of view, they are quite a different fish from wild trout. Not a bad fish, but a different one. The flesh is white, not pink or red, and the flavor is more bland than that of wild trout.

Harvesting

A normal rate of growth for unfed trout stocked as suggested here would be 1 inch (2.6 cm) a month up to about 10 inches (25.6 cm), and slower thereafter. For best results, you should start harvesting when the trout reach 6 inches. To delay harvesting until after that point will only result in unnecessary loss of fish as they die off naturally and a smaller total harvest. Figure II–6–2 graphically illustrates the fate of trout in an unfished pond.

You should count on completely turning over your stock every two years. If there are substantial numbers of trout over two years old in your pond, you should make

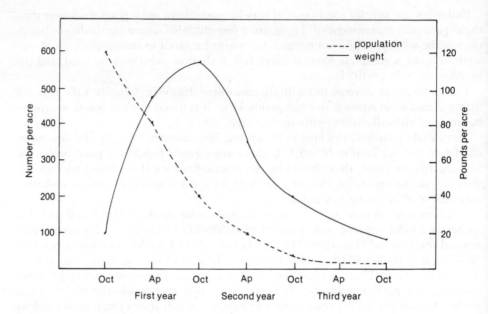

FIGURE II–6–2. *Population and weight of rainbow trout in a stocked, unfished pond over a period of three years.*

every effort to remove them. Large trout grow very slowly, compete for food with smaller trout, and may cannibalize fry and fingerlings.

Trout are among the easiest fish to harvest, individually or *en masse*, from small ponds. See Chapter VI–1 for methods.

Small scale trout culture in flowing water

Trout culture in flowing water for home consumption requires essentially the same techniques as conventional commercial trout culture, but stocking and feeding rates may be less. Commercial trout farming, like channel catfish farming, is well established in North America and elsewhere; there is no need to duplicate the abundant available material in the field. Here I will attempt only to help you determine whether trout farming is feasible in your particular situation.

There is a misconception in some circles that to be successful a commercial trout farm must be big. Certainly there are large trout farms; the world's largest, The Snake River Trout Company, produces 1.5 million pounds (0.7 million kg) of trout annually. But I also know of successful commercial trout farms consisting of no more than three small ponds. Granted that small farmers are unlikely to be able to compete with the giants for the supermarket trade, they can provide fresher, higher quality fish for local markets. (Two other ways for the small operator to make money with trout are by operating a hatchery to produce "eyed" eggs or young trout for sale, and with fish-out ponds. These businesses are discussed briefly in Chapters IX–1 and IX–3.)

Exploring markets

One of your first tasks, as a prospective small-scale commercial trout farmer, then, is to thoroughly research possible local markets, including restaurants, hotels and highway tourist trade. You should also determine the costs of stock and feed in your area.

A small commercial trout hatchery in Wisconsin, showing earthen rearing pond (foreground) and concrete raceways (background). Note aerator in pond. (Photo courtesy Meredith Olson, Wright County Community Action, Waverly, Minnesota.)

Armed with the results of such economic investigations, you should be able to determine how many trout you will need to sell to make a go of it. The next step is to determine whether you can produce that quantity of trout. The answer to this question will be largely determined by the volume of suitable water available.

Water volume factors

In flowing water fish culture the volume of water must be considered differently than in pond culture. The critical factor is not how much water can be impounded, but how much can be passed over the fish per unit of time. There is, of course, a limit of water velocity above which trout cannot maintain themselves in position. At high but less extreme velocities, the fishes' energy which might be used in growth will instead be expended in holding against the current. But up to that point, the density at which trout can be stocked, (and therefore the maximum amount of trout which can be produced for a given area), will increase directly with flow rate. Maximizing flow rate also enables you to take advantage of your water while it is still well oxygenated and before it is heated by exposure to sunlight.

Stocking rule

Experienced culturists gauge their stocking to produce 1.8 lb of marketable size trout per gallon per minute (0.18 kg/liter/minute). Beginners should aim lower: 1.25–1.5 lb/gallon/minute (0.12–0.15 kg/liter/minute).

Determining flow rate

Another useful rule of thumb in planning a trout farm is that the flow rate should be sufficient to provide at least one complete exchange of water per hour at whatever time of year the water is both warmest and least abundant. The rate of flow is determined not only by the size of the trout pools and the volume of water delivered at the source, but also by the shape of the pools and the topography of the site. A long, narrow pool or raceway will produce less eddying and more rapid replacement of water than will an equal area in a wider pool. For the same reason, circular pools are preferable to square ones.

Topographic aspects

The topographic aspect of most concern to trout farmers is slope: the steeper the land the more rapidly can water be passed over it. But on very steep slopes erosion problems may arise and construction costs are higher. The ideal site for a trout farm has a slope of 1% to 3%. On such terrain, pools may be designed both to maintain a satisfactory flow rate and to incorporate a few one foot (0.31 m) drops to reoxygenate the water.

If, after considering the economic and environmental factors outlined above, commercial trout culture seems to fit your situation, you should consult with other trout growers and study books and publications on trout culture. (See references in Appendix IV–1.)

"Pot hole" farming

Small-scale commercial trout culture in standing waters is possible under certain special conditions. For example, consider the success of Canadian farmers in stocking prairie pot holes. "Pot hole" in this context refers to the numerous five to forty acre (2 to 16 ha) lakes which dot the southern half of the prairie provinces of Canada and parts of the states of North Dakota and Minnesota. The principal agricultural crop in this area is wheat, and wheat farmers have generally considered these pot holes simply as nuisances. Since they are not connected to river systems but derive their water from snowmelt they contain no fish. Hunters and conservationists, however, prize the pot holes as perhaps North America's most important waterfowl breeding grounds, and have resisted efforts to drain them.

In recent years the conflict has moved toward resolution, as farmers have also come to value the pot holes—as trout rearing lakes. The rearing of trout in prairie

A typical "pothole" lake used for trout culture in Saskatchewan. (Photo by M.E. Swanson, Saskatchewan Fisheries and Wildlife Branch.)

pot holes involves only two steps: stocking of fingerlings at 200–300/acre (494–741/ha) in the spring, and harvesting of marketable size trout in the fall. Most pot hole farmers stock 2–2½ inch (5.1–6.4 cm) fingerlings and harvest at ½–1 pound (0.22–0.45 kg). But some fish up to 6 inches (15.4 cm) long are stocked, and trout of 1½ pounds (0.68 kg) or more are caught. No feeding is necessary, since the pot holes teem with natural food, notably the fresh water amphipod *Gammarus lacustris*. The flesh of trout which feed largely on *Gammarus* takes on a pink color like that of wild fish, and such trout can be sold at a premium price.

Not every pot hole is suitable for trout rearing, though certainly a high percentage are. Four conditions should be met before commercial trout culture is attempted in a prairie pot hole.

Conditions for commercial pot hole culture

(1) The water temperature at the peak of the summer should not exceed 75°F (23.9°C). Some pot holes less than 10 feet (3.05 m) deep may fail to meet this condition.

(2) There should be a complete winterkill every year or nearly every year, so that uncaught fish will not survive to cannibalize fingerlings stocked the next spring. You can size up the likelihood of winterkill by measuring the D.O. (dissolved oxygen concentration) of the water shortly before the ice breaks up in the spring. If the D.O. is zero, winterkill is certain. A sure indication of zero D.O. is the smell of rotten eggs caused by hydrogen sulfide in surface water. But deoxygenation can occur without this odor, so it is best to test the water. (See Chapter X-2.)

(3) The total concentration of dissolved solids should be less than 2,000 p.p.m.

A tub of pan-sized rainbow trout from a Saskatchewan pothole. Note the full, rounded bodies and small heads, indicating very rapid growth. (Photo by M.E. Swanson, Saskatchewan Fisheries and Wildlife Branch.)

There is a poor survival rate for trout in pot holes with higher amounts of dissolved solids.

(4) The water should not give an objectionable taste or odor to the fish. This has been a problem in some waters which were perfectly satisfactory from the trout's point of view. The only sure way to determine the suitability of a pot hole's water is to stock it and sample the results. It might be a good idea to stock a pot hole for the first time with just a few fish, perhaps in cages. Your taste tests should be carried out on fish caught late in the season, as some lakes will produce a "muddy" taste in the summer but not in the fall.

The foregoing conditions do not guarantee success. For unknown reasons experimental stocking of four pot holes in North Dakota was a failure, due to very poor survival. But generally, it these conditions are met, a pot hole can be stocked about two weeks after the ice goes out, by which time the water should be well oxygenated.

About 24 percent of the harvest (including that of non-commercial growers) is made by angling in the fall, but the major commercial harvest is by gill net just before or after the water freezes. Growers believe that delaying the harvest until as late as possible results in better flavor and less spoilage. (See Chapter VI–1 for details of harvesting methods, including specialized techniques for gill netting under the ice.)

Recovery rates in pot hole farming vary widely—from 0 to over 80%; the average recovery seems to be between 20 and 30%. This may seem very low, but given the lack of operating costs, it is commercially satisfactory. Yields usually reported are in the range of 100 pounds/acre (110 kg/ha). The dollar return of pot hole trout culture is about 2½ times that of wheat farming per unit area.

Another method of producing trout in standing (or running) water is cage culture. Cage culture methods for trout are no different from those used for other fish (see Chapter IV–3), but it may be more difficult to meet the trout's environmental requirements in cages. The dissolved oxygen content should be watched closely; some losses of caged trout have been reported due to oxygen depletion. In order to oxygenate the surface water trout cages should be placed so as to receive maximum exposure to wind and wave action.

Cage culture

Cage culture can be used to produce trout in waters which would otherwise be unusable for trout culture, including lakes too large for intensive management, irrigation canals, and ocean bays. (Trout can readily be acclimated to salt water.) In at least one instance, in Manitoba, cages were used to overwinter trout in a lake which freezes down to several feet in winter, where complete winterkill would ordinarily be expected. The movement of a great number of trout concentrated in a small area kept the water in the cages from freezing.

At the opposite extreme, mention has already been made of the use of cages in winter to make trout culture possible in bodies of water which are too warm for trout in the summer. (See page 54.)

Prairie pot hole culture and cage culture of trout are rather specialized methods of interest to relatively few trout farmers. They have been discussed here, not only for their own sake, but also as examples of the imaginative ways in which "idle" bodies of water can be used to produce food fish.

CHAPTER II-7

Yellow Perch Culture—A Possibility?

This chapter has been included not because I personally advocate cultivation of yellow perch (*Perca flavescens*), but because there has been considerable publicity about perch culture, and some readers will be curious. Most of the publicity has been generated by the work of a team of scientists, economists, food processors, etc., at The University of Wisconsin. Their research was stimulated by the decline of the yellow perch fishery in the Great Lakes. The yellow perch is certainly one of our better food fish and one which people in the Great Lakes basin, and particularly in Wisconsin, have long been accustomed to consume in quantity.

The present situation in Wisconsin in several ways resembles what happened to channel catfish (*Ictalurus punctatus*) in Arkansas in the 1950's: a popular food fish is suddenly in short supply and prices are rising sharply. Due to the price situation, commercial aquaculture, which was not previously considered economically feasible, is the subject of serious study. And like the channel catfish, the yellow perch is high on the food chain. If commercial perch culture is to be a reality, perch farmers will need a high-protein feed similar to those now manufactured for catfish.

Part of the reason for the decline of the perch fishery is the deterioration of the Great Lakes environment, while natural cycles of scarcity and abundance may be involved as well. But overfishing is also implicated. Though any species can be over-fished, a piscivorous fish like the yellow perch will suffer sooner than one lower on the food chain. The root problem in the perch fishery, then, as in the catfish fishery before it, is overdependence on a high order carnivore, when it would be better to use a less carnivorous fish. When the method of exploitation is fishing, the potential result is stock depletion. When the method is aquaculture, the same root problem leads to the high economic and ecological cost of feed, as the catfish culture industry has seen. (See Chapter III-6 for a discussion of fish food materials and costs.)

In the long run, it would be wiser to concentrate on improving environmental quality and intelligent management of perch fisheries, while devoting major aquaculture efforts to biologically more appropriate fishes like carp (*Cyprinus carpio*) or bluegills (*Lepomis macrochirus*). Biologists used this same ecological argument in the 1950's in favor of buffalofish over catfish, but they failed to convince due to the hard fact of the relative prices of catfish and buffalofish. Nor will it do much good to suggest that a hopeful commercial grower consider cultivating bluegills in Wisconsin, where there is presently neither a market nor shortage of them, and where perch are marketable at high prices.

There is one important aspect in which the prevailing method of catfish culture differs from the methods being tested for perch, and it may prove critical in the evolution of perch culture. For whatever reasons, the University of Wisconsin researchers have emphasized "closed system" culture. (See Chapter IV-4 for a critique of closed systems.) It seems clear to me that we have reached a point in history where such energy and technology-intensive systems are inappropriate for fish production. It also seems likely that, even if all the technical "bugs" are worked out, they will eventually become economically unviable.

In the beginning, this opinion was surely not shared by the members of the Wisconsin team. However, despite the generally good quality of their ongoing research, they have failed to demonstrate the feasibility of perch culture for Wisconsin farmers. No one has yet sold a crop of yellow perch, and the early emphasis on promotion of closed system perch culture has been dropped in favor of a more purely research oriented program.

The prospects for perch culture might be altered somewhat if one of two things were to occur. If semi-closed aquaculture systems such as those now in operation at New Alchemy, but more efficient, could be developed, or if methods for open pond culture of yellow perch were perfected, ecologically more gentle forms of perch culture analogous to small-scale trout culture might arise. Nevertheless, the yellow perch would remain a highly carnivorous fish and, where the goal is to maximize the production of fish as food, other species would continue to be preferable as the principal crop.

CHAPTER II-8

Exotic Fishes

Pros and cons of exotic fishes

There are over 8,000 species of fresh water fish in the world; of which more than 700 presently inhabit North American waters. "Exotic," as used here, does not imply a species with peculiar or bizarre qualities, but merely that it is native to another country. In this book only about a dozen of these are given anything approaching extensive treatment as objects of culture for food. Other species may eventually become important for food culture, but fresh water aquaculture in North America need not be limited to our comparatively few native fishes. If we look at agriculture, we will see that the majority of the plants and animals we raise did not originate where they are grown.

Some of the fish we now raise are not native to the places where they are cultured. Of the major species discussed in this book, one, the common carp (*Cyprinus carpio*) has its origins in the Old World, though by now it is firmly established in North America. Among our native fishes, the rainbow trout (*Salmo gairdneri*), the largemouth bass (*Micropterus salmoides*) and the bluegill (*Lepomis macrochirus*) have been introduced just about everywhere there is suitable habitat, and almost all of the other fish mentioned here have been transplanted somewhere outside their native range.

It should not be surprising, then, that certain exotic fishes are suggested for culture from time to time. Whenever this happens, a controversy immediately arises between the advocates of the particular fish and environmentalists who oppose its introduction. There is a good deal to be said for the environmentalists' position. It is virtually certain that any species of fish introduced into an outdoor culture system will eventually find its way into nearby natural waters. When we turn to the history of exotic animals in North American environments, we find that more of these introduced species have been detrimental than were beneficial to the native ecosystems. (A still greater number have simply failed to survive or reproduce, or have had no important effect.) But a smaller number of introductions, including some fish, have been beneficial. For instance, most fishery biologists would agree that the brown trout (*Salmo trutta*), introduced from Europe, has been the salvation of trout fishing in much of the United States.

What it all comes down to is that each case will have to be considered individually—and cautiously. Any species brought in for culture must be introduced only under the most rigidly controlled conditions. And it should be shown not only that it is extremely unlikely to cause serious harm in its new environment, but also that its successful introduction will lead to considerable benefits. This involves careful study. I would strongly discourage the unauthorized (by state or provincial authorities) introduction of non-native fishes anywhere they could have access to natural waters. If you are interested in working with an exotic species, think twice and see if there is not a suitable native species. Then consult your state or provincial fishery authorities both for their opinions and to find out if the proposed importation is legal. (Often it will not be.)

With that said, we can discuss three types of recently introduced exotic fish which, however, are still controversial and not widely distributed in the United States. I refer to the grass carp, the tilapias, and the walking catfish.

Ctenopharyngodon idellus is generally referred to by environmentalists opposed to its introduction in the United States as "grass carp," while its proponents prefer to call it "white amur," after one of its native rivers in Siberia and Manchuria. I prefer to call it grass carp simply because that name was in general use before anyone dreamed up the name white amur. The existing variety of colloquial names for fishes is bewildering. The old names at least have the charm of regional vernacular; let us not add to the confusion by using names created solely for their public relations value. Let the status of the grass carp in North America rise or fall on its merits as a fish.

The particular attractions of the grass carp are the high quality of its flesh and, more important, its virtually insatiable appetite for green plant matter. It is one of the principal components in classical Chinese pond polyculture (see Chapter I–2); Chinese fish farmers feed it great quantities of vegetable tops, grass clippings and the like. In North America, as in China, it is considered desirable to have a high quality food fish which is not only low on the food chain, but will make good use of feeds which are not digestible by most other fish. Perhaps even more attractive from a North American point of view is the prospect of a fish which can act as a biological control against the weeds which choke so many of our waters.

Reports of the effectiveness of grass carp as a weed control vary widely. Table II–8–1 is a summary of the results achieved in weed control with grass carp in various parts of the world.

Grass carp in weed control

There are certain generalizations which can be made about the grass carp as an agent for weed control:

• Above a minimum size of perhaps 2½ pounds (1.14 kg), it is the total weight of grass carp in a body of water, not the number of individuals, that determines its effectiveness. Seen against the figures in the table, some of the advertising claims which have been made for grass carp seem dubious.

• Although the native habitat of the species is in a temperate climate, it is more effective for weed control in warm water. (It can tolerate temperatures up to 95°F

Table II–8–1.   Grass carp's effectiveness in weed control in various locations

State or Country	Age or Size of Fish	Stocking Rate	Results
Alabama	mean 0.29 lb (130 g)	140/acre (346/ha)	Controlled *Mougeotia*, *Zygnema* and *Eleocharis* within 30 days.
	—	19 fish in 0.12 acre (486 square meters)	Complete cleaning of a very weedy pond in five days.
	—	20 to 40/acre (49–99/ha)	Control of *Chara*, *Potamogeton* and *Eleocharis* within 1 month.
	—	685/acre (1,692/ha)	Elimination of *Najas*, *Potamogeton*, *Elodea*, *Chara*, *Spirodela*, *Utricularia*, *Eleocharis* and *Pithophora* within 3 weeks; elimination of *Alternanthera*, *Myriophyllum* and *Eichornia* within 5 weeks.
Arkansas	¼ to ½ lb (0.11–0.22 kg)	16/acre (40/ha)	Did not control *Najas* and *Potamogeton*.
	¼ to ½ lb (0.11–0.22 kg)	100/acre (247/ha)	Controlled *Najas* and *Potamogeton*.
	11.7–15.6 inches (30–40 cm)	20–40 acre (50–100/ha)	Control of *Chara* and other weeds.
	11.7–15.6 inches (30–40 cm)	688/acre (1,700/ha)	Elimination of all submerged plants within 6 weeks.
Bulgaria	—	—	Control of a wide variety of weeds during the summer only.
Czechoslovakia	yearlings	—	Control of 35 species of water plants.
England	2-year olds	267 lb/acre (300 kg/ha)	Reduced aquatic plant growth by 50% in 5 months.
Florida	2.2 lb (1 kg) or more	81–162/acre (200–400/ha)	Control of dense infestation of *Hydrilla*.
	mean 1.05 lb (2.3 kg)	5/acre (12/ha)	Controlled *Hydrilla* reinfestation after herbicide treatment. Did not control *Chara*, *Vallisneria* or *Najas*.
Germany	—	—	Not considered an effective weed control at any population density.
India	—	142/acre (350/ha)	Elimination of *Hydrilla* from "choked" ponds within 2 months.
Japan	yearlings	8/acre (19/ha)	Not effective as a weed control.
	yearlings	22/acre (55/ha)	Seventy percent elimination of extremely heavy weed growth in one year.
	yearlings	20/acre (50/ha)	One-hundred percent elimination of submerged and floating plants in 1 year; no effect on marginal plants.
Mexico	—	202–1,012/acre (500–2,500/ha)	Controlled water lilies in small ponds after 60 days, in the absence of other types of plants.
		81/acre (200/ha)	Controlled submerged plants in two lakes

Table II-8-1. *Continued*

State or Country	Age or Size of Fish	Stocking Rate	Results
Rumania	yearlings	202–324/acre (500–800/ha), followed by 81–202/ acre (200–500/ha) the next year, if necessary	One-hundred percent elimination of aquatic plants.
	3 years	32–61/acre (80–150/ha)	Control of aquatic plants.
	4 years	12–20/acre (30–50/ha)	Control of aquatic plants.
	3 and 4 years	65–97/acre (160–240/ha)	Elimination of aquatic plants.

[35°C], and has been cultured in climates hotter than any listed in the table. At temperatures below 60°F [15.5°C], it virtually ceases to feed.)

• Certain types of plants are eliminated more readily than others. (Food preference experiments have been carried out with the grass carp; it inevitably selects the softest plants first, and prefers submerged to emergent or floating plants. The hard parts of emergent plants such as cattails are eaten as a last resort, and only partially, if at all.)

There are other points in favor of the grass carp:

• It grows rapidly and reaches a large size. Individual fish have been reported to gain 5 pounds (2.3 kg) in a summer. The maximum weight does not seem to be known, but it is probably over 100 pounds (45 kg).

Grass carp's advantages

• In informal, blindfold taste tests, the flavor and texture of grass carp was judged superior to other commonly eaten fresh water fish, and was considered equal in quality to red snapper. (Its food value, particularly when small, is reduced by the large numbers of its intermuscular bones—the small Y-shaped "floating" bones some people find irritating.)

• Another advantage of the grass carp, when used in polyculture, comes as a by-product of its voracious feeding habits: Though it eats great quantities, it is rather inefficient in digesting its food. However, the partially digested material in grass carp feces is digestible by common carp, and probably other omnivorous fishes. As a result, feeding grass carp may increase the amount of food available to other fish with no extra cost or labor. (This has been disputed, however, and will be discussed further.)

There is, then, a lot to be said for the grass carp, but environmentalists point out that it is unclear what effect it would have in the wild in North America, should it become established. There is the possibility that it would eliminate water plants which are beneficial to wildlife and other fish. Or, by feeding selectively, it might bring about the replacement of an existing community of aquatic plants by other, more noxious, varieties. Also, the grass carp's diet is not totally restricted to plant matter, and it is unclear if plants are always its preferred food. Grass carp are known to feed on mollusks and fish, and cultured fish are sometimes fed silkworm pupae. Anglers have caught grass carp on live minnows and artificial lures. It is,

Grass carp's disadvantages

therefore, not unreasonable to fear that it would predate on or compete with valuable native fishes.

Natural spawning of grass carp

Grass carp's proponents counter all this by saying that it is "impossible" for it to reproduce naturally in North America. It is true that its native range and habitat are quite restricted, consisting of a very few rivers in Russia and China, and that pond culturists in China must obtain fry from these rivers or resort to artificial spawning. (See Chapter V–1.) And it does seem extremely unlikely that the grass carp would reproduce in small drainage systems in North America. But it has spawned naturally in Japan, the Ukraine, Taiwan and Mexico—all places outside its natural range. It is possible that suitable spawning conditions might exist in the Mississippi—Missouri system, where the species is definitely present. So, while natural spawning of the grass carp in North America has not been reported and there is a case to be argued that it is not likely, those who state that it is *impossible* seem, in their eagerness to promote the fish, to be confusing fact with wish.

In 1982, one of the most enthusiastic proponents of the grass carp, J. M. Malone, of Lonoke, Arkansas, began to market a sterile "triploid grass carp." Just how this fish is produced has not been revealed, but it is claimed to be the equal of regular grass carp in all respects, but without the risk of unwanted reproduction. The legal response by the various states to this development remains to be seen.

One state (Arkansas) is now actively promoting the grass carp for culture and introduction as a food fish, for weed control, and as a sport fish. The Arkansas Game and Fish Commission has stocked it in some reservoir lakes with weed problems, and a number of fish farmers in the state are growing it. Twenty-six other states, led by neighboring Missouri, have absolutely prohibited its importation, while all but seven others require some sort of permit. For the time being at least, most Americans will find it difficult or impossible to stock it legally.

Other Chinese carps

Students of Chinese aquaculture are often interested in some of the other Chinese carps, particularly those which are plankton feeders. Some of these species are being raised experimentally in Arkansas, and I have heard that they can be found—as fresh fish—in the Chinese markets in San Francisco, indicating that someone is growing them. But all the environmental and legal questions concerning the grass carp apply equally to these fish, and they are not generally available in North America.

Tilapia

The only exotic food fish which have received publicity comparable to the grass carp are the various members of the genus *Tilapia*.[1] Like the grass carp, *Tilapia* spp. have been dubbed with glamorous trade names: "St. Peter's fish" and "African perch." The name "St. Peter's fish" refers to the fact that they undoubtedly formed the major part of St. Peter's catch in the Sea of Galilee. Today they are important as food fish throughout the Middle East and Africa, where they are native, as well as in much of India and southeast Asia and parts of Latin America. They are a high quality food fish; John Hess, a *New York Times* food writer, on sampling tilapia raised at The New Alchemy Institute, called them the best farm-raised fish he had ever tasted.

There are many species of tilapia, with a variety of feeding habits, though all are

[1] In 1977, most of the larger *Tilapia* spp. were reclassified as the genus *Sarotherodon*. In 1981, this was changed to *Aureochromis*—an example of the sort of obfuscation sometimes carried on by taxonomists in the name of clarification—frightening people away from the very useful Latin names.

Blue tilapia (Tilapia aurea). (*Photo by Robert Sardinsky.*)

low on the food chain. Of those commonly available in the United States, *Tilapia mossambica* seems the most omnivorous and adaptable. *Tilapia rendalli* and *Tilapia zillii* prefer leafy plants as, to a lesser extent, does *Tilapia nilotica*. *Tilapia aurea* (blue tilapia) eats both leafy plants and plankton, while *Tilapia hornorum* is thought to be primarily a plankton feeder. "Monosex" hybrids, notably ♂ *T. hornorum* × ♀ *T. mossambicus* and ♂ *T. hornorum* × ♀ *T. niloticus* are becoming increasingly important in fish culture. (See Chapter V–2 for a general discussion of monosex hybrids.)

In most parts of the United States, tilapia will become neither pest nor major crop, since they are not able to survive sustained temperatures below about 55°F (12.8°C). Nevertheless, their import is banned in a number of northern as well as southern states. In Louisiana they were (and possibly still are) grown in polyculture with channel catfish, with very satisfactory results, but they are now illegal. California has taken perhaps the hardest anti-tilapia line of any state, but an exception was made in the Colorado River valley, where *T. mossambica* was stocked to control weeds in irrigation ditches. In Florida, at least one species (*T. aurea*) has become naturalized in some waters.

The chief concern of the environmentalists with regard to tilapia has to do with their prodigious reproductive potential. Tilapia do not produce an exceptional number of eggs, but most species have the unusual habit of incubating the eggs in one or the other parent's mouth and protecting the young there for the first few weeks of life. This leads to a very high rate of hatching and survival. In other parts of the world, tilapia have often overpopulated ponds to an extent that makes our common problems with sunfishes look inconsequential. A number of methods have

Tilapia's reproductive potential

been developed to deal with this problem. (See Chapter V–1.)

The only part of the United States where tilapia have become truly abundant is in the portion of Florida directly east of Tampa. I have observed populations of *T. aurea* in Lake Parker, at Lakeland, Florida, that amount to the densest natural concentrations of fresh water fish I have ever seen. Local attitudes toward this situation are interesting. I talked to a number of apparently poor people who were harvesting tilapia with cast nets; it was their opinion that the new fish were one of the best things that had happened locally. Sport fishermen, on the other hand, grumbled that the tilapia were the cause of the drastic decline in largemouth bass and bluegill fishing.

Tilapia and phosphates

Certainly bass and bluegill fishing was down. In nearby Lake Effie, tilapia, which represented only 1% of the fish biomass in 1968, had reached 93% by 1972. But this change is almost certainly due more to ecological perturbation of the environment than to the introduction of tilapia. That portion of Florida is a major producer of phosphates. A phosphate plant sits on the shore of Lake Parker, and the result is an amazing sustained phytoplankton bloom. Similar problems, related to sewage or agricultural fertilizer, occur in other Florida lakes. In such circumstances, one would expect filter-feeding fish like *T. aurea* to prosper, while visual feeders like bass and bluegill decline.

The phenomenon of Lake Parker suggests the possibility of using *T. aurea* or similar species in the smaller, and equally green, pits and ponds that dot the phosphate-producing regions of Florida. The phosphorus "pollutant" provides free nutrients; all that remains is to stock and harvest tilapia. When blue tilapia were experimentally stocked in phosphate pits at 1,000 per acre (2,470/ha), they grew approximately 7 inches (17.9 cm) in eight months without feeding; annual production was reckoned at 1,810 pounds per acre (1,996 kg/ha/year). A few commercial fishermen are already harvesting some of the Florida lakes. The same approach, perhaps in conjunction with appropriate supplemental feeding, could be used anywhere in the semitropical U.S. that there are eutrophic ponds unconnected to natural waterways. Production could probably be increased by the use of hybrids or other techniques to control reproduction.

Cage culture of tilapia

Another possible use of tilapia is in cage culture in fertile environments like phosphate pits or, with feeding, in ponds. In semitropical environments cage culture has the advantage of preventing reproduction, while in more northerly locations it permits short-term culture during the summer months. Culture methods for tilapia in greenhouse ponds are discussed in Chapter IV–4.

Walking catfish

The "walking catfish" (*Clarias batrachus*) is likely to interest only residents of certain parts of Florida. It escaped from a tropical fish farm in the mid–1960's and has since been establishing itself in much of southern Florida. Despite the belief, based partly on its bizarre appearance and partly on the prejudices and fears of channel catfish farmers, that the walking catfish is inedible, it is an important (and reputedly excellent) food fish in its native country, Thailand. Other species of *Clarias* are similarly important throughout southern Asia and Africa.

In Thailand, fish farmers take advantage of the air-breathing ability of the walking catfish by crowding them into ponds at densities other fish could not tolerate. In this way they are able to achieve what may be the world's highest yields in standing water commercial fish culture—up to 86,427 lbs/acre/year (97,000 kg/ha/year).

Either wild walking catfish fry are captured in hand nets, or spawning methods like those used in the U.S. for channel catfish are employed. After hatching in jars or troughs, the fry are stocked in small pools and fed on zooplankton and boiled fish for several weeks, until they are 0.6 to 3.9 inches (1.5–10.0 cm) long.

Next, the young fish are stocked in ponds with smooth banks, surrounded by a fence to prevent the fish from walking away. They are fed on a 9:1 mixture of ground trash fish and rice bran; the conversion rate is around five or six. They reach marketable size (about ⅓ lb or 0.15 kg) in four months.

If a market for walking catfish could be developed, similar methods could certainly be successful in Florida. At present, legal restrictions prevent such development. A special permit is required to possess live *Clarias* spp. in Florida, and at least twenty other states have similar laws. Permits for outdoor culture would probably be obtainable only in the region where *C. batrachus* is already established. In my opinion, this is as it should be. These hardy, aggressive and highly mobile creatures should not be grown outside their current range.

CHAPTER II-9

Frogs

One of North America's native aquatic gourmet foods is frog meat. The demand for frogs, both for food and as laboratory animals, coupled with continuing inroads on prime frog habitat, has led us to the brink of a critical shortage of large frogs. Not surprisingly, there is considerable interest in frog culture. Despite the periodic appearance for more than forty years of books and pamphlets with titles like "Commercial Frog Raising," "Practical Frog Raising," and "How to Raise Bullfrogs for Fantastic Profits," I have not been able to discover one example of a successful commercial frog farm based on sale of frogs for table use. Somewhat more success has been achieved with culture of laboratory frogs. This type of frog culture is very energy and labor-intensive, and is beyond the scope of this book; I have included appropriate references for those who wish to pursue the matter further. (See Appendix II–4.)

Low intensity frog culture

To attempt intensive indoor or outdoor frog culture competing with wild-caught frogs as commercial food animals seems futile for reasons outlined below. But it may be feasible for some people to harvest frogs as a "bonus" crop from fish culture ponds. Another possibility is the modification and/or management of marshes and swamps not suited to fish culture. The most likely benefit of these approaches would be a few gourmet meals for the frog culturist, with perhaps a small supplemental income. You should *not* expect to make a living, or anything near it, from such low intensity frog culture.

Frog biology and culture

For those readers who are not already totally discouraged, I will discuss the biology of frogs, to explain some of the factors limiting frog culture.

Several species of frogs are eaten and/or used as laboratory animals in the United States and Canada. Except where I indicate otherwise, I will be discussing the bullfrog (*Rana catesbiana*), the largest North American frog and certainly the most important as a commercial food animal.

The bullfrog is native to most of the United States and southern Canada east of the Rocky Mountains. It has been widely introduced outside that range, especially in California. Its popularity as a food animal is based on its size. Several other species are equally good for food, but they are less efficient to harvest and process because of their small size. Large bullfrogs may measure eight inches (20.5 cm) from the tip of the nose to the end of the backbone and weigh as much as two pounds (0.9 kg); the maximum for most other edible frog species is half that length and much less in weight. Two other large frogs which may be of special interest where bullfrogs are not found are the redlegged frog (*Rana aurora*) of the Pacific Coast states and the pig frog (*Rana grylio*) of the deep South. These may reach five or six inches (12.8 or 15.4 cm) respectively.

Bullfrogs lay their eggs in shallow water some time between March and June, depending on latitude. Bullfrog egg masses can be distinguished from those of all other frogs by their great size; they usually cover about five square feet (0.45 square meters). The eggs, which hatch in a period of from four days to three weeks, depending on temperature, produce the familiar larval frog, or tadpole. Tadpoles will eat almost any soft organic matter; in the wild they consume mostly algae, and occasionally dead and decaying plants or animals. Metamorphosis to the frog state takes from five months to two years, depending on temperature and food supply. Unlike the tadpole, the adult frog is entirely carnivorous and will accept only live, moving food. Bullfrogs live and forage mainly along the shoreline, though they will hide in deep water when frightened. The adults are territorial and may defend as much as twenty to twenty-five feet (6.1–7.6 m) of shoreline. In general, frogs grow slowly. It can be said that to reach marketable size (¼ to ½ lb or 0.11–0.22 kg) will require from one to several years after metamorphosis, depending on the food supply and length of the growing season.

By now, the reader will have recognized some of the problems inherent in frog culture. The chief limiting factors are:

• The low growth rate. (The South is favored over the North because of the longer growing season.)

• Adult bullfrogs are very territorial. Their space needs may be reduced in proportion to the amount of food that can be provided.

• Adult frogs refuse any but moving food. This problem has been "solved" in Japan by gently oscillating non-moving food on a tray with a small motor. This seems a technologically extreme solution.

• The optimum environment for tadpoles differs from that for frogs. Tadpoles do best in isolated shallow pools with high algal production and few large predators. But frogs in such places will not find enough food to allow for rapid growth.

• The predators of tadpoles and frogs are legion. They include humans, fish, herons, some ducks, turtles, snakes, cats, raccoons, crayfish, large aquatic insects, alligators and other frogs. Protection against each and all of them in an outdoor situation is virtually impossible.

• Even if you were to succeed in overcoming all the other limitations and could maintain a very high population density of frogs, disease would then enter the picture. Frog diseases are quite rare in nature, but where frogs are kept in great numbers they have presented serious problems. Treatment techniques are almost unknown; intensive frog culture facilities must rely on regular inspection, heavy and frequent disinfectant treatments, and culling of unhealthy frogs.

From all this it becomes clear that few if any natural ponds, or ponds constructed for fish culture, irrigation, recreation, etc., will support dense populations of frogs. There are, however, steps which can be taken to increase the carrying capacity and to enhance the survival and growth rate of frogs in such ponds. Or ponds can be specially constructed for frog farming.

Suitable frog ponds

To maximize frog production in a given space, it would be best to construct two types of pond—one for frogs and one for tadpoles. The latter are easy to construct; they can be any shape and need be no more than a few inches deep. They should be exposed to the sun for a large portion of the day, though provision should be made to prevent total evaporation. A frog pond need not be as deep as a fish pond, in fact much of it should be shallow, but it should be deeper than a tadpole pond to provide food, shelter and a place for hibernation. The deepest water can be one to one-and-a-half feet (0.3 to 0.45 m) in the South and deep enough so as not to freeze to the bottom in the North.

Perhaps the most important consideration in designing a frog pond is the maximization of shoreline. The ideal system would consist of a series of wide ditches (preferably running north and south to take best advantage of the shade provided by plants on the banks). Where topography does not permit this, or when modifying an existing pond, the prospective frog raiser should construct as many inlets, peninsulas and small islands as feasible. Willows or other shade trees should be planted on the banks.

Stocking

Both tadpole ponds and frog ponds can be stocked. Tadpole ponds are stocked by introducing bullfrog egg masses gathered from other ponds during the breeding season. Frog ponds may be stocked with wild-caught or commercial frogs, but you should be careful to get only bullfrogs, and not smaller species. Do not handle frogs during hot weather, as this greatly increases the risk of disease. But, if you stock them in cool spring weather, they may not have enough time to become accustomed to their new surroundings before breeding time. So the best time for stocking is the fall.

Food supply

As in any form of animal rearing, food supply is critical. Tadpoles may be fed indirectly, by fertilizing the water to promote the growth of algae, or directly with any soft organic matter. Some frog raisers have fed the waste from butchered frogs to their tadpoles. In adding organic matter to very shallow ponds you must be extremely cautious not to completely deoxygenate the water.

Feeding adult frogs is complicated by their insistence on moving food. You may culture or mass harvest earthworms, flies, crickets, and other natural frog food organisms (see Chapters III-3 and III-5) but the relatively low population density of frogs leads to inefficiency in feeding, even in specially constructed ponds. Generally, the best that can be done is to provide conditions conducive to the growth of natural food organisms, including small fish, aquatic insects, crayfish, and smaller species of frogs, and to attempt to attract insects to the pond banks. The last can be done naturally, by planting flowers along the banks, or artificially, through the use of lights. (See Chapter III-3.)

Predator control

Perhaps the most difficult aspect of frog farming is predator control. This is especially true when frogs are intended to be a by-product of a fish pond. Any carnivorous fish will eat a frog of suitable size. Among common North American pond fishes, largemouth bass (*Micropterus salmoides*) and chain pickerel (*Esox niger*) are particularly fond of frogs. (In evaluating fish versus frogs, bear in mind that virtually any fish species will produce a greater quantity of meat per area than will frogs.) Terrestrial predators may be discouraged by fencing the pond; a three foot (0.9 m) high fence of close mesh wire topped with an outward angled section one-and-a-half feet (0.45 m) wide should suffice. Birds are another matter; since it is generally illegal to kill herons and other fishing birds, the best that can be done is to stretch a wire net over the pond, or at least the shallows. Often this is either impractical or esthetically unappealing.

Harvesting methods for "cultured" frogs are the same as those for wild frogs: it comes down to a matter of hunting. Frogs may be caught by hand, with a net, by dangling a simulated insect on a hook in front of them, or with a spear. Night harvesting with the aid of a flashlight, or preferably a headlamp, is most effective. I have heard it is possible to take advantage of the winter behavior of frogs by digging holes about three feet (0.9 m) deep in shallow water. Frogs are said to choose these spots for hibernation, and it is possible to harvest by simply digging them out.

Harvesting

Perhaps the most significant measure that could be taken to increase the consumption of frog meat would be to teach people to eat not only the hind legs of the frog but the back, shoulders and front legs as well. The quality of the meat in those parts is exactly the same as in the hind legs, but they are usually discarded. The reason is that, while dressing the hind legs of a frog is extremely simple—just snip them off close to the body and peel the skin back like a glove—extracting the rest of the meat involves a rather laborious gutting and skinning process. But in ordinary frog processing, the result is that almost half the meat is wasted.

Still, even if all the measures suggested in this chapter were widely adopted, it is doubtful that frogs would become important in the North American diet.

Using frog meat

CHAPTER II-10

Crayfish

Importance of crayfish

Among the most widespread aquatic food animals are the fresh water crayfishes (family Astacidae), native to most temperate zone waters. Crayfish (often called "crawfish" or "crawdads" in North America) are esteemed as a gourmet food in some European countries and are a main source of protein for certain tribes in New Guinea, but they may generally be described as underexploited. In the United States, commercial crayfish fisheries exist in the states of California, Oregon and Washington (*Pacifastacus klamathensis*, *Pacifastacus leniusculus*, and *Pacifastacus trowbridgi*), Wisconsin (*Orconectes virilis*) and Louisiana (*Procambarus clarki*, the red crayfish, and *Procambarus blandingi*, the white crayfish). The products of these fisheries are sold locally as food and fish bait, shipped to biological supply houses, and exported to European countries, especially Sweden. By far the largest crayfish fishery is the one in Louisiana, where ten to twelve million pounds (4.5–5.5 million kg) are harvested annually.

Louisiana is also the major and, until recently, only producer of cultured crayfish; sixty million pounds (27 million kg) are produced annually through culture in ponds and rice fields. In 1982 Mississippi became the second cultured crayfish producing state, with 300 acres (120 ha) devoted to the crop. South Carolina and Arkansas are expected to be producing by the time this book is out. Traditionally, crayfish culture in Louisiana has been concentrated in the bayou country of the southern part of the state, but one large, successful commercial crayfish farm has been established in northeastern Louisiana. Louisiana crayfish culture methods are well documented, so if you live in Louisiana, or elsewhere within the ranges of the red or white crayfish, you will first want to study the literature on Louisiana crayfish culture. (See Appendix IV–1.)

Selection of species

The relative importance of the two Louisiana crayfishes does not mean that they are the best or only suitable species for culture. In fact, the introduction of the red crayfish, or any species, into regions where it is not native should be discouraged;

the red crayfish has become a serious agricultural pest in Japan and Hawaii. The preeminence of south Louisiana in crayfish culture is probably as much due to the region's French cultural background as to the qualities of the native crayfishes. There are over three hundred known species and subspecies of crayfish in North America; they are found in almost all types of shallow aquatic environment. Some of these crayfishes are too small to be of importance as food, and there may be some of inferior quality, but careful scrutiny of the North American crayfishes would be sure to turn up some which are promising for culture as food animals.

The species most adaptable to culture for food purposes are likely to be those found in shallow, swampy waters, particularly waters which periodically dry up. Such crayfish are preferable to stream, lake and pond species for the following reasons:

• Methods already developed in Louisiana are likely to be adaptable to them with a minimum of trial and error.

• They may be grown in waters which are temporary and/or too shallow for fish culture. The possibilities of finfish/crayfish polyculture are limited by many fishes' predilection for crayfish as food. When the choice between finfish and crayfish must be made, finfish, which make use of the entire water column rather than just the bottom, will ordinarily win out.

Whatever type of crayfish you consider, bear in mind that crayfish habitat requirements are very specific. A species native to a fast-moving, rocky stream will not succeed in a swamp and vice versa.

If you live within the range of the red or white crayfish, you will perhaps be able to adapt Louisiana methods with appropriate adjustments for the dormant period of your local crayfish. Red crayfish are found in the Mississippi and Ohio valleys as far north as Missouri and Tennessee, in Gulf of Mexico drainages from Florida to Louisiana, and have been introduced to southern California, Nevada and Hawaii. The range of the white crayfish extends up the Mississippi valley to southern Wisconsin, through the Great Lakes to the Atlantic Coast and south to Florida. In the South, dormancy coincides with the hottest weather. In the North, crayfish become dormant when water temperatures fall below 45–50°F (7.2–10.0°C).

Red and white crayfish

If you do not have access to red or white crayfish, you will find yourself with a research project on your hands. Do not expect to become a serious competitor to the Louisiana crayfish industry; all of the established commercial aquacultures have evolved gradually over a period of decades, with contributions from a number of people. But you may be able to provide gourmet fare for your table and perhaps realize some supplemental income. You may even find that your pond or nearby waters support enough crayfish naturally so that you will not need to culture crayfish at all, but can simply fish for them.

Your first task will be to sample the crayfishes in your locality. In some environments, particularly shallow streams, you may be able to capture considerable numbers by turning over stones and logs and grabbing or netting the animals. This process is apt to be more efficient at night. In larger streams and shallow ponds a seine may be more effective. In deep or swampy waters, you will probably want to resort to trapping.

A typical crayfish trap, made of chicken wire, is illustrated in Figure II–10–1. Such traps are set on the bottom, with a buoy attached by a rope when used in deep water. They are usually baited with large chunks of fresh fish. If fresh fish is

Trapping crayfish

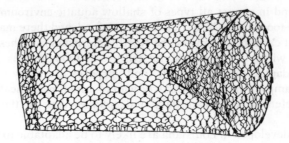

FIGURE II–10–1. *Crayfish trap.*

unavailable, some other sort of meat, for example a punctured can of dog food, will do. Small pieces, or meat which is not fresh, will not work. Traps may be checked within a few hours of setting. Night-time trapping is often best, since crayfish tend to be more active then, and may move into shallow water after dark.

Identifying crayfish species

Once you have a sample of crayfish (assuming that the size and quality of flesh are acceptable), your next task is to identify their species. For this, your first resource is likely to be Pennak's "Fresh Water Invertebrates of the United States." This standard text, which is likely to be found in any university library, has a taxonomic key to all the North American genera and the most common species, as well as a good bibliography to use as a jumping-off point for library research. Your state Fish and Game Department, or provincial Fish and Wildlife Branch, may also be able to help you secure information.

Jay V. Huner, of Southern University, has suggested the following species, in addition to the red and white crayfish, for small-scale culture: the paper shell crayfish (*Orconectes immunis*) of the midwest and mid-Atlantic states, *Orconectes nais* of the Great Plains region, and *Procambarus alleni* of south Florida. However, any species of suitable size, table quality and habitat preference is worth a try.

Reproduction

The most important aspects of crayfish biology, from a culturist's point of view, are reproduction and feeding habits. Given the diversity of crayfish species, you might think there would be great variation in these matters, but in fact there is a surprising uniformity.

Following mating, which may occur at any season, depending on the species and the climate, the female stores the live sperm for a period of from several weeks to several months. After the eggs are laid and fertilized, they remain attached to the female's ventral surface in a sticky mass which is quite visible. Such females are said to be "in berry" and should not be harvested. If the mating season is known, the same prohibition applies to females during the period from mating to egg laying. The incubation period usually ranges from two to twenty weeks, depending primarily on the temperature. The young remain attached to the mother for a period of from a week to a month. By the fall of the year of hatching, or at the latest by the following spring, most of the young are sexually mature.

Harvesting crayfish on a Louisiana farm. The fisherman in the foreground is emptying a good catch from a funnel trap like that shown in Figure II–10–1. Stakes mark trap sites (Photo courtesy B. Glen Ham, Cities Service Co., Fairbanks, Louisiana.)

The culturist should try to select species which produce large hatches, and which go into the dormant period (remaining inactive in a hole) in berry. This simplifies the stocking of ponds. Stocking is done just before the dormant period. Then the pond is drained and planted to any of a variety of cover crops, or to rice. Reflooding is timed to coincide with water temperatures which will bring on crayfish activity. At this time young crayfish emerge. The maximum life span of crayfish is generally no more than two years (though *Procambarus hagenianus,* and perhaps other species may attain seven years), so the young are ready for harvest by the next time the pond is drained. Harvests rarely approach one hundred percent, so enough reproductive females escape to continue the cycle and further stocking is seldom or never necessary. (See Appendix IV–1 for some references on crayfish life cycles.)

Crayfish in general are omnivorous, but not predaceous; their natural food consists mainly of plants. They will also eat dead animals if they are still fresh.

Crayfish diet

Intensive feeding of cultured crayfish has rarely been done. Cover crops or rice stubble may serve as the primary food. Louisiana crayfish culturists plant two aquatic plants known as alligator grass (*Alternanthera phylloxeroides*) and water primrose (*Jussiaea* spp.). Probably any soft aquatic plant would be more or less acceptable, but hard plants, such as cattails or bullrushes, will not be eaten. (Crayfish may be stocked to control weeds in fish ponds. See Chapter VIII–1.) Growers of bait crayfish have successfully fed aquatic plants, sorghum, soybean meal, millet, corn meal, cottonseed meal, rice bran, cut hay, potatoes, commercial

A commercial crayfish culture pond in Louisiana. Note that the water surface is almost completely covered with alligator weed, a food for the crayfish. (Photo courtesy Louisiana Agriculture Experiment Station and Jay V. Huner, Southern University, Baton Rouge, Louisiana.)

fish food and ground fresh fish. Including some animal protein in the diet will probably increase the rate of growth. A mixed diet may be best; there is some evidence that a combination of animal and vegetable foods produces better growth than either type of food alone, and it is believed that if there is not adequate vegetable matter in the diet the quality of the flesh will be inferior.

Feeds should be added sparingly and spread out over the entire pond. In fed crayfish ponds keep an eye out for crayfish coming to the surface—a symptom of oxygen stress. If this occurs, remove as much of the feed as possible and take other emergency measures as described in Chapter VIII-5.

Special problems   There are two other aspects of the biology of any crayfish which should enter into deciding whether or not to attempt its culture. For many species, burrowing is a necessary part of the reproductive cycle. These species should not be stocked where such activity might weaken earthen dams or dikes. Some species are potential crop pests. Problems of this sort will not often arise if a species is used within its natural range, but they could if, for example, a swamp species were stocked in a rice field.

Polyculture with
fish   Polyculture of crayfish with fish has traditionally been discouraged, partly owing to the pest potential of crayfish in fish culture, but also because of the predilection of many fish for crayfish as food. It is true that stocking crayfish in a bass, trout, or catfish pond in mid-summer would more likely result in increased growth of the

fish than in substantial crayfish harvests. Even fish with small mouths, such as sunfishes or common carp (*Cyprinus carpio*), may be serious predators under some conditions. However, there are conditions under which polyculture of food or bait crayfish with fish can profitable.

The safest sort of fish-crayfish polyculture involves non-predaceous fish, such as bait minnows or some of the tilapias. Bait farmers in Arkansas have harvested up to 366 pounds per acre (400 kg/ha) of *Procambarus acutus* from minnow ponds. Nearly as safe is the combination of crayfish with predaceous fish confined in cages. In one such experiment 742 pounds per acre (833 kg/ha) of red crayfish were harvested from ponds containing caged channel catfish (*Ictalurus punctatus*), while control ponds without fish yielded only 419 pounds per acre (470 kg/ha).

It is also possible, though more risky, to culture crayfish together with predaceous fish in an open pond situation. Good crayfish harvests have occasionally been realized in bass ponds, but the chances of success are far greater with fish like channel catfish, which are stocked annually as fingerlings, rather than allowed to reproduce in the pond. The key to success lies in timing; the fish should not be stocked until late enough in the season when the crayfish have matured. In experiments at Louisiana State University's Ben Hur Fisheries Research Center, 445 to 535 pounds per acre (500–600 kg/ha) of red crayfish have been produced with this approach. In the South this scheme of stocking is practical as the crayfish growing season begins earlier than the catfish season. Further north, where growing seasons are shorter for all animals, there is more overlap between crayfish and fish seasons. Here, the added production of crayfish might not compensate for the loss in fish production from delay in stocking.

Whether or not polyculture is attempted, cultured crayfish should be given plenty of cover to protect them against fish, frogs, birds, raccoons, mink and other predators, and to reduce the incidence of fighting among themselves. In France, where crayfish are cultured more intensively than in North America, stacked terra cotta pipes are used as crayfish "hotels." *Predators*

Where mass culture of crayfish other than the two commercially important Louisiana species has been attempted in North America, it has usually been for the production of fish bait. A number of publications on the culture of bait crayfish are listed in Appendix IV-1. Successful bait crayfish culture methods may or may not be appropriate for commercial production of food crayfish, but they are likely to provide useful guidelines for small family projects. Another possibility is the combination of commercial culture of bait crayfish with gourmet meals for the culturist. *Adapting bait crayfish culture methods*

Standing crops of crayfish in nature usually amount to less than 100 pounds per acre (110 kg/ha) but may reach 1,600 pounds per acre (1,765 kg/ha). Louisiana crayfish culturists have harvested as much as 3,600 pounds per acre (3,971 kg/ha); normally they get 300–1,000 pounds per acre per year (331–1,103 kg/ha/year). Though harvesting of cultured crayfish is generally done by trapping, in regularly shaped, shallow ponds that are not too weedy, seining may be effective. Seines used for crayfish should be more heavily weighted than normal fish seines. (See Chapter VI-1 for details on seines and seining.) Production levels in commercial culture of crayfish for bait or food vary widely, from 50 to 1,500 pounds per acre (55–1,655 kg/ha). *Production and harvesting*

Most of the meat on a crayfish is in the tail, though larger specimens also have a substantial amount in the claws. Most cookbooks give few, if any, recipes for crayfish, but Creole cookbooks usually offer a variety. You can also obtain a free pamphlet on crayfish cookery from Cities Service Company, Fairbanks, Louisiana. (See Appendix I.) Or you can try adapting any lobster or shrimp recipe for crayfish. Always cook crayfish thoroughly; in the eastern United States many species are an intermediate host of the parasitic lung fluke *Paragonimus westermani*, which causes the disease known as parasitic or oriental hemoptysis.

One can only guess at the potential of crayfish as a commercial crop and a food resource in North America. Crayfish presently support one major fishery and several minor ones, as well as the most solidly established localized aquaculture system in the United States. With local and export food markets, plus the bait and biological supply house markets, there is undoubtedly room for more commercial crayfish growers outside of Louisiana. Certainly, most rural Americans could enjoy at least an occasional meal of fresh wild or cultured crayfish, and there are many localities where crayfish can be developed into an important food resource.

CHAPTER II-11

Fresh Water Shrimp

Apart from crayfish, the only fresh water crustaceans of aquacultural importance are the giant fresh water shrimp or prawns found in tropical waters around the world. These shrimp, which reach lengths of 9.4 inches (25 cm), not including front legs and claws, are relatively little known in North America, but most people who have eaten them agree that they combine the best qualities of marine shrimp and lobsters. Not surprisingly, they have attracted the attention of professional aquaculturists. Most research has been concentrated on the widespread Pacific species *Macrobrachium rosenbergi*, but work has also begun on culturing *Macrobrachium carcinus* of the Gulf of Mexico and Caribbean Sea.

Though the growth of juvenile *M. rosenbergi*, and their maturation, mating and spawning in captivity were fairly easily achieved, for years biologists were unable to rear the larvae past the first stage. Then, in 1961, researchers at the Marine Fisheries Research Institute, Penang, Malaysia, made a simple discovery: the tiny first stage larvae of *M. rosenbergi*, which in nature are hatched in flowing water, drift down to estuaries where they molt, and then pass the second larval stage. Only after this do they return to fresh water. Following this discovery, biologists at Penang and elsewhere began to replace fresh water with eight to twenty-two parts per thousand saline water at the second larval stage. This opened the door to successful experimental culture of fresh water shrimp.

Schemes have been devised for culturing fresh water shrimp in greenhouses or other heated environments, but to date their feasibility has not been demonstrated. So far the prospects look bright only for outdoor culture in very warm climates. In the United States outdoor culture would seem possible only in south Florida, extreme south Texas and Hawaii.

Commercial culture in
Hawaii

In recent years, the greatest progress in fresh water shrimp culture has been made in Hawaii, where the research of Takuji Fujimura and his associates at the Anuenue Fisheries Research Center has led to the first commercial culture of *M. rosenbergi*. At the Anuenue hatchery shrimp are bred and reared to the juvenile stage, at which time they are 2.0 to 2.2 inches (5 to 6 cm) long. Then they are distributed to private culturists for stocking in 1 to 2 acre (0.4 to 0.8 ha) fresh water ponds (maximum salinity 3 ppt) 2.5 to 5 feet (0.8 to 1.6 m) deep, which are provided with gently flowing water. Juveniles are stocked at 50,000 to 100,000 per acre (123,500 to 247,000/ha) and fed on chicken broiler starter or commercial catfish feed.

Growing ponds may also be fertilized with 500 lb. of lime and 300 lb. of dried chicken manure per acre (550 and 330 kg/ha) or 100 lb. per acre (110 kg/ha) of synthetic 16-20-0 fertilizer. Subsequent fertilization is done as needed to maintain Secchi disk visibility at 5 to 7.5 inches (12.5 to 19 cm).

Harvesting begins seven to eight months after stocking, when the fastest growing 3 to 5 percent of the population will have reached marketable size (0.1 lb or 0.045 kg). Harvesting is repeated every three weeks, using a 1 7/8 or 2 inch (4.8 to 5.1 cm) seine which allows sub-marketable shrimp to pass through. The harvested shrimp are replaced with a comparable number of juveniles.

These methods are still largely experimental but, as of 1977, 11 farmers in Hawaii were involved in commercial culture of *M. rosenbergi* on farms ranging from 1 to 21 acres (0.4 to 8.5 ha) in size. Annual production has been 3,000 to 3,500 pounds per acre (3,318 to 3,871 kg/ha) after the initial seven to eight month growing period. With the present state of the art, a 10 to 20 acre (4.0 to 8.1 ha) farm seems most likely to be profitable in Hawaii. The economics of *M. rosenbergi* culture in Hawaii are well described by Shang and Fujimura, and their paper is recommended to all hopeful fresh water shrimp growers. (See Appendix IV–1.)

Non-commercial
culture

The non-commercial culture of fresh water shrimp is also a possibility. Whether it makes sense economically will depend largely on the price at which hatcheries are able to provide juvenile shrimp. Non-commercial growers can use individual ponds of less than an acre (0.4 ha), without flowing water, relying on organic fertilization rather than feeding. Non-commercial growers could also choose to harvest shrimp at a smaller size, should that prove advantageous.

CHAPTER II-12

Fresh Water Clams

Among the least utilized North American food organisms are the fresh water "mussels." These so-called mussels are, in fact, not mussels but are classified in the order Eulamellibranchia, and are more properly called "fresh water clams." Their life habits are in some respects quite different from marine clams or mussels. So far they have been cultured only experimentally, but their potential is so intriguing and they are so widely distributed that they are worth discussing here.

Most of the native North American fresh water bivalve mollusks belong to the family Unionidae, but in many parts of the United States there are also large populations of the introduced Asiatic clam *Corbicula* (family Corbiculidae). The Unionidae includes many species, all of which have similar life cycles. Critical in this cycle is the presence of a glochidium larva, which parasitizes fishes. This habit is not as undesirable as it might seem, since the larvae only rarely occur in numbers sufficient to injure the host or severely retard its growth. *Corbicula*, however, is unusual among fresh water clams in having a pelagic veliger larva, like marine clams, with no parasitic stages.

Of the two types, a little more is known of the biology and potential usefulness of our native Unionidae. At least one species, *Lampsilis claibornensis*, has been of economic importance since the early years of the century. It is sought primarily for its shell. Formerly this was used in the manufacture of buttons, but now the shells are ground or cut into small pellets and exported to Japan, where they are used as "seed" for cultured pearls. (Many of the Unionidae occasionally produce pearls, but this characteristic has not been much exploited.)

L. claibornensis is eaten in the Tennessee Valley, and perhaps elsewhere, but is not marketed for food. I have had some Unionid clams which were excellent, but more often I have found their flavor non-existent or "lakey." A number of acquaintances, mostly in the South, dispute my evaluation. The problem may lie with the species I have eaten, their habitats, or my culinary skills.

Generally, fresh water clams have only accidently appeared in aquaculture sys-

Taxonomy and life-cycle

N. American native Unionid clams

tems, but they could prove to be of considerable value, not only as a crop but also as "living filters" in fish culture. A fat mucket (*Lampsilis radiata*) weighing 7 ounces (200 g) was observed to siphon 8.8 gallons of water (33.3 l) per day. An interesting experiment in clam/fish polyculture is described in Chapter IV–1.

L. *claibornensis* is by no means the only native Unionid clam which reaches a suitable size to harvest for food. I have read of an instance when a large pond at the United States Fish Farming Experimental Station was drained, uncovering an unsuspected population of large clams, weighing as much as ¾ pound (0.34 kg) each. (A biologist friend guesses they were *Anodonta grandis*.) Despite the presence of several species of clam-eating fish, the population of clams amounted to more than 1,000 pounds per acre (1,100 kg/ha) (including shells). These clams were reportedly delicious, and I understand plans were made for further studies, though the results are not yet available.

In the first quarter of this century, when fresh water clam shells were still in demand for button manufacture, the fat mucket was successfully cultivated in crates floated in the Minnesota and Iowa waters of the Mississippi River. The wooden framed crates, covered with suitable metal screen, were lashed to long narrow rafts formed of barrels. The young clams placed in the crates were protected from predators, and provided with food by the river. By the end of the second summer they reached commercial size. In a variation on the same idea, a similar crate, measuring 10 feet by 10 feet by 8 inches (2.7 m by 2.7 m by 20.5 cm) was sunk in Lake Pepin, Minnesota. Instead of young clams, it was stocked with fish infested with glochidia. By the end of the year of stocking, the cage contained 11,000 fat muckets.

The Asian clam, Corbicula

As there is some confusion in the taxonomy of the genus *Corbicula* I shall refer to these clams by their generic name only. Since 1938, at least one, and maybe more species have been known to be present in the United States. The first specimens were found in the Columbia River near Knappton, Washington. Since that time populations have been found in most of the Pacific Coast drainages from San Francisco Bay north to the Columbia River, much of the Colorado River drainage, the lower Mississippi Valley, the Ohio and Tennessee River drainages, many streams tributary to the Gulf of Mexico from the Mississippi to the Apalachicola River in Florida, various Atlantic Coast drainages as far north as the Delaware River, and perhaps other localities; in many places they have become extremely abundant.

Filtering ability

Corbicula may be even more efficient filterers than *Lampsilis claibornensis*. One worker reported that three small *Corbicula* cleared 500 cc of very turbid water in less than twenty minutes, and I have had similar experiences with these clams in dense algal blooms. The only quantitative study I have seen measured the average laboratory pumping rate of a 1 gram *Corbicula* at 0.13 gallons per day (0.5 liter/day).

Although *Corbicula* run smaller than our edible native fresh water mollusks—the largest size reported in the United States is 3.1 inches (80 mm.)—they seem to me to be a better quality food animal. They are commonly eaten in the Orient, but in this country they have been used widely only as fish bait. (Although if you have enjoyed canned "smoked baby clams" you have eaten *Corbicula sandai*, imported from Japan.) The shell of *Corbicula* is not used for button manufacture or seed pearls.

The potential food value of *Corbicula* may be outweighed by its impact as a pest, and much of the attention focused on these clams in the United States has centered on this aspect. There are two principal problems associated with *Corbicula*. In the Tennessee Valley, where sand and gravel is often taken from river beds for construction use, they may populate gravel to such an extent that setting concrete is weakened by clams rising to the surface in an attempt to escape. (Considering the ecological havoc a sand and gravel operation can raise with a river, *Corbicula* may be an asset to the Tennessee Valley!) A probably more serious problem is the clogging of waterworks pipes and irrigation canals by thousands of *Corbicula* in the Tennessee Valley, the Sacramento Valley of California, and elsewhere. In some environments *Corbicula* also tend to compete with and eliminate native clams, some of which are considered endangered species. For these reasons, I strongly recommend that they not be introduced into river systems which are still free from *Corbicula*.

A few years ago, I tried *Corbicula* culture in fertilized aquaria. They were marvelous filters, but failed utterly to grow in confinement. Clams can often be induced to breed through shock, but mine did not respond to the standard technique of stimulating the adductor muscle with a needle. Even less endearing was their tendency to die off *en masse* for no discernible reason, a habit which has also been observed under natural conditions.

Corbicula's value as a "living filter" in polyculture with fish is discussed in Chapter IV-1.

The use of fresh water clams in aquaculture remains an intriguing possibility. Apart from problems of biology and culture techniques, there is much to be learned about their preparation for the table. If you find an edible clam in your vicinity, or if you are interested in their use as filters or fish food, you might try stocking a few in your fish pond. Don't stock too many at first, or you may be left with an odorous mess of pollution should they all die; some species will not live in standing water. (A note of caution: Clams, being such efficient filter feeding animals, tend to concentrate whatever contaminants are in their water. Do not eat clams from waters polluted by chemicals or human wastes, even though fish from these same waters are considered safe.)

Apart from basic biological information and encouragement, there is little assistance to be offered for fresh water clam culture; these animals have attracted surprisingly little attention from professional aquaculturists. The few studies I know of are all described in this short chapter. Perhaps you will make a contribution in this poorly developed field of aquaculture.

CHAPTER II-13

Other Possible Food Species

Present knowledge on species selection

Of over seven hundred recognized species of fish native or successfully introduced to North American fresh waters, thirty-six (thirty-one food species and five others), plus the bullfrog and four invertebrates are treated at some length in this book. (See Chapters II–1 to II–12 and Appendix II.) These include nearly all species which have been practically applied in food-producing systems in North America, and a few others chosen because of their economic importance, successful cultivation for food in other parts of the world, or potential (to the author) for use in small-scale systems. But this does not mean that the other species have been thoroughly evaluated and found wanting. Almost everywhere, the selection of fish species for cultivation in North America has proceeded erratically, with decisions made primarily on the basis of rather narrow economic factors.

Research on species

Two exceptions are the work of George W. Bennett and his co-workers in Illinois and Homer S. Swingle, E.V. Smith, and their associates, in Alabama, both beginning in the nineteen thirties. By 1943 Bennett's group had tested twenty species of fish for stocking in small fishing reservoirs in Illinois. In 1952 Swingle was able to list twenty-seven species and two hybrids which had been evaluated in experimental ponds at Auburn University. The procedure followed at Auburn is excellent, so I will quote it at length:

> In testing these species for use in ponds, a more or less routine procedure has been followed. First each was isolated to determine whether it would reproduce without special manipulations and to prevent introduction of diseases and parasites. Those species that failed to reproduce were considered for possible use as commercial species and were tested for production in ponds with fertilization and by use of supplemental feeding. If they reproduced under ordinary pond conditions, they were considered of possible value for use both as commercial fish and in combination with other species. Next they were tested in 1- to 2-year experiments to determine the production that might be obtained both alone and in combi-

nation with other species. They were then placed in 3- to 7-year experiments to learn what they contributed to the fishermen's catch and to see if they could maintain themselves adequately in the population. It thus requires at least 3 to 5 years merely to determine whether a species is of any value and 5 years or more to determine how promising species may best be used.

The work of the Illinois and Alabama groups was crucial both for the evolution of the farm pond program and the development of commercial food fish culture in the South Central United States. Neither group, however, could predict the economic and ecological conditions that led to the preeminence of the channel catfish (*Ictalurus punctatus*). Despite the high standard of their work their recommendations cannot be taken as the last word; every set of circumstances and priorities will require its own species and methods.

You and I are not likely to be in a position to undertake the kind of exhaustive investigation just described. But, recognizing that some readers will be thinkers and tinkerers, willing to stray beyond conventional boundaries, I offer an annotated list of additional species which might be grown as food. Some of you who will not be planning to depend on aquaculture for your livelihood or primary source of protein are in some ways freer to experiment with small-scale cultures than either commercial growers or those scientists paid to do aquaculture research; for those perhaps this chapter will inspire some new and original ventures.

I have concentrated on those fishes which seem to me to hold promise of value, together with a few which are likely to occur to the inquisitive reader. Bear in mind that for the animals discussed earlier you can often predict the biological and economic feasibility of a particular culture plan. Most of the creatures on the following list offer no such security. If "success" is crucial for you, I suggest you skip the rest of this chapter, but if you are experimentally minded, read on—at your own risk.

The list is organized by families, in conventional taxonomic order, and a word about aquatic food animals other than fish concludes the chapter.

Polyodontidae (paddlefish): Although the only North American member of this family, the paddlefish (*Polyodon spathula*) is a very large fish, it is a plankton feeder. It could eventually play a role in fish culture similar to that foreseen for the sturgeons. (See below.)

Paddlefish

Acipenseridae (sturgeons): Sturgeon are a favored food fish and the primary source of caviar, but are usually thought of as too big for pond culture. However, juvenile sturgeon are large enough to be eaten, and white sturgeon (*Acipenser transmontanus*) have been reared to a size of 2 to 3 pounds (0.9–1.2 kg) in a year. Not surprisingly, most of the research and development of sturgeon culture has taken place in the Soviet Union, but a number of species, including the white sturgeon, the green sturgeon *(Acipenser medirostris)* and the Atlantic sturgeon *(Acipenser oxyrhynchus)* are under study in the United States. All sturgeons are omnivorous; while they readily accept commercial feeds they also make good use of detritus organisms which are neglected by most cultured fish.

Sturgeons

Salmonidae (trout, salmon and whitefish): The only trout discussed at length in this book is the rainbow trout (*Salmo gairdneri*); some space is also devoted to the brook

Trout and salmon

trout (*Salvelinus fontinalis*). It seems likely that these two species will cover virtually every food fish cultivation situation. The brown trout (*Salmo trutta*), while the most widely distributed trout in North America, is inferior for food cultivation because of its poor tolerance for crowding. It is an important sport fish, however, and may be valuable to commercial hatchery operators. (See Chapter IV–3.) Other important sport trout include the cutthroat trout (*Salmo clarki*), the golden trout (*Salmo aguabonita*) and the Atlantic or landlocked salmon (*Salmo salar*). More or less the same hatchery methods may be used for them as for rainbow trout. The lake trout (*Salvelinus namaycush*) might be included, since it is important to both sport and commercial fishermen, but it is very difficult to rear, due to its specialized environmental requirements.

Another salmonid, the arctic char (*Salvelinus alpinus*) is important as a food and sport fish in the Arctic. So far there has been no incentive for its culture due to the sparse settlement of the Arctic and the slow growth of Arctic fishes. However, if human exploitation of the Arctic increases, there may one day be cause to cultivate this fish.

Wherever salmonids have access to the sea or to large, cold lakes, anadromous strains are likely to develop. The Pacific salmons (*Oncorhynchus* spp.), the Atlantic salmon, the rainbow trout and the Arctic char are particularly noted for this trait. Apart from being outside the scope of this book, cultivation of salmon and trout in sea water appears to be economically best suited to the corporate approach. However, there is the possibility that the "homing instinct" of these fishes could be taken advantage of on a subsistence or small commercial scale. This possibility was first raised by the pioneering work of Lauren Donaldson at the University of Washington. Starting in 1953, Donaldson successfully established and maintained runs of chinook salmon (*Oncorhynchus tshawytscha*) and other species in a small stream on the university campus. It is easy to see how Donaldson's techniques might be applied to production of salmon as food, though to date no one has succeeded in doing so. Even if someone eventually does succeed, the legal problems involved in such a fishery could become formidable.

The other side of the salmon behavior coin is the existence of "landlocked" strains of several species. Particularly interesting is the sockeye salmon (*Oncorhynchus nerka*), known as "kokanee" in its landlocked form. Kokanee are mid-water fish, and are unlike other salmon in obtaining most of their food from zooplankton. They are excellent quality, fast-growing food fish and exploit an ecological niche in which there is little competition. Eventually they might be used in types of cold water fish culture similar to "prairie pot hole" culture of rainbow trout. (See Chapter II–6.)

Whitefish   Another important group of salmonids feeding mainly on invertebrates are the whitefishes and ciscos (*Coregonus* spp.) formerly assigned their own family, the Coregonidae. Whitefishes are among the finest of food fishes and are commercially important in Arctic and subarctic regions around the world. In Canada, the lake whitefish (*Coregonus clupeaformis*) is the single most important fresh water commercial fish, but catches are steadily declining. Though hatchery methods were developed in the 19th century, stocking was subsequently found to have no effect on the catch. In the Soviet Union, several species of whitefish are cultured in ponds and the idea might be worth a try in North America.

Hiodontidae (mooneyes): The goldeye (*Hiodon alosoides*), while said to be dreadful when cooked conventionally, is palatable when smoked. Though rarely eaten in most areas, it is an important fishery product around Winnipeg, Manitoba. Scott and Crossman in "Fishes of Canada" report that the food of goldeyes includes "almost any organism encountered," and "whatever is available predominates" in their diet. These characteristics, and the fact that goldeye seem to prefer turbid water, suggest that they might be useful to some fish culturists.

Goldeye.

Esocidae (pikes): Of the five North American species of Esocids, the chain pickerel (*Esox niger*) seems best adapted, because of its intermediate size, to pond culture as a food fish. However, the larger and much more widely distributed northern pike (*Esox lucius*) may do well in small bodies of water. It is sometimes stocked in European polyculture ponds to control "weed" fish or excess reproduction by cultured species. Though hatchery methods exist, they are costly due to the young pike's stringent dietary requirements and their tendency towards cannibalism.

Pike and pickerel

Cyprinidae (carps and minnows): Only two cyprinids, the common carp (*Cyprinus carpio*) and the grass carp (*Ctenopharyngodon idellus*), are treated as food fish in this book. Both are exotics, though the carp has become widely established in North America. These and several other cyprinids are widely used in Oriental and European fish culture; traditional Chinese pond polyculture relies almost entirely on cyprinids. The North American cyprinids, however, while numerous, are mostly too small to be of much interest for food. There are a few exceptions; at least one of these, the golden shiner (*Notemigonus crysoleucas*, see Chapter IV-2) can be cultivated productively, though its food quality is inferior. I suspect the same would hold true for some of the other large native cyprinids. One which might be of interest is the Sacramento blackfish (*Orthodon microlepidotus*). The Sacramento blackfish, which feeds on plankton and small benthic creatures, and the related hitch (*Lavinia exilicauda*) and hardhead (*Mylopharodon conocephalus*), support a small commercial fishery in a few lakes in California. Another exotic cyprinid cultured for food in Europe, but largely ignored in North America, is the tench (*Tinca tinca*), established in the Columbia River system, the Puget Sound area, the Potomac River, and perhaps in other localities.

Cyprinids (carps and minnows)

Catostomidae (suckers): The bigmouth buffalo (*Ictiobus cyprinellus*) described in Chapter II-4, is the only sucker important for food culture. Two slightly smaller and less prolific buffalofishes, the smallmouth buffalo (*Ictiobus bubalus*) and the black buffalo (*Ictiobus niger*) have also been cultured. However, their subterminal mouths (see Figure III-1-1) make them less well adapted to filter feeding, leading to slower growth.

Buffalofish and other suckers

There are a number of other species of sucker which reach a considerable size. Most, if not all, have sweet, flaky, though very bony flesh, which has been rated superior to the flesh of buffalofish. The bigmouth buffalo has been preferred for commercial cultivation because it is traditionally in demand in the lower Mississippi Valley, and because it is a plankton feeder rather than a benthos feeder like the rest of the family. The other suckers should nevertheless be considered, since they represent most of the potential alternatives (or supplements) to the predaceous salmonids in cold water fish culture. One species, the white sucker (*Catostomus commersoni*), is already cultured as a bait fish, but seems to grow too slowly to be of great value as a cultured food fish.

Catfishes
: Ictaluridae (catfishes): Six catfishes have been discussed so far; three species of bullheads (Chapter II–2) and three larger species (Chapter II–3). Most of the other ictalurids are small or with very limited distribution. One other very large species, the flathead catfish (*Pylodictis olivaris*) has been studied in the South. But as it is extremely piscivorous, and can grow up to one hundred pounds (45 kg), it is of little interest to fish farmers.

Eels
: Anguillidae (freshwater eels): Except by a few ethnic groups, the American eel (*Anguilla rostrata*), has been largely ignored as a food fish. Recently, however, a minor industry has sprung up in collecting young eels (elvers) from Atlantic coast tidal marshes. These are then shipped to Taiwan and Japan, where they are reared to market size in fresh water ponds. This practice, which is due to the depletion of native eel stocks in the Orient, can ultimately lead to a depletion of our own stocks. There are few fish as good for food, and we should be educating ourselves to the joys of eel cookery instead of shipping them abroad. (See Appendix I.) North Carolina State University maintains a demonstration eel culture project that is open to the public at Aurora, North Carolina.

The spawning habits of eels are a subject on which there is much speculation and little information. It is fairly certain that they spawn somewhere far at sea. Duplicating their natural spawning conditions, even if known, would be tricky, to say the least. For the foreseeable future, eel culture will remain a matter of capturing elvers on their return migrations and growing them out.

Though eels are highly carnivorous and not economical to feed, they bring high prices where a market can be located. In Taiwan they are sometimes stocked in small numbers to do "clean-up duty" in complex polyculture systems. Another possibility is using eels to process the offal from other fish culture or fishery operations. Little research has been done on American eel culture, but it should be possible to adapt Oriental methods. Someone in New Brunswick has reportedly begun to successfully grow out locally caught eels, but he does not seem to be eager for publicity.

Serranids: white bass, yellow bass and white perch
: Serranidae (sea basses): Three pan-sized species of this mostly marine family are found in North American fresh waters. In 1943 the yellow bass (*Morone mississippiensis*) was dismissed as being unsuitable for reservoir lakes in Illinois, but the white bass (*Morone chrysops*) and the white perch (*Morone americana*) have not been considered by fish culturists. The yellow bass and white perch are commonly found in small ponds in nature, and there may be situations where they could be used in place of the traditional farm pond carnivores (i.e., largemouth bass, bluegills, redear sunfish and channel catfish). More promising is the so-called "sunshine bass," a hybrid of the white bass with the marine striped bass (*Morone saxatilis*), which has grown very well when fed in fresh water ponds as small as ¼ acre (0.1 ha) in South Carolina.

Black bass
: Centrarchidae (sunfishes and black basses): The centrarchids may be loosely broken down into three groups. The black basses (*Micropterus* spp.) are by far the largest and are piscivorous, as one might guess from their cavernous mouths. The

species commonly referred to as sunfishes (excepting the green sunfish, *Lepomis cyanellus*), have small mouths and, while carnivorous, they consume few fish. Intermediate to these species are a mixed bag of fishes: the crappies (*Pomoxis* spp.), the green sunfish, the rock bass (*Ambloplites rupestris*), the Sacramento perch (*Archoplites interruptus*), the flier (*Centrarchus macropterus*), the warmouth (*Chaenobryttus gulosus*) and a few others not widely distributed.

Among the black basses, the largemouth bass (*Micropterus salmoides*), discussed earlier, clearly dominates the field, and the smallmouth bass (*Micropterus dolomieui*) was also mentioned for stocking cooler waters. The lesser known spotted bass (*Micropterus punctulatus*) is another satisfactory farm pond fish whose environmental requirements fall roughly between the largemouth and smallmouth. Three other smaller *Micropterus* spp. of very localized distribution may also be suitable for cultivation.

Most of the sunfishes, other than those discussed earlier, are too small to be of interest. Possible exceptions are the redbreast sunfish (*Lepomis auritus*) and the longear sunfish (*Lepomis megalotis*). Sunfish connoisseurs consider the redbreast to be the tastiest of all sunfishes. It does well in quite cool water. The longear sunfish consumes a higher percentage of terrestial, floating insects than any other sunfish.
Sunfishes

The crappies were discussed in Chapter II and the rock bass, a fine food fish capable of rapid growth, has been mentioned as a possible candidate for cool water farm ponds. (See Chapter IV–1.) After study the flier and warmouth were both found to be unsuitable for pond culture by Swingle, et al. and Bennett and Swingle. However, it seems that the Sacramento perch has not yet been considered as a fish for pond culture.
Crappies, rock bass, other Centrarchids

Percidae (perches): Most of the perches are very small fishes, known as darters. One exception, the yellow perch (*Perca flavescens*), is discussed in Chapter II–7. Two still larger perches, sometimes called pike-perches, are the walleye (*Stizostedion vitreum*) and the less well known sauger (*Stizostedion canadense*). They are among the finest of food fishes, but are not suitable for small-scale culture. Spawning requires technically sophisticated methods, losses to cannibalism are very high, it is difficult to get the young in captivity to accept any food besides each other, and pike-perches of all ages have a high requirement for animal protein. If anyone did succeed in culturing them to marketable size, the result would certainly be an extremely expensive fish.
Perch, walleye, sauger

Sciaenidae (drums): The fresh water drum (*Aplodinotus grunniens*) is only a fairly good food fish but quite adaptable environmentally. It feeds mainly on benthos, including mollusks, reaches a large size and may be suitable for some culture systems.
Fresh water drum

Cichlidae (cichlids): To aquaculturists the most familiar members of this large and varied family of tropical fishes are the herbivorous and omnivorous tilapias discussed earlier. (Chapter II–8.) Only one cichlid, the Rio Grande perch (*Cichlasoma cyanoguttatum*) occurs naturally north of Mexico—in Texas. Ten exotic species are known to be established in Florida. Some of them are highly predaceous and/or aggressive and competitive with other fishes.
Cichlids

Table II-13-1.   Exotic cichlid species known to be established in Florida.

Common Name	Scientific Name
Blue Tilapia	*Tilapia aurea*
Mozambique Tilapia	*T. mossambica*
Spotted Tilapia	*T. mariae*
Black Acara	*Cichlasoma bimaculatum*
Jack Dempsey	*C. octofasciatum*
Firemouth Cichlid	*C. meeki*
Rio Grande Perch	*C. cyanoguttatum*
Convict cichlid	*C. nigrofasciatum*
Oscar	*Astronotus ocellatus*
Jewelfish	*Hemichromis bimaculatus*

Most cichlids I have eaten were very good food fish. A number of species are used as predators to prevent overpopulation in tilapia culture in Latin America and Africa. These might be appropriate for similar use in Texas or Florida, but only where they are already established. Warnings about intentional or accidental introductions of exotic fishes in North America must be emphasized for predatory cichlids. Some species are already considered pests in Florida, and other cichlids have contributed to the disruption of ecosystems in various parts of the world.

Mullets

Mugilidae (mullets): The herbivorous mullets, though important as food fish, are largely marine animals, and so fall outside the scope of this book. However, one species in particular, *Mugil cephalus*, variously known as gray, black or striped mullet, is found in tropical and semitropical waters around the world and is cultured in places as diverse as Hawaii, Taiwan, Israel and Italy. Mullet culture methods vary in intensity from the simple trapping and rearing of naturally spawned fry in ponds receiving tidal flow to technologically sophisticated techniques.

Most mullet culture takes place in brackish water. However, mullets, including *M. cephalus*, are remarkably tolerant of salinity changes and may even be grown in fresh water. Readers located near a coast where mullet fry can be obtained may want to give them a try in fresh water polyculture systems.

Invertebrates

Three groups of aquatic animals other than fish are discussed in this book. Mention has been made of most, if not all, the frogs likely to be of interest, but there are dozens of species of crayfish and clams which might conceivably be useful; space does not permit listing them here.

Turtles

Fresh water turtles are a final possibility for culture. The snapping turtle (*Chelydra serpentina*), Florida snapping turtle (*Chelydra osceola*), and alligator snapping turtle (*Macroclemys temmincki*) rank among the finest of aquatic food animals. They are marketed to some extent, but to my knowledge their cultivation has never been attempted. Breeding facilities could easily be provided; all that is necessary is well-drained sandy soil not too far from water and exposure to sunlight. Their other environmental demands are few and obvious and, contrary to popular belief, the snapping turtle—though perhaps not the alligator snapper—is herbivorous to a degree.

PART III

Fertilization and Feeding

CHAPTER III-1

Introduction

PART III

Fertilization

and feeding

State of the art

One of the principal concerns of the fish farmer, like any other livestock raiser, is to see to it that all animals are properly fed. But perhaps such worries are a bit less severe than those of some other farmers as it is very unlikely that fish will starve to death. Even in the total absence of food—another unlikely event in any outdoor situation—fish can absorb mineral nutrients directly from the water, warding off starvation for a considerable time. But as it is the business of the fish farmer to maximize the growth rate of his stock, nutrition must be considered.

Technical knowledge

A great deal has been written about the feeding of fish. Casual inspection of the scientific literature on the subject might give the impression that the subject of fish nutrition has been thoroughly investigated—but it has not. To realize the extent of the gaps in our knowledge, you need only scan the most comprehensive technical book on the subject to date, the 713-page tome "Fish Nutrition," edited by John Halver. The great majority of the work of the fourteen contributors to Halver's volume deals with only one family of fish, the Salmonidae (trout and salmon). Even greater weight is given to materials intended as components in prepared commercial fish feeds. Many high quality fish food materials, including most of the important natural foods, are not discussed at all. This is no omission on the part of the editor or authors, but a reflection on the state of the art. Fish nutrition research has had a very narrow focus for two reasons:

Emphasis on salmonids

• The first fishes to be intensively cultured in North America were salmonids. Commercial trout culture has become an important business, and Pacific salmon are cultured both for sale and for stocking to augment commercial fisheries, but there is a much older tradition of salmonid culture for sport fishery purposes. Since the establishment of the first trout hatchery in North America by Seth Green at Mumford, New York, in 1864, the practice of growing trout for stocking in fishing waters has become established in sport fishery management in nearly every state and province where trout are found. The concerns of state-supported hatcheries are in many respects the same as those of a commercial trout farm—including the need to maximize growth by use of high quality feeds. So it is not surprising that by now we know nearly as much about the requirements for protein, vitamins, etc., of trout as we do about our own. As the catfish culture industry in the United States grows in importance, a comparable body of knowledge is being amassed about the channel catfish (*Ictalurus punctatus*). But our knowledge is fragmentary and almost entirely empirically derived concerning the other fish species.

• The emphasis on prepared dry feeds is directly and logically related to the peculiar history of American fish culture. The first purpose of fish culture in this hemisphere was to provide, not food, but recreation, and there was no incentive to grow fish cheaply if faster growth could be realized at slightly greater expense. Commercial fish culture in North America developed much later, during the post-World War II economy, and has emphasized fish which would provide a high economic return rather than those which might be most productive or easiest to grow. It is not surprising, then, that the considerable convenience factor of prepared feeds has won them such great acceptance. Add to this the fact that much fish nutrition research is directly or indirectly supported by the feed industry, and you will see why we are where we are. *Emphasis on dry feeds*

To the person investigating fish culture for the first time, it might appear that commercial feeds are the only way to go. In fact, they represent just one shade in a spectrum of feeding strategies. It is possible, in some places, to raise a respectable amount of fish without any feeding at all, by simply stocking them in an enclosed body of water of sufficient fertility. But results can nearly always be enhanced by supplementation of the fishes' natural diet.

Before embarking on a feeding program, you should know as much as possible about the food requirements of the fish you are attempting to raise. (This may even lead to a reassessment of the choice of species. You should choose not just species to culture, but the most appropriate combination of species and foods, given your particular resources.) There are four sources of such information: *Different species' requirements*

(1) *The experience of other fish culturists.*

If you know other people who are growing the same species you want to raise, by all means consult with them. You will also want to look into the fish culture literature and related scientific publications. What you find there will depend on the species you are interested in. If you wish to raise rainbow trout (*Salmo gairdneri*), you will find much more information than you probably care to deal with. On the other hand, if you have chosen a species of little present commercial importance, you may find next to nothing. *Information sources: experience of others*

(2) *The feeding habits of wild fish.*

Every fisheries student sooner or later has the experience of opening fish stomachs and counting and weighing the contents. You can do the same, but you will have a better understanding if you study the fishery literature, which by now includes an abundance of information on the natural feeding habits of common North American fishes. (See Appendix IV–1.) *Wild fishes' habits*

Bear in mind, though, that most fish are opportunistic feeders. In some natural situations, they may seldom encounter their preferred foods, and so settle for what is available. For example, I have noticed that, in captivity, some so-called herbivorous fishes greedily consume insects, earthworms, etc., when these are offered in quantity. It seems that their feeding habits in nature are the result, not of preference, but of intense competition for food resources. Nevertheless, the examination of stomach contents can give valuable clues as to what a species will or will not eat.

(3) *Trial and error.*

You should not assume that just because a certain substance is not found in the stomachs of wild fish, they will not eat it. It is doubtful that wild carp (*Cyprinus carpio*) encounter many roasted soybeans in their travels, but captive carp are very fond of soy meal. Though it is possible to make a good educated guess about what a species will accept, you never really know until you try. If you have an idea that seems not to have been tested, try a small sample of the prospective food in a situation where you can either watch the fish eat or refuse it, or where you can later see if it has been eaten. Do not overdo making the process easy to watch by confining them too closely, or you may make them too nervous to eat at all. And bear in mind that some species like different foods at different ages.

(4) *Clues gained by inspection of the fish.*

It is possible, in a general sort of way, to infer the feeding habits of a fish by visual inspection. Here are some useful clues:

Size: Obviously, the larger the fish, the more and/or larger food items it will require. Do not, however, conclude that all large fish want their food in large pieces. Some of the largest species of fish are plankton feeders, and even such fish as large trout, which are normally placed very high on the food chain, may sometimes gorge on the tiniest of insects.

Age: Like many other animals, fish have a greater need for protein, particularly animal protein, when they are young. Even the young of largely herbivorous fishes benefit from and prefer animal feeds. Some fishes never "outgrow" this need and are never able to digest much plant material.

Mouth size: This trait is particularly useful in determining the maximum size of food that a fish will consume. Generally speaking, a fish will accept as large a food item as it can swallow. (On rare occasions, fish will overestimate their capacity and choke, particularly if the food is another fish of a spiny-finned variety.) The size of the mouth can also be used to determine the type of food a fish might prefer. As a rule, the larger the mouth of a species, the higher it is on the food chain. This trait should not be interpreted without also considering mouth shape, dentition, and gill rakers, which are discussed below.

Mouth shape: Icthyologists use a number of terms to describe the shape of a fish's mouth. Most of the fish considered here can be classified as having a "terminal" mouth. Fish with terminal mouths tend to be adaptable in their feeding, in that they may feed on the surface, in mid-water or on the bottom. Carp and some other fishes mentioned in Chapter II–12 have a "subterminal" mouth. (See Figure III–1–1.) Fish with subterminal mouths tend to be restricted to harvesting from the bottom or other solid surfaces.

FIGURE III–1–1. *Fish with terminal mouth* (goldfish, Carassius auratus) *and subterminal mouth* (common carp, Cyprinus carpio).

Dentition: Fish do not chew. In feeding, the teeth are used only for seizing and holding prey in preparation for swallowing. A fish may be considered as predatory in proportion to the prominence of its teeth. (Some fishes have pharyngeal teeth, used in grinding food. These are located in the throat, and are not visible unless the fish is dissected.)

Gill rakers: (See Figure III–1–2.) These are small protrusions found on the gill arches, opposite the red gill lamellae. The gill rakers function as a filter, allowing unwanted objects entering the mouth to pass out through the gill slits. What is rejected or retained for consideration as food depends on the size and spacing of the rakers. If a fish's gill rakers are few and widely spaced, you may infer that it is a predatory fish which feeds on sizable animals. If, on the other hand, there are many, closely packed gill rakers, the fish must "filter feed" by straining small items from the water. If, in addition, the mouth is large, it may also eat larger items.

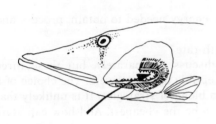

FIGURE III–1–2. *Gill rakers of a predatory fish (*chain pickerel, Esox niger*) and a filter feeder (*blue tilapia, Tilapia aurea*). Note that the former are stout and widely spaced, while the latter are fine and closely packed.*

Digestive system: If you cut open a fish and find a large stomach, clearly distinguishable from the rest of the digestive system, and a short, relatively straight intestine, it is a safe bet that this fish is high on the food chain. But if the stomach is small or indistinguishable from the long, coiled intestine, it indicates that the fish is low on the food chain.

Habitat: A fish is likely to feed at the water level you find it in. For example, a benthic fish (i.e., living on the bottom) will be better adapted to foods found in that environment. However, a cultured fish will not necessarily be subject to the same limitations as a wild one. For example, common carp, which are primarily benthic fish, in captivity quickly learn to accept floating food.

Behavior: Of course much can be learned from watching the feeding behavior of wild fish. Social behavior will also furnish clues to feeding habits. If individuals of a species tend to be solitary and defend a sizable territory throughout the year (instead of only during their breeding season, when many fishes become territorial) this behavior is likely to be related to feeding. Such a fish is likely to be high on the food chain; it needs to defend a territory in order to assure a supply of its relatively scarce food. Schooling fishes, on the other hand, are likely to be lower on the food chain.

I recommend that the beginning aquaculturist makes use of as many of these sources of information as possible. At the outset, refer to Appendix II, for broad outlines of what is known about the feeding and natural foods of the species discussed in this book.

With the requirements of the species you intend to culture firmly in mind, you can then consider possible feeding strategies. I have grouped them in five categories, with increasing degrees of intensity:

(1) Fertilizing the water to promote growth of natural fish food organisms.

(2) Attraction or capture of naturally occurring fish food organisms from outside the fish pond.

(3) Using wastes from fish culture or other agricultural operations as fish food.

(4) Culture of fish food organisms.

(5) Manufacture or purchase of prepared feeds.

All of these strategies have their places, and it is rare for any one of them to be found inappropriate in a given situation. I strongly urge all readers not to be swayed by advertising, or other propaganda, on behalf of any one feed or feeding strategy, but to consider all possibilities. You should always ask these questions:

• Which food is cheapest in dollars?

• Which food is cheapest in terms of the energy needed to obtain, process and feed it?

• Which food will produce the best growth rate?

There are other factors which will be discussed separately, but these three questions must come first, and are not susceptible to pat answers. The choice of a "best" food must finally be a compromise—a blend of strategies. It is unlikely that the best food for promoting growth will also be the cheapest. And how can standards possibly be set for comparing "costs" in terms of money and labor? To make matters more complicated, feeding strategies will sometimes have to change—both within a year as the fish grow and the natural food supply varies seasonally, and from year to year, as the art of fish culture, economic conditions, and the knowledge of the individual fish farmer change.

The art of feeding North American fishes is very young. To enter the field now is to join in a creative process, trying constantly to improve on the last harvest—and, I think, to have fun. Keeping this in mind, let us look at some of the current possibilities. We will begin with fertilization since, no matter what feeding strategy is adopted, the fertility of the water will affect the results.

CHAPTER III-2

Fertilizing the Aquatic Environment

As we have seen, a harvestable size fish represents the end of a food chain which ultimately leads back to non-living nutrients. We fertilize in order to improve food fish production by starting at the bottom of that chain. By adding fertilizers to the pond, we aim to expand the base of the "pyramid of numbers" (see Figure I-2-2) increasing the biomass of organisms at all levels. Pond fertilization is essentially the same as fertilization of agricultural land, although there are more links in the aquatic chain before the effect reaches the human consumer. Another use of fertilizers is to control unwanted plants; see Chapter VIII-1.

*Fertilization:
in general*

There is, in Western culture, a certain resistance to the thought of fertilizing water,—especially with animal manures. When we spread animal manures and other natural materials, as well as synthetic chemicals, on our land we call it "fertilization." When we put the same materials in water, it is generally called "pollution." The word "pollute," when applied to water, implies a value judgment—unless we want to consider only pure distilled water as unpolluted. Where a substance is toxic, acting directly on living creatures, there is little controversy as to what constitutes pollution. But if we are discussing natural organic materials, one person's pollutant is another's fertilizer. Up to the point of severe oxygen depletion, organic materials tend to increase the number and mass of living organisms, and in this way their action is the opposite of toxicants.

"Pollution" defined

Still, the introduction of organic wastes in amounts less than those causing oxygen depletion is not always desirable or "non-polluting." It depends on the planned utilization of a body of water. You would probably not want to fertilize a pond at all if swimming is the primary use, and fish culture a secondary one. Fertilization methods also depend on what type of fish you wish to grow. Levels of fertilization that might be "enrichment" for common carp (*Cyprinus carpio*) could be nearly lethal for trout. In general, fertilization is more suitable in water intended exclusively for food fish production than in multiple use ponds. In my opinion, large natural water bodies should not be intentionally fertilized at all for environmental reasons,—unless human nutritional levels are drastically low.

State of the art

While the Chinese certainly lead the world in fertilizing to enhance aquatic harvests, they have done little to formulate their methods. Their approach is to treat each pond individually, in a manner that is perhaps more art than science. After reading Western "scientific" studies of pond fertilization one is likely to conclude that the Chinese know best. The state of the art has been well summarized by Burwell C. Gooch. In reviewing the North American literature on aquatic fertilization, he was compelled to say that while it is "... relatively safe to predict that fertilization, either with inorganics or organics, will lead to an increase in aquatic productivity of one kind or another, little progress has been made beyond this primitive predictive stage." He also mentions "the inexplicable inconsistent results" of aquatic fertilization studies and "the meagre knowledge now extant about fertilizer dynamics in aquatic systems."

Experience at The New Alchemy Institute has certainly borne out Gooch's statements. Even in small plastic pools where it ought to be possible to control virtually every environmental factor, identical treatments have resulted very differently. As I look out across the New Alchemy farm, I see a battery of fiberglass "solar-algae ponds" which have received similar doses of fertilizer, populations of fish, and feeding. Two are an opaque green, bordering on black; others are bright translucent green, yellowish, or nearly colorless. Something is obviously going on that we do not understand and cannot control.

How fertilizers work

With this caution, we can now discuss in a general way how fertilizers work in aquatic systems and the types of fertilizer which can be used. There are three general rules for aquatic fertilization:

Rules for aquatic fertilization

(1) A pond will respond to fertilization in accord with Liebig's Law of the Minimum, which says that production will be limited by the supply of whatever indispensable nutrient is present in minimal amounts. In other words, it will do no good at all (and possibly some harm) to put large amounts of nitrogenous fertilizer in a pond which is deficient in phosphorus.

(2) The great majority of nutrient material added to a pond is not taken up directly by living organisms, nor does it dissolve in the water. Instead, nutrients are absorbed by the soil on the bottom, from which they are released little by little into the food chain. For this reason, it is important for the pond's bottom materials to be absorptive. The best type of pond bottom is a rich colloidal mud, though in some cases this may be inconvenient for or incompatible with certain aquacultural practices or other uses of the pond. Colloidal mud is formed mainly through the decomposition of soft stemmed submerged plants and algae. Tough stemmed emergent plants are high in cellulose and form a much less nutritive mud.

(3) The pH should be neutral to slightly alkaline for fertilizers to be most effective. Acid conditions reduce the mud's capacity to absorb. For this reason, many ponds should be limed. (See below.)

Natural fertility

Before discussing chemical and organic fertilizers in detail, we need to look at certain natural factors involved in pond fertility, and some forms of fertilization which occur incidentally to the practice of aquaculture. Since these factors are only slightly or not at all under the fish farmer's control, they must be considered before embarking on a fertilization program.

Every soil, and therefore every body of water, has a measurable natural fertility. When first beginning aquaculture in a pond, you would do well to have both the soil and water tested for possible deficiencies, particularly in nitrogen and phosphorus.

But even without testing, you can come to certain inferences about water fertility. Natural ponds become more fertile with time through the process of eutrophication, by which a pond or lake created by some natural process gradually becomes filled with the remains of animals and plants, and with soil eroded from adjacent land. While in the end this results in the elimination of the pond, before that stage it becomes progressively enriched, since much of the material which eventually fills it in is nutritive. Ponds last longer than human beings, so we cannot directly know the age of a natural pond, but the age and fertility of a natural pond can be inferred from the following clues:

- Depth: If there are extensive areas of shallow water, the pond is more likely to be old and fertile than if there are not.   *Clues to fertility*
- Bottom type: Soft, colloidal mud implies high fertility; materials such as sand, gravel and rock connote low fertility if they are not covered by sediments.
- Profile: A pond with banks which drop sharply into deep water is likely to be less fertile than one with a gradually sloping shoreline, and shallow water near shore.
- Water color and transparency: A brownish, reddish or green tinge to the water, often together with high turbidity, suggests an abundance of plankton, which in turn indicates high fertility. You should not confuse this with the opaque white to brown color due to suspended silt particles. This condition, if it persists for more than a few days, is harmful for most aquatic life, and its negative effects may override the positive action of fertilizers. There are many other substances, opaque or transparent, which may color water; some are toxic, while others are harmless, and they may have no relation to fertility. But, the "crystal clear," colorless pond, while "pure" and esthetically pleasing, is bound to be infertile.
- Vegetation: As on land, an exuberant growth of vegetation indicates high fertility. Be sure to look for both macrophytes (plants large enough to be seen as individuals), and phytoplankton (tiny suspended algae). Either type is a good indicator of fertility, but as they are competitive it is rare that both will be abundant in one place.
- Animal populations: The greater the animal populations, the higher the fertility. Try to be objective about this. To the trout fisherman a clear, gravel-bottomed, relatively sterile pond may have "more fish" than a near-bog, but if we consider all species of fish, plus frogs, crayfish, insects, zooplankton, etc., it is easy to see that the trout pond supports less animal life.

You may also infer relative fertility of a pond from the fertility of the surrounding land, though ponds are likely to be more fertile, since they serve as catch basins for the surrounding higher ground.

Artificial ponds are somewhat different. Since their form is up to whoever constructs them, the criteria of depth and profile are of little use in estimating fertility. Also, unlike natural ponds, they may be more fertile when new, due to nutrients in reserve in the soil before it is flooded. These nutrients are rapidly used up until a certain equilibrium is reached. From then on, while the artificial pond is surely affected in the same way as the natural pond, its fertility increases much more gradually than it decreased at the outset.

Two incidental types of fertilization occur in a pond which is managed for fish production. The first is the result of stocking it with fish. Fish culture usually involves a fish population density much greater than is normal in nature. The excreta of these fish, and the bodies of any that die, are important sources of fertilizer. The importance of this kind of fertilization can be more clearly seen just after a major fish harvest. The removal of large numbers of animals from at or near the top of the food chain reduces predation on invertebrates and small fish, which greatly increase their numbers and/or growth rate, until the fertilizer released by the fish which have been removed is used up.

Fertilization of ponds also results as a by-product of feeding, particularly if non-living feeds are used. Even if you avoid over-feeding, there is always some waste with such feeds, and the decomposing material will find its way into the food chain. It has been said that "the most effective known pond fertilizer is fish feed," and certainly its use is usually accompanied by an increase in biological productivity quite apart from the cultured fish.

The types of incidental fertilization described are mainly sources of nitrogen, but nitrogen is not the only valuable element in fertilizing fish ponds. Among the most important elements in pond chemistry is calcium. Though it is very rare to find an aquatic ecosystem or food chain deficient in calcium, it is common to supplement the calcium content of fish ponds by liming. Strictly speaking liming is not a form of fertilization, but a preparation for effective fertilization. Among the specific purposes of liming are the following:

• Liming raises the pH of the water. Most fishes do best in water which has a pH between 7.0 and 8.5. (See Chapter X–2.)

• Liming also raises the pH of the mud, upon which the effectiveness of fertilizers depends.

• Liming "buffers" the pH. There is considerable fluctuation in pH in some bodies of water, but proper use of lime stabilizes pH.

• Liming increases the methyl orange alkalinity of the water. (See Chapter X–2.) This results in more calcium being made available for plants, crustaceans and mollusks.

• Nitrogen from natural processes or fertilizers is often present in the form of ammonia or ammonium compounds, which may be toxic to fish and inhibit their growth. When sufficient calcium is present in the water it is possible for these compounds to be broken down to nitrates or other beneficial compounds.

• Liming accelerates the decomposition and mineralization of organic matter, making it chemically available without deoxygenation of the water.

• Many fish parasites can be killed by liming, with no harm to the fish.

• Liming causes the precipitation of excess suspended organic matter, depositing it on the bottom, where it is useful. Apart from its fertilizing value, this is a good control for the disease known as gill rot.

• If lime is applied a few weeks before a phosphatic fertilizer, the calcium in the lime will displace phosphates from colloidal mud, making them more immediately available.

• In ponds where there is relatively little free carbon dioxide, liming can increase the amount of carbon available for phytoplankton photosynthesis by increasing the amount of bicarbonate in the system.

Liming is not always necessary. Excessive use may even be harmful, by causing precipitation of phosphorus in the form of calcium phosphate, which can "smother" the bottom and take phosphorus out of circulation.

To find out if a pond needs liming, you need to know the methyl orange alkalinity of the water. (See Chapter X–2.) A methyl orange alkalinity of less than 30 ppm shows a definite need for liming, and ponds measuring up to 150 ppm may benefit from it. Methyl orange alkalinity over 350 ppm is considered excessive, though it does not preclude fish culture.

There are three compounds which can be used in liming fish ponds. The naming of these compounds is somewhat confusing. I have listed the chemical formulae, along with the most common vernacular names, but you may still have to confront a variety of trade names:

(1) Calcium Carbonate, $CaCO_3$, is ordinary agricultural lime. It is often sold in the slightly less pure form of powdered limestone, or as marl, which may be quite high in impurities. $CaCO_3$ has very low solubility in water but, over a period of weeks it will be transformed into soluble calcium bicarbonate, $Ca(HCO_3)_2$. The finer the grain of the $CaCO_3$, the more rapidly this will occur. In small systems you may use natural lime sources, such as crushed bivalve shells.

(2) Quicklime, or calcium oxide (CaO) is a stronger form of lime. When moist, it forms calcium hydroxide [$Ca(OH)_2$], sold in powdered form as "caustic," "slaked," or "hydrated" lime. The best place to buy this form of lime is a masonry contractor.

(3) Calcium cyanamide, as sold, amounts to about 60% calcium cyanamide ($CaCN_2$) and 17% quicklime.

These three compounds are listed in increasing order of toxicity. The last two can produce minor skin burns when wet, and should be handled carefully. For many purposes agricultural lime, which has little toxicity and is safe to handle, will be sufficient. To control parasites and for the precipitation of suspended organic matter, you will need quicklime. Calcium cyanamide, while it will accomplish all the purposes of liming, is generally recommended only for treatment of specific parasites.

Liming rates and methods

The rate of liming and the methods of application will vary with the pH of the water and bottom soil, and also with the reasons for liming. The rates suggested below are for $CaCO_3$ in the form of powdered limestone. They can be halved if quicklime or calcium cyanamide is used. For marl or other impure sources of $CaCO_3$ the rates can be adjusted according to the percentage of $CaCO_3$ present. All types of lime should be applied at much lower rates if the pond has little or no mud on the bottom; in such cases it may help to apply organic manure (see below) *before* liming.

The most effective method of liming is to apply the lime directly to the dry pond bottom. But if the pond cannot be drained, the water can be limed instead. Where control of gill rot is intended, this is the only appropriate method.

Lime applied to the water should be spread from a boat. The maximum recommended dose at one time is 440 pounds per acre (485 kg/ha) of $CaCO_3$. If you are using one of the toxic forms of lime, keep watch to see that fish or other desirable organisms are not injured. The pH should be constantly monitored and not allowed to exceed 9.5 under any circumstances.

For liming the dry pond bottom, the following dosages of $CaCO_3$ are appropriate:

• To prepare non-acid soil for other fertilizers: 900–1,800 lb/acre (990–1,980 kg/ha).

• To increase the pH of an acid pond: 500–2,000 lb/acre (551–2,206 kg/ha), depending on the original pH and the texture of the soil. (Heavier soils can handle more lime.)

• To increase the alkalinity of ponds where pH is satisfactory: 200 lb/acre (221 kg/ha) should be enough to raise methyl orange alkalinity by 1 ppm.

• To control parasites: 1 ton/acre (2,206 kg/ha) of quicklime or calcium cyanimide, spread while the bottom is damp.

If the pond has a flowing inlet, you can avoid the necessity of spreading by using it to introduce the lime. On European fish farms, a device known as a lime mill is used for this purpose. It is essentially a water wheel connected to a funnel which releases a certain amount of lime each time the wheel goes around. If the flow rate of the inlet is known, the amount of lime added to the pond can be regulated by adjusting the opening of the funnel. An imprecise but simple variation on this system is to line the inlet stream with crushed bivalve shells.

Essential chemical elements

So far, twenty chemical elements have been shown to be necessary for the maintenance of plant and animal life, and at least seven more may play vital roles in ecosystems. While the functions of some of the twenty essential elements in fish have not yet been described, they are all almost certainly necessary in aquatic ecosystems. Carbon (the keystone of all organic structures) and hydrogen can be considered to be present in sufficient quantities almost everywhere. Another element of major importance is calcium, which has been discussed under liming. Oxygen, next to carbon the most important element for living organisms, is not taken up through the food chain, so I will deal with oxygen problems elsewhere. (See Chapter X–2.) The sixteen remaining essential elements (nitrogen, phosphorus, potassium, iodine, iron, copper, cobalt, sodium, zinc, magnesium, sulfur, fluorine, manganese, chlorine, boron and silicon) are all considered to be nutrients and may be supplied through fertilization or feeding. A definite deficiency syndrome for one other element, selenium, has been described in trout, and the importance in fish diets of vanadium and molybdenum is suspected, but it has not been established that these elements could be limiting in aquatic ecosystems.

Three elements, nitrogen, phosphorus and potassium, have been found to be of primary importance in agricultural soils, and the composition of commercial fertilizers is usually given in terms of the percentage of nitrogen (N), phosphorus (P), and potassium (K). The remainder of the known essential elements, needed in smaller quantities and less well understood, are perhaps less likely to be inadequately supplied in soils, and are lumped together as "trace elements" or "micronutrients." For the sake of convenience, I will keep to this arrangement in discussing pond fertilization and fish diets.

Natural vs. synthetic fertilizers

In popular agricultural writings, fertilizers are often divided into the categories "chemical" and "organic." "Chemical" fertilizers are generally taken to include all that are commercially produced by controlled chemical reactions. In the processing cessing of "organic" fertilizers (which may or may not be purified, filtered, or steril-

of "organic" fertilizers (which may or may not be purified, filtered, or sterilized), the active ingredients are not chemically altered from the form in which they are found in nature. To a chemist, this classification is meaningless, since all substances are chemicals or mixtures of chemicals, and all compounds, natural or synthetic, containing carbon are referred to as "organic." For the sake of clarity, then, I will refer to "synthetic" and "natural" fertilizers.

Synthetic fertilizers, although a rather recent development in agriculture, are by now standard in much of the world. (The first synthetic fertilizer was produced in 1840 by applying strong acids to bones. Research on synthetic fertilizers of many kinds proliferated in the 1880's, but they did not come into general use until this century.) Their use has, however, been rather severely criticized, especially by the "organic" farming movement. Since I am generally sympathetic to these criticisms, I will review them in order to explore whether they are relevant in discussions of aquaculture. The major criticisms of synthetic fertilizers are:

• Natural fertilizers such as animal manures and compost enhance not only the nutrient content of the soil, but also its texture. Chemical fertilizers, which are usually sold as dry pellets or powder, do not. For the farmer, soil texture is important because it is related to the soil's capacity to retain water. Obviously, this quality will not concern the aquaculturist. On the other hand, the presence of a layer of colloidal mud on the bottom of a fish pond is essential to the action of any kind of fertilizer. Synthetic fertilizers will not aid in the formation of mud, and may even cause its decomposition, while most natural fertilizers will help mud formation. One more fact to be considered about the effects of fertilizers on pond soil texture is that the natural accumulation of organic matter in a pond is far greater than its accumulation in an agricultural field. A pond is more like a forest in this respect.

Criticisms of synthetic fertilizers: soil texture

• Disciples of the "organic" method argue that the nutrient elements in synthetic fertilizers are not in forms which are as readily usable by plants as those in natural fertilizers. Their opponents, however, say that one of the reasons for the invention of synthetic fertilizers was to make certain nutrients more readily available. It seems likely that this debate can only be resolved case by case; I leave it to the reader to research the question. The matter is even more problematical for the fish farmer than for the agriculturist since, while the nutrient needs of rooted aquatic plants (and perhaps planktonic plants), are similar to those of terrestrial plants, the chemical breakdown of any kind of fertilizer may proceed very differently in water than on land.

Availability of nutrients

• Synthetic fertilizers are usually less rich in trace elements than natural fertilizers. While little is known about the trace element needs of fishes or aquatic ecosystems, this objection is probably valid both for aquatic and terrestrial farming.

Trace elements

• Synthetic fertilizers may be toxic to beneficial soil organisms on land or to benthic organisms and, though rarely, also to fish in bodies of water. A similar criticism could be levelled at natural fertilizers, which may "burn" soil organisms if applied too heavily. Though rapid heating of the soil by natural fertilizers is scarcely possible under water, their overuse can kill animals through deoxygenation of the water.

Toxicity

• Long-term use of synthetic fertilizers can lead to contamination of waterways or ground water. In one region of California, synthetic phosphate fertilizers have been pinpointed as a cause of metheneglobenemia, a serious disease of infants. Certainly if your aquaculture facilities are connected with natural surface or

Drinking water contamination

ground waters, you must consider the possible undesirable effects of any sort of fertilization.

Breakdown rate

• Natural fertilizers are preferred because they break down and release their nutrients gradually, rather than all at once. This objection may or may not hold in aquatic systems, since rapid, one-time release of nutrients may actually be preferred in certain situations.

High energy costs

• An objection which applies equally to terrestrial or aquatic systems is that the manufacture of synthetic fertilizers requires great amounts of energy, which are usually supplied by burning fossil fuels.

Psychological factors

• To these objections, I would add a psychological one which is also equally valid regardless of where a fertilizer is used. It seems to me that the use of a processed, store-bought fertilizer ("chemical" or "organic") will diminish one's awareness of the relationship between the food-producing process and the larger ecosystem. On the other hand the processing of the "wastes" of one's own farm, as in composting, is likely to enhance this awareness.

Advantages of synthetic fertilizers

Having made my sympathies known, I must admit that I am not religiously opposed to the cautious use of synthetic fertilizers where they can be shown to be economically preferable and ecologically tolerable. With that said, let us look at the respective advantages of synthetic and natural fertilizers. The advantages of synthetic fertilizers include:

• Dry synthetic fertilizers (and some processed "organics") are undeniably more convenient to transport, store, and apply.

• While the chemical composition of manures and other natural fertilizers is bound to be variable, synthetic fertilizers can be bought in confidence that their composition will always be the same. Then, if specific mineral deficiencies are known to exist, synthetic fertilizers can be applied to compensate in the correct amounts.

• Synthetic fertilizers are much less likely to create oxygen deficiencies than decaying natural materials.

• When natural nitrogenous fertilizers decompose, it results in the formation of ammonia and ammonium compounds. These compounds may be toxic to fish or inhibit their growth. This effect can only be remedied by transforming the compounds to nitrates before they enter the food chain. However, you can obtain synthetic fertilizers which contribute only beneficial nitrogen compounds.

• Natural fertilizers may be conducive to disease organisms. At least one fish disease, gill rot, has been shown to be related to the presence of suspended organic matter, such as manure, in the water. No such relationship is known for synthetic fertilizers.

Advantages of natural fertilizers

Several of the advantages of natural fertilizers can be understood from the objections to synthetic fertilizers. To the advantages of soil texture improvement, greater availability of nutrients, higher content of trace elements, lesser likelihood of toxic effects, lesser severity of possible resulting pollution, low energy requirement for production and the psychological effect, we may add three more:

• Animal manures, and perhaps other natural fertilizers, aid the entry into the food chain of naturally present phosphorus and potassium.

• Natural fertilizers aid the growth and action of beneficial bacteria. For instance, in ponds where rooted plants are few, carbon accumulates chiefly in the fish. When a large proportion of the fish are removed at once, the carbon to nitrogen ratio may be abruptly altered, to the disadvantage of nitrogen-fixing bacteria. This effect is less drastic if nitrogen fertilization is carried out with natural materials which contain carbohydrates that may be used by bacteria as an energy source.

• Zooplankton may be able to feed directly on the proteins in some natural fertilizers.

Before discussing fertilization methods, we should consider the role in the food chain of the various essential elements and the known effects of fertilizing with them. I will discuss the specific dietary needs of fishes for these elements in a later chapter. First we will discuss the functioning of the whole aquatic ecosystem.

Nitrogen to phosphorus ratio

I have mentioned that the elements nitrogen, phosphorus and potassium are considered most important in agricultural fertilization. In fish farming, so far as we know, the first two are far more important than potassium. Their actions are intimately interrelated, and it is often found that the N:P ratio is what is crucial, rather than the concentration of either one element. In highly productive natural waters, the N:P ratio will reach 8:1 or usually higher. But the optimum rate for fish culture is lower—about 4:1.

The great majority of research on pond fertilization has involved synthetic nitrogenous or phosphatic fertilizers, manures (functioning primarily as nitrogen sources) or combinations of these three. Most of the published fertilization studies have been done in eastern Europe, particularly Poland, at Auburn University in Alabama, and in Israel, with a few in other parts of the United States, western Europe and various Asian countries. Since the biological results of pond fertilization vary even from pond to pond within a region, we will assuredly find great differences when comparing results from such widely dispersed sites. To minimize the dangers of generalization, I have concentrated here on studies done at Auburn and in Poland, where the climate (if not the soil type), loosely resembles that of the northern United States and much of Canada. The abundant Israeli reports seem less applicable to North American climates and soils.

Nitrogenous and phosphatic fertilizers

There have been changes in fish culturists' views on the subject of nitrogenous and phosphatic fertilizers since the first scientific investigations were made in the 1920's. While the importance of nitrogen in natural systems has never been questioned, it was long thought that there was no need for nitrogenous fertilizers and that phosphorus fertilization alone would always increase production. But in the fifties it was realized that the water at the one fish culture station in Germany where all the early pond fertilization studies were done was atypical in its natural mineral content. So now the usual prescription is for a combination of nitrogen and phosphorus. In very recent years culturists have swung back a little toward the use of phosphorus alone, partly due to the cost of synthetic nitrogenous fertilizers and partly with the discovery that nitrogen levels in old, established culture ponds may stabilize at satisfactory levels without deliberate fertilization. But we can still say that all ponds not used for intensive fish culture for some years, and probably many which have, will benefit from the combination of nitrogen and phosphorous.

It seems that phosphorus is a limiting factor in the productivity of almost all natural waters. Apart from its importance in normal metabolic functioning, phosphorus is necessary to maximize utilization of nitrogen; it may also aid bacteria and blue-green algae to fix atmospheric nitrogen. And organic manures, a natural source of nitrogen, aid in phosphorus metabolism.

Phosphorus

Though nitrogen can be supplied by natural or synthetic fertilizers, the appropriateness of the synthetics is doubtful. In Europe, the application of synthetic nitrogenous fertilizers has resulted in 50% increases in fish production (even without supplemental phosphorus), but their economic feasibility has never been proved. Nevertheless, these fertilizers continue to be studied, and are used in practical fish culture in some places. Among the synthetic nitrogenous fertilizers used in practical or experimental aquaculture are:

• Ammonia liquor (NH_4OH), as used in some European countries, contains 20% nitrogen. It is in a form which, while highly stimulating to phytoplankton, especially green algae, is also quite toxic to fish. It is an inexpensive fertilizer and a "fast" one, in that a high percentage of its nitrogen goes into the food chain soon after application. For these reasons it may become more popular despite its toxicity.

• Sodium nitrate ($NaNO_3$) has given results almost identical to ammonia liquor.

• Ammonium sulfate [$(NH_4)_2 SO_4$] has been extensively used in Israel, but not in North America. It is among the more acidic fertilizers, and may result in lowering pH significantly.

• Norway saltpetre or ammonium nitrate (NH_4NO_3) contains 20% nitrogen in the ammonia form, and 14–15% as nitrate. It is only slightly acidic and may promote growth of blue-green algae.

• Acidic ammonium carbonate (NH_4HCO_3), although it contains only 1% nitrogen, is another fast fertilizer, said to be especially beneficial to green algae. It is one of the least acid and safest of nitrogenous fertilizers, hardly toxic even to small fry. But it does seem to have the effect of reducing bacterial populations, which may not be beneficial, depending on circumstances.

• Urea [$(NH_2)_2CO$] is an inexpensive fertilizer, containing approximately 45% nitrogen in the amide form. As amides do not break down rapidly in water, urea is a "slow" fertilizer, which might give best results in combination with a fast nitrogenous fertilizer. Opinions differ as to the toxicity of urea in aquatic systems; it is certainly not one of the most toxic nitrogenous fertilizers.

In my opinion, there are three situations when synthetic nitrogenous fertilizers may justifiably be used in aquaculture.

(1) When a heavy plankton bloom is needed very quickly. For example, in rearing fry of certain fishes where the growing season is short, a fast synthetic fertilizer may be best.

(2) New ponds which have not had a chance to form natural mud may benefit from an initial dosage of a synthetic nitrogen source.

(3) A single dose of a "fast" nitrogenous fertilizer may be beneficial in early spring, when the water is cool and microbial action is slow.

Apart from these three circumstances, the aquaculturist should definitely consider the possibilities of manures and other natural nitrogen sources. Animal dung, green manures, and seed meals are used as nitrogen sources in aquaculture. Almost any type of animal dung may be used, though you should exercise caution with bird manures, which tend to be more acidic than others. Some researchers believe that dung promotes the growth of both phytoplankton and zooplankton, while green manures are effective only for zooplankton. Zooplankton grows particularly well with seed meals, apparently because it can make direct use of their proteins. Both types of manure are said to be faster acting if properly composted. Certainly compost is less likely to deoxygenate water than raw manure.

This is not the place for a treatise on composting, which is well described in many gardening books. (See Appendix IV–1.) As in the use of synthetic fertilizers, there may be reason to combine a fast natural fertilizer (compost) with a slower one (dung or green manure).

Another rich source of natural nitrogen is sewage, which has been used in Oriental aquaculture for centuries. It is uncertain whether Oriental methods allow bacteria to deal with pathogens, or whether the people have developed a tolerance for certain disease organisms. The experience of travelers who become ill in southeast Asia suggests that the local people do have a higher resistance; certainly it would be best to follow the conservative course here, even if the Board of Health does not speak up.

Nitrogen: sewage

There are also more "Western" ways of using sewage to the benefit of fish culture. Fish are cultured as an integral part of municipal sewage treatment in Poland and Germany, but that sort of fish raising is beyond the scope of this book. However, it is now possible to purchase any of a number of brands of processed sewage sludge for use as a fertilizer. These products are sold in dry soluble form, like synthetic fertilizers, with the N:P:K ratio specified on the label. You need not worry about pathogens associated with the use of processed sewage sludge as it is sterilized.

However, there is another problem, and that is "pollution of the pollutant," largely by heavy metals. This problem is acute for heavily industrialized cities where household and industrial sewers share a common drain. Some municipalities have had to abandon plans to process sludge for sale as fertilizer or to carry out aquaculture as part of sewage treatment, because the sewage or sludge was toxic to plants or animals, or because the human food products did not live up to environmental standards and could not be legally sold. (According to Pietro Ghittino, of the Instituto Zooprofilattico, Torino, Italy, another problem is that "Carp reared in fish farms supplied by sewage waters from towns frequently have a disagreeable phenol taste.")

Polluted sewage

Even in communities which can produce legally acceptable fertilizers or fish, the North American lifestyle would almost certainly lead to at least traces of potentially dangerous heavy metals. There have been virtually no studies of the metabolism of these substances from sewage through algae and invertebrates to fish and ultimately the human consumer, so the prospective users of sewage products in aquaculture must use their own judgement.

Manures can be placed directly on the bottom of dry ponds or in the water. In either case (with the single exception of dry, sterile, pond bottoms), manure should not be spread, but placed in small, scattered heaps. It must be applied repeatedly in small quantities, never in large amounts. Both of these precautions are necessary to avoid the dangers of deoxygenation and high ammonia concentration. (The ammonia problem may sometimes be improved by stocking bottom-feeding fish, such as carp. Their rooting action may cause the adsorption of ammonia on clay colloids.)

Applying manures

When manure is applied with water already in the pond, it is best to keep it near the surface, where dissolved oxygen concentrations are highest. This results in the fastest decomposition of the fertilizer and confines deoxygenation to the zone which can best afford it. Platforms or floats may be built in the pond for this purpose or, in small ponds, the manure heaps may be placed only around the banks. In either case, it may be necessary to weight down green manures. Another method, particularly good for stable dung containing wood chips, sawdust or other insoluble particles, is to float the manure in porous bags. Periodic infusions of a liquid manure "tea" may also be used. This method may be the best for promoting plankton growth, but it also results in much ammonia being released.

One of the chief objections to manuring is the considerable labor of procuring, transporting, and applying manures. Instead, the aquatic farmer should consider the more direct application of natural nitrogen. You can obtain green manures by cutting plants growing in or close to the water. Animal manure can be provided by confining livestock so that their wastes fall into the water or will easily be washed in. This method works particularly well with waterfowl. (For a full discussion of the integration of waterfowl and terrestrial livestock with aquaculture see Chapter IV-2.)

Natural and incidental nitrogen inputs

Before applying any kind of nitrogenous fertilizer, you should carefully consider the natural and incidental inputs of nitrogen. Fish, stocked at the usual high densities, are one of the best sources of nitrogen, especially when fed on prepared feeds. However this should not be construed as an excuse for overfeeding. (Apart from the expense involved, it has been shown that the biological oxygen demand [B.O.D.] of commercial fish feed is ten to sixty times that of manures, which in turn have much higher B.O.D. than synthetic fertilizers.) You can create further "natural" nitrogenous fertilization by drying the pond bottom between growing seasons, which may result in the mineralization of naturally occurring nitrogen. (Some, however, contend that drying reduces fertility. This is just one example of the contradictions which abound in reports of aquatic fertilization.) Seeding the dry pond bottom with plants, especially legumes, and plowing them under can further increase the benefits of drying.

Still another method of adding nitrogen to an aquatic system without the use of chemicals is to stock the floating aquatic fern, *Azolla*. *Anabaena azolla*, a blue-green alga associated with *Azolla* is able to fix atmospheric nitrogen in aquatic systems. In experiments in Denmark, it fixed as much as 85 pounds per acre (94 kg/ha) of nitrogen in a summer. *Azolla* must be thinned periodically to prevent the formation of an excessive "cover," shutting off light and atmospheric oxygen. (In semi-tropical or tropical climates, *Azolla* can become a pest. Do *not* introduce it in such climates.)

You may well find, after a while, that no nitrogenous fertilizer need be applied. Experiments at Auburn University indicated that when no nitrogenous fertilizer was placed in ponds fertilized for the previous five years with commercial 8-8-2 fertilizer, there was no decrease in fish production over the following six-year period.

Phosphorus fertilizers

Phosphorus is found in all natural ecosystems, and in most natural fertilizers, but the Belgian fish culture expert Marcel Huet says "It seems that nearly all waters lack phosphorus and consequently mineral phosphate fertilizer is nearly always worthwhile." While nitrogen is usually applied to fish ponds by manuring, phosphatic fertilization is generally done with synthetics. Unlike synthetic nitrogen fertilizers, the use of phosphates has been found economically appropriate, not only in Europe and the United States, but in somewhat less affluent countries, such as Taiwan. Huet lists four types of synthetic phosphorus fertilizers: superphosphate (16–20% P_2O_5-, some inert $CaSO_4$), basic slag (an industrial by-product, of variable composition), Rhenian phosphate (25% P_2O_5-, 42% CaO), and di-calciumphosphate (35% P_2O_5-, 28% CaO). Others mentioned by early culturists include colloidal phosphate (24% P_2O_5-, 25% CaO) and Thomas meal (13–20% P_2O_5-, 40–50% CaO).

In North America, by far the most commonly used are superphosphate and triple superphosphate (superphosphate minus the $CaSO_4$), which are also the most soluble.

The use of the superphosphates requires adequate amounts of lime (i.e. calcium), but they should not be applied immediately after liming, or the precipitation of insoluble calcium phosphate [$Ca_3 (PO_4)_2$] may result. Liming is not necessary with basic slag, which contains 40–50% calcium oxide (CaO) as well as abundant trace elements. The other phosphatic fertilizers (mentioned in the previous paragraph) are said to be intermediate with respect to solubility and calcium content.

The organic farming movement generally uses ground phosphatic rocks, sold as "rock phosphate." This material contains approximately thirty percent phosphoric acid, but most of it is not readily available to agricultural plants. Whether the same is true for aquatic systems is not known.

Organic gardeners also consider basic slag acceptable as a phosphorus source. The next richest "natural" sources of phosphate are bone meal and other types of processed bone (22–30% phosphoric acid), and processed sewage sludge (2.5–4.0% phosphoric acid).

There is much difference of opinion about how often to apply phosphatic fertilizer. Clearly the need for phosphorus is not as constant as the need for nitrogen, i.e., ponds can go for a time without available phosphorus with no effect on fish production, while a nitrogen deficiency will have almost immediate negative effects. Of course a chronic shortage of phosphorus will be limiting, while overdoses of phosphate may result only in phosphorus being "locked up" in the mud. Equipped with these facts, Huet maintains that phosphatic fertilizers need be applied no more often than annually, while the Polish researcher Stanislaw Wrobel advises frequent small applications. It is not known how much these recommendations were influenced by the character of individual ponds, or how they might be altered if "slow," natural phosphorus sources were used instead of synthetic compounds.

Frequency of phosphatic fertilization

I mention potassium next, not because of its relative importance in aquaculture, but because it is the third major nutrient in many commercial fertilizers. In fact, there has been no clear demonstration that potassium is a limiting factor in productivity of water bodies. Maybe extremely infertile ponds would benefit from fertilization with potassium, but only if nitrogen and phosphorus were abundantly supplied.

Potassium

I have mentioned that carbon is present in profusion almost everywhere. But it may happen that less than optimal quantities of carbon are present in the form of carbon dioxide (CO_2), a critical compound for many biological processes, not the least being the respiration of plants. As CO_2 enters the environment chiefly as a product of oxidation, attempts to produce it in water are apt to lead to deoxygenation. The brilliant Hungarian fish culturist Elek Woynarovich makes frequent small applications of manure—spread, not heaped—to stimulate CO_2 production, but this method is only recommended for those who clearly understand water bodies' biochemical cycles, and also have the means to monitor CO_2 and oxygen constantly.

Carbon dioxide

Carbon in the form of bicarbonate may also be limiting. The availability of bicarbonate may be increased by liming, as discussed earlier in this chapter.

Some of the trace elements have scarcely been looked at in aquatic ecosystems.

Trace elements
Some have been studied primarily with respect to their importance in fish physiology. If deficiencies of any of the elements known to be physiologically important to fish are found, they are perhaps best treated by supplementing the diet of the fish. (See Chapter VIII-2.) Other trace elements have been studied to determine their effect on organisms lower on the food chain, principally the plankton. These are properly discussed here, under the heading of fertilization.

Iron
Iron (Fe), in the few cases when it was studied as a fertilizer component, showed no effect on plankton production. Furthermore, large amounts of iron are generally considered undesirable in ponds, since iron tends to precipitate in insoluble forms. There may be forms of aquatic life which benefit from supplementation with iron. One such possibility, the larvae of chironomid midges, is discussed in Chapter III-5.

Silicon
Silicon (Si) is contained in high percentages in diatoms, one of the common groups of planktonic algae, and can be limiting in their growth. In one series of experiments conducted in Taiwan, application of zeolite, which contains 73% silicon, at 10 pounds per acre (11 kg/ha), resulted in 18–40% increases in fish production. However, Taiwanese aquaculturists do not presently consider zeolite to be effective in increasing primary productivity in most ponds.

Manganese
Manganese (Mn) is also limiting for diatoms. In experiments conducted in India, phytoplankton production was increased by applying elemental manganese at 0.9 pounds per acre (1 kg/ha) in conjunction with phosphatic fertilizers.

Magnesium
In one set of experiments at Auburn University, heavy liming (2,936–5,826 lb/acre or 3,295–6,539 kg/ha of $CaCO_3$) resulted in three- to six-fold increases in concentrations of Magnesium (Mg), together with increases in phytoplankton production.

Boron
Boron (B) is essential both for diatoms and for higher plants, but its use in fertilizing aquatic systems has not been studied.

Cobalt
Cobalt (Co) is an important constituent of Vitamin B_{12}, which is found in considerable quantities in phytoplankton. In the Soviet Union, application of cobalt chloride to a fish pond at 4.5 pounds per acre (5.0 kg/ha) stimulated the growth of phytoplankton and resulted in an 87% increase in fish production.

Copper and Sodium
Copper (Cu) and sodium (Na) have been studied in aquatic ecosystems, but have not been shown to limit phytoplankton production.

Specifics of fertilization
So far this discussion of fertilization has avoided mention of specific fertilizer dosages because the results of fertilization of aquatic systems are so unpredictable. Wilhelm Schaperclaus, in his classic early text *Lehrbuch der Teichwirtschaft* (Handbook of Pond Husbandry), suggested beginning by applying various fertilizers to small containers of "parent" water to see which gave the best algal production. But he was the first to admit the limitations of this system, not the least of them being the need to judge qualitatively, as well as quantitatively, the results of fertilization. Finally, the only real test is fish production.

Nevertheless, it would be remiss of me to altogether omit examples of fertilizer dosages used in practical aquaculture. Therefore, bearing in mind the limitations of our knowledge, I will offer some examples of fertilization schemes which have been successful in achieving specific goals, or which have been recommended by fish culturists or fishery biologists. First, I will discuss methods which have successfully enhanced the production of particular types of fish in the United States.

The original concept of the American "farm pond" included fertilization, but as long as we continue to underfish these ponds, fertilization will be superfluous. As I write, I have before me a number of farm pond booklets put out by various states. Generally, in the northern states fertilization is not recommended except as a method of weed control (see Chapter VIII–1), while in the South fertilization is still recommended as a basic management technique. This is usually explained on the basis of climatic difference; the chance of winterkill in fertilized ponds is certainly greater than in unfertilized ones. However, I suspect it may also reflect the greater utilization of pond fish as food in the South.

The Department of Conservation of the state of Alabama recommends the use of synthetic fertilizers in amounts sufficient to provide 8 pounds of available nitrogen, 8 pounds of available phosphorus and 2 pounds of available potassium per acre (8.8, 8.8 and 2.2 kg/ha, respectively) in ponds which have not been fertilized previously. After two years, the nitrogen portion may be eliminated if plankton blooms are satisfactory. The fertilizer is normally applied by placing opened bags on a single submerged platform. The first three applications are made at two-week intervals, beginning with the first warm weather of late winter. Subsequent applications are made every three to five weeks, or whenever Secchi disk visibility (see Chapter X–1) exceeds 12 inches (30.7 cm), until the start of winter weather. This normally comes out to 10 to 12 fertilizer applications per year, and is said to increase largemouth bass and bluegill production from three to seven times.

Bass-bluegill farm ponds

The Conservation Commission of Missouri, a state which is climatically neither strictly northern nor southern, recommends regular fertilization only for heavily fished ponds. They suggest a completely soluble high phosphorus fertilizer, such as ammonium phosphate, for clear, alkaline ponds. A practical application method is shown in Figure III–2–1. Fertilizer should be applied in five or more cans at 100 pounds per acre (110 kg/ha), and replaced whenever Secchi disk visibility exceeds 2 feet (0.6 m).

The nutritional basis of the channel catfish (*Ictalurus punctatus*) culture industry is intensive feeding, not fertilization. While maintenance of plankton blooms is useful both as a supplemental food source and to reduce weed problems, in conven-

Channel catfish

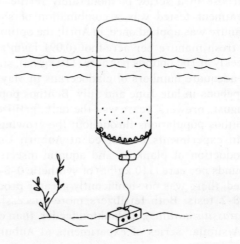

FIGURE III–2–1. *Device for gradual application of synthetic fertilizers made from a can.*

tional catfish culture adequate fertility is almost guaranteed by high fish population densities and the use of commercial feed. Still, many catfish farmers like to get a jump on the season by fertilizing their rearing ponds once a year, about two weeks before stocking. Recommended treatments include 50 pounds of 16–20–0 and 50 to 100 pounds of 20–20–5 per acre (55–110 kg/ha). Sometimes fertilization is done in the winter, at dosages of 150–225 pounds per acre (165–248 kg/ha).

Buffalofish   Cultured buffalofish, unlike catfish, get most of their food from "natural" sources, so fertilization is critical. The first application of fertilizer must be done immediately after spawning if the fry are not to starve. In Arkansas, a dosage of 100 pounds per acre (110 kg/ha) of commercial 8–8–2 fertilizer, applied monthly after the initial fertilization, produced adequate amounts of diatoms and zooplankton.

Trout   Fertilization is not normally practiced in trout culture. However, the U.S. Department of Agriculture does recommend fertilization for those trout ponds intended to produce fish primarily for home consumption. They suggest eight pounds of available nitrogen and eight pounds of available phosphorus per acre (8.8 kg/ha), with the optional inclusion of two to four pounds per acre (2.2–4.4 kg/ha) of available potassium where the soil is known to be deficient in that element. Application of fertilizer on platforms or floats begins when the ice breaks up in the spring, or when the crocus blooms, and is repeated weekly until Secchi disk visibility is 18 inches (46 cm) or less. The treatment can be repeated as needed until early summer, when fertilization should cease. Some trout farmers have reported trout deaths from heavy use of commercial fertilizer.

The aforesaid fertilizer treatments were designed to increase production of specific fish species. Fertilizer recommendations would be more generally useful if the initial effect of the fertilizer on plankton, benthos, rooted plants, etc., were specified. A few such examples follow.

Zooplankton   One of the most difficult of American fresh water food fish to raise is the walleye (*Stizostedion vitreum*), as young walleye require large amounts of zooplankton and benthos throughout their first spring and summer. The difficulty of providing adequate amounts of these organisms during the summer months has been one of the main constraints on hatchery rearing of walleyes. The Minnesota Department of Conservation has tested a variety of synthetic and natural fertilizers for this purpose in a series of moderately fertile alkaline ponds. The most satisfactory treatment tested was a combination of sheep manure and brewers' yeast. The manure was applied once, in April; the optimum rate of application was 248 pounds of fresh manure per acre-foot (0.091 kg/m³) of water. Brewers' yeast was applied weekly at 100 pounds per acre (110 kg/ha), starting June 15. This treatment resulted in adequate numbers of cladocerans in May and June, followed by a good crop of copepods in late June and July. Benthos populations were high from May through August; brewers' yeast was the only fertilizer tested which produced satisfactory benthos populations throughout the growing season.

In experiments conducted at Auburn University, Alabama, some increase in production of plankton and aquatic insects occurred in ponds treated with 100 pounds per acre (110 kg/ha) of synthetic 0–8–2 fertilizer. When 8–8–2 fertilizer was used, there was no significantly greater production of these organisms than in the 0–8–2 tests. Both fertilizers more successfully enhanced production of goldfish (*Carassius auratus*), a plankton feeder, than common carp, mainly a benthos feeder.

A similar series of experiments at Auburn, using Mozambique tilapia (*Tilapia mossambica*) as the test fish, gave the same results: 8–8–2 fertilizer was not significantly better than 0–8–2 for production of fish, zooplankton, or chironomid larvae. Benthic forms other than chironomids did not respond to either fertilizer.

Still other experiments at Auburn dealt with phytoplankton. Indications were that nitrogen was unnecessary in synthetic fertilizers, and could even be harmful, in that it favored the growth of blue-green algae, some of which may be toxic to fish. When triple superphosphate was applied at 17.33 pounds per acre (19.45 kg/ha), an initial bloom of blue-green algae appeared, then died off, to be replaced by more desirable green algae. When 23.5 pounds per acre (26.4 kg/ha) of ammonium nitrate was applied with the superphosphate, the blue-greens persisted. *Phytoplankton*

Perhaps the most difficult group of organisms to grow with fertilization are benthic animals. Fertilization may actually reduce their numbers for while they need nutrients like any organism, they also need oxygen. One of the first results of conventional fertilization is a dense layer of phytoplankton which, although it oxygenates the upper layers of the water, shuts off light and greatly reduces photosynthesis on the bottom. Perhaps benthos-feeding fishes would actually benefit most from fertilizer doses much lower than those normally used. And perhaps animal manures, though less effective in promoting plankton blooms than the more readily soluble synthetic fertilizers, would more effectively aid the growth of benthic animals. Two researchers, Meehean and Marzulli, observed that "organic fertilizers may be digested by bacteria and the fragments produced in the process, along with the bacteria, may be utilized by a flourishing population of microcrustaceans closely associated with the bottom, but which never become part of the plankton." (See Appendix IV–1.) *Benthic (bottom) animals*

Another difficult situation occurs when fish culture is attempted in peat bogs. Peat soils are typically acid and poor in nutrients. However, they hold water well, and conversion into ponds is one of the most logical uses of excavated peat bogs. The fertility of peat soil may be made satisfactory by combined use of heavy doses of lime, manures and synthetic fertilizers, but only at considerable expense. A less expensive method of making peat-bottomed ponds suitable for fish culture was developed in the Soviet Union. It involves releasing nitrogen and other nutrient elements which occur naturally in peat, but in forms unavailable to plants. This is achieved by application of a common Russian peat-based fertilizer known as TMAU. TMAU is prepared by treating peat millings (25–30% decomposed peat containing 60–65% water) with enough ammonia liquor to provide 0.6–0.9% ammonia nitrogen in one hundred percent dry peat. The resulting fertilizer is first applied to the dry pond bottom at 4,455 pounds per acre (5,000 kg/ha), following which the bottom is harrowed. Beginning thirty-five to forty days after filling with water, an additional 4,455–6,237 pounds per acre (5,000–7,000 kg/ha) is gradually applied to the water surface. Where the peat soils are deficient in phosphorus, 22 pounds (10 kg) of superphosphate are added to each 2,200 pounds (1 metric ton) of TMAU, but in some ponds this is not necessary. It was calculated that costs for nitrogenous fertilizer were decreased in this way by factors of 2.8 to 3.0 and costs for phosphates by factors of 1.7 to 1.9. At the same time, production of common carp is increased by 73 pounds per acre (80 kg/ha) or more. *Peat soils*

It will come as no surprise to most readers that in Western society much more funding is available for the study of processed feeds and fertilizers than for natural ones, and that is why most of this discussion has dealt with synthetic fertilizers. In *Chinese methods in polyculture* recent years the success of the Chinese in "organic" pond fertilization has been paralleled by Israeli fish culturists. In polyculture systems, they have been able to produce the same weights of fish with either fresh or fermented animal manure as with conventional processed fish feeds. It was not until 1975 that the first attempt was made to test a North American counterpart to traditional Chinese pond fish culture.

Nourishing via manure

In that year, Homer Buck, R.J. Baur, and C.R. Rose stocked two 0.3 acre (0.12 ha) ponds at the Sam A. Parr Fisheries Research Center, Kinmundy, Illinois, with a mixture of seven kinds of fish and tried to nourish them solely through fertilization with manure. The fish populations were not only comparable to Chinese polycultures, they included three Chinese species: the grass carp, silver carp (*Hypophthalmichthys molitrix*), a phytoplankton feeder, and bighead carp (*Aristichthys nobilis*), traditionally considered a zooplankton feeder. These three species accounted for seventy-eight percent of fish production in the ponds. Nutrients were provided by penning pigs next to the ponds and washing the pens directly into the ponds. The two ponds were provided with five and eight pigs, respectively, or about sixteen and twenty-seven pigs per acre of water surface (39 and 66 pigs/ha).

The chemical aspects of the pond environment, including dissolved oxygen concentration, pH, methyl orange alkalinity, ammonia concentration and B.O.D. (see Chapter X–2) were monitored and found consistently satisfactory. Over a growing period of one hundred and seventy days, the total production of fish in the

Hogs penned beside a polyculture pond at the Sam A. Parr Fisheries Research Center, Kinmundy, Illinois. Wastes from the hog pen are washed into the pond as fertilizer for algae and zooplankton to be eaten by the fish. (Photo courtesy Homer Buck, Illinois Natural History Survey.)

pond receiving the manure of five pigs was 2,647 pounds per acre (2,971 kg/ha), while that of the pond receiving manure from eight pigs was 3,416 pounds per acre (3,834 kg/ha). Subsequently Buck obtained a maximum production of 4,585 pounds per acre (5,043 kg/ha). Despite the fact that the hogs' feed contained corn, soy meal, minerals and vitamins, these results are impressive when compared to conventional North American fish culture.

The most highly developed Western form of food fish culture in standing water is channel catfish monoculture. (See Chapter II-3.) Catfish culture is traditionally carried out in the South, where water temperatures are more consistently favorable for growth than in Illinois. The fish are fed on a commercial ration more complex than the one fed to the hogs in the experiment; it contains substantial amounts of animal protein in addition to grain products. The techniques of culture are the product of some twenty years of experience, and catfish growers benefit from a strong program of scientific research and technical aid. Catfish farmers usually produce less than 3,000 pounds per acre per year (3,309 kg/ha). Buck and his co-workers were able to surpass this figure in less than a half-year on their first try with a Chinese type of polyculture system based on manure and without using any commercial feeds or synthetic fertilizers.

Species combinations for polyculture are discussed in Chapter IV-1. But whether you give greater importance to the polyculture aspect of the experiment or the fertilization technique, the message as to the appropriateness of traditional North American fish feeding methods is clear.

U.S. catfish culture compared with Chinese pond polyculture

CHAPTER III-3

"Trapping" Fodder Animals

Bug lights

While fertilization makes the water more suitable for the growth of aquatic organisms, the "catch" for the plankter or insect is that it will very likely be eaten by a fish. To feed our fish we can also play the classic spider and fly game, by attracting and trapping animals which do not live in the water. The fatal attraction of the moth for the flame is well known, so it is not surprising that this principle should be applied to trap insects for purposes of pest control and, still more recently, for feeding fish.

Early models

Any source of light will attract some insects, but ultraviolet or "black" lights are much more effective than other natural, incandescent or fluorescent lights. Early models of ultraviolet "bug lights" have a killing grid of closely spaced electrified wires. When an insect hits the grid it closes the circuit and is electrocuted. Such lights are still sold and used and can be adapted for aquaculture use either by attaching a tray to collect insects which fall off the grid, or by placing them directly over the water. However, there are a number of objections to these lights. The continual noise they make on a "busy" night can be annoying, and it may be that partially "fried" insects are not as nutritious as live ones. While the killing grid is not powerful enough to injure humans, touching it can be unpleasant and frightening. I know of one instance where a bird lit on the grid, panicked and was killed.

U.S.D.A. bug light experiment

A quieter and safer sort of u-v light insect trap was developed in the 1950's by the Plant Pest Control Division of the U.S. Department of Agriculture, for the control of European chafer beetles. This trap uses a metal baffle and a funnel. Insects are attracted to the light, strike the baffle and fall into the funnel, which leads to a collecting jar. The u.s.d.a. extensively tested various aspects of trap design and concluded that, for European chafer beetles, the best trap was made with a single 30 watt black light bulb, suspended perpendicular to the ground, with a four-winged baffle over a funnel with a diameter of 12 inches (30.7 cm) (Figure III-3-1). The best material for the baffle is anodized aluminum, which reflects ultraviolet light. Increasing the wattage or number of tubes or the size of baffles and funnel did not improve the performance of the trap. You can easily build such traps at home. If they are to be used safely over water, they should be provided with marine type connectors for extension cords.

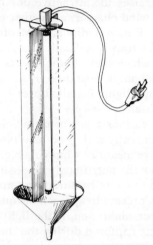

FIGURE III–3–1. *Ultraviolet bug light provided with aluminum baffles, for use as a fish feeder.*

A chafer beetle type of trap was used in the first serious study of lights as a tool in fish feeding, by Roy Heidinger of Southern Illinois University. Heidinger grew bluegills (*Lepomis macrochirus*) in floating cages, some of which were provided with 15 watt ultraviolet lights equipped with baffles and funnels, but no collecting jar. Instead, the lights were suspended 5.9 inches (15 cm) over the water surface so that insects striking the baffle fell directly into the water. To induce bluegills to feed at night, a 25 watt incandescent bulb was added, though it is not clear to me that this was necessary.

Researching bug lights for use with fish

Heidinger found that the animals attracted were of three sorts:
(1) Flying insects metamorphosing and emerging from the pond. (It has been estimated that 75 percent of such insects never return to the water.)
(2) Terrestrial flying insects lured out over the pond.
(3) Aquatic insects and zooplankton which swam into the cages.

Analysis of the stomach contents of the caged bluegills (which received no other food during the experiments) revealed that 74 percent of their food, by dry weight, was aerial insects, 14 percent was aquatic insects and 1 percent was zooplankton. ("Miscellaneous food items" accounted for the remainder.)

The average weight of bluegills stocked in the cages was 1.8 ounces (50 g). Up to 9.3 fish/ft² (100 fish/m²) were stocked, but the supply of insects exceeded the capacity of the fish even at the highest population densities tested. Growth rates as high as 147 percent in sixty days were recorded. While Heidinger's research used caged fish, the technique of using ultraviolet lights to attract insects can also be used in open pond culture. Heidinger estimated that 25 ultraviolet lights with baffles, evenly spread around the periphery of a 0.4 acre (1 ha) pond in southern Illinois would increase bluegill production by 160 pounds (351 kg) in 180 days, at a cost of four cents per pound of fish produced (assuming use of a conventional power source).

It might be possible to further reduce this cost by using battery power. United States Department of Agriculture reports say that their "European chafer beetle survey traps" are "wired to operate on either a.c. or d.c. circuit from storage batteries." This raises the attractive possibility of a pond-side or floating battery platform with a wind charger, altogether eliminating the cost of public power.

More recently, Elmer Hedlund, of Medford, Wisconsin, has developed an ultraviolet light insect trap specifically designed for use in feeding fish. This type of feeder has a fan which sucks in the insects, which are then deflected into the water. Where current cannot be sent directly to the pond, a collecting bag can be attached to the trap, though lights installed over land are usually less effective than those operated over water.

We have used various types of ultraviolet light insect traps and fish feeders at New Alchemy, generally with good results. It seems to be one of the least expensive and labor-intensive means of supplying high protein food to fish. (See Table III–7–1, pages 167–168 for the nutritional composition of insects.)

The bulk of the insects captured on Cape Cod are midges and moths. The only beneficial insects present in appreciable quantities are lacewings (usually less than three per light per night) and, if the lights are left on long past dawn, bees. The quantity of insects captured during the "bug season" (June through September) is extremely variable, ranging from almost nil to nearly one half pound (227 g) per light for lights mounted over land and provided with trays or bags. The best results

"Will-o-the-Wisp" bug light fish feeder installed on a farm pond in Wisconsin. Nocturnal flying insects are attracted to the u-v light tube, sucked in by an impeller fan and blown down through the chute at the back of the unit and into the water. (Photo by Elmer Hedlund, Hedlunds of Medford, Inc., Medford, Wisconsin.)

are obtained on still, sultry nights. We are certain that our light traps would be much more effective in the Midwest or South, where such nights are the rule, while on Cape Cod most nights are cold and/or windy.

A number of other types of insect traps are used in pest control. Those which capture sufficient quantities of insects can be used as sources of fish food, but some of the hard-bodied insects, like the Japanese beetle, may need to be ground up before feeding. It is easy to imagine other applications of the principle of attraction, but to date only lights have been tested in fish culture. Some successful insect pest control programs have used attractant pheromones (the chemicals which attract adult insects to other insects of the opposite sex). New Alchemy's Earle Barnhart has suggested that such pheromones could be placed on posts or floats in ponds to lure insects over the water. Probably some types of plants planted along the shore, or even on floats, would increase the number of insects available for food. Another possibility is to increase the production of insect larvae and other benthic organisms by placing some sort of structures in the pond to increase the substrate surface area.

Other trapping methods

CHAPTER III-4

Use of Waste Materials as Fish Food

Plant wastes

Agricultural wastes include both plant and animal materials. Most North American fish farmers will find little use for green plant materials as fish food, since our continent is poorly supplied with large herbivorous fishes. At New Alchemy, we do use garden weeds (purslane, sheep sorrel and certain grasses), as well as "waste" portions of crop plants (carrot tops, marigold blossoms) as food for tilapia. In Chapter III–5 I discuss how these and other plants may be cultivated as food for herbivorous fishes, and water weeds are discussed in Chapter VIII–1. Grain brans and other grain wastes can be fed to carp, buffalofish and bullheads. But the cultivation and processing of grains, as normally carried out in North America, do not result in quantities of these materials being made available.

Meat and fish wastes

A rich potential source of high protein fish food occurs wherever meat animals or fish are butchered. To get the highest nutritive value from meat and fish waste, it should be fed fresh, or frozen and prepared when needed. The meat of warm-blooded animals must be boned and cleaned of sinew and fat. The meat is then passed through a hand operated or power mincing machine, set so as to produce the size of particles appropriate to the fish to be fed. When using fish as food only large individuals need be boned, and then only the larger bones need be removed, as carnivorous fish are well adapted to digest fish bones.

Nutritionally it is preferable not to cook meat and fish, as cooking lessens the availability of proteins and vitamins. It also will require the use of fillers to prevent inflammation of the intestines. Cooking is advised when the feed is not fresh, when using the carcasses of sick animals, and especially to prevent the transmission of diseases and parasites from feed fish to crop fish. This risk can be lessened if only salt water fish are used as food for fresh water fish, but many times fish farmers want to take advantage of the carcasses of their own harvests. If cooking is necessary, it should be done in a pressure cooker at 248°F (120°C) for ten to fifteen minutes.

Another good food is blood from slaughterhouses. It is used as a dry meal, usually mixed with other feeds, and can be fed raw or lightly cooked. Cooked, lightly salted blood can be kept for several days in a cool place with no loss of nutrients. Kept in this way it will congeal to a semi-solid state in which it can be fed.

Most fresh fish and meat products need to be minced before use. If some sort of binder is added prior to mincing there will be more defined particles and less loss of fluids. About ten percent of sawdust is sometimes used for this purpose, but it is better to use a substance such as white cheese or vegetable meal, which has some nutritive value.

Five percent of yeast is usually added to such feeds, and if a "complete" feed is desired, synthetic vitamins can be added as well. (See Chapter III–6.) Thiamine is absolutely necessary where herring are used, since they contain an enzyme called thiaminase, which breaks down thiamine.

Readers who do not have access to agricultural, fishery, or slaughterhouse wastes have probably thought of household waste. Unfortunately, I know of no fish comparable to the chicken as a garbage omnivore, though bullheads might be fed on carefully sorted garbage. In water the effects of environmental contamination by uneaten "food" are potentially much more serious than on land. All in all, probably the best application of household garbage in aquaculture is to use it to raise earthworms to be fed to the fish. (See Chapter III–5.)

Household wastes

CHAPTER III-5

Cultured Fish Foods

Fresh vs. processed feeds

The fertilization of a body of water could be seen as a form of culturing fish food organisms. You might suppose, then, that from fertilization a more intensive culture of particular species of forage animals and plants would logically have evolved. But, for the most part, the evolution of feeding in fish culture has passed directly from fertilization to the use of prepared "dry" feeds. Most aquaculturists still consider live foods to have limited applicability. For instance, Huet says "A distribution of small natural fresh water fauna rarely pays and can only be considered for young fish of high value such as salmonids and pike."

Advocates of live foods value their assured freshness and high vitamin content. We are beginning to accumulate some evidence that a live food diet produces growth superior to that obtained with the best dry feeds. But the feed industry can counter with some formidable arguments for the use of their product. Compared to dry feeds, live foods are extremely high in water content, which involves the handling of far greater weights of food to provide a similar amount of nutritive material. The industry also points out that when you compare the costs of purchasing, handling and storing dry feed with the time, labor and space requirements for culturing fish food organisms, the processed feeds are, in fact, cheaper. While in all but a few special cases, the feed company arguments will win in North America at this time, there are some good reasons to consider the use of natural foods:

Advantages of natural foods

• Most published discussions of fish farming are aimed at the large-scale commercial producer, but the economic advantage of dry feed becomes debatable with a reduced scale of operation.

• In general, the prices of dry feeds and their ingredients are rising, and it appears that this rise will continue. Some fish farmers also tell me that the quality of commercial feeds is declining. The costs of culturing natural foods remain more stable, and of course the quality does not change (excepting, possibly, as a result of environmental contamination).

• Processed dry feeds are preeminent in fish culture in North America and a few other highly developed countries only because of the peculiar economic situation the world found itself in after World War II. Specifically, it is the availability of cheap petroleum-based energy which makes it possible to grow fish at a profit using these feeds. If you take a more global view of the process which terminates on an American fish farm, the ultimate limitations of our prepared feeds become apparent. Consider the harvesting of fish from the oceans of the world and reducing them to meal; then, the growing, harvesting, roasting and milling of soybeans; and then synthesizing vitamins from petroleum in a factory. Consider bringing together all these ingredients and more, processing them, and shipping the product to the feed store. Now you can see that, in terms of fuels, human energy and resources, the dry feeds are produced at greater cost than some foods which may be cultured by the fish farmer. Insofar as we are concerned about world food and fuel supplies, we should be thinking about this expense now. Eventually it will be passed on to us and we will be forced to think about it.

• The most expensive ingredients of dry feeds are the sources of protein and vitamins—precisely the classes of nutrients best supplied by live foods. Eventually it should be possible to work out practical fish diets in which the carbohydrates, fats and some of the proteins and vitamins needed by fish are provided by dry feeds much cheaper, and ecologically more sound, than the present day "complete" feeds. At the same time the bulk of the animal protein and vitamin needs could be supplied by relatively small quantities of live foods.

From a practical point of view, I advise beginning fish growers to use whatever combination of foods is most advantageous at the moment, but to be prepared to change strategy, as circumstances dictate.

Certainly the culture of fish food organisms has not been left unexplored. Literally thousands of species of small animals potentially edible by fish have been cultured; if you add plants to the list, it becomes almost infinite. I cannot describe culture methods for each possible fodder organism, or even list them all. But I will mention some of the more useful and promising ones, and provide some background to help you make choices, together with appropriate references.

Sources of information

With regard to food animals, writings worth investigating have been produced by workers in three fields.

• Food and sport fish culturists: As implied by the quote from Huet, practical aquaculturists have grown fodder organisms only where there was a clear economic payoff. Their emphasis on young fish is logical; young fish have both the highest percent need for animal protein and the lowest absolute need for quantities of food. It may be feasible to feed them a high proportion of live animal material when to provide adult fish with the same percentage would be unfeasible.

• Research biologists other than aquaculturists: A tremendous variety of invertebrates are cultured for use in biological and medical research, or as food for other laboratory animals. While many of these culturists do not work within the same economic limitations as the fish farmer's, many of their methods are quite efficient and may be adapted to aquaculture needs.

• Aquarists: Public aquarium keepers and aquarium hobbyists share a common concern with the fish farmer: healthy, well-nourished fish. While their quantitative requirements and economic priorities are in a completely different ballpark from the fish farmer's, they have studied fish foods for many years and much of their work has been published, though in anecdotal form.

In perusing the list of food animals, the reader is sure to think of many organisms which are not included, some of which are important foods for fish in nature. In considering an animal for culture as fish food, you should consider not only its acceptability by fish and its nutritional value, when known (nutritional values of fish feeds and ingredients are listed in Table III-7-1, pages 167–172), but also its life cycle and ecology. As in selecting a species of fish for culture, you must reckon with the food chain. An example of an animal which has been omitted due to food chain considerations is the dragonfly. While dragonfly larvae are excellent food for fish and fairly easy to culture, they are predators and it is doubtful that they could be reared efficiently.

Culturing of fish food plants involves somewhat different considerations, to be discussed later.

The fodder animals are discussed in conventional systematic order, rather than any order of particular use to the fish culturist. For the reader's convenience, in Table III-5-1 these animals are roughly divided into three groups on the basis of size—one of the most critical criteria for the fish culturist. Those marked with an asterisk are already being cultured for use in growing food fish. The others are included because they are known to be good fish food; many of them are collected from the wild by aquarists or fish culturists. Though culture methods for some of these animals have been developed, they have not yet been used to produce the large quantities needed for practical food fish culture schemes. Some species should prove fairly easy to adapt to practical mass culture. Others will require considerable modification of existing techniques, or may eventually be rejected for mass culture. Full taxonomic classifications are given in the text to aid the reader interested in delving into the biological literature.

Table III-5-1.
Animals presently or potentially usable as foods for cultured food fish.

Large animals, suitable for adult fish, but which may have to be cut or ground for young fish	Medium-sized animals	Small animals, best suited for feeding very young fish, unless very large quantities can be obtained
*Earthworms	*Enchytraeid worms, white worms, or potworms	*Nematodes, including micro-worms and vinegar eels
Cockroaches		
Grasshoppers		
Crickets	*Amphipods, scuds, or beach fleas	Tubifex worms
*Silkworms	Fairy shrimp	*Artemia or brine shrimp
*Mealworms	*Midge larvae, Chironomid larvae, or bloodworms	*Daphnia or water fleas
*Bees		Cyclops
Snails		Ostracods, or "hardshell daphnia"
Clams or "fresh water mussels"	*Fly maggots	Mosquito larvae
*Minnows	Soldier fly larvae	

*Presently cultured as foods for food fish.

NEMATODES
> Nemathelminthes
>> Nematoda
>>> Cephalobidae
>>>> *Turbatrix aceti*
>>>> *T. ludwigii*
>>>> *T. silusiae*
>>>> *Cephalobus*
>>>> *Panagrellus silusiae*
>>>> *P. redivivus*
>>> Rhabditidae
>>>> *Rhabditis*
>>> Diplogasteridae
>>>> *Diplogaster*

The nematodes (order Nematoda) include many species parasitic on humans and other animals as well as plant pests. We are here concerned with the small (1–2 mm long) free-living species, which include the "vinegar eel," *Turbatrix aceti*, and a number of species raised by aquarists and collectively known as microworms. You can buy starter cultures of microworms from aquarium stores.

The main use of cultured nematodes has been in the Soviet Union, where all the listed species of *Panagrellus* and *Turbatrix* are cultured indoors on a medium made from oats or barley and used as food for very young fishes. Daily production of *P. redivivus* has varied from 0.03 to 0.25 oz/ft² (10–75 g/m²), depending on the nutrient source, temperature and age of the culture. Harvest of a culture may begin ten or eleven days after setting up, and cultures may be kept going for six months or more.

Nematodes

TUBIFEX WORMS
> Annelida
>> Oligochaeta
>>> Plesiopora
>>>> Tubificidae
>>>>> *Tubifex*
>>>>> *Limnodrilus*

These animals are mentioned because they are among the live foods most commonly used by aquarists, who refer to them collectively as "tubifex." Tubifex worms are most often found in the mud below sewage outfalls, so their culture would seem feasible. However, according to the late Charles O. Masters (see Appendix IV–1) "For all practical purposes, tubificids cannot be reared in quantities large enough to make it worthwhile." If this is true for the aquarist, it is certainly true for the food fish culturist.

Tubifex

Tubifex worms occur in nature in enormous numbers and, if one cares to undertake the odious task of collecting and washing them, they make an excellent food for small fish. But wash both the worms and yourself well, as their environment is a breeding ground for pathogens of fish and humans alike.

ENCHYTRAEID WORMS
> Annelida
>> Oligochaeta
>>> Plesiopora
>>>> Enchytraeidae

> Enchytraeus albidus
> E. buccholzi
> E. capitatus
> E. minutus

White worms   Another group of animals long favored by aquarists and first successfully mass cultured in Russia are the enchytraeids, known in the aquarium trade as white worms or potworms. Cultured enchytraeids run 1.4 to 1.8 inches (35 to 45 mm) in length; some Russian fish farms produce from 1,100 pounds (500 kg) to several tons per season. Their culture is carried out indoors in boxes filled with soil and stocked with 0.3 oz of worms per square foot (100 g/m²). Boxes are harvested 45 to 50 days after stocking, yielding 0.9 to 1.5 pounds (400 to 700 g) each. To achieve this production requires 0.55 to 0.66 pound (250 to 300 g, dry weight) of proteolyzed yeast, 1.98 pounds (900 g, dry weight) of meal or groats or 6.6 to 8.8 pounds (3,000 to 4,000 g, wet weight) of green plants. Stocks for starting white worm cultures can be bought from aquarium stores. Some aquarists believe that eating too many white worms causes fish to become constipated.

EARTHWORMS
Annelida
> Oligochaeta
>> Ophistophora
>>> Lumbricidae
>>>> Lumbricus rubellus
>>>> L. terrestris
>>>> Eisenia foetida
>>>> Eudrilus eugenie
>>>> Pheretina sp.

Earthworms   No animal is more firmly associated with fish food in people's minds than the earthworm, the archetypal fish bait. And for good reason; few fresh water fish, even those which rarely encounter an earthworm in their natural habitat, will refuse this nutritious morsel. In the only experiment on earthworms as a fish food to come to my attention, channel catfish fed diets in which twenty-five and fifty percent of their commercial ration was replaced by worms, grew faster and converted feed more efficiently than catfish fed a control diet of commercial feed alone.

The earthworm is also an animal for which culture methods are well developed. Yet it is not widely cultured as a fish food; cultured worms are intended primarily for bait or for agricultural use. The most commonly cultured species are the red worm, Lumbricus rubellus, the manure worm, Eisenia foetida and, to a lesser extent, Pheretina sp. A fourth species, the African nightcrawler, Eudrilus eugenie, is considered a superior bait worm, and is sometimes cultured for that purpose, but it has no advantage as a food worm, and is more difficult to raise. The American nightcrawler, Lumbricus terrestris, is our largest native earthworm. It may sometimes be harvested (by stalking on lawns and golf courses at night with a flashlight) in sufficient quantities to justify gathering it to be cut up as food for small fish. But the nightcrawler, along with many other native species, has not been intensively cultured successfully.

Writings on earthworm culture are increasing (see Appendix IV–1), and stock of superior varieties can be bought to start cultures. Earthworms are cultivated in "beds" (usually wooden boxes) loaded with manure or compost. They may be fed on virtually any biodegradable material, including household garbage, manures, paper and cardboard and sewage sludge.

Since most worm farmers are more concerned with the number of worms produced than their size, it is difficult to get an accurate estimate of the weights of worms which can be produced to feed fish. At the New Alchemy farm it has been our experience that uncultured red worms, harvested as they come, without selecting for size, average about ¾ gram. Combining this figure with the production figures reported by Gaddie and Douglas (Appendix IV–1), I calculate that an initial stock of 100,000 worms (165 lb or 75 kg) could produce 3,200,000 worms (over 7,000 lb or 3,182 kg) or more in a year.

Apart from the quantities of worms which can be produced and their high value as fish food, there are still other aspects of earthworm culture which suit it for fish farming. If the fish farm is to be integrated with vegetable or tree farming worms can be used to the benefit of the crops, and worm feces or "castings" make excellent potting soil or fertilizer. Further, worms and worm castings represent a potential cash crop of great value. A combination commercial worm farm and subsistence fish farm might be a very good arrangement for someone with limited land and/or water to farm. Or worms could make a good quick cash crop to defray expenses of developing a commercial fish farm over a longer period of time. (Earthworm culture has received quite a bit of bad publicity, due to the shady sales practices of some unscrupulous wholesale dealers. It is unfortunate that this has reflected poorly on the feasibility of small scale commercial earthworm culture.)

In feeding earthworms (and other types of worms) it is usually best to use a worm

FIGURE III–5–1. *Earthworm feeder.*

feeder of the type used by aquarists, only larger. This is easily made of styrofoam or any other buoyant material (see Figure III–5–1). Use of such feeders permits the worms to enter the water one by one, preventing waste or the "hogging" of most of the worms by a few dominant fish.

Crustaceans

AMPHIPODS
Arthropoda
    Crustacea
        Amphipoda
            Gammaridae
                Gammarus
            Talitridae
                Hyalella azteca
            Orchestidae
                Orchestia sp.

Amphipods   There are a number of kinds of amphipods, or scuds, native to fresh and salt water. They run from about $^1/_{16}$ inch to ½ inch (16–128 mm) in length, and all look pretty much like Figure III–5–2. Amphipods are an excellent food for almost all fish, though young fry of some species may find their hard shells difficult to digest.

Amphipods are often found in nature in very great numbers, and populations in excess of 929 per ft² (10,000 per m²) have been reported. Very little has been written on the culture of amphipods, but in 1975 Masters wrote the following anecdote about *Gammarus*: "European fish culturists have produced them in ditches 120 feet (36.6 m) long by 12 feet (3.7 m) wide and 5 feet (1.5 m) deep. Approximately 650 to 900 lb (295 to 409 kg) of scuds were produced in such a 'tank' per month. Animal manure and plant refuse from local markets were used as food, and the water was allowed to flow slowly through the ditches, which probably kept the dissolved oxygen content sufficiently high."

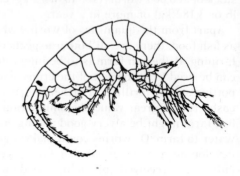

FIGURE III–5–2. *Amphipod, or scud* (Gammarus palustris).

One word of caution on fresh water amphipods: they often serve as intermediate hosts for Acanthocephalan parasites of fishes. Sometimes the larvae of these parasites are visible as small red spots on the amphipods.

"Beach flea"   At the New Alchemy farm on Cape Cod, we have stumbled on another crude form of amphipod "culture" which may be of interest to readers who live near the coast. On the Cape, gardeners customarily mulch with "seaweed" (mainly eel grass and *Codium*), which can be gathered from the beaches in abundance after storms. These piles of plant material often contain great numbers of the marine amphipod or "beach flea" (*Orchestia* sp.), which may survive in the garden for months if mulching is heavy or if a few piles of material are left. We have taken advantage of this by making heaps of "seaweed" around our greenhouse-ponds. The material is turned over daily, causing the amphipods to leap about; many of them fall into the water where they are eaten by the fish. As long as the material is kept moist and not allowed to become too hot, there is an abundance of amphipods.

Incidentally, we have found these beach hoppers quite palatable when gathered in numbers and deep fried; perhaps the *Gammarus* culture system mentioned by Masters could be applied directly to human nutrition, instead of only as food for fish.

FAIRY SHRIMP (BRANCHIOPODS)
Arthropoda
　　Crustacea
　　　　Eubranchiopoda
　　　　　　Chirocephalidae
　　　　　　　　Eubranchipus vernalis
　　　　　　　　Streptocephalus torvicornis

The Branchiopods are small (approximately ⅛ inch to 1 inch, or 0.3–2.6 cm) animals, and among the most characteristic inhabitants of temporary woodland pools. *E. vernalis*, known in North America as "fairy shrimp," is eagerly sought after by aquarists as a fish food. It is considered a better food than other crustaceans because of its thinner, softer shell. It is the species primarily responsible for the excellent growth of rainbow trout stocked in "pot holes" in the prairie provinces of Canada.

To date, no methods have been developed for mass cultivation of fairy shrimp and their relatives. However, Ivleva discusses laboratory methods for culturing the Russian fresh water branchiopod *S. torvicornis* and writes that "the satisfactory results obtained in the development of the basic procedures for the cultivation of these crustaceans hold great promise for future success." Her methods involve maintaining a small stock of breeders, and the separate hatching and rearing of the young, which are raised on cultures of bacteria and yeast. (See Appendix IV–1.)

BRINE SHRIMP
    Arthropoda
        Crustacea
            Eubranchiopoda
                Artemiidae
                    Artemia salina

On the whole, the Branchiopods are a poorly studied group of animals; an exception is the brine shrimp (*Artemia*), an inhabitant of highly saline waters throughout the world. It is very easy to culture and has been used as an experimental animal and fish food for many years.

Most aquarium hobby stores carry brine shrimp eggs to be hatched at home. Eggs are collected for sale from a number of bodies of water, but the best North American source seems to be the San Francisco area. For reasons which are not clear, but which may be connected with pesticide pollution, brine shrimp hatched from eggs from other localities have been associated with high mortalities of young fish in some hatcheries.

Most aquarists are content to purchase eggs, hatch them to obtain the tiny young shrimp, or nauplii, and feed these to the fish. However, some hobbyists and dealers are beginning to rear *Artemia* to maturity and reproduce them. This involves careful monitoring of the temperature and salinity of the water and maintenance of algal cultures as food. Masters describes good methods for both hatching and harvesting the nauplii, and the more complex maintenance of cultures. (See Appendix IV–1.)

Larger scale mass culturing methods which make use of fertilization and yeast cultures, have been described by Ivleva. (See Appendix IV–1). Data on the sustained productivity of such systems are scarce, but she gives one instance of a culture maintained for 1½ months in three tanks 11.7 inches (30 cm) deep, totalling 27.9 ft² (2.59 m²) in surface area. During this time an average of nearly 6 pounds (2.7 kg) was harvested daily.

Brine shrimp, like fairy shrimp, have thin, soft shells and are considered excellent food. Due to the high salinity of their culture environment they should be washed in fresh water before being fed to fresh water fishes in confinement. The Indians living around Great Salt Lake, Utah, are said to have used dried brine shrimp as food at one time.

DAPHNIA
　　Arthropoda
　　　Crustacea
　　　　Cladocera
　　　　　Daphniidae
　　　　　　Daphnia
　　　　　　Moina
　　　　　　Bosmina

Daphnia (water fleas)　　Daphnia, or water fleas, were the first live food used in quantity by aquarium hobbyists, and they are a good one. The term "daphnia" commonly refers to a number of Cladoceran species of the genus *Daphnia*, as well as representatives of some other, similar genera, especially *Moina* and *Bosmina*. The species most commonly collected and cultured are *D. pulex*, *D. longispina* and *D. magna*. In nature, they are found most abundantly during cool weather, in small, sometimes temporary pools, especially those "polluted" by organic matter. Masses of daphnia give a rusty to dark brown color to the water. If you look closely the individual animals 0.008 to 0.12 inches (0.2 to 3.0 mm) long can be seen swimming along in a jerky fashion.

Many more or less effective systems for culturing daphnia indoors have been developed, most of them relying on feeding with yeast. Some of these are reported by Needham and his co-workers. (See Appendix IV–1.) In the same book is a report by Embody and Sadler describing a system for culture and mass harvest of daphnia in fertilized outdoor pools. They were able to harvest better than two pounds (0.9 kg) of daphnia from an 8 by 12 foot (2.4 by 3.7 m) pool every three weeks. Under optimal conditions, this rate could surely be improved on. Apart from Embody and Sadler's work, the reports of daphnia culture are regrettably short of precise information about standing crops or sustained harvests. As a rough index of the incredibly high numbers in which they sometimes occur, I will give figures based on a commercial daphnia culture operated in conjunction with a sewage oxidation pond in Arcata, California. This pond was harvested by using a low pressure irrigation pump which delivered about 100 gallons (424 liters) per minute. Collections averaged about one pound (0.4 kg, wet-drained) per minute, with peaks up to twice that.

Both natural and cultured populations of daphnia have somewhat unpredictable population peaks and "crashes." Populations also usually decline in the hottest weather. So the fish farmer growing daphnia should have several culture tanks if possible, and should be prepared to use alternate foods when daphnia are not available.

CYCLOPS
　　Arthopoda
　　　Crustacea
　　　　Eucopepoda
　　　　　Cyclopidae
　　　　　　Cyclops viridis
　　　　　　C. serrulatus
　　　　　　C. vernalis

Cyclops　　This tiny crustacean usually less than 0.08 inches (2 mm) long is suitable only as a food for small fry. It is mentioned because it is one of the zooplankters most commonly found by daphnia hunters, and because it is quite easy to culture, being more tolerant of environmental changes than daphnia. You can recognize Cyclops most easily by looking for the characteristic egg sacs of the female. (See Figure III–5–3.)

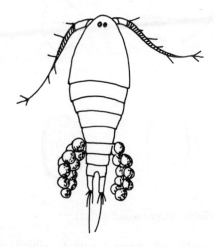

FIGURE III–5–3. *Female* Cyclops.

Cyclops can be cultured in jars fertilized with manure; methods are described by both Needham and Masters. (See Appendix IV–1.) No data are available on yields of cyclops culture. Some aquarists say that eating too many cyclops causes diarrhea in fishes.

CYPRIS AND OTHER OSTRACODS
>Arthropoda
>>Crustacea
>>>Podocopa
>>>>Cypridae
>>>>>*Cypris*
>>>>>*Cypridopsis*

These creatures, like cyclops, are mentioned because they are often found by the daphnia hunter. They are rather featureless (see Figure III–5–4); at rest they may be mistaken for tiny mollusks, although their vigorous swimming should help you to distinguish them. Aquarists' opinions of the food value of "hard shell daphnia," as they are sometimes called, differ. Some consider them excellent, while others maintain there is too high a proportion of shell.

"Hard shell daphnia"

Ostracods, like daphnia, are found in a great variety of environments, and may sometimes be collected in incredible numbers. There would seem to be good potential for culturing them, but though suggestions for culture have been published, I have not heard of any successful sustained mass cultures.

The creatures so far described are aquatic or, if they are not aquatic, sink when placed in water. We are about to consider a number of members of the large class Insecta, or insects. All the insects discussed here are terrestrial as adults, while the larvae inhabit a variety of environments. As a rule, adult terrestrial insects will float when placed in water, while larvae of all kinds normally sink. Some of the fish suggested for culture in this book are natural surface feeders (trout, sunfishes, young tilapia); others (carps, catfishes, buffalofishes) are not. This behavioral factor may limit the fish farmer's choice of insects as food. However some fish, such as

Insects

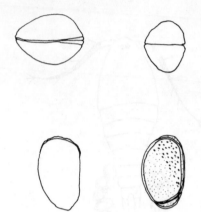

FIGURE III–5–4. *Ostracod* (Cypris subglosa).

common carp, bullheads and channel catfish, which are not normally surface feeders, can learn to feed at the surface when food is abundantly available there. Such learning is particularly rapid if there are surface feeding fish present to "teach" them.

COCKROACHES
  Arthropoda
     Insecta
        Orthoptera
           Blattidae
              Periplaneta americana
              Blattella germanica
              Blatta orientalis

Cockroaches   Considering the large numbers of cockroaches which all too often occur where they are not wanted, it would seem simple to mass culture these unpopular insects. Instructions for their care and feeding can be found in a number of publications, but to date, there are not methods for true cultivation. (Any attempt at culture of cockroaches should be carried out well away from dwellings or food storage areas.)

GRASSHOPPERS
  Arthropoda
     Insecta
        Orthoptera
           Acriidae
              Melanoplus
              Romalea
              Dactylotum
              Encoptolophus
              Chortophaga
              Chloealtis, and others

Grasshoppers   The grasshoppers are a common and widespread group of insects sometimes cultured by researchers and tropical fish hobbyists, but they have not been mass cultured as fish food. There are many genera and species of North American grasshoppers; those listed here can be recommended on the basis of their hardiness, availability, small size and high reproductive potential. Most other flightless species can be raised, as long as a continuous supply of acceptable food plants can be provided.

The essentials for grasshopper culture are cleanliness, warmth, sunlight, a dry atmosphere and plenty of food. Cultures are normally maintained only during the warm months. Maintenance of cultures throughout the year in the North will ordinarily require either a refrigerator for resting eggs and an incubator, or a constant temperature room.

There is no way to accurately estimate what the productivity of mass grasshopper cultures might be, but Masters reported that a cage measuring 2 feet by 1½ feet by 1½ feet (0.6 × 0.46 × 0.46 m) will maintain several hundred newly hatched grasshoppers or about thirty adults, and that females of *Melanoplus* and *Romalea* may produce as many as three hundred offspring per year.

CRICKETS
    Arthropoda
        Insecta
            Orthoptera
                Gryllidae
                    Gryllus domesticus
                    G. assimilis

Crickets are used as fish bait more often than grasshoppers, and are sometimes sold for that purpose. Methods have been developed for culturing bait crickets, but they need to be improved if they are to be cultured as fish food. Using current methods, a can 24 inches (61.4 cm) in diameter can be made to produce about four hundred crickets every three months during warm weather. About 8 pounds (3.6 kg) of laying mash is required as food for that quantity of crickets.

Crickets

SILKWORMS
    Arthropoda
        Insecta
            Lepidoptera
                Bombycidae
                    Bombyx mori

Apart from bees, perhaps no insect has been so widely cultured as the silkworm moth, *Bombyx mori*. It was first applied as a food for cultured fish in Japan, where culled pupae are fed to a variety of fishes. The methods used in large scale silkworm growing are adaptable to North American conditions, provided there is a source of white mulberry, *Morus alba*, leaves. These methods have been described in a number of publications. (See Appendix IV–1.) Commercial silkworm culture is a fairly intensive proposition, however, requiring selection of mulberry leaves, frequent cleaning, and careful tending of small groups of larvae. Whether it could economically be adapted to the production of larvae and pupae for use as fish food is not known. A possible drawback is that silkworm larvae have been implicated in "off" flavors observed in carp cultured in Japan.

Silkworms

MOSQUITOES
Arthropoda
Insecta
Diptera
Culicidae
Culex pipiens
C. restuans

Mosquitoes   All of us are only too familiar with adult mosquitoes, but not everyone recognizes the pupae and larvae, or "wigglers." Figure III–5–5 is a generalized drawing of the larva and pupa of a culicine mosquito. The less common anopheline mosquitoes are similar, but rest horizontal to the water surface. Mosquito larvae are a favored food of many fishes and, while they have not been cultured by fish farmers, some aquarists have been able to produce *Culex* in numbers sufficient to offer them for sale in frozen packages. All that is necessary, given warm weather, is a number of wooden or plastic pools containing 6 to 8 inches (15.4–20.5 cm) of water, and the addition of a handful of dried yeast every two weeks or so. Adult mosquitoes will be attracted to such pools to lay eggs.

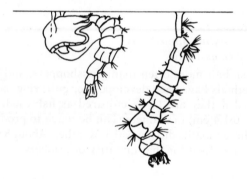

FIGURE III–5–5.   *Pupa and larva of culicine mosquito*

Harvesting the larvae is simple, since they tend to congregate at the surface, around the edges of the pools, and can easily be netted. The extent of the harvest will vary somewhat with the weather; the culturist should keep a close watch on the larval population and be sure to crop it heavily, both for the sake of efficiency, and to prevent the emergence of large numbers of adults. Apart from personal discomfort, no one wants to be known as a source of biting mosquitoes. This is perhaps the most serious drawback of mosquito culture. You may find that, despite your best efforts, you are blamed for every local mosquito population peak, whether or not there is any basis in fact. The best policy on all counts is to rear mosquitoes well away from residential areas or anywhere that people congregate.

MIDGE LARVAE
Arthropoda
Insecta
Diptera
Chironomidae (= Tendipedidae)
Chironomus (= *Tendipes*) *tentans*, and others

Larvae of chironomid midges, often called "bloodworms" because of the brilliant red color of some species, are among the favorite natural foods of fresh water fishes; carp in particular seem to be fond of them. The adults are the small, mosquito-like (but non-biting) insects you may have noticed dancing in swarms over fields or lawns in the late afternoon at all times of the year, even on warm winter days when few other insects are active. The family Chironomidae (also called Tendipedidae) is distributed world-wide, wherever there is fresh water, so there are certainly some species in your locality.

Some fish culturists have long felt that midge larvae might have a particularly beneficial effect on the growth of fish. A few years ago Joe Levine, Marcus Sherman and I carried out the first controlled experiments to determine the growth-promoting value of chironomid larvae in fish diets. We found that when young blue tilapia (*Tilapia aurea*), fed on rolled oats, roasted soymeal and fresh green algae, received a dietary supplement of midge larvae amounting to 2 percent of the total diet (wet weight) or 0.28 percent (dry weight), the rate of growth was increased significantly. Essentially the same results were obtained with young lobsters (*Homarus americanus*) fed a standard diet of frozen *Artemia* supplemented with live midge larvae.

The quantities of larvae fed were too small for the increased growth to have resulted from only the increase in crude protein; it must have been due to the addition of specific proteins, amino acids or vitamins to the diet. It may be that other live foods will eventually be found to have the same effect. It is known that vitamins, in particular, are quite rapidly lost from "complete" dry diets and frozen foods during storage. It may be that the tropical fish hobbyists, who have long maintained a sort of mystical faith in the value of live foods, know something which has been largely overlooked by fish farmers. Considering the small quantities of midge larvae involved in our experiments, the results justified the effort we at New Alchemy have put into a new method of midge culture. Further, they suggest the feasibility of providing at least some live or fresh food to all cultured fish.

The culture methods we have developed are primarily for the *Chironomus* (= *Tendipes*) *tentans* group[1]; they are less productive than those previously developed for these insects in Israel and the Soviet Union. The late Dr. A. Yashouv in Israel produced 0.81–1.22 oz/ft² (250–375 g/m²) per day, while A.S. Konstantinov in Russia did even better—1.30 oz/ft² (400 g/m²) per day. Our production during the peak season is about 0.3 oz/ft² (100 g/m²) per day. However, most readers will probably prefer our technique from a labor/technology standpoint.

Yashouv's system involves constant circulation of the water and provision of a "natural" bottom layer of soil and sand for the larvae. With such a substrate it is necessary to sieve or sort out the larvae at harvest time. Konstantinov's technique eliminates the substrate, but also uses a constant flow of water. He has further complicated matters by growing his midges in a battery of aluminum trays in a room kept at constant temperature.

[1] The taxonomy of the Chironomidae, or Tendipedidae, according to which authority you consult, is quite confused and identification to species is difficult. The *Chironomus tentans* "group" (also referred to as the *Tendipes tentans* group) includes a number of virtually indistinguishable forms. Their identity is of little importance to the fish farmer, and I will refer to all of them, plus related species, as "midges" or "chironomids."

The Israeli and Russian methods are suited to their own economic and biological contexts. In Israel, midge larvae are used primarily in the experimental culture of mullet (*Mugil* spp.). At the time Yashouv was doing his chironomid culture work, relatively small numbers of these commercially valuable, but hard to feed fish were being grown. But as mullet culture intensifies, it may become necessary to improve the harvesting methods in Israeli midge culture systems. In Russia chironomid larvae are fed to young sturgeon to be released into rivers to supplement an endangered fishery. Given the small size of these young sturgeon, the great economic value of the fishery and its imperiled status, the technological expense of Konstantinov's methods is probably justified.

New Alchemy midge culture

At New Alchemy we have developed a year-round outdoor midge culture system without water circulation, and without the labor of separating the larvae from the substrate. Our basic technique can be described very simply: Construct a series of small, broad, shallow pools in a sunny spot, preferably but not necessarily near a natural pond. Fertilize them with a small amount of manure or other organic matter, and wait for adult midges to come and lay their eggs. Within one to three weeks, depending on temperature, you will have your first crop of larvae ready to be harvested or to emerge as adults. From then on, all you need do is harvest and replenish the fertilizer.

Ordinarily, you would have to laboriously remove the larvae from the bottom and sides of such pools, but we bypass this step by providing a removable substrate in the form of a piece of cloth. At harvest time the substrate is taken out and suspended in the fish pond. The fish do the rest of the work by picking off the larvae and other attached food organisms—including algae, if you are growing herbivorous fish.

Over the years, we have experimented with a number of variations and embellishments on this method. So far, here is what we have learned:

The best fertilizer is Milorganite®, a commercially available processed sewage sludge, applied at about one pound (0.45 kg) per m² of pool surface. It has definitely outperformed horse manure and chicken manure, perhaps because of its high iron content (4.5%). (The brilliant red color of most of the cultured chironomid larvae is due to the large amounts of hemoglobin they contain, and they need iron, just as human blood does. There are many species of chironomids lacking this red color, and therefore with perhaps less hemoglobin. Whether they would be as nutritious to fish or if they could be similarly cultured is not known.) Some of the fertilizer is ingested directly by the larvae, but much of it goes to support green algae which the larvae use both as food and as material to construct the tubes in which they live. The more algae is growing directly on the cloth substrate, the more larvae will settle there, instead of on the pool's sides.

Realizing this led us to abandon our original system of hanging the cloth pieces in the pools, like clothes on a line, and to adopt instead the simpler method of laying them on the bottom. We fertilize each substrate piece individually. An algae-fertilizer mixture with an ideal texture for the larvae is produced by preparing a mixture of Milorganite, sand and pond mud in a ration of 7:2:1, adding water to make a slurry, and distributing this evenly over the surface of the cloth.

Burlap is the most effective substrate material we have tested, presumably because of its rough texture. We would like to find a less expensive material with the same texture, but with greater durability in water.

Slow or intermittent circulation of the water increases production, but is not necessary. Aerating the water—we used a windmill powered air compressor—had a negative effect in our experiments; the aerated pools developed a dense growth of filamentous algae, which seemed to discourage midge larvae.

Indoor greenhouse pools produced more larvae during the winter than outdoor pools.

In terms of the quantity of food that can be provided culture of midge larvae is not comparable to, say, earthworm culture. (It has been calculated that it would take nearly 700 *Chironomus* larvae per day to produce normal growth of young carp in nature.) Fish farmers who can feed their fish a great deal of live or fresh food may therefore choose to ignore it. However, for the culturist who must rely on dry foods, it may be a very effective way of improving fish growth with little expense in money, space, time and labor.

FLY MAGGOTS

Arthropoda
> Insecta
>> Diptera
>>> Muscidae
>>>> *Musca domestica*
>>>> *Phaenicia sericata*
>>> Sarcophagidae
>>>> *Sarcophaga bullata*, and many others

Fly maggots

There are any number of flies with small, worm-like larvae, often called maggots, which might be reared as food for fish. I mention the three which have commonly been cultured for various purposes, but there may be others as good or better to raise as fish food.

Bill Gressard, of Trail Lake Fish Hatchery, Kent, Ohio, has developed an extremely simple, low-intensity form of maggot culture to supply food for bluegills and other panfish. Spoiled meat, fish or roadkilled animals, wrapped in paper and thoroughly moistened, are placed in 17 by 17 inch (43.5 by 43.5 cm) hardware cloth baskets suspended a few feet over the surface of a pond. In the natural course of things, maggots appear. As the time for pupation approaches, maggots instinctively move away from their food source, leading them to fall into the water. According to Gressard, three or four pounds (1.4–1.8 kg) of meat or fish will produce five thousand maggots a day for a week before the "bait" has to be replaced. In addition to providing food, this method gathers fish in one area for efficient angling. And if the fish caught are cleaned on the spot, the carcasses can be used to replenish the fly bait.

The esthetic drawbacks of such a maggot "culture" technique are fairly obvious. I have heard of one chicken farmer who provided maggots for his birds in the same way. But he also constructed a small metal "chimney" over the platform. In addition to shielding the rotting carcass from squeamish eyes, the chimney was described as drawing the foul odors upward.

As several types of maggots are used for various medical and biological purposes, a number of tidier laboratory culture methods have been developed, some of which might be adaptable for aquaculture. Culture methods for a variety of flies may be found in Needham et al. (See Appendix IV–1.)

I know of only one instance of controlled fly culture being incorporated in an aquaculture program. This is the Amphibian Facility of the University of Michigan, where a self-perpetuating colony of several species of frogs is maintained for laboratory purposes (but not for use as food). The director of the Facility, George W. Nace, has found that optimal growth and rapid sexual maturity occur only if the frogs are fed on live insects. After studying fly culture he settled on two species, the flesh fly (*Sacrophaga bullata*) and the greenbottle fly (*Phaenicia sericata*) as the most suitable. The same techniques are used with both species: adult flies are maintained on water, sugar, and a sugar solution, and allowed to deposit their eggs on a moistened mixture of sawdust and dog food, topped with several thin slices of raw liver, placed in a plastic tray. After twenty-four hours in a breeding cage, each such tray is placed in a 12.1 × 11.0 × 3.2 inch (31.0 cm × 28.5 × 8.1 cm) deep stainless steel pan, lined with paper toweling. Maggots are kept from escaping by a nonlethal "electric fence," created by running a ten volt current through a copper strip mounted on insulation affixed around the lip of the pan. When fully grown, the maggots migrate from the food tray into the pan, where they pupate. Pupae can be stored at 39.2°F (4°C) for as long as three to four months, then warmed to 86°F (30°C) for hatching. After hatching, the flies can again be chilled or anesthetized with CO_2 and fed to the frogs while in a torpid state. This entire operation, which produces 25,000 flies daily, is confined to a 7.9 × 9.8 ft (2.4 × 3.0 m) room.

The disadvantages of the system from a food fish grower's point of view are the rather expensive foods for the maggots and the need to maintain constant high temperatures. Perhaps you could bypass the temperature problem by growing the house fly (*Musca domestica*) or some other northern native species. However, the temperature needs of flesh flies and greenbottle flies have the advantage that escapees of these species will not survive to create a serious local pest problem in Michigan.

SOLDIER FLY LARVAE
    Arthropoda
        Insecta
            Diptera
                Stratiomyidae

Soldier fly larvae   The soldier flies (Stratiomyidae) are not included in any of the texts dealing with insect culture, but I have included them on the strength of the experience of a correspondent, Stanley Tinkle, of Anaheim, California. I take the liberty of quoting from one of his letters:

> A year ago I began feeding red worms a diet of household garbage. They thrived in 45 gallon (170*l*) garbage cans without bottoms — until the level of the contents exceeded about 14 inches (35.8 cm). Then anoxia caused some souring, with predictable die-off. Last August we returned from two months' absence to find the cans crawling with fat maggot-like grubs. They have shouldered aside the worms and all competing house-fly larvae, and by themselves now consume the 1½ gallons (5.7*l*) of food-type garbage which our family of four discards daily. The level in the garbage can does not rise!
>
> The adult stage finally appeared in October. Our Anaheim Health Department's "Environmental Health Officer" identified them as "Soldier

Flies," or in Latin, "Stratiomyidae." He says they're common, harmless, found in soggy compost heaps. The larvae may be predaceous, because other insect larvae seem to avoid them. Adults look like wasps but are harmless and are thought to feed on nectar. When the larvae are ready to pupate, they change from tan to coffee-brown and climb out of the covered can, seeking a quiet spot. They then shrink inside their skin, which toughens and forms a protective shell. Within days or perhaps three weeks by my observations, the adult appears.

The balmy temperature beneath a plastic dome should ensure an endless supply of grubs. Ours have prospered from 105° to 45°F (40.6–7.2°C) so far. You could simply suspend a garbage receptacle above your pool, allowing a ramp for grubs to flee the madding crowd—whereupon they'd fall into the water below. The only hitch in the plan is that the grubs flee sunlight. A black plastic shroud would serve, or the darkness of evening might lure them out.

Tinkle later offered a few more observations on his soldier flies. He discovered that when the lids were left off his garbage cans, soldier fly larvae populations went down, but substantial numbers of much smaller maggots, probably house fly larvae, developed in their place. The approximate average weight of the maggots was 0.04 g, while the soldier fly larvae averaged 0.4 g. Perhaps a combination of lighted, open and dark, covered pails could be used to provide different sizes of food for different sizes of fish without otherwise changing the culture method. He also found that, while soldier fly larvae will survive under water for some time, they try to escape when their culture medium is drenched.

These observations were the basis for Tinkle's proposed "Zero Labor Fish Feeder." Unfortunately, it has not, to my knowledge, been constructed and tested. (See Figure III–5–6.)

Zero Labor Fish Feeder

Crop: Larvae of Soldier Fly

Medium: Household garbage mixed with sawdust or compost

Containers: Tomato crates or nursery flats lined with discarded door screen (see local door and window company)

Racks: 1 inch (2.6 cm) angle iron or 2 × 2 inch (5.2 × 5.2 cm) fir

Catch screen beneath racks is of 1 × 4 inch (2.6 × 10.2 cm) pine, lined with screen. Inside is edged with plastic sheet to trap escaping larvae.

Method: Each container holds about one week's garbage for family of four. Once daily (or as needed), pour a bucketful of water onto top container. Larvae become active under slushy conditions, migrate downward, finally reach catch-screen, head for darkened corner and through black plastic pipe, fall into pool.

As top container is vacated or exhausted, remove it and move others upward, placing new container on bottom rack. Larvae and adults should populate it, without help.

Drain water would be a rich broth, could be poured onto a compost pile, or earthworm culture, or perhaps back onto top container.

Larvae are most active above 70°F (21.1°C), do little below 50°F (10.0°C). Would

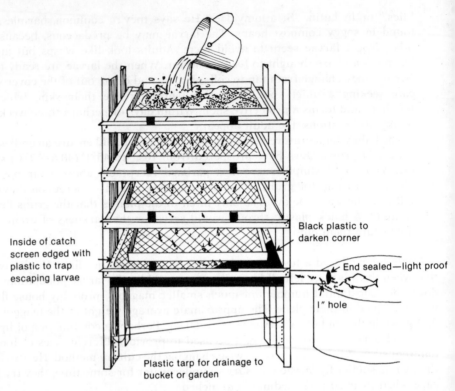

Inside of catch
screen edged with
plastic to trap
escaping larvae

Black plastic to
darken corner

End sealed—light proof

1″ hole

Plastic tarp for drainage to
bucket or garden

FIGURE III–5–6. *Stanley Tinkle's "Zero Labor Fish Feeder" for producing soldier fly* (Stratiomyidae) *larvae on a diet of household garbage.*

thrive under plastic dome beside pool. Racks should be draped with heavy canvas or ten mil black plastic to keep out sun, which larvae instinctively avoid.

MEALWORMS

   Arthropoda

      Insecta

         Coleoptera

            Tenebrionidae

               Tenebrio molitor

               T. obscurus

               Blapstinus moestus

               Tenebroides mauritanicus

               Alphitobius diaperinus

               A. piceus

Mealworms   The term "mealworm" refers to a variety of beetle larvae often found in stored grain and cereal products. Perhaps the best known are the yellow mealworm (*Tenebrio molitor*) and the dark mealworm (*T. obscurus*), both of which are commonly reared as food for captive birds, fish and reptiles. There are also a number of smaller species which are more suitable for aquarium fish, but they are not as commonly used by aquarists as you might imagine. I have not heard of any significant use of mealworms by food fish growers. Table III–5–2 lists the maximum length of the various species of mealworms.

Table III–5–2. Maximum length of common species of mealworms.

Tenebrio molitor	½ inch plus (12.8 mm)
T. obscurus	½ inch plus (12.8 mm)
Blapstinus moestus	⅜ inch (9.6 mm)
Tenebroides mauritanicus	⅓–½ inch (8.5–12.8 mm)
Alphitobius diaperinus	¼ inch (6.4 mm)
A. piceus	¼ inch (6.4 mm)

Culture techniques are similar for all species. All life stages are kept together in large, shallow wooden or galvanized iron trays or drawers, stacked on top of each other. The surface of the containers must be smooth to keep the adult beetles from crawling out. The best and cheapest culture medium is chick-growing mash, which may occasionally be enriched with sliced potatoes, vegetables, grains or cereals. The mash is placed in thin layers, with layers of burlap between, and sprinkled daily with water.

Starter cultures can be bought from pet shops or biological supply houses. If kept indoors (80°F or 26.7°C is an optimal temperature), larvae can be harvested beginning about three months after stocking. If cultures produce more larvae than you need, you can slow down their development by refrigeration.

There are many other beetles whose larvae are found in stored foods; Masters has offered a sizable list of species. Culture methods for these animals have not been developed as fully as for mealworms, but similar techniques should be workable. Some of these larvae have the probable advantage of being softer and more thoroughly digestible than mealworms.

BEES

Arthropoda

Insecta

Coleoptera

Apis mellifera

The world's most commonly cultured insect, the honey bee, may be of use as a fish food. Not that anyone would want to raise bees specifically for this purpose, but there normally is a daily mortality in beehives which could be taken advantage of.

Bees

In a serendipitous experiment at the Farallones Institute, in California, beehives were placed over a fish pond. The bees disposed of dead bees by pushing them out of the hive and down into the pond, providing over one hundred bees a day for the fish. At New Alchemy we placed a tarpaulin under a bee hive on the ground so as to be able to measure potential contribution of a bee hive to a fish pond, but we did not find nearly the numbers of bees reported at the Farallones Institute. Either our mortality rate was lower, or the bees were carrying away their dead, and not simply dumping them. If they were carrying them off, the factors involved might be the strain of bee, climate, or elevation of the hive. Certainly, if you are growing fish and bees, and hives can be conveniently placed over ponds, you might as well try to take advantage of the "free" food.

Molluscs

SNAILS

Mollusca

Gastropoda

many species

Snails   Eleven families of snails are found in North American fresh waters; nearly all types of environments contain at least a few of them. In some waters they are rare, but under certain conditions they may be found in enormous numbers. All are edible by fish, but only a few species make much use of snails in nature, probably because of the obstacle of their shells. Among the fishes of interest to readers, yellow perch (*Perca flavescens*), redear sunfish (*Lepomis microlophus*) and pumpkinseed sunfish (*Lepomis gibbosus*) seem to be particularly fond of snails. Methods for culturing fresh water snails on a larger than aquarium scale have not been developed but, under certain circumstances, it might be possible to encourage their growth in the fish pond.

The most critical environmental factor for snails seems to be the amount of dissolved salts, especially calcium carbonate. Abundance of carbonates is nearly always found together with a high pH, so do not look for dense populations of snails where pH is low. Most snails also need rather high levels of dissolved oxygen.

Snails may eat some foods of animal origin, usually decaying flesh, but they are mainly herbivorous. Some work has been done on feeding snails in aquaria; the two best foods seem to be head lettuce and dried maple leaves. It would obviously be impractical to try to provide lettuce in large outdoor systems, but it might be possible to supply substantial amounts of maple leaves.

CLAMS

Mollusca

Pelecypoda

Eulamellibranchia

Corbiculidae

Corbicula

Unionidae

Sphaeridae

many species

Clams   There are two major families of native North American fresh water clams, the Unionidae and the Sphaeridae, both represented by many species. Some Unionid clams, along with the introduced Asiatic clam, *Corbicula*, are of value as human food. In the next chapter I will discuss their potential use in polyculture, including the possibility of grinding up excess clams for use as fish food.

Most species of fish find the shells of large intact Unionidae and Corbiculidae too stout for their tastes, though channel catfish are a notable exception. But the Sphaeridae, which include the fingernail clams and seed shells, rarely exceed 0.03 in (10 mm) in length and have very thin shells. In nature, they are readily eaten by yellow perch, catfishes, some sunfishes, and buffalofish, but culture methods for these small clams have not been developed.

Forage fish   "MINNOWS"

Pisces

Chondrichthyes

Cyprinidae

Pimephales promelas
Notemigonus crysoleucas
Poeciliidae
Poecilia reticulata, and others

Many popular North American food fishes are more or less piscivorous; among *Fathead minnows*
the species included in this book they include the trouts, the catfishes, the black
basses, the yellow perch and the chain pickerel (*Esox niger*). Providing live small
fish for these species is ordinarily not economically feasible. But in the United
States, the practice of stocking the fathead minnow or tuffy (*Pimephales promelas*)
in channel catfish ponds has been adopted by some fish farmers as an offshoot of
the bait minnow culture industry.

Catfish farmers use fatheads largely for growing catfish beyond the standard
market size of one pound (0.45 kg). The adult minnows are stocked in catfish ponds
at one to five pounds per acre (1.1–5.5 kg/ha). As the catfish get used to eating
minnows, the feeding rate of commercial feeds can be reduced from the normal
daily rate of three percent of the total weight of catfish to only one percent. This
represents a real saving to some farmers, particularly as many catfish farmers are
also in the minnow culture business.

Dr. H.S. Swingle at Auburn University experimented with adding fatheads to a
polyculture of largemouth bass and three species of tilapia. Although the bass were
supposed to be acting as predators on excess young tilapia, introducing the
fatheads led to a five-fold increase in bass production. After seven months the bass
had completely done away with the minnows.

Several other species of fish are widely cultivated for use as bait in the United
States (see Chapter IX–2), and one can imagine numerous uses of these and other
species in food fish culture.

One such fish, the golden shiner (*Notemigonus crysoleucas*) is sometimes recom- *Golden shiners*
mended for stocking in northern farm ponds in place of the bluegill. While this may
result in superior sport fishing for largemouth bass, a properly fished bass-bluegill
pond will produce a much greater quantity of food fish.

Fodder fish have to be restocked periodically. With the high population density
found in all but the least intensive food fish culture operations, it would be unlikely
for fodder species to form a permanent breeding population. One way around the
need to restock might be to grow some species of fodder fish which will reproduce
more or less naturally in a pond adjacent to the main culture pond, but screened off
from it. When populations of the fodder fish reach a certain level, the screen could
be removed and some of the small fish flushed or driven into the main pond. This
method might be particularly well suited for semi-tropical waters where small
fishes of the livebearer family (Poeciliidae), which reproduce constantly as long as *Livebearers*
temperatures are suitable, could be used as the fodder fish.

A similar tactic was adopted at New Alchemy by Stewart Jacobson for the
unlikely purpose of feeding non-piscivorous tilapia. He found that, despite their
reputation, blue tilapia in confined situations would eat guppies (*Poecilia reticulata*)
less than 0.6 inches (16 cm) in length. There are also reports in the literature of
other tilapias, including *Tilapia hornorum*, *Tilapia mossambica* and *Tilapia rendalli*, preying on small fish. Jacobson suggested that guppies or similar fish might
be grown in separate small pools, in the biological filters sometimes used in small
closed fish culture systems (see Chapter IV–4) or in floating cages. In this way the

large individuals would be retained while the young could be released into the tilapia culture pool. Assuming that tilapia are relatively inefficient predators, it might even be possible to polyculture guppies in the tilapia pool.

When stocking "fodder" fish, you should take care not to upset the ecological balance of food fish cultures. If for some reason the "predators" do not prey significantly on the fodder fish, and these then turn out to be competitive or otherwise harmful, your only recourse might be to poison the pond.

Plants as fish food

Plant materials fed to fish may be divided into two broad categories: low protein green or leafy plants (including both aquatic and terrestrial types) and high protein grains and seeds. North American fish farmers will find the former of small value. (Two possible exceptions are alfalfa and comfrey, discussed below.) Of the animals included in this book only grass carp (*Ctenopharyngodon idellus*), crayfish and some of the tilapias will consume large amounts of green plant material. Mirror carp will take some but are not primarily herbivorous.

Though the remains of aquatic plants have been found in the stomachs of predatory fishes, they appear to be swallowed only incidentally while feeding on attached animals. Experiments have shown that aquatic plants have almost no nutritive value for channel catfish and bluegills, and the same is probably true for other predatory fishes.

High protein grains and seeds are another matter. These materials, which fish seldom encounter in nature, can be substituted for a portion of the animal matter normally required by predatory fishes, and are an important ingredient of prepared feeds. Some fishes, notably carp, bullheads and tilapia, will also take them "straight," though with most fish it is necessary to roast and reduce them to meal.

This is not the place for a lengthy discussion of agricultural methods. However, some readers may be interested to know which plants can be cultivated on a small scale for use as fish food, or for ingredients in prepared feeds. Little attention has been paid to the cultivation of plants for feeding fish, and the following comments on agriculture and processing plant materials for fish are drawn largely from experience accumulated at New Alchemy. We have barely begun research in this area; much more remains to be learned, even about those few plants we have already tested.

Aquatic plants

I will begin by discussing those aquatic green plants we have tried at New Alchemy. Though it seems more logical to grow aquatic green plants than to grow terrestrial ones for fish, in practice there are great difficulties. If you try to feed herbivorous fish with plants grown in the fish pond, the result will usually be virtual eradication of the plants. On the other hand, if aquatic plants are grown separately, the difficulties of harvesting from the water, and the tremendous amount of water which clings to the plants and must be carried or disposed of generally makes use of the plants impractical. I can imagine some sort of "pasturing" system, in which fish are moved from one pond, or part of a pond, to another, in relation to crops of aquatic plants. There is some doubt, though, as to whether this would be an efficient use of space, and moving numbers of live fish from place to place is much more difficult than moving land animals.

The two most promising aquatic plants we have found so far are both tropical: water hyacinth (*Eichornia crassipes*) and the aquatic fern known as azolla (*Azolla* spp.). Both have tremendous growth rates, making it possible to crop them quite heavily, and both are floating plants, a factor which greatly simplifies harvesting.

We have found that water hyacinth can be grown in pools with the herbivorous blue tilapia, which will not feed on them in their natural state. But if individual plants are turned upside down, the tilapia will browse the leaves (though other greens are preferred). This is not as convenient as it sounds, for a mat of floating plants in a fish culture system has its drawbacks. The roots may shelter young fish, protecting them from predatory fish which could otherwise control overpopulation. The plants can also create tremendous mechanical interference with harvesting, feeding and other management operations. And the shading effect of the floating mat, while it may be beneficial in some cases, is just as likely to be harmful by cooling the water, hindering phytoplankton blooms, and reducing the opportunity for night-time absorption of oxygen from the air. (For further discussion of water hyacinth, see Chapter VIII–1.) *Water hyacinth*

Azolla cannot be grown with herbivorous fish, but it has the advantage of producing large amounts of vegetation in very small spaces—wading pools and the like. In nature it is found in assocation with a nitrogen-fixing blue-green alga, *Anabaena azolla*. This, plus azolla's ability to root and grow in moist soil, suggest its agricultural value as a fertilizer or mulch. The total impact of the *Azolla-Anabaena* association on aquacultural ecosystems can only be guessed at. *Azolla*

The duckweeds (*Lemna* spp. and *Spirodela* spp.), found in most parts of North America, are similar in appearance, though unrelated to azolla. The reproductive capacity of duckweeds is staggering; it is estimated that, left undisturbed, a patch of *Lemna minor* initially one square inch (645 mm²) in area would grow to cover well over an acre (0.45 ha) in fifty-five days. Potential productivity of *Spirodela oligorhiza* and *Spirodela polyrhiza* has been estimated at 445 pounds per acre per day (500 kg/ha). *Duckweeds*

Duckweeds are relished by herbivorous fish and waterfowl. Though these plants cannot fix atmospheric nitrogen, they can very effectively take up nutrients from water. As a result, the nutritive value of duckweeds, as with water hyacinth, varies according to the water's degree of enrichment. The duckweed *Spirodela*, grown at Louisiana State University in lagoons used in treatment of wastes from swine and dairy cattle, contained 37–45 percent protein (dry weight).

Other than the phytoplankton grown in our fish culture systems (see Chapter IV–4), the only submerged plants we have used are filamentous algae which grow unbidden in fish pond filters. Blue tilapia eat them with gusto, but we have not tried to cultivate them, analyze their nutritive value or measure the harvest. *Submerged plants*

Alligator grass and water primrose cultivation are mentioned in the chapter on crayfish culture. (See Chapter II–10.)

All aquatic plants are potential pests. We cultivate water hyacinth (already a major weed in Florida and many tropical countries) and azolla only because we know that they could not overwinter in our region—Cape Cod. Since it is so much easier to accidentally introduce an aquatic plant into an ecosystem than an animal, I urge you to grow *no* aquatic plant in a region where it is not already established if there could be the slightest risk of its "escape" into the natural environment. *Aquatic plants as pests*

We have used the following terrestrial greens: *Terrestrial green plants*

Alfalfa (*Medicago sativa*), although a common ingredient in trout and catfish feeds, in the green state proved unattractive to blue tilapia, grass carp and mirror carp. From a fish culturist's point of view perhaps it should be treated as a "grain." *Alfalfa*

Carrots (*Daucus carota sativa*) need no introduction as a vegetable, but it is not

Carrots

well known that their leafy tops rank quite high in protein among greens. Blue tilapia like them at least as well as any green available at New Alchemy. Though it would scarcely make sense to cultivate large amounts of carrots specifically for the tops, if you are gardening and growing herbivorous fish they should not be wasted.

Comfrey

Comfrey (*Symphetum officinale*) is a perennial plant which seems to be known mainly as a medicinal herb. However, it is exceptionally high in protein for a green plant, and is one of the few natural sources of Vitamin B$_{12}$, so that it has tremendous potential as a livestock feed. Comfrey leaves are among the favorite foods of all our herbivorous fishes, and seem also to benefit their growth.

Comfrey, which is usually planted from root stock, seems to grow anywhere except in very acid soils. Good strains have been reported to yield up to 115 tons per acre per year (254 metric tons/ha); 25 tons per acre (55 metric tons/ha) is considered minimal. Harvesting is best done just before flowering by cutting the entire plant about two inches (5.2 cm) above the ground. The plant benefits from frequent cutting.

Hairy vetch

Hairy vetch (*Vicia villosa*), also known as winter vetch and sand vetch, is a popular nitrogen-fixing plant used agriculturally for forage and as a cover crop for farms and roadsides. When planted densely, large quantities can easily be cut with a scythe or machete for use as a fish food; yields of up to twenty tons per acre (44 metric tons/ha) have been reported. Hairy vetch is at its best in late spring and early summer; after it flowers it tends to be "viney."

There are several other species of vetch; some with toxic alkaloids in the leaves. Fish appear to avoid these varieties. Our experience at New Alchemy has been that tilapia readily consume hairy vetch, but will not touch crown vetch (*Coronilla varia*). Some vetches are also grown as seed crops. (See the discussion of fava beans below.)

Marigolds

Marigold (*Tagetes patula* and *Tagetes erecta*) foliage is not eaten by herbivorous fish, but we have found that tilapia like to nibble the petals off the blossoms. Their yellow to orange color suggests that they may be a source of Vitamin A.

Purslane

Purslane or pusley (*Portulaca oleracea*), considered to be either a potherb or a garden weed, is distributed worldwide. In 1977 we obtained some seeds of a French "golden" garden variety purslane. Though its larger, more succulent leaves may mean that it is superior to the wild variety as human food, it is inferior as a fish food. Not only is it less robust, it produces less foliage per plant.

Purslane produces an abundance of tiny round black seeds in late summer and early fall. It should be a simple matter to collect these seeds by shaking the plants. Think twice, however, about encouraging such a persistent plant in your garden. But if you have a garden already blessed with purslane and are raising herbivorous fish, you will certainly want to take advantage of it.

Sheep sorrel

Sheep sorrel (*Rumex acetosella*) is another edible common weed which is relished by herbivorous fish. It usually appears in late summer, often colonizing sandy patches where few other plants grow. We have never made any effort to cultivate it, but some of its relatives, notably French sorrel (*Rumex scutalus*) and garden sorrel (*Rumex acetosa*), are grown by French gardeners.

Squash

Squash of several varieties were grown around the periphery of the pond in our Back Yard Fish Farm, originally to produce squash for human consumption. But we later discovered that blue tilapia will graze the squash leaves which trail down into the water.

Most important with any green grown for fish culture, after learning its protein and vitamin content and palatability to fish, is the percentage of edible matter. Table III–5–3 shows the amount of edible material (mainly stems, stalks and main leaf veins) and the dry matter content of four green plants grown as fish food at New Alchemy.

Table III–5–3.   Percent of edible matter in four
greens grown as fish food at The New Alchemy Institute.*

Plant	Percent edible matter, as harvested (total weight minus portions not consumed by fish)	Percent dry matter of edible portions only
Azolla	100.0	2.0
Comfrey	79.5	8.9
Hairy vetch	28.4	8.35
Purslane (domestic)	—	8.9
Purslane (wild)	51.0	9.1

*All plants, except Azolla, harvested by cutting the entire plant 2 inches (5.2 cm) above the ground. Azolla was fed whole.

Tuberous plants

Leaves of a number of tuberous plants have been used as fish foods in the tropics. They include the taros (*Colocasia* spp.), cassava or manioc (*Manihot dulcis* and *Manihot esculenta*) and sweet potato (*Ipomoea batatas*). Of these, the taros produce by far the most green matter; some varieties produce dozens of giant "elephant ear" leaves, measuring up to several feet in length, per plant. (Caution: Some ornamental "elephant ears" are extremely high in oxalic acid.) All these plants can be cultivated in warmer portions of the United States. They can be recommended for diversified farming because, in addition to greens for fish, they produce starchy tubers edible for humans or livestock.

In choosing greens to be grown as fish food, you should give preference to those which can be grown near the pond, perhaps even on the banks. It may not seem like much work the first few times you cut a supply of greens and carry them to the pond, but in the long run you will find it to be a great labor saver if you can simply "cut and throw."

One task connected with feeding green plants which is generally unavoidable is retrieving the uneaten portions with a net (occasionally one may encounter a pond where they are better left as fertilizer.) This task can be simplified by tying the plants in bundles before feeding.

Grains and seeds

The list of cereal grains and seeds which could be used for feeding fish includes all those now used for human food or livestock, plus others yet to be applied. Here I will discuss only those plants which we have grown at New Alchemy specifically for testing as fish food crops. 1977 was our first year in this endeavor. We began with plants which are relatively little known or which we felt might be particularly suitable for small-scale cultivation. Though grains as a group may be easy enough to grow, they are not easily processed on a small scale. As a prospective small-scale fish farmer/grain grower you would be well advised to look into the technology of low cost threshing equipment.

In addition to the cereal grains and seeds discussed here, there are many mass cultivated, commercially available grains which are often incorporated in commercial fish feeds. Some of these are included in Table III–7–1 which gives the nutritional values of various feeds (pages 167–172). The reader interested in the use of plant proteins in prepared fish feeds should also consider at least two green plants, azolla and comfrey.

Amaranth

Amaranth is a common name used for the entire family Amaranthaceae. Many will be familiar with the various edible "weed" amaranths known by such common names as pigweed, goosefoot, lamb's quarters, good King Henry, and strawberry spinach. The one we are interested in is grain amaranth (*Amaranthus hypochondriacus* and *Amaranthus cruentus*), the grain crop of the Incas.

Apart from its nutritional value and historical interest, we were intrigued by grain amaranth because the tiny size of the grain suggested that it might not need to be ground before feeding or blending with other materials. There was also the possibility that the abundant foliage (edible by humans) could be harvested throughout the growing season as food for herbivorous fishes. However, our tilapia did not care for leaves of grain amaranth or its wild relative, lamb's quarters (*Chenopodium album*), and we have not tried feeding grains.

In our short experience, we have found grain amaranth very easy to cultivate, drought-resistant and amenable to close planting. Our only real problem has been in distinguishing the young plants from the many wild amaranths which colonize our gardens.

Corn

Corn (*Zea mays*) appears to be one of the most suitable grain crops for cultivation on a small scale as a fish food. Cultivation methods are well known, and the grain is among the easiest to harvest and process. The "field" varieties are more productive and vigorous than "sweet" corn. Corn which has not ripened and hardened may be fed to fish as whole grain kernels; crp are particularly fond of it. Dried corn needs to be soaked or ground. Corn is the favored grain for "finishing" fish to improve their flavor, though some European authors say it gives a distinctive corn taste which masks the true flavor of the fish. (See Chapter III–7.)

Fava beans

Fava beans or broad beans (*Vicia faba*) are a traditional legume crop of several European ethnic communities in the United States. They are said to be a high yielder, but we were disappointed in our first trials, as the plants grew poorly and were subject to a variety of pests. Harvest and processing was easy only up to the stage of grinding in a hand grinder, which the hard, lima-sized beans resisted.

A common problem with fava beans is that they can cause the disease fabosis, through an alkaloid which is present in the green bean. Whether fabosis, which affects both humans and livestock, would affect fish, is not known. The responsible alkaloid is absent in the fully matured bean, and cooking will destroy it at any stage.

Flax

Flax (*Linium usitatissimum*) seed has been an important ingredient in livestock feeds in North America since the nineteenth century, and is used as a fish food in Europe. In our first trials flax was easy to cultivate and harvest, but relatively low yielding. It proved difficult to separate the seeds from the pods and processing was also complicated by the light weight of the seeds, which made winnowing inefficient.

Lupine

Lupine (*Lupinus* spp.) is grown primarily as an ornamental in North America, but in Europe and Latin America lupines are regarded as food and fodder plants.

Lupine has a light seed, and ground lupine seed is more likely than other grain feeds to blow away when fed on windy days. You will have to be very careful in choosing lupines for cultivation. Some lupines, including such ornamental varieties as Quaker bonnet (*Lupinus perennis*) and bluebonnet (*Lupinus subcarnosa*) contain toxic alkaloids. Generally, it seems that you might characterize the small-seeded varieties as toxic and large-seeded ones as safe, but there may be exceptions. In 1977 we planted several varieties of lupine but all germinated poorly and did not withstand four weeks of drought.

Sorghum

Varieties of sorghum (*Holcus sorghum*) are grown both as a livestock feed (grain sorghum) and a raw material for molasses (cane sorghum). It is not difficult to cultivate, but may be marginal for northern areas in terms of the length of the growing season. The yield is perhaps the best of the grain crops we have planted, but we have found it difficult to extract the seeds from the seed heads.

Soybeans

Soybeans (*Glycine max*) are the legume most widely used in feeds for livestock, including fish. They are high in protein, but cooking is required (roasting at 424°F [218°C] for fifteen minutes is suggested) to neutralize natural toxins and to make that protein available to warm-blooded animals. In any case, most, if not all fish will reject ground, unroasted soybeans. It has recently been found that soy meal is best digested by rainbow trout after it has been roasted so that the interior temperature of the bean reached 349°F to 383°F (175–195°C).

Of all the grains and seed plants discussed, soybeans are probably the best adapted to small-scale cultivation. They are easy to grow and harvest, high yielding and fairly simple to process, though roasting is essential. By far the most commonly available variety of soybean seed is Canrich, but small specialty seed companies sometimes offer varieties better suited for small-scale cultivation in northern climates.

Sunflowers

Sunflowers (*Helianthus* spp.) yield a nutritionally superior seed for consumption by humans, livestock and fish. In addition to containing twenty-five percent protein, they are high in vitamins A, B and E, calcium, potassium, magnesium, and iron. The common ornamental sunflower (*Helianthus annus*) is usable as a fodder plant, but the giant sunflower (*Helianthus giganteus*) is still better. Good varieties are Mammoth Russian and Manchurian.

Cultivation is easy, but it seems that some gardening books over-rate the "pest-resistance" of sunflowers. Though the growing plant may have few insect pests, we find that the drying heads quickly become infested with what looks like a type of weevil. And the seed heads, both on and off the plant, are extremely susceptible to theft by birds, as well as rats, which will climb the stalks to get at them. There may be further losses by the heads rotting during wet fall weather.

The first stages of harvesting and processing sunflower seeds are simple. Just cut the heads and hang in a well ventilated place until thoroughly dry. Seed removal is easily done by scraping the heads back and forth over a piece of chicken wire covering the top of a barrel.

The next step, however, is shelling the seeds—always a tedious task. I have described an instance of carp solving that problem for us, but you can not depend on most fishes to be so resourceful. (See Chapter III–7.)

CHAPTER III-6

Prepared Feeds

History

After reading what has been published on commercial fish culture or visiting fish farms or hatcheries, you might conclude that commercial prepared dry feeds are essential for fish culture. Yet in actual fact they are only one aspect of a spectrum of feeds and feeding strategies. I have already covered the pros and cons of prepared feeds; here I will discuss the composition, cost and use of commercial dry feeds, plus some alternate types of prepared feeds, including some which are suitable for on-site preparation.

Historically, prepared dry feeds are the descendants of the familiar commercial feeds used in raising poultry and other livestock. In the United States they were first developed for federal and state hatcheries raising trout to be stocked as sport fish. They were soon adopted by the commercial trout culture industry throughout the world. Subsequently, research programs were begun with the aim of developing similar feeds for use in the burgeoning catfish farming industry. Since about 1960, prepared diets tailored for rainbow trout (*Salmo gairdneri*) and channel catfish (*Ictalurus punctatus*) have been generally available and almost universally used.

Supplemental vs. complete feeds

Early prepared feeds were "supplemental" feeds, in that they were intended to augment the growth of fish in ponds stocked too densely to allow for good growth with only the available natural foods. But the trend in commercial fish culture has been toward a "complete" diet in the form of a single feed. For trout, salmon and catfish in such environments as cages, concrete raceways and recirculating systems, where there are few natural foods, the feeds used probably approach nutritional completeness.

Growers of fresh water fish other than trout and catfish are less fortunate. While common carp (*Cyprinus carpio*) nutrition has been studied in Europe and Japan, we know very little compared to what we know for trout and catfish, and there is no commercially available feed for carp. Recently, prepared feeds for eels, shrimp and tilapia have been offered for sale in North America, but their development has not been as well supported by research as that of trout and catfish feeds. As for other North American fresh water fishes, knowledge of their nutritional needs is almost non-existent. Nor is it likely, given our society's present priorities for applied research, that there will be much change in this situation, unless and until culture of one or more of these fishes becomes commercially important.

This is not to say that you should never take advantage of the convenience of packaged dry feed by, for example, feeding trout feed to bluegills, or that such feeding will not result in improved production. But while on the one hand the protein content of trout feed is higher than is probably necessary for sunfishes, on the other hand, trout feed will not provide a complete diet for them. So the grower of sunfishes, carp, etc., who wishes to use a commercial dry feed is forced to buy a product which is more expensive than one tailored for those fish probably would be, while the effect on growth is less than it could be.

The important constituents of commercial dry feeds include sources of protein, carbohydrates, fats, vitamins, minerals, medicines, preservatives, coloring matter, and binders.

Important constituents of commercial dry feeds

Protein is the most critical ingredient. While the percentage of protein required for mere maintenance of fish is much lower than that found in commercial feeds, the fish culturist wants to maximize growth. This is generally equated with providing maximum protein, though, as we shall see, this may not always be true. The minimum feasible protein level in commercial culture of channel catfish and rainbow trout is generally considered to be about twenty-five or thirty percent. Optimum levels for common carp vary from ten to thirty-eight percent. These are of course very approximate figures, which vary with many factors, but especially with the age of the fish, the quality of the protein, and water temperature. (More protein is needed at higher temperatures.)

Protein

Fish are like most other animals in that the young need more protein, and particularly more animal protein, than older individuals. Feed for trout fry, for instance, should contain about fifty percent protein. The value of animal protein seems to hold for the young of all fish; even young herbivorous fish like grass carp (*Ctenopharyngodon idellus*) and tilapia will benefit from the addition of animal protein to their diets.

The single most important source of animal protein in commercial feeds is fish meal. Other common sources are meat scraps, slaughterhouse offal, feather meal, blood meal, liver meal and dairy products. A certain amount of the protein requirement of carnivorous fishes can be supplied by grains and seed meals, among them soy, corn, cottonseed, alfalfa, wheat, peanuts and rice. But if the percentage of vegetable protein is too high, it may not only be wasted, but may actually be harmful. It is generally recommended that at least fifty percent of the protein in channel catfish diets be of animal origin. The corresponding figure for trout is sixty to seventy-five percent depending on the age of the fish. It is likely that for many other fishes, including some carnivores, the animal protein percentage requirement would be found to be considerably lower.

As amino acids are the "building blocks" of protein, any discussion of protein quality leads to the question of amino acid requirements. So far, these have been determined for six species of fish: channel catfish, rainbow trout, common carp, two of the Pacific salmons and Japanese eel (*Anguillia japonica*). All were found to need

Amino acids

the same ten amino acids, though not necessarily in the same amounts. (See Table III–6–1.) (A requirement for Phenylalanine has been proved in Chinook salmon, but not in the others.)

Table III–6–1. Amino acids known to be essential to fish, and their minimum daily requirement for rainbow trout.

Amino Acid	Requirement (as percentage of total diet)
Arginine	2.5%
Histidine	0.7%
Lysine	2.1%
Methionine	0.5%
Cystine	1.0%
Tryptophan	0.2%
Threonine	0.8%
Valine	1.5%
Leucine	1.0%
Isoleucine	1.5%

Knowledge of these requirements opens the way for developing feeds with less animal protein by substituting plant materials known to be high in certain amino acids. In recent years a number of grain varieties have been developed as substitutes for animal flesh. Other possible alternative sources of amino acids are oxidation pond algae from sewage treatment plants and artificially produced single cell proteins.

Carbohydrates   The main role of carbohydrates and fats in food is to provide energy, and they are usually economically and ecologically cheaper than proteins. There is considerable controversy among fish culturists about the use of carbohydrates in "protein-sparing." Research in the 1940's at the U.S. Bureau of Sport Fisheries and Wildlife's Eastern Fish Nutrition Laboratory at Cortland, New York, showed a very low tolerance for carbohydrate in rainbow trout. Inclusion of more than nine percent of *digestible* carbohydrate in trout diets led to deposition of glycogen in the liver resulting in death. (In trout feeds, the carbohydrate in most of the grains is considered virtually indigestible. Carbohydrate is generally fed to trout in the form of sugars such as glucose and lactose.) This has become the accepted view on carbohydrates in trout diets, but after further study, the research at Cortland has been questioned. Other researchers have given as much as 61 percent carbohydrate to salmonids without ill effects, and protein-sparing has clearly been shown with channel catfish fed an 18.6 percent carbohydrate diet as compared to diets with half that amount. It is almost certain that fish lower on the food chain than trout and channel catfish can tolerate greater carbohydrate content in their food, and they may be able to utilize some of the carbohydrates in grain and cereals.

Fats   Fats, or lipids, are also used in sparing protein in fish diets. Before dry feeds were developed, trout diets contained ample amounts of fat, owing to the use of horse meat, liver and meat scraps. However, present-day diets are comparatively low in fat, due to the greater cost of these materials and the difficulty of storing feeds containing oily and fatty materials.

Since fish are cold-blooded animals, their ability to digest fats is primarily governed by the environmental temperature. If the melting point of a fat is above this temperature, it will solidify in the fish's gastrointestinal tract and be poorly digested. Because of this, diets for cold water fishes (trout and salmon) do not contain saturated fats such as tallow, but vegetable oils instead. (Cottonseed oil is sometimes used, but in fish it may produce bad effects similar to those familiar to poultry farmers, so corn or soy oil is preferred.) *Temperature and fat absorption*

On the other hand, in the normal growing range for warm-water fish above 68°F (20°C), fish utilize saturated fats better than unsaturated ones. For example, channel catfish digest olive, safflower or hydrogenated corn oil less efficiently than beef tallow or menhaden oil, but more efficiently than unsaturated corn oil.

Only sketchy findings are available on optimum lipid levels for fish. The recommended level for rainbow trout is five to ten percent of the total diet; greater amounts may lead to fatty degeneration of the liver and kidneys. Commercial diets for common carp in warm water contain ten to fifteen percent fat, but lower levels are recommended when temperatures are below 68°F (20°C). The relation of temperature to dietary lipid level holds true for other species. For example, the generally recommended fat level in channel catfish diets is four to eight percent, but fish reared at 82.4°F (28°C) were reported to efficiently utilize up to twelve percent fat.

One important aspect of fat metabolism in fish is the relative value of the various types of fatty acids. For rainbow trout, it was found that linolenic acid and fatty acids of the linolenic series are essential, while acids of the linoleic series are not. It is worth noting that most insects are high in linolenic fatty acids, while the vegetable oils commonly used in trout feeds are low in these acids. For cultured trout, this suggests the value of insects, and perhaps other natural foods, even when they are receiving "complete" diets. *Fatty acids*

Vitamins in fish diets have been comparatively well studied, particularly for trout and salmon. Eleven of the water-soluble vitamins and four fat-soluble vitamins have been found essential for salmonid nutrition. Some research has also been done with carp, catfish and other fishes, and their needs appear to be essentially the same. *Vitamins*

Most of the essential vitamins in commercial trout and catfish feeds are incorporated as petroleum-derived synthetic compounds. This has become a bit of a problem, as their cost and availability fluctuates with the economy and politics of petroleum. Actually, vitamin supplementation is usually needed only for fish grown in cages, raceways or other environments where very little natural food is available.

Of all the ingredients of dry feeds, the vitamins deteriorate most rapidly; even the best feeds, if stored too long, may be vitamin-deficient. Fish consuming a variety of live foods never experience vitamin deficiencies, so even if you depend on dry feeds, you would do well to see that your fish have access to some natural food.

Quantitative vitamin requirements have been determined for rainbow trout, but only in a few cases for other fish. Table III-6-2 lists those vitamins which to date have been determined to be essential, and their minimum daily requirement for rainbow trout. Most often, vitamin problems involve deficiencies, but for the fat-soluble vitamins, particularly Vitamin D, hypervitaminosis—over-dosing—may also occur. *Vitamin needs of rainbow trout*

Table III–6–2. Vitamins known to be essential
for all fish, and the minimum daily requirement of rainbow trout.

Vitamin	Minimum Daily Requirement of Rainbow Trout
Water-Soluble Vitamins	
Thiamine or B_1	0.150–0.2 mg/kg of fish
Riboflavin or B_2	0.5–1.0
Pyridoxine or B_6	0.25–0.5
Pantothenic Acid	1.0–2.0
Niacin or Nicotinic Acid	4.0–7.0
Biotin or "Vitamin H"	0.04–0.08
Folic Acid or Folacin	0.10–0.15
B_{12} or Cyanocobalamin	0.0002–0.0003
Ascorbic Acid or C	450–500
Inositol	18–20
Choline	50–60
Fat-Soluble Vitamins	
A	8,000–10,000 IU/kg of feed
D	1,000 IU/kg of feed
E	125 IU/kg of feed
K_3	15–20 mg/kg of fish

Minerals   The mineral elements known to be essential for growth of fish have already been mentioned in the discussion of pond fertilization. (See Chapter III–2.) Since natural foods contain many minerals, and since fish can absorb dissolved minerals directly from the water via the gills, mineral deficiencies are rare. If mineral deficiency in fish is suspected, or if a body of water is known to be deficient in minerals, you then have to ask whether the answer is to fertilize the water or to add the missing minerals to the diet. The first is likely to be simpler, but in flowing water, the second method may be more effective. If, rather than lacking one or two specific minerals, the water is generally deficient in minerals, the commercial mineral supplements fed to cattle may be used.

The minerals most commonly found to be limiting in fresh water fish are iodine and phosphorus. Rainbow trout are known to need 0.00027 to 0.0005 mg of iodine/lb (0.0006 to 0.0011 mg/kg) of body weight daily. Iodine deficiency can be corrected or prevented by mixing sea salt with the food in proportions as high as two percent of the total. Most commercial feeds contain adequate amounts of phosphorus for trout and catfish. Supplemental phosphorus may be necessary for less predaceous fish, such as carp, which are not adapted to efficiently metabolize phosphorus of animal origin.

The benefits of adding minerals to the diet (or the water) are not limited to the prevention and treatment of deficiency problems. In experiments in Russia, when cobalt chloride was added to common carp diets at a rate of 0.04 mg/lb (0.08 mg/kg) of fish, the growth of fingerlings was increased by thirty percent and that of two-year old fish by fifteen to twenty percent. Supplements of copper and manganese also aided fishes' growth, though to a lesser degree. Other mineral elements sometimes incorporated in commercial feeds are calcium, potassium and iron.

Drugs   Drugs have been added to fish feeds both to prevent and cure diseases. Antibiotics have also been added because of their alleged growth-promoting qualities, but they have been found to have either zero or negative effects as often as positive ones. In the United States the use of antibiotics and other medicines in fish diets (except for treatment of specific diseases), has declined due largely to federal restrictions on the use of chemicals in culturing commercial food fish. Considering the experience of poultry and cattle farmers with the development of drug-resistant strains of disease microbes, it is probably just as well.

Despite controversy about the effects of preservatives in human foods, their presence in fish feeds has been found to be benign. Anti-oxidants such as ethoxyquin and BHT are included to retard the breakdown of vitamins, which is one of the principal problems with dry feeds. *Preservatives*

The "finishing feeds" used on commercial trout farms during the last four to eight weeks of rearing contain natural or synthetic carotenoids resulting in orange, pink or red color of the flesh in imitation of wild trout. These substances also bring out the red spots of brook trout and brown trout. The best trout feeds use shrimp meal —also an excellent source of protein, vitamins and minerals—for this purpose, but the more common practice is to mix in two percent paprika. The color of the flesh brought on by paprika or synthetic carotenoids fades with cooking, unlike the color of wild trout or trout fed on shrimp. *Coloring matter*

A more practical use of coloring agents in dry feeds is to make them more attractive to fish. This is useful for sight feeders (trout, bass, sunfish) but probably superfluous for chemosensory feeders (catfish, carp). Commercial feeds are colored red or brown, but in experiments rainbow trout have shown preferences for other colors. Probably many factors, including lighting and background color, are responsible.

Nutritionally inert binders are essential to retard the disintegration of dry feeds in particles larger than meal size. Early trout feeds used soft wood sawdust, but this has been replaced by bentonite clay, gypsum, or various starches. These components can also serve a beneficial role in overly rich diets which might otherwise cause inflammation of the intestines. *Binders*

Dry feeds are made in floating and sinking varieties. Nutritionally, floating and sinking feeds of the same manufacture are usually similar, the main difference being in the expansion process used to reduce density in floating feeds. I have found that some "floating" feeds, particularly in the smaller sizes, are considerably less than one hundred percent buoyant when dropped onto the water from any distance or directly among feeding fish. The advantages of a feed which floats are: *Floating vs. sinking feeds*

• Overfeeding is much easier to detect with floating feed, as uneaten food remains visible.

• In many situations, use of floating feed provides the only routine opportunity for the farmer to see and inspect his stock.

• Floating feeds are stable in water up to twenty-four hours, while sinking feeds last only a few hours.

The advantages of sinking feeds are:

• They cost less.

• In ponds exposed to strong winds, floating feed may be blown onto shore before fish can eat it.

• Certain species of fish are reluctant to take floating feed, although most eventually learn to accept it. (But the redear sunfish, *Lepomis microlophus*, is said to persistently refuse floating feed.)

Particle size

Commercial feed is available in particles from fine meal up through pellets measuring ³⁄₁₆ inch (48 mm) in length. The size of food particles is determined by the size of the fish's mouth, which of course varies with species and age. Feed manufacturers list recommended particle sizes for trout, salmon, and channel catfish. With other fish, the fish farmer is on his own. Some fishes, like the bullheads, are "all mouth" and can take large pellets early in life. Others, like most of the sunfishes, can scarcely handle ⅛ inch (32 mm) pellets at harvest size. In general, the bigger the pellet, the less waste. Where growth is uneven or there are several species or age groups of fish, it may be necessary to mix two or more sizes of feed. If there is a problem with large fish consuming small food particles meant for smaller fish, this can sometimes be remedied by putting the smaller size feed in an "ad-lib" feeder with a hole too small for the larger fish, or some sort of fine mesh swim-through feeder. (See Chapter III–7 and Figure III–6–1.)

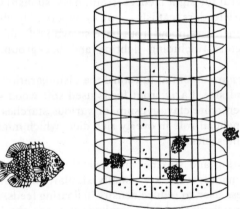

FIGURE III–6–1. *Swim-through feeder for simultaneous use with two or more sizes of feed and fish.*

Feed composition for trout and catfish

Table III–6–3 contains the formulas for three types of dry feed used in U.S. government hatcheries which are very similar to commercially available feeds. The nutritional value of typical commercial feeds is given in Table III–7–1 (pp. 167–172), along with some natural feeds and single ingredients of prepared diets. The purpose of these tables is to help the reader to decide where it might be possible to improve fish diets or save money by substituting natural or fresh foods for prepared ones.

Further information on commercial feeds and recommendations for their use can be obtained by writing to the feed manufacturers.

Table III–6–3.   Composition of typical dry fish diets.

Composition of a Typical Feed Used for Young Trout[1]

Ingredients	Percent (pounds)
West coast Canadian or Alaskan herring meal, maximum fat 10.5%, minimum protein 70%, immediate past season.	49.8
(a) Lecithin (blended with fish meal).	0.2
Corn gluten meal, 41% protein, 1% fat, 6% fiber. (Blend of 60% prime and 25% feed grade can be substituted.)	5.0
Toasted, defatted soy flour, maximum fat 0.5%, minimum protein 50%.	5.0
Steam dried brewer's yeast, minimum protein 40%, minimum fat 0.7%, maximum fiber 3.0%.	4.0
A and D feeding oil, nonsynthetic, stabilized.	6.0
Condensed fish solubles, dried on wheat middlings (equivalent to 100% condensed fish solubles).	5.0
Soluble dried blood flour.	2.0
Kelp meal.	2.0
Wheat standard middlings	7.0
Dried skim milk	5.0
Unextracted liver meal.	5.0
Vitamin premix 4–C. (See below for ingredients.)	4.0

[1] National Fish Hatchery diet specifications for 1970–1971. Courtesy of Bureau Sport Fisheries and Wildlife, U.S. Dept. of Interior. (Starter and No. 1 granule sizes only.)

Vitamin Premix No. 4–C[2]

Ingredients	Guaranteed potency per pound premix (mg)
D-Calcium pantothenate	600.0
Pyridoxine	500.0
Riboflavin	1,750.0
Niacin	6,250.0
Folic acid	100.0
Thiamine	250.0
Inositol	6,250.0
p-Amino benzoic acid	250.0
Biotin	5.0
Vitamin B_{12}	0.25
Menadione sodium bisulfite	125.0
Vitamin E from α-tocopherol acetate (d or dl) in beadlet form	2,718 IU
Vitamin D_3 activity	16,000.0 IU
Vitamin A activity (from Vitamin A palmitate in gelatin beadlets)	76,600.0 USP
BHT antioxidant	800.0
Ferrous carbonate or ferrous oxide	225.0
Copper sulphate	22.0
Choline chloride[3]	40,000.0
Ascorbic acid[3]	3,000.0

[2] One or more diet ingredients as carrier.
[3] These items will not be part of a packaged premix but are to be blended with one or more cereal ingredients, and added to the mixture at the time of manufacture of feed.

Table III–6–3.—*Continued*

Composition of a Typical Feed Used for Larger Trout[1]

Ingredients	Percent (pounds)
West coast Canadian or Alaskan herring meal, maximum fat 10.5%, minimum protein 70%,immediate past season.	34.8
(a) Lecithin (blended with fish meal).	0.2
Corn gluten meal, 41% protein minimum, 1% fat, 6% maximum fiber. (Blend of 60% prime and 25% feed grade can be substituted.)	18.0
Wheat standard middlings, 13% minimum protein, 9.5% maximum fiber.	16.0
Soy bean oil meal, solvent extracted, toasted, and dehulled, minimum protein 50%.	10.0
Steam dried brewer's yeast, minimum protein 40%, minimum fat 0.7%, maximum fiber 3.0%.	4.0
Delactosed whey, minimum protein 16%, maximum sugar 50%.	4.0
Dehydrated alfalfa meal, 17% protein, reground pellets.	3.0
Trace mineralized salt.	2.0
Vitamin Premix No. 14. (See below for ingredients.)	4.0
A and D feeding oil, nonsynthetic, stabilized.[2]	4.0

[1] No. 2 and larger granules and all pellet sizes.
[2] Contractor is authorized to use a lesser quantity of oil to control the total fat content of the finished product. Total calculated fat in the finished feed shall not be less than 7% or greater than 8%. Contractor will exercise such control by adjusting the amount of oil added during mixing operations.

Vitamin Premix No. 14[1]

Ingredients	Guaranteed potency per pound premix (mg)
D-Calcium pantothenate	600.0
Pyridoxine	250.0
Riboflavin	1,750.0
Niacin	6,250.0
Folic acid	100.0
Thiamine	750.0
Biotin	5.0
Vitamin B_{12}	0.25
Menadione sodium bisulfite	125.0
BHT antioxidant	250.0
Vitamin E from a-tocopherol acetate (d or dl) in beadlet form	675.0 IU
Vitamin D_3 activity	8,000.0 IU
Vitamin A activity (from Vitamin A palmitate in gelatin beadlets)	40,000.0 USP
Choline chloride[2]	12,500.0
Ascorbic acid[2]	3,000.0

[1] One or more diet ingredients as carrier.
[2] These items will not be part of a packaged premix but are to be blended with one or more cereal ingredients and added to the mixture at time of manufacture of feed.

Table III–6–3.—*Continued*

Thirty-Six Percent-Protein Catfish Formula (Pelleted or Extruded)[1]
(Auburn Number 4)

Ingredient	Amount in Diet (%)
Soybean, seeds, meal solvent extracted, 44% protein[1]	45
Wheat, grain, ground	22
Wheat middlings, lt 9.5% fiber	10
Fish, meal mechanically extracted, 60% protein	9
Corn, distillers solubles, dehydrated	7.5
Fat, animal	2.5
Organic pellet binder[3]	2.5
Dicalcium phosphate	1.0
Vitamin premix	0.5
Trace mineral premix[4]	0.08

[1] For extrusion processing, the pellet binder should be replaced with an equal amount of cereal grain, and the fat- and heat-labile vitamins should be added onto the surface after extrusion.
[2] 6.25 × percent nitrogen.
[3] Hemicellulose or lignin sulfonate products.
[4] Trace mineral premix should contain the following (mg per kg): Mn, 115; I, 2.8; Cu, 4.32; Zn, 88.6; Fe, 44.

Pond Fish Vitamin Premix[1]

Ingredients	Guaranteed potency per ton of feed
Vitamin A activity (from palmitate in gelatin beadlets)	5,000,000 IU
Vitamin D_2 activity	1,000,000 IU
α-Tocopherol acetate (d or dl) in beadlet form	20 gm
Menadione sodium bisulfite	20 gm
Choline chloride	1,000 gm
Niacin	50 gm
Riboflavin	10 gm
Pyridoxine	5 gm
Thiamine	5 gm
D-Calcium pantothenate	20 gm
Biotin	200 mg
Folic acid	1,000 mg
Vitamin B_{12}	20 mg
BHT antioxidant	10 gm
or Ethoxyquin	136 gm

[1] Premix on finely ground soy bean meal carrier.

Growers of fish with relatively low protein requirements may be able to use other types of commercial animal feeds, which contain little or no animal protein and are usually much cheaper than trout or catfish feeds. At New Alchemy we have had good success feeding commercial rabbit pellets to tilapia. Such feeds are *not* under any circumstances recommended for trout, catfish, bass and other predaceous fishes. *Low-priced livestock feed*

Both the composition and the prices of commercial feeds vary constantly. Trout *Locally milled feeds* and catfish feeds are produced to meet certain minimum specifications, and these may be met by an almost infinite number of possible combinations. Most feed companies now use computers to calculate the effects of changes in the availability or cost of ingredients, and alter their product composition accordingly.

Under these circumstances, you might not think it possible for fish farmers to find better feed or lower prices than those offered by the major producers. However, this is just what has happened in one of the major catfish producing centers—Humphreys County, Mississippi. The catfish farmers of Humphreys County have banded together to create Producers Feed Company, a feed mill totally owned by catfish producers.

Any feed mill can produce a fish feed according to specifications, but there are some special considerations. Conventional pelleted feeds used for land animals disintegrate completely after ten minutes in water. The earliest prepared feeds used sawdust or bentonite clay to get around this problem. Today, manufacturers of water-stable feeds select fibrous ingredients to increase compression during pelleting. These ingredients are ground separately, added to the feed mix, and the entire mix reground through the ⅛ inch (32 mm) screen of a hammermill. The soft feed is passed through dry steam and immediately pelleted in a thick die rotating at 305 rpm. Upon extrusion the feed is cooled to room temperature to retard nutrient loss, but some loss of vitamins inevitably occurs.

Floating pellets are produced by a controlled expansion process involving moisture, high temperature and pressure. This may result in further loss of nutrients. To compensate for this loss, feeds are oversupplied with vitamins before milling, and selected vitamins, amino acids and fats may be sprayed onto the feed after processing.

"Home-made" feeds

The feeds discussed so far can be produced economically by local mills only where there are large numbers of fish farmers. However, another avenue for innovative fish farmers to explore is the development of low cost, relatively low protein feeds, especially for fish less carnivorous than trout and catfish. Such feeds could be primarily grain mixtures, to be used together with live or fresh foods, the fresh foods supplying most of the animal protein and vitamins.

There is also sometimes the possibility of producing low-cost, relatively "complete" feeds by taking advantage of locally available sources of inexpensive animal protein. Two such extremely simple feeds were developed and tested at Southern Illinois University. Ingredients of the two feeds are listed in Table III–6–4. (page 165). Both feeds, when fed to yearling channel catfish, produced growth equal to that achieved with a top quality commercial feed. Preparation of the experimental feeds was quite simple:

Home-made feed preparation

Whole soybeans were roasted for 15 minutes at 424°F (218°C) and then passed through a food chopper using a 3 mm plate. We used a 1–horsepower Toledo Chopper, manufactured by Toledo Scale Company, Toledo, Ohio. After being preground, the soybeans were passed through the chopper with whole (4.6 to 7.0 inch or 12 to 18 cm) frozen gizzard shad (*Dorosoma cepedianum*), 1% vitamin mix and 2% salt mixture. The preparation was then ground three more times to insure complete mixing. The chopper extruded a 3 mm strand of material, which, when placed in a plastic bag and shaken, produced a pellet 0.2 to 0.4 inches (56–100 mm) long and 3 mm in diameter. This preparation was frozen until it was to be fed.

Preservation of these feeds required freezing. (Frozen commercial feeds for salmon and trout are available, but due to their high cost are not discussed here. A similar dry feed would be better, but it would be technologically more difficult to produce.)

Table III–6–4. Composition of two low cost, relatively complete fish feeds developed at Southern Illinois University.

	Ingredient	Percent of total (wet weight)	Percent of total (dry weight)
Feed 1	Ground roasted soybeans	48.5	79.5
	Fresh or frozen gizzard shad	48.5	16.0
	Vitamin mix[1]	1.0	1.5
	Salt mixture U.S.P. XIV	2.0	3.0
	TOTAL	100.0	100.0
Feed 2	Ground roasted soybeans	32.3	52.5
	Corn meal	16.2	27.0
	Fresh or frozen gizzard shad	48.5	16.0
	Vitamin mix[1]	1.0	1.5
	Salt mixture U.S.P. XIV	2.0	3.0
	TOTAL	100.0	100.0

[1] Obtained from Nutritional Biochemicals, Cleveland, Ohio.

Storage

Dry feeds should be kept in a cool, dry place and protected from light. It is best to transfer from the original bags to containers with tight-fitting lids, such as garbage pails. Failure to carry out these cautions can lead to rancid feed, insect infestation or, worse, development of invisible molds, which produce toxic substances (aflatoxins) which can kill fish. The commonest problem arising from poor or prolonged storage of dry feed is the loss of vitamins. Severe vitamin deficiencies in cultured fish have often been traced to old or improperly stored feeds. The *absolute maximum* storage time for dry feeds at optimum conditions is six months; some authorities advise no more than a month.

CHAPTER III-7

Feeding Rates, Techniques and Strategies

Nutritive value of
feeds

Before deciding what and how much to feed, you should have an idea of the feeding habits of the fish you wish to grow and their nutritional needs. (See Part II and Appendix II.) You will also need to know the composition of available feed materials. Table III–7–1 (pages 167–172) is a summary of information on the nutritive value of fresh and prepared fish feeds and ingredients.

The information available on vitamins, minerals and amino acids in fish feeds is fragmentary and confusing. Rather than make Table III–7–1 even more cumbersome, I will give the information on these nutritional factors separately and less precisely. Table III–7–2 (pages 173–74) lists the main feeds with known high concentrations of important vitamins, minerals and amino acids. Included are four categories of feed: animal, grain, green plant and miscellaneous feed items and supplements. Within each category the items containing the highest percentage of a given nutrient are listed. Remember that certain nutrients will be higher in one or other type of food. For example, you would not expect to find concentrations of amino acids, which are protein constituents, in green plant material comparable to the amounts you would encounter in grains or animal products, nor would you expect any animal to rival green plants as a source of ascorbic acid.

Note: The information in Table III–7–1 is drawn from a great variety of sources, but I do not claim to have done a thorough search of the available literature, nor have I reported all that I have found. I have included the data that could be found on those fodder organisms and feed components most likely to be both available and to interest readers of this book. The figures represent the lowest and highest amounts I have found for each item, but they may not represent actual minima and maxima. The alert reader may note omissions or errors, and may find other data which fall outside the ranges listed in the table. Possible sources of variation and error are:

Table III-7-1. Nutritional composition of natural and cultured fish foods, feed ingredients and prepared feeds.

I. ANIMAL FOODS

A. Soft-bodied invertebrates

Food Item	Water (%)	Crude Protein (%)	Fat (%)	Carbo-hydrate (%)	Ash (%)	Fiber (%)	Kcal/g
Limnodrilus	81.8	8.6	4.4				
Tubifex worms	87.15	8.06	2.0	1.88	0.91		
Enchytraeus albidus	82.31	12.4	3.0	2.0	1.0		(0.187%, dry wt)
Earthworms	85	20.3	4.4				449
Aquatic snails (various species, minus shells)	73.0–78.4	10.6–19.1	0.55–1.4	4.3–8.7	3.7–6.95		
Clams (various species, minus shells)	81.7–82.5	9.6–12.6	1.0–1.6	2.0–3.1	1.8–2.1		760
Fingernail clams (with shell)	56.02–75.75	3.06–3.41	0.24–0.26	1.62–2.89	18.04–38.71		281–314

B. Chitinous invertebrates (crustaceans and insects)

Food Item	Water (%)	Crude Protein (%)	Fat (%)	Carbo-hydrate (%)	Ash (%)	Fiber (%)	Kcal/g
Gammarus	14.1–87.5[1]	6.4–24.7	0.8–13.6	2.69	2.7–34.6	2.39	845
Carinogammarus	77.63–79.8	11.1–11.25	1.1–1.73	2.3–4.77	4.62–5.6		994
Brine shrimp							
nauplii		50.6[2]	23.2[2]	6.0[2]	14.7[2]		520[2]
adults	83.8–86.6	5.0–7.0	2.0–4.0	2.0–3.0	2.0–3.0		
Daphnia spp.	89.4–93.0	2.0–6.0	0.6–6.0	0.1–4.07	1.6–3.0	1.47	371–535.0
Moina rectirostris	90.6	6.6	1.5		1.0		
Mixed zooplankton		46[2]	6[2]	23[2]	25[2]		
Crayfish (*Orconectes limosus*) flesh only[3]	80.9	17.2					
Grasshoppers (various species)	10.5–70.6[4]	15.3–46.1	25.0–28.3	6.8–7.5	0.8–5.0		347
Silkworms							
newly hatched larvae							163.0
full grown larvae							115.0
pupae	64.6	19.1–23.1	12.8–14.2	2.3	1.2		216.0
Chironomid larvae	83.3–88.28	6.21–8.2	0.51–1.40	2.42–3.08	0.9–1.50	1.77	549–654
House fly pupae	3.9	63.1	15.5		5.3		

Table III-7-1—Continued

I. ANIMAL FOODS

B. Chitinous invertebrates—continued

Food Item	Water (%)	Crude Protein (%)	Fat (%)	Carbo-hydrate (%)	Ash (%)	Fiber (%)	Kcal/g
Mixed insect larvae	81.8	12.3	2.7	1.9[2]	1.0		
Mixed flying insects[5]		60[2]	6[2]		8[2]	5[2]	

C. Fish products

Food Item	Water (%)	Crude Protein (%)	Fat (%)	Carbo-hydrate (%)	Ash (%)	Fiber (%)	Kcal/g
Sardines							
fresh	70.3	21.4	6.7				449
meal		61.5	10.8	6.1	15.8	1.0	
Shrimp							
fresh	17.0	55.5	5.5				295
meal	10	44.8	3.0	3.7	27.1	11.3	
Carp (*Cyprinus carpio*)							
fresh	71	19.0	7.6		2.3		150
meal	10	52.7					
Catfish (boiled)	40	11.1					
Miscellaneous fresh water fish	72.0–78.0	12.5–15.75	1.26–6.3	0.91	2.0–4.21	(0.57% chitin)	104
Alewife (*Alosa pseudoharengus*)							
whole	76	19.4	4.9		1.5		129
meal	10	65.7	12.8		14.6	1.0	466
Anchovy meal (*Engraulis ringen*)	8	65.7	4.1	6.2	3.76	1.0	461
Herring							
fresh	72.2	19.0	6.7		2.1		142
meal	8	72.2	8.5	0.0	10.5	0.7	493
Menhaden meal (*Brevoortia tyrannus*)	8	61.1	9.7	0.8	19.1	0.8	442
Miscellaneous salt water fish	81.5	10.6–12.5	0.4–4.1		1.1–1.2		
Fish meal	8.0–14.0	48.4–63.0	2.1–11.6	1.6	16.0–32.6		243–350

D. Meat products[6]

Food Item	Water (%)	Crude Protein (%)	Fat (%)	Carbohydrate (%)	Ash (%)	Fiber (%)	Kcal/g
Bone meal	3	11.2	9.2		5.4	1.6	83
Blood							
fresh	81.0	18.3	0.2		0.5		
meal	8–9	75.3–83.9	1.6–2.5	8.5	4.2–5.4	1.0	355–523
Carcass meal	7.0	50.3	17.0	1.0	22.0	2.7	319
Feather meal	7–9.0	80.0–85.4	3.2–5.0	1.2	3.5	1.2	236
Heart	71.5	17.5	10.0		1.0		151
Horse meat	74.2	21.2	3.4		1.2		123
Liver							
fresh	72.4	20.0	4.0		2.0		114
meal	7	66.5	15.7	2.9	6.3	1.4	550
Lung	80.4	15.0	2.6		2.0		82
Meat meal	7–10.8	54.3–72.3	8.2–13.2	2.9	2.9–3.8	2.4	209–418
Meat meal with blood	8	59.5	9.0		21.7	2.2	286
Meat meal with bone	7	50.5	9.9	2.2	28.4	2.0	214
Poultry by-product meal	7	57.8	12.3	5.6	15.2	2.3	283
Spleen	17.8	17.8	3.9				116

II. PLANT PRODUCTS

A. Grains and pulses

Food Item	Water (%)	Crude Protein (%)	Fat (%)	Carbohydrate (%)	Ash (%)	Fiber (%)	Kcal/g
Amaranth	12.3	12.9	7.2	65.1	2.5		358
Barley	10.0–17.8	8.5–27.0	0.7–2.4	60.8–86.8	1.4–6.5	5.0–6.3	286–474
Corn	6.9–22.5	6.2–16.9	4.0–6.0	56.2–81.0	0.7–5.0	2.2–2.9	346–527
Corn gluten meal	9	43.1	3.9	38.3	3.3	4.5	295
Oats							
whole	4.8–21.3	7.4–23.2	2.4–9.9	56.2–76.2	1.5–9.6	10.3–10.8	254–501
groats	11	15.6	6.3	62.5	2.3	2.6	317
Cottonseed							
bran	7.4–10.5	3.4–7.9	0.9	43.6–51.3	2.9–6.2		
cake	8.7	3.6–5.8	5.0	33–41	8.0		
meal	7–11.6	34.4–48.0	0.6–4.6	27.9–32.1	6.0–7.1	11.2–13.6	265–464

I. PLANT PRODUCTS

Table III-7-1 — Continued

A. Grains and pulses—continued

Food Item	Water (%)	Crude Protein (%)	Fat (%)	Carbo-hydrate (%)	Ash (%)	Fiber (%)	Kcal/g
Horse chestnuts (unpeeled)	49.2	4.3	1.5	40.9	1.6	2.5	139
Linseed meal	9–10	34.3–34.9	1.6–5.6	36.2–38.7	5.7–5.8	8.7–8.9	416–446
Lupine	10.0–14.0	29.5–38.3	4.4–8.5	25.4–39.9	2.9–4.2	11.2–14.1	322
Millet	11	12.1	4.1	61.1	3.6	8.3	393
Milo (maize)	12.0	9.2	3.0		2.0	2.0	337
Peanuts							
whole	6.0	26.8	44.9	17.5	2.2		
cake	7.8–11.5	36.0–40.9	3.5–10.0	32.0–35.7	5.5–19.0	25.3–26.1	443–462
meal	8–10	44.0–48.9	1.4–7.3	25.3–26.1	5.3–6.3		
Peas	13.2	22.7	1.9	53.2	3.0		
Rape seed cake	10.0	33.1	10.2	27.9	7.7	11.1	284
Rice							
whole	0.4–14.7	6.5–11.6	1–2	68.6–92.4	0.1–7.2		357–412
bran	9–13.5	6.2–16	2.7–13.9	35.1–47	0.1–12.2	11.6	423
polishings	10	12.1	1.5	54.7	7.6	3.2	313
Rye							
whole	3.1–15.6	8.7–18.2	1.5–1.7	69.5–85.9	1.5–2.8	2.2	253–391
bran	12.5	16.7	3.1	58.0	4.5	5.2	246
Sesame meal	7	8.8	7.9	23.7	11.2	5.7	338
Sorghum	3.7–29.0	8.7–16.8	2.8	65.3–85.3	1.2–7.1	2.4	333
Soybean							
cake	10.1–13.5	40.9–50.0	8.5–11	26.0–35.7	5.5–6		
meal	7.0–13.5	29.5–53.2	0.5–20.0	25.0–31.1	4.7–6.1	3.6–6.1	
Sunflower seed[7]							
whole	7.0–7.5	14.2–46.3	2.7–32.3	14.3–25.5	3.4–6.5	11.0–25.1	239–420
cake	9.2	36.4	11.0	22.9	8.5	14.0	205–221
Vetch seed	13.3–14.0	26.0–27.2	1.6–1.7	49.5–49.8	1.9–3.2		316

Food Item	Water (%)	Crude Protein (%)	Fat (%)	Carbo- hydrate (%)	Ash (%)	Fiber (%)	Kcal/g
Wheat							
whole	6.0–15.9	8.0–21.9	1.6–2.5	66.4–88.8	0.4–4	2.0–2.5	326–502
flour	12.0–15.0	7.5–11.8	0.8–2.0	71.0–87	0.3–1.7	1.0	333–382
bran	11–14.8	14.3–19.9	3.9–4.8	50.7–62	5.9–6.2	10.2–10.3	406
middlings	11–12.0	13.0–16.7	2.0–4.6	56.5	2.0–4.5	7.0–10.0	414

B. Green plants

Food Item	Water (%)	Crude Protein (%)	Fat (%)	Carbo- hydrate (%)	Ash (%)	Fiber (%)	Kcal/g
Chlorella (unicellular algae)	7.0[2]	55.5[2]	7.5[2]	17.8[2]	8.25[2]		520[2]
Duckweed	91.2–93.2	1.5–2.1	0.3–0.7	2.8–5.9	0.5–1.6		
Water hyacinth	80.5–95.7	1.0–32.1	0.1–0.6	2.4–12.2	1.2–4.5		
Alligator grass (*Alternanthera*)	77.5	3.2	0.3	11.9	4.4		
Ceratophyllum	85.7	2.4	0.3	6.0	3.2		
Myriophyllum	86.4	2.4	0.2	6.8	2.5		
Water primrose (*Jussiaea*)	87.5	2.5	1.0	7.9	1.1		43
Alfalfa							
hay	0.6–24.9	9.3–33.7	2.0	25.0–53.2	4.4–17.6	18.6–25.9	380–502
meal	7.0–9.0	15.4–22.0	2.0–3.9	37.1–39.0	8.8–10.4		389–407
Purslane	86.5–91.2	1.0–2.0	0.2–0.4	5.0	1.4–1.9		26
Sweet potato leaf	85.0–87.0	1.6–3.6	0.4–0.8	6.0–9.2	1.2–1.6		48.0
Comfrey	87.6–88.5	2.5–3.4	0.3	4.9–5.0	1.5–2.3	1.5–1.7	

C. Brewery and distillery wastes

Food Item	Water (%)	Crude Protein (%)	Fat (%)	Carbo- hydrate (%)	Ash (%)	Fiber (%)	Kcal/g
Beer mash	9.0[2]	25.5[2]	7.0[2]	42.8[2]	2.9[2]	12.8[2]	265[2]
Corn mash	8–8.6[2]	25.5–27.3[2]	8.4–10.4[2]	40.1–44.2[2]	2.2–7.3[2]	5.2–10.2[2]	252–288[2]
Potato mash	10.0[2]	24.3[2]	3.7[2]	40.8[2]	11.7[2]	9.5[2]	
Rye mash							
fresh	92.2	1.7	0.4	4.6	0.4	0.7	
dried	10.0	18.5	8.2	47.8	1.3	16.2	

Table III-7-1 – Continued

III. MISCELLANEOUS FOODS

Food Item	Water (%)	Crude Protein (%)	Fat (%)	Carbohydrate (%)	Ash (%)	Fiber (%)	Kcal/g
Copra meal	7-8	21.2-21.3	4.1-6.5	46.1-46.9	6.6-6.8	11.5-13.6	149-160
Molasses	10-25	0.1-3.9	0.1-0.9	62.4-65.3	7.7-11.1	4.5	187-197
Potatoes							
fresh	75.0-76.0	1.2-2.1	0.1	21.0-21.5	1.1-1.2	0.7	
dried	9	7.9	0.5	73.7	6.9	2.0	295
Poultry litter	15.0	20.0	2.0		23.0	20.0	
Whey	7[2]	13.1-16.5[2]	0.6-1.1[2]	59.2-70.0[2]	9.4-15.8[2]	0.2[2]	196[2]
White cheese		20.9	1.0	4.3			
Yeast	7[2]	45.1-48.0[2]	1.0-2.4[2]	32.3-37.8[2]	6.8-8.0[2]	2.4-3.1[2]	202-211[2]

IV. COMMERCIAL FEEDS[8]

Food Item	Water (%)	Crude Protein (%)	Fat (%)	Carbohydrate (%)	Ash (%)	Fiber (%)	Kcal/g
Dry trout feed	10	38-50[9]	4.0-8.0	4.0-8.0	13.0-15	4.0[10]	260
Dry catfish feed	10	32-40[9]	4-8	20	8	8-20	54
Moist pellets	30-40	30	4-13	13			
Rabbit feed		20[11]	3.0[11]			12.0[10]	
Dry eel feed[12]	8.73	51.91	5.36	16.03	17.97		

1 The great variation in the water content of *Gammarus* may be because the heavy, thick shell of this animal comprises a greater percentage of the weight of small specimens.
2 Based on dry weight.
3 76% of live weight is shell.
4 The source consulted (Taylor, 1976; see Appendix IV-1) gives no explanation for the great variation in water content. Taylor lists "sun-dried grasshoppers" as containing 5.0% water.
5 Insects captured at night by a U-V "bug light" feeder. The majority were chironomid midges.
6 Beef products, unless otherwise noted.
7 Data may include dried and undried, shelled or unshelled seeds.
8 Specifications for commercial fish feeds are based on permissible maxima and minima plus a cursory examination of some available products.
9 The lower figures correspond generally to ordinary growing foods; the higher protein feeds are used principally for fry and fingerlings.
10 Maximum.
11 Minimum.
12 Used in Taiwan.

• The difference in water content of fresh and "dried" materials due to weather and handling, which in turn affects all the other figures.

• The difference in diet (animals) and soil or water fertility (plants) reflected in the nutritional content of the organisms.

• Varietal differences in cultivated plants.

• The different processing techniques to which widely grown crops are subjected. For instance, to make any sense of the nutrition literature on wheat, you must deal with at least eight processing categories in additon to data for the whole grain.

• The lack of standardization in the writings on nutrition, both in what is reported and how it is presented.

• Errors arising from my own lack of technical training in the field of nutrition.

Table III-7-2. Good sources of amino acids, vitamins and minerals in fish feeds (best sources are in italic)

A. AMINO ACIDS

Amino acid	Animal feeds	Grains	Green plants	Miscellaneous feeds and additives
Arginine	*Feather meal, herring, poultry meal*	*Sesame, peanuts*	Water hyacinth	Copra meal, yeast
Histidine	*Blood*, alewife, sardine	Sesame, soy, sunflower seed, cottonseed, corn gluten meal		Yeast
Lysine	*Blood*, sardine, herring	Soy		Yeast
Methionine	*Herring, sardine, anchovy, alewife*	Sunflower seed	Comfrey	Yeast
Cystine	*Feather meal*, blood	Cottonseed		Whey, yeast
Tryptophan	*Blood*, herring, anchovy	Sesame, soy, sunflower seed, flax, peanuts	Comfrey, alfalfa	Yeast
Threonine	*Feather meal, blood, alewife*	Soy, sesame, sunflower seed, cottonseed		*Yeast*
Valine	*Blood*, herring, liver	Sesame, cottonseed, soy, corn gluten meal		Yeast
Leucine	*Blood*, feather meal	Corn gluten meal, soy, sesame		Yeast
Isoleucine	*Feather meal, liver, sardine, alewife, herring, anchovy*	Soy, corn gluten meal		Yeast

B. VITAMINS

Vitamin	Animal feeds	Grains	Green plants	Miscellaneous feeds and additives[1]
Vitamin A	Clams, shrimp	Corn, wheat, barley	*Sweet potato leaf, water primrose*	Molasses

Table III–7–2.—*Continued*

B. VITAMINS—*continued*

Vitamin	Animal feeds	Grains	Green plants	Miscellaneous feeds and additives[1]
Vitamin B$_1$	Daphnia, bone	*Peas*, wheat germ, wheat bran, rice bran	Alfalfa	*Yeast*
Vitamin B$_2$	*Liver, midge larvae*, herring, poultry meal	Wheat bran, corn mash	Alfalfa	*Yeast, whey* *Yeast, whey*
Vitamin B$_{12}$	*Liver*, herring	Corn mash	Alfalfa	Whey
Niacin	*Insect larvae*, liver	*Rice polishings, rice bran*, wheat bran, sunflower seed	Alfalfa	*Yeast*
Vitamin C	Clams	Amaranth, corn	*Water primrose*, sweet potato leaf, purslane	
Vitamin D (All foods mentioned here are low in Vitamin D; known sources are listed in this table.)	Sardines, midge larvae	Rice bran, corn		
Pantothenic acid	Insect larvae, liver	Peanuts, rice polishings	*Chlorella*	Yeast

C. MINERALS

Mineral element	Animal feeds	Grains	Green plants	Miscellaneous feeds and additives
Calcium	*Bone*, menhaden, sardine	Sesame	Comfrey, duck-weed, water primrose	*Oyster shell, calcium carbonate, limestone, phosphate, mono-dicalcium phosphate, dicalcium phosphate*
Phosphorus	Bone, meat meal, fish	Oats, corn	Purslane, sweet potato leaf	*Sodium tripoly-phosphate, mono-dicalcium phosphate, dicalcium phosphate, phosphate, yeast*
Potassium	Midge larvae	Amaranth, sorghum, oats, cottonseed bran, barley	*Alfalfa*, comfrey	
Iron	*Blood, midge larvae, daphnia*	Oats, cottonseed bran	Alfalfa, water primrose	*Phosphate, monodicalcium phosphate, limestone*

[1] All of the vitamins mentioned are commonly added to prepared feeds in synthetic form.

Conversion ratio

A central factor in feeding is the concept of conversion ratio. The conversion ratio (sometimes called feed quotient, growth coefficient, or nutritive ratio) is commonly used as a measure of the effectiveness of livestock feeds. Next to crude protein content, it is the statistic most commonly discussed in relation to fish diets. In very simple terms:

$$\text{Conversion ratio} = \frac{\text{weight of food}}{\text{increase in weight of fish}}$$

The lower the conversion ratio, the better the food. Of course, whenever food is consumed, less than 100% of it goes into growth. Usually some of it is indigestible waste or roughage, and some is burned to provide energy for life processes. The very best food, fed under optimum conditions, must then have a conversion ratio greater than 1. If you inspect some of the advertising literature for commercial fish feeds you will see claims for conversion ratios of 1.1 or even lower, and it would seem as though fish feed technology has approached perfection. It has not.

There is nothing really dishonest about the feed companies' advertising, but it is confusing, and it can be misleading if you try to compare different types of food. There are three points to clarify before published conversion ratios can be considered intelligently.

(1) Increased rate of crude weight gain does not always indicate a proportional gain in the formation of edible flesh. Some feeds cause fat deposits in internal organs, or water in the flesh which, while they certainly add weight to the fish, add nothing from the consumer's point of view. Very little is known about this aspect of fish feeding. *Edible flesh gain*

(2) In practice we deal with *relative* conversion ratios, where one food may be compared with another used under similar circumstances. Absolute conversion ratios are obtainable only under the most strictly controlled laboratory conditions, with no known source of nutrients other than the feed being tested. When feeding trials are done under field conditions, what is known is the amount of feed given to the fish and their weight gain. But it is unknown how much natural food is also being consumed. As a result aquaculturists often use the term "S conversion" (supplementary feed conversion) rather than the "C conversion" (complete feed conversion) used in livestock feeding. It is even possible to have an S conversion of *less* than 1. In such a case, if we could separate the effect of the feed being studied from that due to natural feeds, the C conversion would be considerably greater than 1. *Absolute and relative conversion ratios*

(3) "Dry" feeds contain only about 10% water, whereas live fish are about 75% water. The conversion ratio of dry feed to fish is thus a ratio of dry weight to wet weight. But the water content of fresh fish foods is higher: 10–15% for grain, 70–75% for animal, and 75–95% for green plant material. So, conversion ratios for the last two are ratios of wet weight to wet weight. *Dry and wet weights*

Almost inevitably, then, a processed dry feed is going to exhibit a better conversion ratio than a fresh food, if we in each case consider the total weight of feed to be handled by whoever is doing the feeding. In one way this is a fair comparison; it *is* more work to transport and feed 500 pounds (225 kg) of live food than 100 pounds (45 kg) of dry concentrate. On the other hand, *someone* had to process the water and other impurities out of the fresh ingredients that went into the dry feed and, speaking in terms of economics or energy, someone has to pay for that processing.

To further illustrate the hidden complexities in using conversion ratios as yardsticks for fish feeds, let us see what might happen in three hypothetical carp ponds. (These are imaginary ponds since, to my knowledge, no actual experiment of the type I am about to describe has been done.) We will compare feeding strategies in these ponds in two ways; first, in the usual manner, with a conversion ratio based on the total weight of feed used, and second, based only on the dry weight of feed. Feeds will be chosen from those known to be acceptable and digestible by common carp (*Cyprinus carpio*). Water content and conversion ratios will be based on those reported in scientific studies. *"Experiment" with various feeds*

Let us suppose that in each pond we will produce the same amount of fish: 1,000 pounds (450 kg). In Pond A we will feed a good commercial dry feed, containing 10% water, with a conversion ratio of 1.5.

In Pond B, we will feed equal portions (fresh weight) of daphnia (90% water), earthworms (85% water) and lupine seed (14% water). Conversion ratios have been reported ranging from 4 to 6.4 for daphnia, 8 to 10 for earthworms and 3 to 5 for lupine seed. Let us suppose, then, that the conversion ratio for these three fresh feeds used together is 6. (In practice, a mixed diet is usually better balanced than a diet of one food, and the conversion ratio will be better than the average conversion ratio of foods fed singly.)

In Pond C, we will feed a grain mixture: equal parts of corn (12% water), wheat bran (15% water), and lupine seed (14% water). The conversion ratio ranges for these foods, fed to carp, are 4–6 for corn, 4.2–7.3 for wheat bran, and 3–5 for lupine seed. Used together in our experiment, we can assign them a conversion ratio of 5.

So, to produce 1,000 pounds (450 kg) of carp we need 1,500 pounds (675 kg) of commercial dry feed, 5,000 pounds (2,350 kg) of mixed grains or 6,000 pounds (2,700 kg) of the live food/lupine seed mixture. Using conventional conversion ratios as a criterion we would conclude that: the live food/lupine seed mixture fed in Pond B was the least effective of the three feeds; the grain mixture (Pond C) was slightly better; the commercial feed (Pond A) was four times as effective.

But what happens if we express the weights of the three feeds in terms of *dry* weight? (For easy comparison, we will continue to express fish production in wet weight.) We find that the actual weights and conversion ratios of *food material*, minus water, supplied to the three ponds would be as shown in the last two columns of Table III–7–3.

Table III–7–3. Hypothetical example illustrating the difference between conversion ratios based on wet and dry weights of natural and prepared feeds.

Pond	Feed	Lb of fish produced	Raw (wet) weight of feed (lb)	Conversion ratio (based on wet weight of feed)	% dry matter	Dry wt. of feed (lb)	Conversion ratio (based on dry weight of feed)
A	Commercial dry feed	1,000	1,500	$\frac{1,500}{1,000} = 1.5$	80 %	1,350	$\frac{1,350}{1,000} = 1.35$
B	Daphnia		2,000		10	200	
	Earthworms		2,000		15	300	
	Lupine seed		2,000		86	1,720	
	Total Pond B	1,000	6,000	$\frac{6,000}{1,000} = 6.0$	36.7	2,220	$\frac{2,220}{1,000} = 2.22$
C	Corn		1,667		88	1,467	
	Wheat bran		1,667		85	1,417	
	Lupine seed		1,667		86	1,434	
	Total Pond C	1,000	5,000	$\frac{5,000}{1,000} = 5.0$	86.3	4,318	$\frac{4,318}{1,000} = 4.32$

Evidently, the carefully balanced, expensive dry feed (Pond A) with its fish meal, package of synthetic vitamins, etc., was only about 1.6 times as effective as the live food/lupine seed mixture (Pond B). And the live food/lupine seed mixture was nearly twice as effective as the unfortified dry diet of grains (Pond C).

All this is oversimplified, and does not take into account the economic factors you would have to deal with in a real fish culture situation. It does show the limitations of conversion ratios. They are very useful for comparing the effectiveness of similar feeds—say different brands of commercial feed—but if you try to compare different types of food, together with data involving different environments or species of fish, the picture becomes fuzzy.

Usually, the choice of a particular feeding strategy is basically an economic one. Perhaps you have no intention of purchasing feed or selling fish, but there are still equipment, time, and labor involved, and you will want to compare the cost of raising fish with the cost of raising or buying other protein foods. (Even if you are going to raise fish just because you like fish, you are probably interested in not spending money unnecessarily. So read on.) We can set up four broad categories of feeding strategy, in ascending order of intensity and, usually, cost. Of course there are any number of intergrades and hybrids between these categories:

(1) No effort at all is made to provide feed for the fish beyond what occurs naturally. Management is limited to stocking, guarding against predators, and harvesting.

(2) Food is not directly provided, but the water is fertilized to increase production of fish food organisms. (Methods of attracting and trapping naturally occurring food organisms, for example the use of lights to attract insects, fall somewhere between this and the following category.)

(3) Prepared and/or cultured feeds are fed only to supplement naturally occurring foods or those produced through fertilization.

(4) Feeds, usually though not only commercial, are used as the main source of growth material for the fish, in most cases with an incidental input of "natural" foods.

At the lower end of this grouping you will rarely find a small body of water which will provide fish over and above the occasional needs of a family without any feeding or fertilization. On the other end, if you contemplate intensive feeding of fish in a fair-sized body of water, you probably have a market in mind. Clearly, the degree of feeding intensity is directly related to the eventual economic use of the fish. You must ask yourself which of the following you are most interested in:

• An occasional meal, perhaps as a by-product of recreational fishing.

• A steady supply of fish for home use.

• A small-scale commercial enterprise (or a steady supply of fish for a small community), perhaps as part of a diversified farming program.

• A major commercial venture.

Table III-7-4 shows how to set up an economic analysis of feeding strategies, using data drawn from carp culture in Israel. The example omits the final step—comparison of the total market value of fish produced by each method and the profit (or saving, if the fish are to be used where grown) in each case. But note that, while the use of feeds increased the cost of the operation per unit area of water, in this instance it also decreased the cost per unit weight of fish.

Many prospective fish farmers in North America will not be able to come up with such precise data as that available for a firmly established business like carp culture in Israel. But the effort to anticipate economic realities will generally be rewarded.

Intensities of feeding strategy

Economics of feeding

Table III–7–4. Costs per hectare and per ton
of common carp culture in Israel with and without feeding.

Costs	Yield with feeding		Yield without feeding	
	per ha 2,100 kg	per ton – 0.47 ha	per ha 1,000 kg	per ton – 1.0 ha
Charges for capital invested in ponds and fishing gear	$360	$169	$360	$360
Water	210	99	210	210
Fertilizers	87	41	87	87
Maintenance	106	50	106	106
Feed	370	174	–	–
Labor	286	116	150	150
Marketing costs	32	15	12	12
Interest on working capital	5	2	2	2
General and overhead expenses	10	5	8	8
Total	$1,466	$671	$935	$935

Source: Tal and Hepher (1966).

Feeding rates: general considerations

Of course the economics of feeding cannot be assessed without considering the rate of feeding, which in turn depends on the conversion ratio. If you have decided to grow catfish or trout using a commercial feed, you will be able to take advantage of the manufacturer's recommendations. These should be followed, with adjustments as noted below or in the catfish or trout culture literature. Otherwise, you will need to determine your own feeding rates.

A rule of thumb in feeding fish is to not feed at one time more than the fish will consume in a half hour. Holding closely to this rule almost demands hand feeding and the use of floating feeds. It is not necessarily appropriate for live foods which will survive for a time in the pond environment.

Most feeding schedules are calculated on the basis of the known or estimated standing crop of fish (total weight of fish in water at given moment). Commercial dry feeds are usually fed at rates of two or three percent of the standing crop daily. Live and fresh feeds are fed in greater quantities to compensate for the water content.

A number of situations call for lower than normal rates of feeding, including the following:

Occasions for lowered feeding rates

• Cold water (usually occurring near the beginning or end of the growing season): In North America, outside of a few subtropical regions, fishes' metabolism is greatly slowed in the winter, feeding is reduced, and almost no growth occurs. In the dead of a northern winter, a single, small, weekly feeding may suffice for warm water fishes. Trout may be fed regularly in all but the very coldest of waters.

• On very hot, still, or cloudy days, or whenever there is reason to believe that dissolved oxygen levels may be below normal. At such times there is acute danger of deoxygenation from the decomposition of uneaten food. If you discover any signs of organic pollution, whether from feeding or any other source, you should stop feeding until the situation can be remedied.

• When fish are sick, they usually go "off feed." (Bullheads are sometimes the exception to this rule; I have seen them eat heartily and die later the same day.) Reluctance to feed may also indicate environmental problems, including a shortage of dissolved oxygen.

• When only a few fish are seen to be diseased, and medication is applied, all the fish may go off feed for a day or two. Not all disease treatments have this effect; the farmer must rely on direct observation.

• When there is an unusual abundance of natural feed. Recommendations for commercial feeds make the assumption that an "average" amount of natural food will be present. If your waters are unusually fertile, or if you supply some natural or fresh food, you will want to adjust the feeding rate of dry feed downward. Remember, when trying to calculate the value of live or fresh foods, that "dry" feed contains only about 10% water, while most animals and plants are 70% to 95% water. (See Chapter III–5 for suggestions on substituting natural foods for a portion of a dry feed diet.)

• When fish have been recently disturbed or alarmed as, for example, after a partial harvest or sampling.

• Whenever you notice that the fish are not promptly consuming all the food you give them. A submerged feeder is a good aid. (Figure III–7–1); if substantial amounts of food remain on such feeders an hour or more after feeding, you are overfeeding.

• During the few days immediately before handling or transport fish should not be fed at all. Handling is safer with empty stomachs, and they will not vomit feed or excrete and pollute their transport water.

Heavier than normal feeding may be called for in an environment without natural foods, for example, concrete or plastic pools. It is also permissible to fatten fish for sale, in the last few weeks before a mass harvest.

Perhaps the most problematical aspect of devising feeding schedules is calculating the standing crop. It is possible to skip such calculations and feed by "feel." One of the most effective fish feeders I have known was a man with an eighth grade education who worked in a government trout hatchery. He delighted in throwing away his college-educated bosses' carefully prepared feeding charts and feeding according to what he saw, with phenomenally good results. However, he had forty years experience with fish feeding; most of us will have to rely on weighing and calculating for the first twenty years or so.

FIGURE III–7–1. *Submerged feeder.*

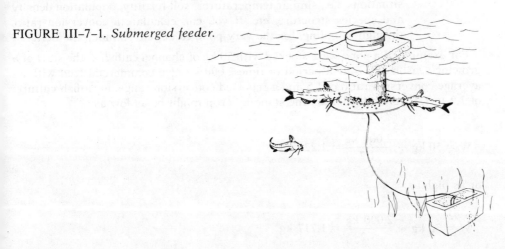

Biologists' formulas to estimate standing crop can be rather complicated; you can find examples in fishery biology texts. (See Appendix IV–1.) Calculation of standing crop in fish culture systems is usually easier than in nature, since fish culture ponds tend to be smaller, biologically less complex, and may be designed to simplify the capture of fish. Another advantage where reproduction does not occur is that the culturist can know precisely the initial numbers and weights of fish stocked.

Basically, determining standing crop in natural systems involves capture and weighing of samples of fish. In systems where reproduction occurs, an estimate is needed of the numerical population, and perhaps its age distribution, before the weight of fish can be assessed. This is not only quite complicated mathematically, but so fraught with error as to be of dubious value to the fish farmer. Even in ponds where there is no reproduction, there are two sources of error in such estimates.

• Death or other losses of fish can occur undetected, causing overestimation of the standing crop.

• Since we cannot assume that all fish are equally easy to catch, we have no way of knowing if a sample catch represents the population. Small fish are generally more catchable than large ones, leading to underestimation of the standing crop, but the opposite may also occur.

Another objection to estimating standing crop by sampling is that it inevitably entails considerable disturbance and handling of fish, which can lead to decreased feeding and growth and raise the incidence of injury and disease.

Most North American commercial fish farmers rely little on sampling and compute growth using the known initial weight, the number of fish, and the reported conversion rate for the feed used. Of course this task is simplified when a single feed is fed to a widely cultured type of fish. The formula for this method, simply stated, is:

Standing crop (W) = I + F/S

where I = the initial weight of fish stocked or known to be present at the beginning of a certain time interval

   F = the weight of feed added in that time

and S = an average S conversion for the feed used, based on similar situations, i.e., similar temperatures, soil fertility, population density and species structure, etc. If you can calculate a conversion ratio, through sampling, for the very pond to be fed, so much the better.

For example, if a pond were stocked with 50 kg of channel catfish at the start of a growing season and over a period of time 2,000 kg of a commercial feed with an average conversion ratio of 1.5 and a reported conversion range, in catfish culture, of 1.2 to 1.7 were added, then the standing crop might be as low as:

$$W = 50 \text{ kg} + \frac{2,000 \text{ kg}}{1.7} = 1,226 \text{ kg}$$

or as high as:

$$W = 50 \text{ kg} + \frac{2,000 \text{ kg}}{1.2} = 1,717 \text{ kg}$$

but estimated as:

$$W = 50 \text{ kg} + \frac{2,000 \text{ kg}}{1.5} = 1,385 \text{ kg}$$

This system may appear inaccurate, but in experiments where standing crops were estimated by both methods, and then the water drained and the actual standing crop measured, it was found to be as accurate as estimation by sampling.

The estimated standing crop may simply be multiplied by a suggested percentage to arrive at a new feeding rate. A more sophisticated approach is to calculate the rate of increase in weight expected in a given time period (R_T) with a specified feeding program:

Determining optimum feeding rates

$$R_T = 1 + DP/S$$

where   D = the number of days during interval T when fish are fed,

T = equal intervals at which the feeding rate is readjusted (e.g., two week intervals, four week intervals)

P = rate of feeding expressed as a percentage of estimated standing crop

and   S = S conversion.

For example, if channel catfish are fed six days a week with the same feed as in the previous example, the weight of feed being calculated at 3% of the estimated standing crop, and the feeding rate is adjusted every four weeks, then D = 24 (six days × four weeks), T = 4, P = 0.03 and S = 1.5. Then:

$$R_T = 1 + \frac{(24)(0.03)}{1.5} = 1.15$$

That is, at the end of the four week period T, the fish are expected to weigh 1.15 times their weight at the start of the period. This formula, together with economic data on the cost of food, value of fish, etc., can be used to plan an optimal feeding schedule.

A related logarithmic formula which you can use to determine the amount of time, in terms of number of "T" intervals (N), necessary to produce a desired weight of fish is:

$$N = \frac{\log W - \log I}{\log R}$$

where W = the estimated standing crop at some point in the growing season

I = the initial weight of fish present, and

R = the rate of weight increase, as calculated above.

The calculations just described will not be necessary for many non-commercial fish farmers. One rule that all fish farmers must heed is: *do not overfeed*. Uneaten feed may act as a fertilizer, but it is just as likely to be a pollutant, resulting in

Time and frequency of feeding

deoxygenation, the spread of bacterial diseases and poor utilization of the next batch of feed. And wasting feed costs money—there are cheaper fertilizers.

On the other hand, you do not want to underfeed. Your goal, after all, is to maximize the production of fish flesh. You may be able to buffer your margin of error in feeding by a discriminating use of polyculture. (See Chapter IV–1.)

Most commercial fish farmers feed six days a week. Besides giving the farmer a day off, this gives the ecosystem time to deal with the effects of any unintended overfeeding. Catfish farmers have shown that feeding seven days a week can add 12% to 15% to annual production, but more caution must be exercised.

One feeding a day, usually at dawn, is the normal practice, except for young fry, which may be fed up to eight times daily. Most cultured fish will feed at any time of the day (or night, for such nocturnal feeders as bass, crappies, catfishes and brown trout, *Salmo trutta*), but there are a number of reasons for dawn feeding:

Dawn feeding
• Dawn feeding forces you to look at your fish at the most critical time of day when dissolved oxygen is usually lowest; if anything is wrong you are more likely to be able to detect it then.

• Some kinds of fish feed more enthusiastically or will accept greater quantities of food at dawn (or dusk). I have noticed this particularly with sunfishes.

• There is now some evidence that the time of day of feeding can affect growth rates and conversion ratios. This evidence suggests dawn as the best time for feeding.

Perhaps a regular feeding schedule is as important as determining the precise time of day; most fish learn to expect food at a certain hour. In large ponds this can be important; having the fish in the right place at feeding time can save feed.

We may eventually learn that a single daily feeding is not always best. I have read reports suggesting that two daily feedings to satiation produced the best growth rate in rainbow trout. Another source reported that channel catfish grew faster when fed to satiation every other day. Still another approach to fish feeding schedules is to use demand feeders. (See Figure III–7–2.)

Feeding site
You should always feed in the same place. If you cannot feed at the same time daily, you can train fish to come to a certain place. If the sight of a person on the bank or the sound of food hitting the water is not enough to bring the fish, you can "ring the dinner bell" by driving a pipe a few feet into the ground by the bank and beating on it with a stick. Though this may alarm the fish the first few times, they will soon learn to associate it with food.

In selecting a feeding site you should consider the following:
• It should give you a good vantage point to observe feeding.

• Ordinarily, the water at that point should not be over four feet deep. This increases the likelihood of any food that sinks to the bottom being picked up eventually.

• The water must not be too shallow, or it will be unduly roiled during feeding and fish, particularly the larger ones, will be inhibited from entering the area.

• The bottom at the feeding site should be clean and hard, if possible. This will prevent roiling and make it easier for fish to find food which sinks. If a pond has no hard bottom areas, you may want to gravel one spot. (Clean bottoms are more important for some fish than for others. At one extreme, trout generally refuse to feed off a soft bottom. On the other, carp not only will root in mud, but should be encouraged to do so.)

• If there is a flow through the pond you will want to feed near the upper end to prevent loss of feed.

When feeding by hand, broadcast the feed as gently as possible to take maximum advantage of floating feeds. You will want to scatter the feed a bit, so that all the fish get a chance at it, but you do not want the feed to scatter too widely. To prevent wide dispersal, some sort of floating boom or feeding ring may be helpful.

It is not necessary to feed by hand; alternatives are ad-lib, automatic, or demand feeders. An ad-lib feeder (suitable for use only with sinking feeds) may be made from a plastic drum or other container by simply cutting a hole larger than the largest fish to be fed in the top. The drum, loaded with no more than one day's ration, is then sunk upright in the pond. Ad-lib feeders have the disadvantage of reducing your opportunity to watch the fish feed, but once the fish learn that food will constantly be available in them, they reduce competition for food, which may then lead to more uniform growth. I have no personal experience with ad-lib feeders, but am quite certain that some fishes would not learn to enter them; they have been used successfully with carp and tilapia.

Automatic feeders are set to provide a certain amount of feed at specified times. They are certainly a convenience, but they have not found great acceptance. In practice, fish feeding requires too much daily inspection and adjustment to allow for long-term setting of automatic feeding devices.

Demand feeders (purchased or constructed) operate on the principle that a fish knows best when it is hungry. (See Figure III–7–2 and Appendix IV–3.) Certainly fish quickly learn to depress a plate and release a small quantity of food when they want it. But more research is needed to determine whether demand feeding guarantees optimum feeding rates, or whether a few dominant fish can control the food supply more easily with a demand feeder than with hand feeding.

FIGURE III–7–2. *Demand feeder.*

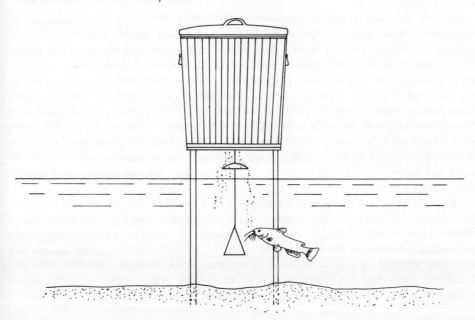

Feeding devices

At present it seems that, though there may be some economic gain for the large-scale aquaculturist who is using automatic or demand feeders, their only real value for the small-scale grower is to allow short absences and to eliminate guesswork in winter feeding. Dependence on these devices can have a negative effect, particularly for the novice. Having to meet a feeding schedule forces the farmer to inspect the aquaculture system daily—a practice I heartily recommend, even if *no* feeding is going on. One of the greatest pleasures of fish culture is to daily observe your healthy, growing fish enthusiastically feeding. Why miss it?

Farm ponds

In concluding this section, I will offer some feeding suggestions for the various species and groups of fish discussed in this book, with reference to the different intensities and economic goals listed above.

Direct feeding of fish is not generally practiced in traditional farm ponds, though sometimes such ponds are fertilized. But there is good reason to break with tradition, particularly if your goal is increased production of food fish, and not just recreational fishing. While fertilization affects the whole food chain, and should benefit all species, direct feeding is usually aimed at a particular species. So I will discuss the fish in the farm pond ecosystem separately. Of course, this information also applies to cultures of these species outside the traditional farm pond framework.

Bass

The largemouth bass (*Micropterus salmoides*) is at the very top of the farm pond food chain; it feeds naturally on frogs, crayfish, large insects, and other live animals of comparable size. But above all, it is supposed to feed on the young of bluegills and other panfish, and to control their populations. For reasons already discussed, this does not always work out as well as planned. Almost from the beginning of the farm pond program, biologists have tinkered with the idea of stocking bass, with or without bluegills (*Lepomis macrochirus*) in conjunction with other forage species. The most comprehensive research of this nature is still Swingle's work in Alabama in the 1940's.

Stocking combinations

Swingle stocked fertilized experimental ponds with various combinations of largemouth bass, bluegills, goldfish (*Carassius auratus*), golden shiners (*Notemigonus crysoleucas*), gizzard shad (*Dorosoma cepedianum*) and mosquitofish (*Gambusia affinis*). He concluded that, while some of the combinations involving goldfish, golden shiners, or gizzard shad gave higher yields of bass over a two-year period, the traditional bass-bluegill combination promised the highest sustained yield of bass. Swingle was primarily interested in bass, rather than total food fish (bass plus bluegill) production. In going over his figures, it appears that, at least in the short run, total food fish production was greatest with a combination of bass, bluegills and golden shiners. Goldfish and gizzard shad were not suitable substitutes for the golden shiners; gizzard shad appeared to compete with the bluegills, while bluegills are serious predators on goldfish eggs.

The main objection to the use of golden shiners was that after about two years the bass drastically reduced their numbers, and mosquitofish were virtually eliminated within a few months. I am not sure that this objection need be a serious one where producing food fish is the goal. Golden shiners, mosquitofish, or other forage fish could be stocked as catfish farmers sometimes stock fathead minnows, with the *intention* that they be eliminated and artificially replenished. A combination commercial bait fish/subsistence food fish farm might be an attractive option.

In a reasonably fertile pond, the growth rate of very young bass should be satisfactory. If their growth is poor, most of the live foods suitable for bluegills (discussed immediately below) should be satisfactory as supplemental feeds.

It is generally believed that juvenile and adult largemouth bass eat only live, moving foods of fairly large size. Of the cultured food animals discussed in this book, earthworms may be most suitable for bass culture. However, I have occasionally caught largemouth bass on dead bait, and in recent years some researchers successfully trained largemouth bass to take prepared feeds. Still, the nutritional and economic importance of this accomplishment is doubtful.

I have mentioned two other fishes, the smallmouth bass (*Micropterus dolomieui*) and the chain pickerel (*Esox niger*), which might play the largemouth bass' predatory role in some ponds. (See Chapter II–1 and Appendix II.) My comments on feeding largemouth bass apply largely to them. However, the chain pickerel is one of the few fishes which no one has been able to induce to accept artificial feeds.

Feeding sunfishes appears more promising than largemouth bass feeding. *Sunfish* Whether fed by fertilization, trapping of food animals, culture of live foods, or processed dry feed, they are likely to return more pounds per acre than bass.

By far the most commonly stocked sunfish is the bluegill. All feeding strategies, from the least intensive to the most, apply to bluegill culture. Fertilization methods for bluegill ponds are discussed in Chapter III–2. Moving half a step up the intensity scale, the use of bug lights to attract and trap flying insects as fish food may be more suited to bluegills than to any other fish. This method will very likely add a significant number of pounds per acre to bluegill production, whether in an otherwise unmanaged recreational fishing pond or an intensive commercial facility, and it should be economically feasible in most circumstances.

The live food "culture" most often used with bluegills is the growing of fly *Culturing live food* maggots on platforms over the water. All of the insects and smaller invertebrates whose culture I have described are potentially good bluegill food. Earthworms are relished, but they may have to be chopped for smaller fish.

Bluegills easily learn to take floating or sinking dry feed, but they do not feed well from the bottom, so be especially careful not to overfeed. Some sort of bottom feeding fish could be stocked with the bluegills, but almost all will eat bluegill eggs, and could lead to losses. A clever solution to this problem has been developed by the staff of *Farm Pond Harvest Magazine*. They use a specially designed bluegill feeder, consisting of an anchored floating ring, suspended from which is a tray inclined at an angle of about 20 degrees. Apparently, bluegills are uncomfortable tilting the axis of their bodies at 90 degrees or more to the substrate, which they must do to pick food off the bottom, but they accept 20 degree tilt. Used with a combination of floating and sinking feed, this device has virtually eliminated wasting of feed. (The feeder, together with a supply of feed and detailed feeding instructions are for sale as a kit. See Appendix IV–3.)

The role of prepared feeds in the development of bluegill culture is still unclear. *Prepared feeds* While bluegills will accept commercial catfish and trout feeds, we do not know their precise nutritional needs. The dry feeds sold as "pond fish" feed are usually ordinary catfish feed. I think it likely that in matters such as protein requirement they will fall somewhere between trout and catfish. Due to the bluegill's small mouth, the bluegill farmer need consider only feeds with the smaller particle sizes; even

A bluegill feeder developed at the Farm Pond Harvest research center, Momence, Illinois. Floating feed is retained by the ring. Feed which sinks is caught on the submerged screen, which holds it at the proper angle for bluegill feeding. (Photo courtesy Leo Pachner, Farm Pond Harvest Magazine.)

a large bluegill will have difficulty handling anything larger than a ⅛ inch (32 mm) pellet.

Present-day commercial feeds should be fed to bluegills only as a supplement. If they are used in cage culture or other systems restricting access to natural foods, there should also be a generous supply of fresh feeds such as insects, earthworms, minced meat or fish scraps. Francis Bezdek suggests salt water fish offal passed through a meat grinder as an optimum diet for sunfishes. You can combine the resulting product with dry feed if it is too "soupy."

Feeding amounts   The quantitative aspect of bluegill feeding is poorly understood, for the following reasons. Intensive bluegill farming is comparatively undeveloped, so that we simply do not know very much about feeding bluegills yet. Also calculation of bluegill standing crops is complicated by their high rate of reproduction and the difficulty of capturing adequate quantities for sampling. Finally, since bluegills usually reproduce in any pond, all sizes and ages of fish are likely to be present, and not all of these will respond the same to, or benefit equally from, a given feed.

Where you can estimate the weight of fish present, a daily dry ration feeding of 2% of total body weight is appropriate during the peak growing months. If you do not know the weight of fish, try 2 to 3 pounds of dry feed per acre (2.2–3.3 kg/ha) or its equivalent.

There are two reasons for feeding bluegills in addition to the straightforward one of producing more pounds of fish. One is to improve hook and line fishing. Bug lights, platform culture of maggots, and feeders such as the *Farm Pond Harvest* device mentioned earlier in this chapter are particularly effective for this purpose. These methods tend to concentrate bluegills in one spot, looking for food. Whether your goal is sport or just increased efficiency of harvesting, better angling is bound to result. It is also likely that some of the bluegills attracted to feeders are socially dominant individuals which would grow more rapidly than others regardless of feeding. Feeding unmanaged bluegill populations at stations in this way serves to "trap" large fish as well as to enhance their growth. (Where there are bullheads, they will also be attracted; don't neglect to try bottom fishing under the feeder.)

"Station" feeding

If your goal is not full-scale commercial production, it matters little if the average weight of bluegills in a pond is low, or if there are many old but small fish, as long as you know where there is a concentration of large fish to be harvested at will with hook and line. "Station" feeding creates this situation. In a pond producing some natural food, such feeding, together with regular fishing, can result in harvests of fish weighing more than the amount of feed provided by the farmer. Concentrating and "finishing" fish in this way seems to me one of the least objectionable uses of commercial feeds.

So far I have dealt exclusively with bluegills. Other sunfishes are generally not as suitable for cultivation as food fish. Many are simply too small, but even among the larger sunfishes only the longear sunfish (*Lepomis megalotis*) is as prone to surface feeding as the bluegill. Redear sunfish (*Lepomis microlophus*), in particular, are said never to surface feed.

In rare instances you may want to stock a species other than the bluegill. For example, the pumpkinseed (*Lepomis gibbosus*) is fond of snails, and might be stocked to take advantage of a natural abundance of that food. The redear sunfish, sometimes called "shellcracker" is an even more avid consumer of all kinds of mollusks. The predaceous food habits of the green sunfish (*Lepomis cyanellus*) are given away by its unusually large mouth. The same is true of most of the centrarchids outside the genus *Lepomis*; except for the black basses, they are poor choices for culture, with exceptions as noted in Chapter II–1 and Appendix II.

At least two hybrids, the "hybrid bluegill" and "hybrid redear" are said to convert dry feed better than pure bluegills or redear sunfish. The hybrid bluegill has one other attractive characteristic: its mouth size falls between that of its parents. While not likely to become a predatory pest like the green sunfish, hybrid bluegills can ingest larger food items than pure bluegills of the same size. This facilitates the task of feeding them. In most respects the factors involved in feeding hybrid bluegills and hybrid redears are the same as for bluegill and redear sunfish.

A problem associated with all types of sunfish culture, and which you should consider carefully if you are contemplating investing substantial amounts of labor or money in sunfish feeding, is the difficulty of harvesting these fishes. Though "station" feeding of otherwise unmanaged populations with sporadic harvests may be a convenient means of providing sunfish for home consumption, in commercial culture other factors come into play. It simply will not do to construct elaborate culture facilities for food animals or purchase large amounts of prepared feeds if you can only harvest 500 out of every 1,000 sunfish you feed.

Bullheads

Probably no fish mentioned in these pages are easier to feed than the bullheads. They will consume practically any material of animal origin and many plant materials (though not acid fruits, leaves or stems). Though we do not know the effects of various foods in terms of nutrition and growth of bullheads, it is logical to expect their requirements to be similar to those of channel catfish, though less restricted. Bullheads are naturally nocturnal and benthic, but they quickly learn to feed at any level in the water, at any time of day. From an early age their large mouths equip them to handle large pieces of feed.

Bullheads are guided to their food primarily by taste and smell. They are perhaps the best suited of all fresh water fishes to take advantage of offal from slaughter of fish or other animals. Any sort of commercial feed, or any of the live foods discussed above should be acceptable to bullheads, though large, sinking items like earthworms or crushed clams are probably best. In close confinement, for example in cage culture, they are effective predators on small fishes, which they track down in the dark, but in open water they are relatively handicapped in catching other fish.

Channel catfish

Conventional feeding methods for channel catfish (*Ictalurus punctatus*), involving mainly the use of commercial dry feeds, are well covered in the catfish farming literature and scarcely need discussion here. (See Appendix IV–1.) But, as commercial feeds have been so heavily emphasized, I want to present a few possible alternatives and supplements, lest the reader think there is only one way to feed channel catfish. Most of what I have to say applies equally to the less commonly grown blue catfish (*Ictalurus furcatus*) and white catfish (*Ictalurus catus*). White catfish are said to be better converters of dry feed than channel cats. Blue cats are the poorest converters of the three commercially cultured species of *Ictalurus*, but are easiest to train to feed at the surface.

Stocking forage fish

The most common alternative or supplement to dry feeds in commercial catfish culture is the stocking of forage fish, which is described in Chapter III–5. Worms are accepted by catfish more readily than small fish, but to date no one has tried to use them in commercial catfish culture. However, I am familiar with one homestead-scale successful attempt to culture channel catfish using earthworms as food. (See Appendix IV–1.) Fresh water clams are another natural food of catfish which might be applied in a culture situation.

Channel catfish, while somewhat more selective in their feeding than bullheads, share their chemosensory feeding tactics and will eagerly consume meat and fish offal. Some culturists use meat scraps as the primary food for catfish fry, which are gradually switched over to dry feed. Such fresh feeds can save you money, but they should be used with extreme caution, as they can easily create serious pollution at normal catfish growing temperatures.

One does not normally think of channel catfish as insect feeders, but some catfish farmers have reportedly realized savings by installing U-V light insect traps over ponds containing fry or fingerlings.

Buffalofish

Buffalofish in experimental culture, particularly spawners and young fish, have successfully been fed sinking dry feeds. The description of one such feed, developed at the U.S. Fish Farming Experimental Station, is given in Table III–7–5. It is a good example of an effective dry feed which does not incorporate expensive materials of animal origin.

Table III–7–5. Feed formula for young buffalofish, developed at U.S. Fish Farming Experimental Station, Stuttgart, Arkansas.

Rice bran (not hulls)	31.6%
Cottonseed meal	31.6%
Dehulled soybean meal (fine ground)	15.8%
Poultry by-product meal	10.5%
Wheat shorts (fine grade)	2.6%
Ground yellow cornmeal	2.6%
Dry whey or other dried milk product	5.3%
Pure vitamin B-12	25 mg/ton
Pure vitamin A	22,000 units/kg (finished feed)
Pure vitamin D	9,900 units/kg (finished feed)
TM–10	1.8–9.0 kg/ton (finished feed)[1]
Salt	18.0 kg/ton
Di-calcium phosphate	18.0 kg/ton

[1] Sometimes fed only when disease outbreak is feared.

In practice, however, buffalofish are nearly always expected to obtain their food through fertilization or, when stocked with other fishes, by cleaning up uneaten small fragments of dry feed. Undoubtedly you could provide buffalofish with acceptable cultured live foods, but, given the respectable level of production you can achieve with fertilization alone, and the current market price of buffalofish, I would not encourage investment in cultured or prepared feeds mainly for these fish.

Carp

Perhaps no fish has been cultured so intensively as the common carp. In Japan and Germany carp have been raised in tanks so densely stocked that they cannot turn around for bumping into each other. Such rearing requires the circulation of great volumes of filtered water and dictates that you provide one hundred percent of the diet in the form of high-protein fresh or dry feed. Apart from my own objections to this approach to carp culture, it is of little interest in North America. Here low prices and the poor acceptance of carp as food dictate that they be reared at low cost for home use or for small, specialized markets.

In North American aquaculture the common carp's primary role is and probably will continue to be as a minor component in polyculture systems. (See Chapter IV–1.) Under such circumstances there will be enough naturally occurring food, fertilizer (added deliberately or as a by-product of high density fish stocking), and waste feed to take care of the carp's needs. But there are possibilities in intensive carp culture for those who would like to learn to enjoy carp as a table fish, or who have identified a suitable market. The following comments on carp feeding are primarily for such readers.

The popular game and pan fishes (bass, trout, bluegill, etc.) are generally better suited than carp to harvesting from unfed and unfertilized systems (except in naturally very fertile waters), so the carp farmer will be interested primarily in the second and third levels on our list of feeding intensities.

Waste material as food   Among temperate climate fish the common carp is probably best adapted to realizing substantial weight gain on the basis of fertilization alone. In Indonesia, carp are reared at high densities in cages placed in rivers which amount to open sewers. In Germany and Poland the same concept is applied in more sanitary fashion by growing carp in secondary sewage treatment ponds, realizing 400–1,000 pounds per acre (441–1,103 kg/ha), while benefiting the sewage treatment process as well. Fish culture in sewage oxidation ponds is a matter for consideration by municipalities, not small farmers and homesteaders, and few of us have (or want!) access to sewage streams. I mention these practices only to suggest the feasibility of organic fertilization in carp culture. A particularly attractive approach is the combination of waterfowl and carp. (See Chapter IV–2 for details on waterfowl in fish farming.)

When looking to further supplement carp nutrition, you should first look around to see what sort of "waste" materials are available for use as combination fertilizer/feeds. In Czechoslovakia, effluents from dairies, sugar mills, slaughterhouses and starch mills have all been used for this purpose.

No significant research on carp nutrition has been carried out in North America. But there is a body of information from Europe, Israel and Japan, which at least provides the North American culturist with a place to begin. In the last fifteen years, European carp growers have followed trout farmers in using the convenient prepared dry feeds—but with a different approach to nutrition. Rather than trying to develop a "complete" ration from costly protein and vitamin sources, they have tried to increase production by feeding an abundance of inexpensive carbohydrate and grain protein. The carp's protein and vitamin needs are rounded out by the "natural" foods in the highly fertile ponds. This is the same strategy long employed in traditional European carp farming, where grains and other starchy materials are fed in whole form.

Adaptability in feeding   Though carp are poor at capturing fish and other fast-moving animals, there are few fishes more adaptable or resourceful in their feeding. There appear to be no grain or seed materials which carp will not eat. At New Alchemy, we have even successfully fed unshelled sunflower seeds to mirror carp. The fish would rise to the surface, seize a few seeds and disappear into the depths of the pond. A few seconds later, up would come the empty shells, one by one. Common carp may be less efficient in digesting green plant material than grass carp and some of the tilapias, but the mirror variety in particular will eat most soft leaved plants. Carp are also capable of filter feeding on plankton. Other feeds which have been used by carp culturists include potatoes, chestnuts, silkworm pupae and terrestrial insects.

Table III–7–6 lists the carp's conversion ratios of some food materials used in Europe and Japan. Table III–7–7 offers examples of some diets used in practical and experimental carp culture.

Table III–7–6.   Reported conversion ratios of certain foods by common carp.

Food	Conversion ratio (based on wet weight)
Barley	2.6–3.7
Corn	4.8
Lupine	2.6–3.7

Table III–7–6—*Continued*

Oats	2.6
Peas	2.7–2.8
Potatoes	33.9
Soy beans	5.0
Wheat	3.2–5.7
Silkworm pupae (dried)	1.3–2.1
Silkworm pupae (fresh)	5.0

Table III–7–7.   Diets used in culture of common carp.

Country	Ingredients—Percent	Nutritional Composition—Percent	Results and Comments
Germany	Fish meal ....15.5 Oats ....10.0 Corn............39.5 Rape ....14.2 Wheat 20.8	Crude protein ....25 Fat ....3 Fiber 6	Gave fair results with carp grown in aquaria, when combined with a vitamin mix containing Vitamin A—two million I.U., Vitamin D_3—300,000 I.U., Vitamin C_3—20 g, riboflavin—4 g, pyridoxine—5 g, pantothenic acid—2 g, and protein chloride—10 g/ton.
Israel	Ground wheat, ground milcorn, ground corn, soybean oil meal, fish meal, wheat gluten, beet molasses, bentonite clay and gypsum (See Hepher & Chervinsky, 1965.)	Crude protein— 26–30	—
U.S.S.R.	Oil cakes...... 47–55 Vegetable products..... 10–20 Grains......... 9–25 Wheat bran.... 10–20 Animal meals....3–10 Oyster shells....... 1	Protein.......23–28 Fat............ 4–5 Carbohydrate. 27–33	Conversion ratio was 3.5. It is customary to add 10 g of cobalt chloride and 1 g of chloramphenicol per ton of feed to combat infectious dropsy.
U.S.S.R.	Sunflower seed cake....... 35 Soybean cake......25 Shorts...........25 Hydrolyzed yeast... 5	—	These three feeds were designed for use in floating cages where carp have less access to natural foods than in pond culture. Thus, one would expect that artificial diets would need to be higher in protein and vitamins. The first diet gave "fair" growth. The second gave growth 14% above "normal". Addition of terramycin at 5,000–10,000 I.U. per kg of food resulted in 10–16% increases in growth with any of these diets.

U.S.S.R.	Grain waste.... 30–40
	Sunflower
	  seed meal.... 15–20
	Peanut meal.... 15–20
	Flax seed meal.....10
	Silk worm pupae...15
	Fodder yeast....... 5
	Paste prepared from
	  aquatic plants... 20
		—	

U.S.S.R.	Broken grains..... 80 Duckweed........ 20	—	

Use of higher quality feeds may be feasible to speed up the growth of young carp. The standard technique for feeding young carp on European and Asian fish farms is to fertilize the rearing pond prior to stocking. Fry ponds may also be drained and planted to cover crops. Flooding just before stocking results in an abundance of zooplankton. In Japan daphnia are cultured in the pond before the fish are introduced. In India, when nineteen carp fry diets were compared, it was found that a mixture of backswimmers (an aquatic insect), fresh water shrimp and cowpeas in a ratio of 5:3:2 produced the best growth. Almost any of the small and medium sized cultured foods listed in Table III–5–1 (page 128) could be fed to young carp. If these were properly combined with North American grains no doubt a similarly successful diet could be produced.

Water temperature and feeding

With carp as with other fishes, the rate of feeding must be adjusted according to temperature, but recommendations vary considerably. This may be because carp were long ago introduced widely outside their native Russian waters, and by now there are strains genetically adapted to both hot and cold climates. The lowest temperature at which common carp are known to feed is 39.2°F (4°C). Experienced carp raisers therefore recommend that you supply a small weekly maintenance feeding whenever the water temperature is 36°F (3.5°C) or higher. You can begin light daily feedings at around 42.8°F (6°C). The minimum temperature reported for growth in temperate climates varies from 55.4°F to 68.0°F (13 to 20°C), while optimum temperatures reported for feed utilization range from 73.4°F to 86.0° (23 to 30°C).

It appears that the North American carp grower will have to "play it by ear," and adjust the daily rate of feeding according to how much the carp will eat. Theoretically, every 18°F (10°C) increase in water temperature above the minimum for growth will double or triple food consumption. Carp are often fed more heavily than other fishes, and daily rations above 7% of the fishes' weight may be biologically and ecologically justified.

Duration of feeding

In determining feeding rates of carp, keep in mind that they feed more slowly than fishes commonly cultured in North America, and our rule of thumb to feed only as much as the fish will consume in a half hour may not apply. You should allow at least an hour, and up to a full day for carp to clean up their food. It may be difficult to find or manufacture dry feeds which are stable enough in water to optimize their nutritional value to such slow eaters. To minimize the leaching of nutrients, wheat gluten is added to pelleted carp feeds in proportions as high as 50 percent. Another way to approach the problem is to place carp together with efficient filter feeders such as buffalofish which would take full advantage of the "wasted" carp feed.

A point to remember in feeding carp is that they do not like to feed over a hard bottom. Feeding carp on sand, gravel or submerged platforms may lower feed consumption and/or injure the carps' soft, delicate mouth parts.

Carp flavor and texture

An interesting aspect of carp feeding is how the diet affects the flavor and texture of the fish. The quality of any fish is affected by its diet, but this may be particularly true for carp. Certainly, opinions as to the gustatory value of carp vary more than for other fish. Wild or cultured carp from polluted waters often have a "muddy" flavor, while carp reared in sewage waters in Europe are said to frequently have a disagreeable phenol taste. Farm-grown carp's flavor is generally rated superior to, or at least more consistently good than that of wild or sewage-grown carp. This may be partly due to feeding with grains; but the control of pollution on commercial carp farms and/or the superior genetic stock used in European carp culture may also be factors.

As for texture, carp consuming primarily natural foods are apt to be more lean, while carp fed on potatoes or bread are said to have wet and fatty flesh. Barley, corn, rice, rye, soybeans and wheat are considered good foods for achieving an agreeable texture between these extremes.

You can correct poor texture or objectionable flavors in carp by maintaining the fish for a while in clean water on selected feeds. You may need as little as a day, or more than ten days to "finish" carp in this manner. Some maintain that corn is a particularly desirable food for this purpose, while others disagree, saying that heavy feeding with corn will substitute a corn flavor for the true flavor of carp.

The conventional feeding methods for trout, as for channel catfish, are already well described in print. (See Appendix IV–1.) However, with trout there is even more room for alternatives in feeding than in catfish culture. As long as the water is suitable, trout are amenable to all sorts of culture, from the most casual to the most intensive, and a variety of feeding methods can be used.

Trout

The first two levels of feeding intensity will not concern us here. But we are interested in feeds less expensive than commercial dry feed for the subsistence grower wanting to supplement his fishes' natural diet, and also alternate feeds to reduce feeding costs for the small commercial grower and others using prepared dry feeds. Alternate feeds for trout must be high in protein, since protein is almost always the major limiting factor in trout growth.

The use of offal from slaughterhouses, commercial fisheries, or the trout farm itself is a method of trout feeding which predates the use of dry feeds in commercial fish culture. Not everyone is located near a major source of such wastes, and up through the late 1950's in the United States, growers generally purchased meat scraps and liver to feed trout. Few can afford to purchase large quantities of these feeds any more, and large-scale trout growers have had to turn almost entirely to prepared dry feeds. Far too many small trout growers have unquestioningly followed their example, when they might be able to save money by using meat and fish wastes in quantities insignificant to the big operator. As an indication of the potential of such feeds, consider Denmark, until recently the world's leading trout producing country, where the entire industry is based on scrap fish and other by-products of the Danish fishing fleet. (Methods of processing and feeding meat and fish for fish feed are described in Chapter III–4.)

Low cost food sources

Flying insects are an excellent and cheap source of protein for all sizes of trout during the warm months, and can be obtained by using U-V "bug lights." (See Chapter III–3.) Since most trout culture is done in flowing water, and since trout (with the exception of the brown trout) may be reluctant to feed at night, you should install some sort of floating ring or boom under or downstream from the light to hold the insects. An alternative is to install powerful incandescent lights which will induce the trout to feed at night.

Almost all of the cultured foods discussed in Chapter III–5 could be fed to trout, but given their natural feeding behavior, it is probably best to use those which float, swim, or can survive indefinitely on the bottom. It is best not to try to feed small fish to trout. Trout are normally harvested before they reach a size at which they are efficient predators on other fish, and introducing live "forage" fish could establish a competitor species.

If you plan to culture aquatic invertebrates as trout food, grow them separately in small pools which do not require the high flow rates and low temperatures needed for trout. In fact, such conditions often result in greater production. But portions of the trout culture facility itself may also be used for food production. The most fertile water on any trout farm is the tailwaters. Many trout farmers find that they have to pass their effluent through gravel and sand-bottomed, unstocked and unfed tailraces in order to comply with pollution laws. You could perhaps make use of this space by stocking it with amphipods and harvesting them periodically to feed the trout.

You can use the water upstream of the growing area in a similar way by installing rotting logs as a habitat for insect larvae and crustaceans, some of which will wash down into the trout ponds. Some natural food will even be produced in the culture ponds if you can construct them with a dirt or gravel bottom. Unfortunately, in the United States, many trout farmers find it necessary to use concrete ponds and raceways, owing to the prevalence of trout disease, which cannot be successfully treated in earthen ponds.

Apart from increased growth and economic savings, there is another reason to provide trout with some live food. Cultured trout fed on live food, particularly crustaceans, tend to be more colorful and have the orange, pink or red flesh characteristic of "wild" trout, rather than the white flesh of typical cultured trout.

Yellow perch

Yellow perch are piscivorous, but also consume insects, mollusks, zooplankton, etc. So far, experiments in perch culture have involved commercial feeds like those fed to trout and channel catfish.

Herbivorous fishes

While it seems reasonable to grow plants specifically for such herbivorous fishes as grass carp and tilapia, you should first look around to see if there are any weeds, harvestable cover crops, or waste plant materials on the site. If you do plant fish food plants, remember that conversion ratios may be as high as 50 due to the extremely high water content of green plants. And be sure to test a sample on the fish before proceeding. A good way to provide both fish food and pond fertilizer is to plant a legume on the bottom of a dry pond and, instead of plowing it in, flood the growing plants.

Perhaps there are no totally herbivorous fish. But certainly a few species are far more herbivorous than the rest. None of the medium-to-large size fishes native to North America make much use of green plants, but the grass carp and some of the tilapias are voracious consumers of higher plants and/or phytoplankton.

Protein needs

Most of the tilapias are what I would call opportunistic feeders, and some think the same of the grass carp. The pattern of young fishes needing more animal protein holds for these fish. Grass carp under 9.8 inches (25 cm) long consume considerable amounts of zooplankton, and blue tilapia naturalized in Florida lakes eat aquatic animals in large amounts only when less than 6 inches (15.4 cm) long. However, in captivity blue tilapia and perhaps other tilapias will continue to consume considerable amounts of animal matter at all ages. At New Alchemy, in addition to algae, leafy plants and roasted soy meal, we have successfully fed tilapia with earthworms, amphipods, midge larvae, daphnia, mosquito larvae, mixed flying insects and commercial dry trout feed. We found that the addition of very small amounts of midge larvae to a diet of algae and grain significantly improved the growth rate of young tilapia. The effect on adults, which appear to be physiologically more completely adapted to a vegetable diet, is unknown. With high population densities, blue tilapia will consume any invertebrate of suitable size. (See Chapter III–5 for a discussion of providing small fish for adult tilapia.)

Since my experience with tilapia is interwoven with investigation of small, closed aquaculture systems at The New Alchemy Institute, the reader interested in tilapia will probably want to have a look at Chapter IV–4, on closed systems. You will find further discussion of herbivorous fishes and their feeding in Chapter II–8.

You will find feeding methods for frogs, crayfish, fresh water shrimp and clams in Chapters II–9, 10, 11 and 12, which are devoted to culture of these animals.

PART IV

Aquaculture Systems

CHAPTER IV-1

Toward a North American Polyculture

Natural compared to artificial systems

So far I have discussed only two essential aspects of aquaculture—the cultured animals themselves and the techniques used to feed them. Before proceeding with such important matters as harvesting, breeding, disease control, etc., it will pay us to look at some aquacultural "systems"—the ways in which cultured animals, their foods and their physical environment are put together.

Compared to natural systems, aquacultural (and agricultural) systems are greatly simplified. Almost all natural environments contain a great variety of plant and animal species, only a few of which are nutritionally or economically useful to humans. We manipulate nature by eliminating or controlling predators, parasites and pathogens, and by trying to divert as great a proportion of the available nutrients as possible into those few forms which we wish to harvest.

This simplification carries risks. Natural systems include a complex array of checks and balances, so that a species is rarely eliminated. Artificial food-producing systems which are simplified may also be less stable; they often create situations in which it is easier for a pathogen to infect an entire population, or a "weed" species to run rampant. It is well known that, as our agriculture has moved more and more toward monoculture, pest problems have become more severe, and control measures more drastic. What is true for agriculture holds for aquaculture. A channel catfish or rainbow trout monoculture is much more likely to be wiped out by disease than if the same species were in a natural environment. Yet some simplification is unavoidable if human needs are to be served. Clearly a well-tended garden will produce more human food than a natural forest. So too, an aquaculture pond will produce more *useful* biomass than its more complex natural counterpart.

It does not follow, however, that the more simplification, the greater the useful productivity. In order to maximize aquacultural production, and to keep the ecosystem stable, it is generally best to aim for a biological complexity somewhere between that of natural systems and that common to North American commercial aquaculture.

The simplest aquaculture systems are the monocultures so commonly found in "aqua-business." To set up a monoculture you need do little more than choose an animal, learn its physical and nutritional needs and proceed to provide them. You might, for example, create a common carp monoculture simply on the basis of reading the appropriate pages in Parts I and II of this book. Conventional monocultures, then, need no further discussion in this section.

What I do want to discuss are some of the ways in which more complex, artificial aquatic food-producing systems can be put together, incorporating a variety of animals and plants and/or integrating the aquatic environment with aspects of the adjacent terrestrial one. In Chapters IV–3 and 4 I will discuss fish culture methods where fish are grown in structures more or less shut off from the "outside world," making the culturist more directly responsible for the maintenance of the total system than in open pond culture.

In earlier chapters I have introduced the concept of polyculture, explained its basic principles and contrasted it to the monocultures typical of North America. While advocating a polyculture approach, I must acknowledge that many readers will have good reason to practice monoculture, at least in the beginning. Some may wish to skip this chapter and read only those portions of the book concerned directly with their chosen fish. But you do so at the risk of missing information which might enhance the production of your favorite, as well as give you another crop species. (Trout culturists are excused, as polyculture has little to offer them.)

Prevalence of monoculture

Readers who are mainly interested in maximizing their crop of edible fish, and who are not set on a particular species, will probably want to read on. I must confess, though, that compared to most of the rest of the book this chapter is long on theory and short on "nuts and bolts," for reasons which will become clear.

I have already mentioned that the traditional American "farm pond," with its two-species fish community, represents a primitive form of polyculture. Clearly, our largemouth bass (*Micropterus salmoides*)—bluegill (*Lepomis macrochirus*) polyculture, which yields at best 300 pounds per acre (331 kg/ha) of fish per year, is no competition for the Chinese, who have produced fish at twenty times that rate for thousands of years. In fact, normal bass-bluegill polyculture is much less productive than conventional North American fish monocultures, for instance channel catfish (*Ictalurus punctatus*) culture.

This raises two questions:
• Why are North American fish monocultures more productive than our one established polyculture?
• If polycultures really are more productive, why haven't we developed them?

Before I can justify the title of this chapter, I shall have to answer these two questions, and the answers are bound up with what I call the "dollars per acre mentality."

Traditionally, agricultural and aquacultural yields are measured in pounds per acre, or some similar unit. But in commercial farming, what finally determines the choice of crops and methods of cultivation is dollars per acre or, to be more precise, profit per acre. At present, the use of this criterion favors aquatic monoculture for the following reasons:

• In America, aquaculture is a new business; from a propaganda point of view it is more "efficient" to promote one or two species of fish to consumers who may not be accustomed to them, than to familiarize them with a variety of fishes.

• Agribusiness-scale aquaculture is based on mass harvests and bulk sales, which often requires draining ponds or capturing all the fish. Where there are two or more species or sizes of fish, time (= dollars) is lost in sorting. Even if all the species can be harvested at the same time, they may need to be processed, packed and/or marketed differently. Under such circumstances, it is easy to see how the farmer with the most pounds of fish to harvest might not be the farmer with the largest profit. Of course, there is ample precedent for this sort of situation in our commercial agriculture.

In North America the great majority of the research, agricultural extension work, propaganda, and development in commercial aquaculture has focussed on the large-scale monoculture of a very few species. Small-scale commercial aquaculture and aquaculture for home consumption are grossly underdeveloped. Where they exist they have usually followed the lead of the commercial fish farming industry, largely for lack of visible alternatives.

So, to answer the questions introduced above:

• Monocultures of catfish or trout are economically more productive than farm pond polyculture, because they are intensive commercial operations which are based on high quality feeds, considerable expertise, and backing by well organized research and extension programs. At present there are no comparable polycultures; farm pond polyculture is commercially unimportant, usually totally unmanaged, and little incentive is offered to alter the situation.

• Though polycultures are often more productive, in the absence of publicized economic incentives they have not been developed.

• The lack of developed polycultures also reflects the general neglect, by government agencies, universities, public and private funding agencies and the media, of subsistence and small-scale commercial growers of warm water fishes — those who could benefit most from development of North American polyculture systems. systems.

Benefits of polyculture

I hope that this book will shed light on the incentive for small farmers to begin with the slow development of sophisticated polycultures based on North American fishes and adapted to North American conditions.

Recently, the major fish producers in the South have even shown an interest in polyculture. To explain why this is so, and, more important, to strengthen the case for polyculture as a tactic for the small fish farmer, I shall have to mention a further advantage of polyculture.

Synergistic effect in polyculture

Clearly, if you have a pond in which you are growing only one species of fish, and if in that pond there is unused space or uneaten food, then by putting in another species of fish which will use that space or food, you may get an additional crop. Less obviously the introduction of additional species of fish may have a synergistic, rather than an additive effect. That is, if you are raising a certain number of pounds of Species A in monoculture and add species B, you may get, not just the same amount of A plus some pounds of B, but *an increased amount of A*, plus some pounds of B.

You cannot automatically assume this will happen any time you add a species. Many species are competitive, and combining them may lead to elimination of one species or even reduced total production of fish. In the following pages I will describe the combinations that are likely to be positively synergistic.

It might be easier to explain synergistic polyculture with examples from China, where stocking combinations and ratios have been established for centuries. But this information, which is readily available elsewhere, would not be directly applicable in North America. (See Appendix IV–1.) So I will begin by offering some North American examples.

Experimental ponds at Auburn University, Alabama, were stocked with channel catfish (*Ictalurus punctatus*) at 2,000 fish per acre (4,940/ha) and fed with commercial dry feed in the usual manner. In test ponds, an additional 568 tilapia (*Tilapia mossambica*) per acre (1,403/ha) were stocked, while the control ponds held catfish only. The rate of feeding was the same in both test and control ponds. The control ponds produced an average of 1,260 pounds per acre (1,390 kg/ha) of catfish, while the ponds with tilapia produced 1,411 pounds per acre (1,556 kg/ha) of catfish, plus 240 pounds per acre (265 kg/ha) of tilapia, for a total fish yield of 1,651 pounds per acre (1,821 kg/ha). The feed conversion ratio for catfish production alone was 1.7 in both test and control ponds. If the tilapia are included in computing the conversion ratio, it comes out to 1.45 in the test ponds. In other words, adding tilapia to a conventional catfish monoculture led to a 12 percent increase in catfish production and a 31 percent increase in total fish production, with equally good or better feed conversion.

Further experiments showed that tilapia could be stocked at numbers up to one third of the catfish population with the same kind of results, and that *Tilapia nilotica* worked as well as *T. mossambica*. Experiments elsewhere with *Tilapia aurea* showed similar results. These findings were put into practice by catfish farmers in Louisiana with good results, at least until tilapia stocking was made illegal in that state.

More recently a similar experiment was carried out with catfish and buffalofish at Rockefeller Wildlife Refuge, Grand Chenier, Louisiana. Six 0.1 acre (0.045 ha) ponds were each stocked with 200 eight-inch (20.5 cm) channel catfish; three of the six also received 10 six-inch (15.2 cm) bigmouth buffalo (*Ictiobus cyprinellus*). The experiment had to be terminated after 143 days of feeding, due to a severe storm. At that time the catfish in the catfish-only ponds averaged 1,657 pounds per acre (1,708 kg/ha), with 89 percent of the fish considered to be marketable size (¾ lb or 0.35 kg or more). The catfish-buffalo ponds had an average of 1,902 pounds per acre (1,961 kg/ha) of catfish, plus 148 pounds per acre (163 kg/ha) of buffalo; 96 percent of the catfish were marketable.

In another experiment at Auburn, monocultures of channel catfish and blacktail redhorse (*Moxostoma poecilurum*), a benthic feeder, were compared with a polyculture of the two species. Results of the experiments in fertilized ponds which were fed in proportion to the estimated weight of catfish only are shown in Table IV–1–1.

Table IV-1-1.   Monoculture and polyculture
production of channel catfish and blacktail redhorse.

		Catfish Monoculture	Blacktail Redhorse Monoculture	Polyculture
Stocking Rate:	Catfish	2,000/acre (4,940/ha)	—	2,000/acre (4,940/ha)
	Redhorse	—	1,000/acre (2,470/ha)	1,000/acre (2,470/ha)
	TOTAL	2,000/acre (4,940/ha)	1,000/acre (2,470/ha)	3,000/acre (7,410/ha)
% Survival:	Catfish	87	—	97
	Redhorse	—	95	75
Production:	Catfish	982 lb/acre (1,083 kg/ha)	—	1,563 lb/acre (1,724kg/ha)
	Redhorse	—	159 lb/acre (175 kg/ha)	219 lb/acre (242 kg/ha)
	TOTAL	982 lb/acre (1,083 kg/ha)	159 lb/acre (175 kg/ha)	1,782 lb/acre (1,966 kg/ha)
Mean Weight:	Catfish	0.56 lb (0.25 kg)	—	0.80 lb (0.36 kg)
	Redhorse	—	0.17 lb (0.08 kg)	0.29 lb (0.13 kg)
Conversion Ratio:		2.3	(unfed)	2.3 (catfish only) 2.0 (both species)

How to explain such polycultures? It is easy to imagine what the tilapia, buffalo or redhorse might eat. Particles of dry feed too small for or missed by the catfish, catfish feces containing partially digested food, and plankton or benthos produced largely via fertilization by catfish feed or metabolites (wastes), are probably the main food items. Since the supply of all these types of feed is due to the presence of catfish, it is easy to explain the greater production of blacktail redhorse in poly-culture ponds with catfish than in fertilized monoculture ponds.

But why the increases in catfish production? Tilapia and buffalofish definitely will eat dry catfish feed, and so, presumably, will blacktail redhorse. You might assume that this competition, however slight, would tend to decrease catfish production, even while total fish production was increased. Two possible explana-tions have been advanced.

The most frequent explanation for this type of effect is that the filter or benthos feeding fish act as "garbage collectors," improving water quality and therefore the health and growth of other fishes. At Auburn the highest acceptable level of dry feed in catfish culture in still water is considered to be 35 pounds per acre (38.6 kg/ha); sustained feeding at higher rates causes oxygen depletion. In the blacktail redhorse polyculture trials, catfish were fed at 30 to 83 pounds per acre (33.1–91.5 kg/ha) for a period of 235 days with no unusual mortality and no apparent reduction in dissolved oxygen concentration.

Another possible explanation for the increased catfish production arises from research in Poland. There, when common carp ponds are stocked with young grass carp (*Ctenopharyngodon idellus*), big head carp (*Aristichthys nobilis*) or especially silver carp (*Hypophthalmichthys molitrix*), the zooplankton's composition alters to contain a higher percentage of the larger animals. This in turn may lead to increased growth of common carp (*Cyprinus carpio*), since they can utilize the large plankton better than the smaller zooplankton, which are used, (along with phytoplankton) by the other carps.

Cropping of zooplankton

The fresh water clams, also plankton feeders, may have more potential for enhancing production of non-filter feeding fishes. Swingle carried out the only experiments I know of using such clams in polyculture. He stocked a 1.9 acre (0.77 ha) pond with a few of the clam *Lampsilis claibornensis*, along with 2,000 bluegills, 1,000 redear sunfish and 200 largemouth bass. The clams multiplied rapidly and were harvested annually along with the fish, yielding an average of 560 pounds (255 kg) of shelled clams (295 lb/acre or 325 kg/ha). The sixth year after stocking, the pond was drained; 42,968 clams were recovered, amounting to 704 pounds (320 kg) of shelled clams (360 lb/acre or 397 kg/ha). Bear in mind that this was in a pond containing redear sunfish, which feed heavily on small mollusks. Where predation was not serious I have seen fresh water clam populations that literally paved the bottom of the pond.

Fish and clams

The standing crop of fish in Swingle's pond was 817 pounds or 371 kilos (418 lb/acre or 461 kg/ha), to which can be added 256 pounds or 116 kilos (125 lb/acre or 138 kg/ha) of fish harvested during the year. In a control pond, stocked at the same time with the same numbers of fish and harvested in the same way, the standing crop of fish was only 557 pounds or 253 kilos (285 lb/acre or 314 kg/ha). Swingle attributed the difference to the clams' filtering action. The productivity of such fish-clam systems might be improved still further by grinding up excess small clams and feeding them to the fish.

Polyculture affecting the quality, rather than quantity, of fish flesh produced was seen in 1976 at New Alchemy. We had stocked a variety of indoor and outdoor pools, primarily with blue tilapia, but also with small numbers of mirror carp, brown bullheads (*Ictalurus nebulosus*) and/or frogs. One pool was operated for the first time as a pure monoculture. The weight of fish produced in that pool was satisfactory but, while our other tilapia were as tasty as always, the monocultured fish had an offensive "muddy" flavor. The harvest amounted to 75 pounds (34 kg) of cat food. When the "cat food pool" was cleaned, we found some large patches of blue-green algae in the mud at the bottom. Blue-green algae are often found together with "off flavor" in fish and anaerobic (oxygenless) conditions. Since the various pools were otherwise very similar, we concluded that the natural behavior of the carp, bullheads and tadpoles kept the bottom mud "turned over" preventing the anaerobic conditons which lead to the development of blue-green algae.

Polyculture and fish quality

Piscivorous fish have a place in some polycultures, especially if one of the cultured species reproduces naturally in the systems. (Sunfishes, tilapia and bullheads are the worst offenders.) It may be hard to get used to the idea that when bass eat bluegills it is good for the bluegill population, but it is, at least from the point of view of the pond owner wanting pan-sized bluegills, not runts.

Population control

You can also stock predators where small fish compete for food with larger, more desirable food species. This sort of stocking means taking great care in choosing the

Weed Control

"Teacher" fish

Different age groups of a species

Poor combinations

size of the predators and/or stocking them late enough in the year so that they will not prey on the other cultured species.

Another type of polyculture, often not thought of as such, occurs when herbivores are stocked in a pond overgrown with weeds. The object is to make the pond more suitable for growing and harvesting other fishes, but at the same time a crop of the herbivore is produced. (See Chapter II–8.)

A peculiar phenomenon is the use of "teacher" species. This is seen most often when bluegills "teach" young largemouth bass to take dry feed, but there are other examples. Some catfish farmers claim better production when they stock 10 percent blue catfish (*Ictalurus furcatus*) with their channel catfish. This may have to do with the blue cats' greater willingness to accept floating feed, which perhaps helps to overcome the more nervous channel cats' initial resistance. I have seen young mirror carp learn to feed on floating insects after a few days in a pool with young tilapia.

Polyculture is usually thought of as a multi-species proposition, but you can also combine different age classes of the same species. This is often done with carp in Europe. Yearlings, which make good use of zooplankton, are stocked with older fish which are primarily benthic feeders. At New Alchemy we feed young tilapia on insects and grain, while older, larger fish in the same pool dine primarily on phytoplankton.

I have not said much about the more obvious sorts of polyculture, involving species with very different feeding habits. I should mention, however, that the species involved in such systems can be quite similar in appearance. For example, bluegills and redear sunfish (*Lepomis microlophus*), while closely related and superficially similar, are sometimes stocked together. This works well as the bluegill likes to feed on the surface, while the redear rarely does, resulting in better feed conversion and cleaner water, particularly when commercial feeds are used over deep water.

This discussion would be incomplete without mentioning a few species combinations which may be harmful. Most obvious are those where direct competition occurs between fishes of unequal value. If, for example, you were growing largemouth bass, you would probably not stock another large-mouthed predator like green sunfish (*Lepomis cyanellus*). The result might or might not be lower total fish production, but the value of the combined crop would certainly be less than that of a crop of bass alone.

Whether polyculture combinations are beneficial or not may be related to the numbers or sizes of fish stocked, environmental factors, and the species ratio. For example, I have described experiments at Auburn University in which blue tilapia stocked together with channel catfish seemed to cause an increase in catfish production. But in another study at the same site, blue tilapia stocked with channel catfish at lower densities appeared detrimental to catfish production.

Some promising polyculture combinations involving predatory fishes like channel catfish fail only because the predator eats the other species. You may be able to get around this problem by putting one or the other animal in cages. (See Chapter IV–3 for details on cage culture.) I discussed one example of this method in Chapter II–10.

Benthic fish like the common carp and bullheads which root vigorously in the bottom may be able to use nutrients which would not otherwise enter the food chain. Their activities may assist in the metabolism of ammonia nitrogen, make food available to other fishes, and discourage the growth of anaerobic plants which can cause "off flavor" in fish. But they also present special problems. Such fish can roil the water to the extent that visual feeders are handicapped, and/or phytoplankton blooms are retarded or eliminated. They often eat the eggs of other fishes. Neither of these types of behavior is always detrimental, but they can be. Culturing benthic fishes usually means that you have to pay particular attention to the rate of stocking. A fish which is an asset in small numbers can sometimes be the ruination of a pond if it is overstocked.

As promised, there has been a minimum of "nuts and bolts" in this chapter. If you are seeking to establish a polyculture system you will have little more than hints to go on. The first step is to identify a species of fish which is of primary interest to you, and set up a successful monoculture of that species (or possibly a bass-bluegill polyculture, since that combination is readily workable and quite well understood). Then consider what other fishes are available and capable of living in your water. Consider their habitat and feeding preferences. If one or both of these are different from your primary species, you may be able to plug them in and obtain an additive or, still better, a synergistic increment in fish production. Chapters II–1 through 13 on individual types of fish and Appendix II should help you in making choices. As a further aid, Figure IV–1–1 shows the major types of positive interaction known to occur between fishes with different feeding habits. (The figure says nothing about rates of stocking, since they depend on the natural fertility of your water and the feasibility of providing various types of food.)

The composition of Chinese polyculture systems was discussed briefly in Chapter I–2. The "ideal" North American system, while not identical, would incorporate animals in the same general niches. These niches are listed below, along with some comments and suggested animals. The animals mentioned are drawn exclusively from those discussed in this book, but you need not be limited exclusively to these.

Filter feeders represent the most productive niches in most Chinese ponds. Many fish culturists make a distinction between phytoplankton and zooplankton consumers, but some biologists believe that this distinction does not stand up well in nature. The phytoplankton and zooplankton feeders in classical Chinese polyculture are, respectively, the silver carp and the bighead carp. These fishes are not currently widely available in North America, though they have been used in some experiments. They are illegal in several states, but it may prove difficult to find their counterparts. It happens that of all the continents, evolution has least favored ours with large filter feeding fishes. Almost all of our native filter feeders are too small to interest most people as food fish. The most notable exception is the bigmouth buffalo, which should definitely be considered for ponds within its range. Probably even more efficient use of phytoplankton will be made by fresh water clams. In those few areas where they are already established, I would also recommend tilapia.

Apart from buffalofish, the best bets for cropping zooplankton seem to be the young of fishes which are benthic feeders or predators as adults. Carp in their first year of life, bluegills weighing less than about 3 grams each and small yellow perch

FIGURE IV-1-1. *Major known types of positive interaction between different kinds of animals in polyculture systems, with examples.*

(Arrows indicate which types are benefitted by the interaction, e.g., Example G indicates that "learner" fish benefit from "teachers," but the reverse has not been shown.)

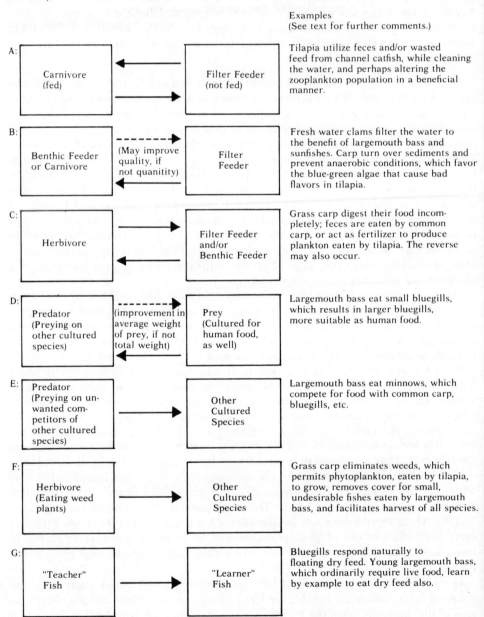

Examples
(See text for further comments.)

A: Carnivore (fed) → Filter Feeder (not fed)

Tilapia utilize feces and/or wasted feed from channel catfish, while cleaning the water, and perhaps altering the zooplankton population in a beneficial manner.

B: Benthic Feeder or Carnivore → (May improve quality, if not quanitity) → Filter Feeder

Fresh water clams filter the water to the benefit of largemouth bass and sunfishes. Carp turn over sediments and prevent anaerobic conditions, which favor the blue-green algae that cause bad flavors in tilapia.

C: Herbivore ⇄ Filter Feeder and/or Benthic Feeder

Grass carp digest their food incompletely; feces are eaten by common carp, or act as fertilizer to produce plankton eaten by tilapia. The reverse may also occur.

D: Predator (Preying on other cultured species) → (improvement in average weight of prey, if not total weight) → Prey (Cultured for human food, as well)

Largemouth bass eat small bluegills, which results in larger bluegills, more suitable as human food.

E: Predator (Preying on unwanted competitors of other cultured species) → Other Cultured Species

Largemouth bass eat minnows, which compete for food with common carp, bluegills, etc.

F: Herbivore (Eating weed plants) → Other Cultured Species

Grass carp eliminates weeds, which permits phytoplankton, eaten by tilapia, to grow, removes cover for small, undesirable fishes eaten by largemouth bass, and facilitates harvest of all species.

G: "Teacher" Fish → "Learner" Fish

Bluegills respond naturally to floating dry feed. Young largemouth bass, which ordinarily require live food, learn by example to eat dry feed also.

(*Perca flavescens*) certainly fall into this group. Perhaps those culturists able to harvest and willing to eat fish much smaller than those usually preferred by North Americans will open the door to more productive uses of zooplankton in our food fish culture.

Benthic feeders are stocked in proportion to availability of nutrients, but you must also consider the texture of the bottom soil. The harder the bottom, the less likelihood of roiling water becoming a problem. On all but very hard bottoms, carp are probably the best exploiters of the benthic biota. Bullheads and crayfish are perhaps better at scavenging foods which might otherwise be wasted. Where there are large numbers of mollusks not of high priority as a crop, redear sunfish, pumpkinseed sunfish (*Lepomis gibbosus*), channel catfish or bullheads can be used. *Benthic feeders*

Herbivores (other than phytoplankton feeders) can be stocked where aquatic weeds are a problem (see Chapter VIII–2) but ordinarily they will need supplemental feeding with green matter. Almost certainly the most effective herbivore is the grass carp, but many readers will not be able to use it. A good second choice is the mirror carp; certain species of crayfish or tilapia may also be used. *Herbivores*

Surface feeding insectivores are not represented in Chinese fish culture, but they should be considered in North America. Most such fishes are adaptable feeders, able to benefit from many kinds of cultured or processed feeds. Almost certainly the best of these fishes is the bluegill. Young of largemouth and smallmouth bass (*Micropterus dolomieui*) also fit in this niche. *Surface feeding insectivores*

Piscivorous fishes are used only where one or more of the cultured species has a high reproductive potential, or where non-food fishes are present. The classical North American fish for this niche is the largemouth bass, but smallmouth bass, crappies, rock bass (*Ambloplites rupestris*), chain pickerel (*Esox niger*), yellow perch and channel catfish are all more or less appropriate. *Piscivorous fishes*

Additional fishes may be added according to the farmer's resources. For example if you have access to substantial quantities of meat or fish scraps, you should consider one of the catfishes. If it is a simple matter to bring electricity to your pond, you are in a good position to use bug lights to enhance insectivore production over what the pond would normally support. Or if your resources include money you may want to invest in commercial feed for no reason other than to support a personal favorite fish. *Additional fishes*

You can vary stocking percentages for the various niches and species to an almost infinite degree according to your pond's characteristics and your own resourcefulness. The change in feeding habits which some species experience with age, and the fact that animals generally do not strictly respect the niches we outline for them, but overlap into neighboring niches, complicate the task of quantifying stocking by niches. Table IV–1–2 is intended as no more than a collection of "ballpark figures" to be used as a guide. *Existing and possible systems*

Table IV-1-2. Percentage (by weight) of animals representing various niches stocked in polyculture ponds. (Most important or suggested species in parentheses.)

Niche	Chinese Intensive Fish Culture Pond	American Analog of Chinese Pond[1]	Typical "Farm Pond"	Tentative North American Polyculture Based on Farm Pond
Filter Feeder:	30–95	50	0	25–45
Phytoplankton Feeder	20–65 (silver carp)	33 (silver carp, hybrid buffalo)	0	10–45 (clams, bigmouth buffalo)
Zooplankton Feeder	0–30 (big head carp)	17 (big head carp, young mirror carp)	0	0–20 (young common carp or mirror carp)
Benthic Feeder:	5–60 (common carp, black carp[2], mud carp[3])	45 (mirror carp, channel catfish)	0–25 (channel catfish, redear sunfish)	15–25 (bullheads, channel catfish, redear sunfish)
Herbivore:	0–30 (grass carp)	3 (grass carp)	0	0–5 (grass carp, mirror carp)
Surface Insectivore:	0	0	40–90 (bluegill)	30–50 (bluegill)
Piscivore:	0–10 (sea perch[4])	2 (largemouth bass, channel catfish)	10–40 (largemouth bass, channel catfish)	10–20 (largemouth bass, chain pickerel, channel catfish)

[1] Based on the system described in Buck, Baur and Rose (1977). (See Appendix IV-1.)
[2] *Mylopharyngodon piceus*, not available in North America.
[3] *Cirrhina molitorella*, not available in North America.
[4] *Lateolabrax japonicus*, not available in North America.

CHAPTER IV-2

Integration of Fish with Plants and Terrestrial Systems

In preceding chapters I have, for convenience sake, often treated aquaculture ecosystems as though they were isolated from the outside world, when of course they are not. There is the obvious intervention of the fish farmer, who stocks, feeds, and otherwise manipulates both fish and environment. And there are other, less obvious, interactions between aquatic and terrestrial systems. In this chapter I will discuss some potentially beneficial linkages between the aquaculture ecosystem and the rest of the farm, along with some linkages within the aquatic system involving plants as well as animals. It will perhaps be helpful to keep in mind that the boundaries of an "ecosystem" are largely determined by where we choose to draw the lines. In this chapter, the ecosystem will be the farm as a whole.

Aquaculture linked with larger ecosystems

It is not customary in North America to think of bodies of water used for intensive aquaculture as parts of a total farm ecosystem. A commercial fish farm is a *fish* farm—period. Yet integration of terrestrial and aquatic crops and by-products is part and parcel of many traditional Oriental food-producing systems. The Oriental approach, which is more ecologically sensitive and less dependent on technology, may be difficult to adapt to a large-scale commercial situation in North America, but the small diversified farmer, particularly the farmer for whom fish is a subsistence crop, would do well to study it.

Let me illustrate with an example I have often used. Figure IV-2-1 is a diagrammatic sketch of a type of aquaculture-agriculture system used in rural Singapore. The hill may be seen as a fertility gradient. Minerals, humus, compost, fertilizer—all types of nutrients—eventually work their way down toward the fish pond. Additional nutrients enter the system at the chicken coop, the farm house, and the pig sty, and waste from the latter is periodically washed into the pond. Certain waste portions of crop plants are fed to the fish. In time the pond becomes a dilute

Oriental examples

fertilizer solution, which can be fed to plants. Every few years, the pond can be drained and the sludge at the bottom dug out and hauled back up the hill. In this manner, fish yields as high as 7,000 pounds per acre (7,721 kg/ha) can be achieved, plus pigs, chickens and eggs, vegetables, fruits, and cash crops, with no outside sources of food or fertilizer, almost no technology, and very little nutrient loss from the system.

A similar farm in Bandung, Java, was described to me by John and Nancy Jack Todd. (See Figure IV–2–2.) In this system fish are incorporated at three levels:

(1) A few goldfish are used in primary sewage treatment. These are probably not used as food.

(2) Nursery ponds and intensive rearing ponds are placed midway in the system.

(3) At the lowest level of the farm, in the catch basin below the rice paddies, less intensive fish culture is practiced.

Possible linkages

As usual, the practice is more difficult than the theory. It is quite easy to throw such a system out of tune. Asian farmers have the benefit of many years of tradition to help them avoid mistakes. We should not expect to quickly duplicate their results. But Asian models can show us how to exploit interrelations between aquaculture and the other components of a diversified farm. These interrelations are further illustrated in Figure IV–2–3.

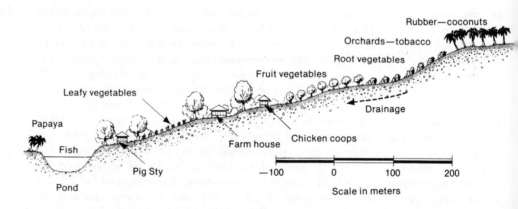

FIGURE IV–2–1. *Profile of integrated aquaculture-agriculture system used in Singapore. (After Ho, 1961.)*

The more of the linkages in Figure IV–2–3 that you can make, the less you will have to depend on materials from outside the farm. I shall not discuss those linkages not directly involving the aquatic components, as they are well covered in organic agriculture publications. The fish pond receives inputs of food for the fish and fertilizer, and contributes outputs of food, fertilizer and water. There are also important nutrient cycles involving both foods and fertilizers which operate entirely within the aquatic system. (See Chapters I–2, III–1 and IV–1.)

Available inputs to the pond

Food for fish may come in the forms of waste portions of human or livestock food, offal from the slaughter of livestock (including other fish), and plants from

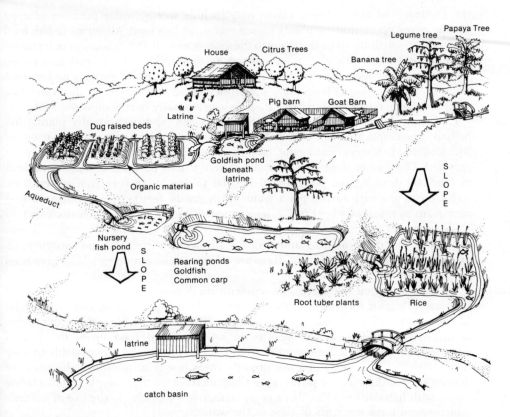

FIGURE IV–2–2. *Aquaculture-agriculture farm observed in Bandung, Java, Indonesia.*

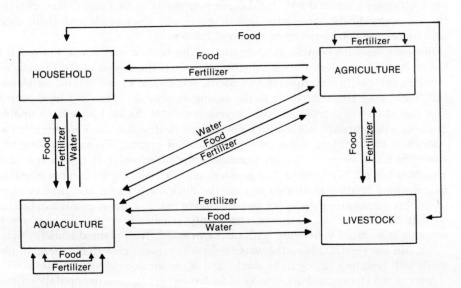

FIGURE IV–2–3. *Potential linkages between the four major components of a diversified aquaculture-agriculture farm.*

the gardens and fields. These plants may include weeds, waste portions of crop plants, cover vegetation, or plants raised mainly as fish food. All forms of fish food inevitably contribute some fertilizer to the system as well. Other sources of fertilizer include controlled run-off from fertilized agricultural lands, animal and (when suitable sanitary measures are taken) human feces, household garbage and various plant materials. Animal feces may sometimes be added directly to the water (see Chapter III–2), but other fertilizer materials are usually best composted first.

The pond's outputs   The most obvious output of the fish pond is fish (and occasionally plants) for human consumption. Fish, especially the waste portions, may also be used as livestock feed. Mineral nutrients tend to accumulate in the pond and may be recycled. This can be done by composting small fish or portions of fish not desired as food, by draining the pond and using the sludge or, in very rich systems, by irrigating crops with fertile pond water. Fish ponds, fertile or otherwise, can be important as reservoirs for crop irrigation and drinking water for livestock and, in some instances, humans.

There is an almost infinite number of ways in which terrestrial and aquatic food-producing systems can be linked. Here are a few specific examples which have been put into practice.

Waterfowl   We have seen that the success of aquaculture depends in part on the fertility of the water. Organic fertilizers, which can be produced on the farm, and commercial fertilizers as well, were discussed in Chapter III–2. Here I will mention an organic fertilization technique which results in an additional food crop—the use of waterfowl. All livestock are potential sources of fertilizer, and it is possible to keep pigs, chickens, etc., near, or even directly over, ponds to minimize the effort of getting the fertilizer to the water. (See Chapter III–2 for an example of integrating swine with fish culture.) But there is an attractive simplicity to the use of animals which spend large amounts of time *in* the water.

Apart from the obvious contributions of manure as fertilizer, and eggs and meat for the farmer's table, ducks can benefit fish ponds in two ways: they effectively control many kinds of aquatic weeds, especially duckweed, and their digging activity releases nutrients from the pond bottom.

Possible problems and solutions   However, as with benthic fish, digging in the bottom may cause too much turbidity. Another problem associated with all birds is the acidity of their feces. When experimenting with waterfowl in fish ponds, you should keep a careful watch on the pH. Ducks may also contribute to the caving in of banks. Though I have been told that this is less of a problem with certain breeds of ducks, I have been unable to discover which breeds. All should be presumed destructive until proven otherwise.

Much of the problem is due to ducks' constant grazing. Perhaps grazing of the shoreline could be prevented by planting some plant which is repellent to ducks. This idea has not been tested. The problem may be reduced, though not eliminated, by providing floating platforms so that the ducks will spend less time resting on shore. Platforms also help in the distribution of manure. Some ponds can be fenced so that ducks have access to the shore only at certain points. Perhaps the best strategy is to build your pond with ducks in mind. Well planted banks which are steep, but not vertical above the water line, will prevent growth of emergent plants while still resisting caving in by ducks and other animals of similar size.

Duck stocking rates   Most of the virtues and drawbacks of waterfowl are related to population density. Recommended stocking rates for ducks depend on whether you read Asian or

European publications. In Asia, stocking rates as high as 1,100 ducks per acre (2,717/ha) are advocated, while European publications generally report figures from 90 to 180 ducks per acre (222–445/ha). However, Elek Woynarovich claims that "at least 500 ducks can be kept on a one hectare pond" (200/acre). According to Woynarovich, "A duck will produce about 6 kilos (13.2 lb) of droppings a month, and 100 kilos (220 lb) of duck manure will result in an increase of 4 to 6 kilos (8.8 to 13.2 lb) of fish." This works out to an annual increase of about 1,267–1,901 pounds per acre (1,398–2,094 kg/ha) of fish. Other European publications report annual increments in carp production of 90 to 450 pounds per acre (99–496 kg/ha) with ducks.

Most reports are not very clear about feeding the ducks, but it is my impression that, even at the lowest stocking rates reported, it is necessary to feed them according to conventional duck-rearing practice. But smaller quantities of ducks may be kept entirely on natural foods, (though with reduced growth of both ducks and fish). Even densely stocked ducks will find some natural food, allowing the farmer to reduce both the quantity and protein content of the duck feed ration. This economy may be multiplied by the fact that some fish will consume any duck feed which gets by the ducks.

Feeding ducks

There are still other advantages to rearing ducks on fish ponds as compared to growing them in pens. Pond-reared ducks are said to be freer of parasites and diseases, to produce higher quality meat, and cleaner, more readily salable feathers.

While domestic ducks are not generally considered piscivorous, they will eat very small fish when they can catch them, and should be kept out of ponds containing fry unless fish population control is needed.

For low-intensity fish-waterfowl culture, geese or swans may be preferable to ducks. Unlike ducks, they are one hundred percent herbivorous, and so more effective in weed control.

There are a number of aquatic mammals which might conceivably contribute manure plus food or salable products, but all of them seem either to eat fish or damage pond banks and dams. Some, like the nutria (*Myocastor coypus*), a good quality food animal and valuable fur bearer which has been introduced in some parts of the United States, can also become serious crop pests. Any experiments with aquatic mammals in aquaculture ponds should be confined to native species and begin with very small numbers.

Mammalian livestock

With the exception of waterfowl, if you are trying to integrate higher animals into fish culture systems, it is best to stay with terrestrial stock. Swine are particularly suitable, since they will eat a variety of foods, and profit from the inclusion of whole uncooked excess fish in the diet. Swine can be penned over ponds as shown on page 118, so that it is a simple matter to hose the manure into the pond. This technique has been widely used in the Orient, but only in 1975 was a similar approach finally tried out in North America. (This was discussed in Chapter III–2.)

The use of nitrogen-fixing aquatic and terrestrial plants to fertilize filled and drained ponds, respectively, has been discussed in Chapter III–2.

Ponds can get *too* fertile. The obvious way to deal with this problem is to drain off some of the water, preferably from the bottom, and replace it with unfertilized water. But the same water which is too fertile for the good of fish may be a rich source of nutrients for plants.

Pond water for agriculture

At New Alchemy we did a series of experiments in irrigating vegetables with water from small fish-growing pools. This water, applied at normal watering rates, seemed to help in the growth of shallow-rooted leaf crops, notably leaf lettuce, but had no effect on other types of plants, at least not over the course of one growing season.

Bear in mind that we are speaking of tremendously fertile water. Although we made no chemical analyses, the water amounted to an opaque green algal "soup," containing high population densities of fish. (See Chapter IV–4 for details on these fish culture systems.) In addition to the fish metabolites and food wastes in the water, we provided our lettuces with millions of decaying one-celled plants. Fish ponds which do not contain algal blooms or very high densities of fish, may be useful as irrigation reservoirs, but it is doubtful that water from such ponds would have any measurable effect on plant growth as compared to "pure" water. Our research was done on Cape Cod, which has fairly wet summers. There might be a more marked effect of such fertile water on plant growth in arid climates, where more frequent watering is necessary.

In using small fish culture pools as irrigation reservoirs, they can be conveniently placed above the land to be irrigated, so that water can be siphoned directly onto the plants.

Aquatic plants to purify water

Much has been written about the possible use of water plants, mainly water hyacinth (*Eichornia crassipes*), for removal of excess nutrients from the water. Promising tests have been made on sewage lagoons, where water hyacinth has shown a great capacity to remove nutrient elements, particularly nitrogen, potassium, and phosphorus. Other plants were shown to be equally effective in some instances, but water hyacinth is favored because of the ease of harvesting a large floating plant.

Water hyacinth

Production rates of water hyacinth in rich environments are prodigious. In a sewage lagoon at Bay St. Louis, Mississippi, 8 to 16 tons per acre (17,648–35,296 kg/ha) per day have been harvested. Even allowing for the fact that 80 to 95 percent of the total amount is water, that is a tremendous harvest of potentially usable, mineral-rich plant material. Of the possible uses which have been explored, the most promising are as compost and livestock fodder. (An aquatic plant with similar characteristics to water hyacinth, plus value as a human food, Chinese water chestnut, is discussed later in this chapter.)

Does this have any value for the small farmer involved in aquaculture? Probably not at this time. If you have to use a body of water which is overrun with weeds, you will certainly want to remove as much of them as possible. If you can compost them or feed them to livestock, so much the better. But, after the initial harvest, you will probably want to concentrate on keeping weeds out of the aquaculture system. There are no known methods of growing and harvesting large quantities of fish together with major crops of large water plants, and there are no aquatic plant cultures which can begin to rival fish culture in nutritional or economic value.

If North American fish culture eventually develops to include highly fertile systems similar to Chinese polyculture ponds, there may be a need for water hyacinth and other aquatic plants to deal with nutrients. The only North American aquaculture waters as fertile of which I am aware, are experimental sewage plant systems and very small semi-closed systems such as those at New Alchemy. (See Chapter IV–4.) We have used water hyacinth in filter tanks with some success, and

*Water hyacinth, restrained by a floating boom, in a fish pond in southeastern China.
(Photo by Robert Sardinsky.)*

they have the added value of being highly ornamental. We compost the excess
plants or feed them to herbivorous fishes, but the quantity of vegetation produced
in such small systems is not great.

A drawback of floating plants in aquatic nutrient removal is that the water under
dense mats of plants becomes anaerobic, offsetting their positive contribution to
some degree. In filters, it is simple enough to aerate the water after it passes
through the water hyacinth chamber. In open waters it is necessary to use rafts or
booms to strictly confine the plants, or entire bodies of water may go anaerobic.

*Drawbacks of floating
plants*

Another potential drawback in open waters, depending on the climate, is the high
water loss associated with dense mats of floating vegetation. While they do block
surface evaporation, floating plants make up for it by losing even more water
through transpiration. It has been estimated that loss of water from surfaces covered with water hyacinth is 3.2 to 3.7 times that from exposed water.

These are just two of the ways in which floating plants may pose problems for
water bodies. Water hyacinth, in particular, is one of the world's most serious economic pest plants. It has been found responsible for clogging irrigation canals,
interfering with hydroelectric power production, obstructing boat traffic, destroy-
ing food and recreational fisheries and fish culture, smothering rice fields, causing
floods and encouraging the spread of water-borne diseases. With this in mind I
would discourage introduction of aquatic plants where there is any chance of their
becoming established in natural waters.

Certain aquatic plants can be grown as food crops for human consumption. Part of their best future use may be achieved technologically through processes such as leaf protein extraction, but these have no applicability to the small farmer and homesteader. There is, however, the possibility of growing aquatic plants for home table use or, perhaps, for sale. The only fresh water plant with major commercial or nutritional value at this time is rice (*Oryza sativa*). There is a small but lucrative market for wild rice (*Zizania aquatica*) (actually not a rice at all). Other aquatic food plants with potential commercial value are watercress (*Nasturtium officinale*) and Chinese water chestnut (*Eleocharis dulcis*). Two potentially cultivable and productive sources of starchy food are the cattails (*Typha* spp.) and arrowheads (*Sagittaria* spp.). There are also a number of edible aquatic greens, some of which are quite widely used in southeast Asia. Their use is virtually unknown in North America, though I have grown the popular aquarium plant known as water fern or water sprite (*Ceratopteris thalictroides*) in indoor pools and used it as a lettuce substitute. It is also eaten in the Philippines.

Potentially the productivity of some aquatic plants is high, but on this continent they are neglected, due to unfamiliarity, the wide availability of traditional food plants, the difficulty of harvesting aquatic plants, and the apparent lack of distinctive qualities in most edible aquatic plants. Their use in aquaculture is further discouraged by the fact that vascular (non-algae) plants are completely detrimental to most conventional forms of fish culture.

In this book, I will only discuss plants and culture systems which can be combined with fish culture. The only plants so far cultured with fish in North America with any degree of success are rice and Chinese water chestnut. Beginnings have been made in the cultivation of watercress and wild rice, but growers of these plants have not integrated them with fish farming. Readers interested in these or other edible aquatic plants are referred to Appendix IV-1.

Chinese water chestnut

Chinese water chestnut, a staple of Chinese cooking, is still little grown outside Asia. But it may represent a potentially valuable cash crop in the southern United States, and the U.S. Department of Agriculture has developed varieties for that purpose. In China, it is planted in flooded fields, often in rotation with rice. Small tubers are planted in nursery beds, then transplanted to the field, which is soon flooded and left. Six months later, the field is drained and the tubers harvested, yielding as much as 17,820 pounds/acre (20,000 kg/ha) of edible material. Presumably this technique could be combined with fish culture in the same manner as for rice culture.

The only trials of Chinese water chestnut with fish of which I have heard were made at Clemson University, South Carolina, by Harold Loyacano and Richard Grosvenor. They grew the plants in a vermiculite substrate on 4.03 foot (1.23 meter) square rafts floating in 10 foot (3.05 meter) plastic pools which also contained channel catfish (*Ictalurus punctatus*) being fed a conventional commercial diet. A comparison was made of catfish yields from pools with rafts and Chinese water chestnuts, with rafts alone, and with no rafts or plants. Table IV-2-1 shows mean survival and production of fish in the three treatments.

The differences in channel catfish survival and production were found statistically not significant (though further trials might have established significance). These differences may have been due to the shading effect of the raft, rather than to the plants removing excess nutrients and competing with unwanted phytoplankton.

Table IV–2–1.   Mean survival and production of channel catfish
in pools with and without floating rafts, and with Chinese water chestnut.

Treatment	% Survival	Production of Catfish in lb/acre (kg/ha)
Neither raft nor Chinese water chestnut	56	1,096 (1,230)
Raft only, no Chinese water chestnut	86	1,667 (1,871)
Raft planted with Chinese water chestnut	87	1,739 (1,952)

Chinese water chestnut production averaged 5.1 pounds (2.5 kg) per pool, which can be extrapolated to 2,836 pounds per acre (3,128 kg/ha).

Rice-fish farming

Far and away the best developed integrated combination is that of fish and rice farming. Unfortunately, both in Asia, where such culture is a traditional food-producing method, and in the United States, where it is relatively new, the practice is declining, due to the increasing dependence of rice farmers on chemical pesticides lethal to fish. At present, production rates of "organic" rice growers are not nearly so high as those of chemical farmers, and their production costs are higher. This may be compensated by the premium price of organic rice, and the added food and income from the production of fish. I leave it to the individual farmer to work out the economics—and ethics—involved. (In China, it has been possible to greatly limit pesticide use by stocking ducks in the rice fields to feed on insects and weeds.)

A further restriction on rice-fish culture, from the point of view of this book, is that modern rice farming in the United States is carried out in areas where the farms are large, and it is not a small farmer's game. On the other hand, small, labor-intensive rice farms are the rule in much of the tropical world; it is possible that small farmers here could successfully emulate such methods, producing rice and fish for home use, or maybe even commercially. Once again, I leave it to the reader to do the necessary economic homework.

History of U.S. rice-fish farming

The original design for rice-fish culture in the United States was conceived in the South Central states, notably Arkansas, during the 1950's. At that time, federal restrictions on the amount of land that could be planted to crops encouraged farmers to look into fish farming as a legal and profitable alternative use for their land. The idea was particularly appealing to rice farmers, who already practiced crop rotation to maintain soil fertility: here was a way to raise a crop in the off years while increasing the fertility of their rice fields.

The principal fish crop raised in rice fields in the early years was buffalofish, and undoubtedly they are among the fishes best suited for the practice. Some catfish were also raised. By the late 1960's, the more profitable catfish had almost replaced the buffalofish and rice-fish rotation had yielded to chemical rice farming and catfish monoculture. (See Chapters II–3 and 4.)

In the short run, rice-buffalofish culture was an economic failure. But I will quote from *Aquaculture* (by Bardach, Ryther and McLarney) to describe the methods used in some detail, because they were successful in terms of total fish and rice production (and could surely be improved, if there were incentive to do so), and also because they may become economically feasible in the future.

Fields used in rice-fish rotation were divided by inside levees to create ponds of suitable size. Some compromise always had to be reached in this regard since fields 35.2 acres (16 ha) or more in area are best for rice farming whereas buffalofish culture proved to be most efficient in ponds smaller than 35.2 acres (16 ha). In practice, ponds of all sizes from 17.6 to 176.0 acres (8 to 80 ha) were used. Whatever their surface dimensions, productive ponds were designed to be no less than 1.6 ft (0.5 m) deep at the shallowest point, and mostly 3.3–6.6 ft (1–2 m) deep, to discourage rooted aquatic plants, provide lower bottom temperatures, and reduce predation by birds.

Production ponds were stocked at various times from late summer to spring. One of the most efficient stocking methods took advantage of the heavy fall rains which usually occur in the South Central states. Prior to the rains, ponds were partially filled, preferably with well water. Fingerlings were then stocked in the borrow ditches leading into the catch basin and automatically released into the rest of the pond when the rains filled it.

Recommended stocking rates varied from 10 to 100 fingerlings/acre (25 to 250 fingerlings/ha) in unfertilized ponds. If potential competitors such as minnows were suspected of being present, largemouth bass fingerlings were sometimes stocked at 24 to 50/acre (60 to 125/ha). Direct feeding of buffalofish in growing ponds was never done, and fertilization was only occasionally carried out. Not only did effective dosages of fertilizer vary greatly from pond to pond, but the farmer had to consider the effect on the forthcoming rice crop as well as on the fish.

The system was based on a two-year growing period, since local consumers are used to buying buffalofish weighing five pounds (2.3 kg) or more. It was possible to raise fish half that size in 18 months and, in experiments at Auburn University, 2½ to 3 pound (1.1–1.4 kg) buffalo were produced in six months in fertilized ponds. The implications are that if people can be persuaded to eat smaller buffalofish, and/or if economically feasible methods of pond fertilization suitable for rice farming are developed, rice-buffalofish farming could become economically viable.

Rice grown with other fishes

Buffalofish are not the only fish which can be raised in conjunction with rice. I have already mentioned the use of catfish. In Taiwan, mullet (*Mugil* spp.)— normally salt water fish, but possessing a wide salinity tolerance—are occasionally grown with rice; it might be possible to follow this practice along the American Gulf Coast. Other possibilities for American rice farmers are carp (see Chapter II–5), some species of tilapia (not those which consume large plants; see Chapters II–8 and 13) and crayfish. Table IV–2–2 lists stocking rates and yields of some of the fishes cultured with rice in various parts of the world.

Rice-crayfish farming

A form of rice field aquaculture which deserves a separate description is Louisiana's rice-crayfish farming. Although rice field culture accounts for a minority of the state's total crayfish production, it is important and becoming more so. The scheme used is a two-year rotation as outlined in Table IV–2–3. Most Louisiana rice fields contain enough crayfish to make stocking unnecessary, but in some cases five to ten pounds per acre (5.5–11.0 kg/ha) of adults may be stocked as broodstock. The crayfish take care of two chores for the farmers by feeding on "stubble" left after the rice is harvested and by consuming aquatic weeds. Flooding and draining of the rice fields coincide with the natural periods of activity and inactivity of the crayfish.

Table IV–2–2.   Stocking rates and yields of fish-rice
culture with different species in different parts of the world.

Fish	Location	Stocking Information	Yields in pounds per acre per year (kg/ha)
Bigmouth buffalo	South Central United States	10–100 fingerlings/acre (25–200/ha) in unfertilized ponds, plus optional 25–50 largemouth bass fingerlings/ acre (62–125/ha)	85–450 (94–496) (Buffalo only. Based on harvests after a *two-year* growing season.)
Bigmouth buffalo	Auburn University, Alabama	500 fingerlings/acre (1,235/ha) in fertilized ponds	1,200 (1,324) (Based on six-month growing season.)
Channel, blue or white catfish	South Central United States	5–75 fingerlings/acre (12–185/ha)	—
Common carp	Japan	1,400–7,000 fry/acre (3,458–17,290/ha)	600–1,000 (662–1,103)
Common carp	Ukraine	40,000–320,000 fingerlings/acre (98,800–790,400/ha)	125–200 (138–221)
Crayfish	Louisiana	5–10 lb of adults per acre (5.5–11.0 kg/ha). (More often adequate broodstock is present naturally, so that no stocking is necessary.)	350–1,000 (386–1,103)

Table IV–2–3.   Schedule of procedures
for rice field culture of crayfish in Louisiana.

Activity	Dates
Plowing	March 1 – April 30
Replowing and planting rice	April 15 – May 15
Flooding, and stocking if necessary	May 1 – May 31
Draining	July 15 – August 15
Harvesting rice, followed by reflooding	August 1 – September 1
Harvesting crayfish	December 1 – June 15
Drain and use as pasture	June 15 – March 1

Given the ubiquity and diversity of crayfish in the United States, they could probably be produced wherever rice is grown. Whether they could be grown commercially outside of Louisiana is another question. Care should be exercised in selection of species, as some crayfishes may damage rice field levees with their burrowing.

One more possible application of food plants in aquaculture systems needs to be mentioned, and that is hydroponics. This is the growing of plants without soil, in a liquid medium containing nutrients in solution. Advantages of hydroponic farming that have been cited are the far greater precision which is possible in "feeding" essential nutrients to the plants, the possibility of growing food plants in places

Hydroponics

completely unsuitable for conventional agriculture, and the tremendous yields which have been achieved. A common criticism of hydroponic farming concerns the total dependence of conventional hydroponics on synthetic chemical fertilizers. It is possible, however, to carry out successful "organic" hydroponics. Though less precision can be obtained in determining nutrient dosages than in chemical hydroponics, much greater is still possible than with soil rooted plants. Presumably, vegetables grown hydroponically in fish pond water would perform some of the same functions as aquatic plants, by removing excess nutrients.

Such thoughts have started many a fish culturist's head working. Unfortunately, most attempts to grow vegetables hydroponically in aquaculture waters have been disappointing. Apparently such water does not usually make a complete nutrient solution for plants. For example, at the U.S. Fish Farming Experimental Station, a number of vegetables were grown in effluent from a raceway complex containing 10,000 pounds (4,545 kg) of channel catfish fed on commercial type feed. Of seventeen garden plant varieties tested, only peas and cucumbers produced "out-

standing" yields. Cantaloupes produced "above average" yields (as compared to nearby gardens). Production of strawberries, okra and swiss chard was "average," while bush beans, cabbage, cauliflower, corn, eggplants, lettuce, onions, radishes, squash, tomatoes and turnips produced crops of "very poor quality and quantity." Chemical analysis revealed, surprisingly, that the limiting element was nitrogen.

Of course fish pond water can be used as a base for hydroponic solutions. The De-Korne family of El Rito, New Mexico, have grown most common garden vegetables in their "Survival Greenhouse" using equal parts of rabbit manure, chicken manure, earth-worm castings and wood ashes in fish tank water. However, they have not found any difference between plants grown in a fish pond water-based solution and those grown in a similar solution in well water.

A more sophisticated hydroponic system was developed by Carl Baum at the New Alchemy Institute, to take advan-

Hydroponically grown tomatoes and cucumbers in the Cape Cod Ark at the New Alchemy Institute, Hatchville, Massachusetts. Fertile water from solar-algae ponds (left background) is passed through the trough (center foreground). Plants are rooted in a tray containing vermiculite (an inert rooting medium) set in the trough. (Photo by Robert Sardinsky.)

tage of the enriched water in indoor solar-algae ponds. A single 600 gallon (2,280*l*) solar-algae pond stocked with blue tilapia (*Tilapia aurea*) produces about 20 ounces (567 g) of excess nitrogen over a 20 week growing period. Approximately half of this is eliminated during periodic removal of solid waste. The remaining 10 ounces (283 g) are theoretically convertible to about 250 lb (104 kg) of green plant material.

This was the rationale for the device shown in Fig. IV–2–4. Water is pumped from pond to pond, and ultimately into the filter tank, by an air-lift system. It then flows by gravity down the trough and back into the head pond. Plants are rooted in perlite in trays so that their roots trail down into the enriched water in the trough. The result has been what is believed to be the most productive aquaculture-based hydroponic system ever devised. Yields of cucumbers have been especially high—as much as 30 to 65 lb (12–26 kg) per plant. And cucumbers can be planted at about 6 times the density of normal garden planting.

However, any benefits to the fish culture system arising from the filtering action of the vegetables have been offset by the fact that we cannot use salt in the system. Fish production has in fact been slightly lower than in solar-algae ponds without the hydroponic component.

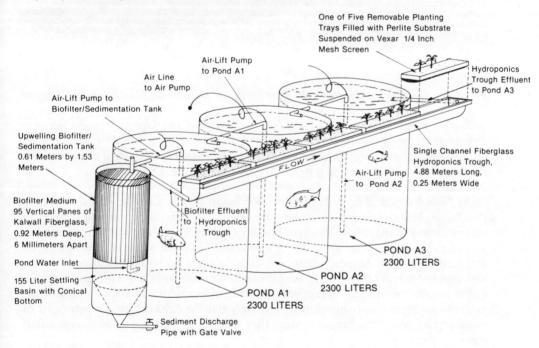

FIGURE IV–2–4. *A recirculating fish culture system combining three solar-algae ponds, sediment removal, biofiltration, and hydroponic removal of excess nutrients.*

So far the discussion of integrating aquatic and terrestrial food-producing ecosystems has centered on outdoor "open" systems. But two of the most ambitious attempts at integration involve "closed" systems. (See Chapter IV–4 for definition and comparison of open and closed system aquaculture.) These are New Alchemy's

Integration in closed systems

The "Cape Cod Ark" at the New Alchemy Institute, Hatchville, Massachusetts. Approx-imately one-third of the 1950 square feet (181 square m) of floor space is occupied by solar-algae ponds like these seen outside in the courtyard. In addition to producing up to 55 pounds (22 kg) of tilapia, carps, and other fishes annually, each indoor pond doubles as a passive solar heater. Outdoor ponds in the foreground are used to grow carp and trout, as well as tilapia hatched in the Ark. (Photo by Robert Sardinsky.)

"Arks," one located at Spry Point, Prince Edward Island, Canada, and the other in Hatchville, Massachusetts. Here we have begun to apply, with varying degrees of success, many of the tactics described in this chapter. Plants grown in the structures are fed to the fish or used as mulch or compost. Fish pond water is used to irrigate and fertilize plants. Aquatic plants aid in processing potential pollutants. Insects with aquatic larval stages aid in pest control. We have tinkered with cultivation of edible aquatic plants and hydroponic growing of conventional vegetables. But the most unique aquaculture-agriculture interaction in the Arks is New Alchemy's use of "solar-algae ponds," in climate control. (See Chapter IV–4 for a discussion of solar-algae ponds.)

New Alchemy's goal with the Arks was to create an indoor aquaculture-agriculture facility able to produce significant amounts of fish and vegetables year-round in northern climates, with little supplemental heat and no inputs from fossil or nuclear fuels. During their first winter, the severe winter of 1976–1977, both Arks performed admirably. The Cape Cod Ark has 1,950 square feet (181.4 m²) of floor space, 650 square feet (60.4 m²) of it devoted to agriculture. (Additional crop space is provided by raised, vertical and hanging growing structures.) Most of the remaining space (370 ft² or 34.4 m²) is given over to a battery of nine 5 foot diameter by 5 foot deep (1.5 by 1.5 m) solar-algae ponds, plus an 8 by 13 foot (2.4 by 4.0 m) sunken reservoir. During the first year, despite a leak leading to ice cold water coming in at ground

A battery of solar-algae ponds in "The Cape Cod Ark," at the New Alchemy Institute, Hatchville, Massachusetts. During the day the solar greenhouse structure heats the ponds. At night, the ponds give off heat to the greenhouse. (Photo by Ron Zweig, New Alchemy Institute.)

level, and the fact that a roof-top vent blew off during a storm, the only supplementary heat needed to maintain vegetables was a wood stove, used only during the single most severe week of the winter. During the equally severe winter of 1977–1978, no supplemental heat was used. Similar results were obtained in the Canadian Ark, which boasts 38 solar-algae ponds and 2,144 square feet (199.4 m²) of agricultural space plus a 1,812 square feet (168.5 m²) one-family living area.

We cannot say to what extent the moderate internal climates were due to solar heating, insulation, heat storage facilities, etc., but certainly the aquaculture area helped to maintain temperatures well above outdoor levels. It is common knowledge that volumes of standing water within a greenhouse will moderate the air temperatures, particularly at night, but our solar-algae ponds go one step further. Due to their transparent walls and dense phytoplankton blooms, during the daytime they trap much more radiant energy than below-ground or clear water pools, functioning as passive solar heaters.

I hope that this chapter has focused the reader's attention on the compartmentalized thinking which has so far limited most of North American aquaculture and agriculture. By referring to Figure IV–2–3 (p. 210) and with the examples of aquaculture-agriculture integration I have described, perhaps you can help build some of the missing links in our food-producing ecosystems.

CHAPTER IV-3

Cage Culture

History

Raising fish in cages is an aquacultural technique which might be applied at all economic levels. As we shall see, in North America it seems to offer particular advantages for homestead fish culture. Like so many things aquacultural, cage culture is Oriental in origin. Perhaps the earliest cage culture of fish occurred in Cambodia, where a variety of river fishes are kept in the "basements" of floating dwellings and fed on fish scraps and rice. Cage culture is also traditional in parts of Indonesia, where cages are anchored in streams which amount to open sewers. Standing water cage culture probably originated in Japan, and this is the form which has been adopted in North America.

Oxygen needs

At first thought it might seem that only a few fish could be placed in a small cage. In fact, fish can be stocked in cages at densities hundreds of times greater than is possible in pools or ponds. To see why this is so, consider a fish's relationship to oxygen. Confined in a small space, such as a sealed bottle, a fish will use all the dissolved oxygen in the water, then die of asphyxiation. In nature, this rarely occurs: oxygen is constantly being replenished by exchange with the air and through the action of aquatic plants, and natural populations are ordinarily limited, by food and other factors, to numbers well below those which would result in depletion of dissolved oxygen. However, it is possible to overstep natural bounds and stock so many fish in a small pool or pond that dissolved oxygen is used up too rapidly to be replaced by natural processes.

It is much more difficult to overstock small cages in a larger body of water. Here, the amount of water to which the fish have access for breathing purposes is equivalent, not to the small volume of the cage, but to virtually the entire volume of the larger body of water. Even in the most stagnant of ponds, wind and convection currents will bring about some circulation through a floating cage. And when the cage is stocked, circulation will be increased, since the breathing and swimming motions of the fish act as a pump. Thus there is a constant exchange of deoxygenated water for fresh water with adequate dissolved oxygen.

The most obvious advantage, then, of cages as fish culture enclosures is that, where there is a large body of water, great numbers of fish can be confined and grown in a very small area. (The same is true of certain "closed" aquaculture systems. Potential fish growers for whom space is an important consideration should also read Chapter IV–4 on closed systems.) Not only is a small amount of space utilized, but caged fish are easily accessible. They can be fed, inspected, observed or harvested in part or totally with greater ease and convenience than in any other type of fish culture.

Cages open the possibility of at least small scale aquaculture in almost any unpolluted body of standing water, and in some flowing waters as well. A partial list of situations in which cage culture may be the only feasible or desirable method of raising food fish follows:

• Cage culture is the most suitable method for many bodies of water intended for multiple use. Recreational fishing, for instance, is really not compatible with conventional intensive fish culture. But cage culture may even enhance some types of angling by attracting wild fish to the shelter and occasional food offered by the cages. Heavily stocked and fed ponds are often unattractive for swimming, but cage culture concentrates dense fish populations and their side effects in a small portion of the pond.

• Cage culture can be carried on in natural ponds without compromising them esthetically or as wildlife habitat. Many types of wild animals and plants are drawbacks for conventional aquaculture, as are irregularities in profile and shoreline. Yet these are the very features which make many ponds esthetically and ecologically desirable. A few cages will disrupt the natural setting no more than a boat dock or swimming raft.

• Some readers may have access to a body of water, but without control of it. For instance, if you live beside or own land on a pond shared by other landowners you probably will not be able to make major modifications of the shoreline or water level, fertilize, or stock and mass harvest fish. But with the understanding of your neighbors and permission from local conservation authorities you may be able to float a few cages off your shoreline.

• Many natural and artificial waters otherwise suitable for fish culture prevent efficient harvesting owing to great depth, brush, weed beds, sediments, etc. Quarries, strip mine pits and ponds in swamps often fall into this category. Use of cages eliminates this problem.

• There are instances where it is not feasible to protect cultured fish from predators such as larger fish, birds, snakes, etc. Covered cages will keep such animals out.

Among the disadvantages of cage culture are the following:

• The close confinement of fish offers an ideal situation for the spread of infectious disease, and treatment of disease in cages may be difficult. (Disease treatment techniques for caged fish are described later in this chapter.)

• Should a hole somehow develop in a cage, the result will ordinarily be close to 100 percent loss of fish. This may not be so serious for the large-scale grower, but for the subsistence farmer with only a few cages, the loss of one cage can be critical.

• In many areas, floating cages are a serious temptation to theft or vandalism.

• Because caged fish have far less access to natural feeds than free-swimming

Cage culture in a natural pond at the New Alchemy Institute, Hatchville, Massachusetts. Two cages are in place in the foreground, while a third is being harvested by lifting it out of the water. Note how unobtrusive the cages are in the natural landscape. (Photo by John Todd, New Alchemy Institute.)

fish, the culturist must provide them with both more food and a more complete diet.

• At high densities cage culture can be a considerable source of pollution. No data are available for family scale cage culture, but the size, rate of water circulation, chemical condition and uses of any body of water should be considered before a commitment is made to cage culture in it.

Territoriality and aggression

Another apparent disadvantage often mentioned by visitors to aquaculture workshops at New Alchemy is the supposed tendency for territorial fishes such as catfishes and sunfishes to become excessively aggressive with each other in close confinement. Yet this does not appear to be a cause for concern. During research on the social behavior of yellow bullheads (*Ictalurus natalis*) John Todd and his coworkers, including myself, observed territorial behavior at low population densities with only occasional, and rarely serious, conflict between individuals. At higher densities, when there was not enough space for each fish to maintain a full-sized territory, fighting became frequent and severe. But at *still* higher densities,comparable to those normal for cage culture, territoriality and aggression disapeared and what what we called the "love-in effect" (amicable crowding) took over. In every experience I have since had with cage culture and in every account I have read of it, the same thing was seen. There may be fish species which fight at very high population densities, but apparently no one has yet attempted to culture them in cages.

One type of floating cage will serve for all species of fish, so I will begin by discussing cage design and construction. There are a number of commercial manufacturers of fish cages or cage kits and their products are generally satisfactory. (See Appendix IV–3.) But commercial cages are also fairly expensive (about $120 to $180 for a 50 to 75 cubic foot or 1.4–2.1 cubic meter cage as of 1982), so here is construction information suitable for those who wish to make their own cages. (A third alternative available in some places is to rent fish cages.) Figure IV–3–1 illustrates the basic features of a fish culture cage. A heavy-duty cage, for raising up to 1,000 pounds of fish, but said to cost less than $50 (in 1980) to build is shown in Figure IV–3–3.

Cage design and construction

Most cages are made out of some sort of mesh material, although a slatted construction is possible. In South America, I have seen excellent and durable slatted cages made of a water-resistant type of bamboo. Bamboo is also used as a cage material in southeast Asia. I am not aware of any North American natural material which lends itself as well to the construction of slatted cages, though redwood might serve as a durable substitute.

Cage materials

Early fish cages in the United States were made of galvanized or aluminized wire, but when immersed in water, the durability of these materials was marginal. Applying asphalt paint to the wire was a slight improvement, but users of such cages still suffered from that insecure feeling. More than one fish grower had the sour experience of pulling up a cage, only to watch the year's crop go through the bottom.

Today cages are manufactured from plastic coated wire or nylon or plastic webbing. I have not worked with the coated wire, but it is my impression that there may be some problem with corrosion whenever the ends of the wire are exposed. Plastic or nylon netting also has the advantage of being lighter; the specific gravity of some of these materials is somewhat less than that of water, so that cages made of them are slightly buoyant. Other aspects of working with the two types of netting are covered in the following discussion.

FIGURE IV–3–1. *A 64 cubic foot (1.8 cubic m) commercial fish culture cage, made of Vexar® nylon netting with aluminum frame. Flotation is inside cage. Below is same cage folded flat for storage.*

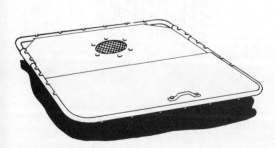

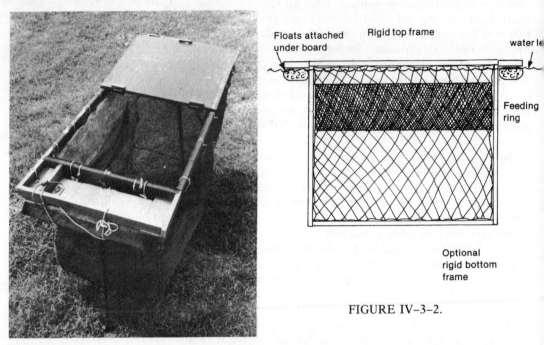

Floats attached under board

Rigid top frame

water le

Feeding ring

Optional rigid bottom frame

FIGURE IV–3–2.

A 54 cubic foot (1.5 cubic m) fish culture cage made of Vexar® nylon netting at the New Alchemy Institute. The frame is treated wood. Flotation is made of blocks of polystyrene foam enclosed in canvas bags. Note feeding ring (band of fine netting around top portion of cage). (Photo by Robert Sardinsky.)

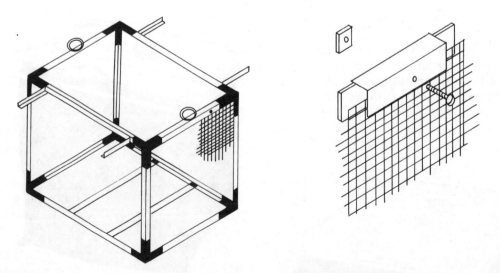

FIGURE IV–3–3. *Design of a heavy duty fish cage. (Inset: method of attaching webbing to frame.)*

As a rule, you should choose the mesh size that is the largest possible yet still will prevent escape. The smaller the mesh the greater the chance of blockage by algae and the more frequent the need for cleaning. Quarter-inch mesh is a good size for most operations beginning with fry or fingerlings, although you may need smaller mesh for small fry of slender fishes like trout.

Cages may be made in any size; the smallest commercial cages I have seen have a volume of about 50 cubic feet (1.4 cubic meters). Under good conditions, a cage only a little over half that size can be used to raise 100 pounds (45 kg) of fish or more. The largest commercial cage I know of has a volume of nearly 1,000 cubic feet (28.3 cubic meters). Most small-scale fresh water fish culturists will find their needs are satisfied by cages in the range of 30 to 300 cubic feet (0.85–8.50 cubic meters). If in doubt, choose smaller cages, as water circulates faster through them. Trout need larger cages; 12 × 8 × 4 feet (3.7 × 2.4 × 1.2 m) seems to be the accepted minimum size. More or less these same proportions should be maintained whatever size cage is used, since trout do not like to be crowded during feeding, and need their food spread out over a fair area.

Cage sizes and shapes

The conventional shape for fish cages is a rectangular solid. Small cages may be cubical, but an oblong shape is better for large ones, to facilitate water circulation and netting of fish. Some fish farmers believe that well defined corners encourage a harmful "bunching up," and prefer a cylindrical cage. (Bullheads appear to be particularly prone to this behavior.) The cylindrical shape is faster to construct, but more difficult to make partial harvests from.

All fish cages must have a rigid top frame. Side and bottom frames are optional, and many commercial cages are made without them, so that they can be folded for storage. I personally prefer a side frame. On several occasions, I have seen the walls of a nylon cage buckle while it was in the water. This in itself was not harmful, apart from reducing the volume of the cage, and may even have provided a bit of cover for the fish. But over a period of weeks, a crack developed in the angle formed by the buckled side. The crack expanded to form a hole and fish were lost. This could have been avoided with a rigid side frame.

Cage frames

Cage frames may be made of wood topped with a coat of non-toxic paint, but aluminum tubing, which you can buy at most hardware or building supply stores, is lighter and more durable. You can staple or nail netting to wooden frames, but corrosion can be eliminated by using plastic fasteners, or nylon electric cable straps, available from electrical supply stores. Panduit Corporation manufactures special plastic ties for use with wood or aluminum frames. (See Appendix IV–3.) When attaching coated wire to cage frames make sure no ends of wire protrude into the cage. These can injure fish or damage nets. We have successfully constructed nylon cages without side or bottom frames by sewing together pieces of netting with nylon rope, but this is quite time consuming. Whatever construction method is chosen, think ahead to the harvest. Estimate the best possible crop of fish, and plan the cage to support that weight when it is lifted from the water.

The other essential component of all cages is flotation. Polystyrene foam is ordinarily used for this purpose, though oil drums or various types of tubing, adaptations of marine buoys, balsa wood, etc., will serve. If you use polystyrene foam, encase it in a nylon or plastic bag so that if it breaks it does not scatter. I have found about one cubic foot (0.03 m³) of polystyrene foam more than adequate to support a 36 cubic foot (1.1 m³) Vexar® nylon cage with a wooden top frame. DuPont suggests about 2 cubic feet (0.06 m³) of polystyrene foam for a 128 cubic foot (3.6 m³) Vexar®

Cage flotation

cage with a full aluminum frame; wire cages may require more. Flotation should be placed so that the top of the cage rides at least a foot (30.7 cm) above the surface of the water. Vexar® cages will float at or near the level of the bottom surface of polystyrene foam flotation. (DuPont has discontinued manufacture of Vexar, but if you can find some, I recommend it.)

"Feeding ring"   If you will use floating foods, it will help to incorporate a "feeding ring." The conventional feeding ring is nothing more than a band of very fine mesh about a foot (30.7 cm) wide, fastened around the perimeter of the cage at the water line to prevent floating food particles from passing out of the cage before they are consumed. You can simplify construction by using a separate, floating feeding ring. (See Figure IV–3–4.)

Cage tops   A hinged top is optional in some circumstances, essential in others. If cages are set where large waves are a possibility, a lockable top is a wise precaution. Tops are also necessary to prevent escape of strong leapers, like trout and tilapia. When growing non-leaping fish, such as sunfish or bullheads, in waters not subject to strong wave action, you will need a top only if you anticipate problems with predators such as snakes, birds and people.

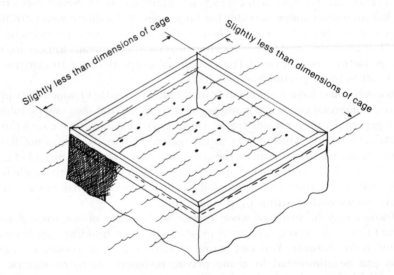

Slightly less than dimensions of cage

Slightly less than dimensions of cage

FIGURE IV–3–4. *Unattached floating feeding ring for use in fish cages.*

Tops hinder the entry of aerial insects to the cage, interfering with the use of such feeding devices as "bug lights." On the other hand, channel catfish (*Ictalurus punctatus*) have been found to use feed more efficiently in cages with opaque tops than in those with transparent tops. Whether this is true for other, less "nervous," species is not known. Most fish like to have some cover, but this can be provided by floating boards in the cage as well as by putting a top on it. Many cage culturists like a small door in the top to facilitate feeding.

Placing cages   You can fasten cages to some other structure, such as a floating platform or dock, but do not attach cages to a stationary dock if you expect extreme fluctuations in water level. You can also anchor cages in open water, where cinder blocks lend themselves for use as anchors. It is best to use two anchors, or four in very windy locations. When more than one or two cages are used, it may be convenient to secure them between parallel cables. The various methods of securing cages are illustrated in Figure IV–3–5.

If cages are to be tended from a boat, you may want to attach some device for snapping or tying the boat to the cage, leaving both hands free to work.

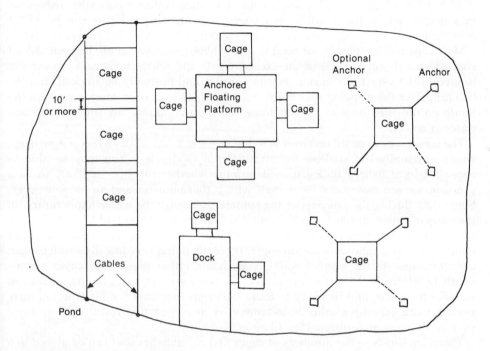

FIGURE IV–3–5. *Four methods of securing floating cages.*

Cages should be placed so that at low water their bottoms are at least a foot (30.7 cm) off the bottom. While the fish might acquire some food from having access to the natural bottom, it would not be enough to make up for the loss of circulation created by resting cages on the bottom.

The precise location of cages should be chosen so as to maximize circulation. While the fish themselves will induce some water circulation, you should plan to take advantage of currents or wind whenever possible. There is, of course, a limit to the amount of current or wind which is compatible with stable placement of cages and convenience in working with them.

Because warm water "floats" on colder water, a special problem may affect cage siting in a few bodies of water where deoxygenation is associated with "turnover." In many ponds and lakes, particularly small, shallow ones, there is a constant slow circulation and exchange of heat so that temperature variation from top to bottom is gradual, and water level stratification by temperature does not develop. But in some bodies of water, particularly deep ones where wind circulation is a minor factor, thermal stratification may occur. At certain times, especially in the summer, there will be distinct layers of warm water on the surface (the hypolimnion) and cold water on the bottom (the epilimnion), with a narrow transition band (the thermocline) in-between. If a change in air temperature or any influx of cold water

"Turnover"

causes cooling of the surface, the hypolimnion may become colder than the epilimnion. When this happens, the pond or lake "turns over," bringing the bottom water to the surface, where caged fish live.

In many bodies of water, this will present no problem. But in some waters, particularly fertile ones, the bottom water may contain almost no dissolved oxygen. Wild fish can escape the effects of turnover in such waters by moving inshore or into deeper water for a while, but caged fish are trapped and can be killed overnight.

Most cage culturists will not need to worry about turnover at all, but you should check to see if you are one of the exceptions. In the North, turnovers are usually limited to the very early spring, before cages would normally be stocked, and the late fall, after the growing season for most fish species is over. Many waters in the South do not turn over at all, but those that do may do so any time the surface water is cooled.

The surest way to avoid turnover is to put cages in water about 6 feet (1.8 m) deep, which is usually too shallow for stratification to develop. You may be able to consult a local fishery biologist to determine whether turnover is likely to be a problem, or you can check it yourself with a thermometer and an oxygen meter. Some time during the hot part of the summer, measure the water temperature, in the early morning, at one foot (30.7 cm) depth intervals. If you find a band of water where the temperature changes very sharply over a few feet (the thermocline), that pond or lake is a candidate for turnover. If you then find very low dissolved oxygen concentrations in the bottom water, you should either place your cages permanently in relatively shallow water, or keep a close watch on water temperatures and weather forecasts, and be ready to move the cages in advance of an expected turnover. A more expensive solution to temporary deoxygenation resulting from turnover is mechanical aeration. (See Chapter X–2.)

Concentration of cages

There are limits to the numbers of cages and caged fishes that can be placed in a body of water, but there is very little danger of reaching these limits except in very small bodies of water or very large commercial operations. Of somewhat more concern to most growers is spacing between cages. No research seems to have been done on this subject, but obviously there must be some space between cages if there is to be effective replacement of dissolved oxygen in the cages. The amount of space needed is determined by the numbers of fish in the cages and the quality of water. In the absence of experimental data, I would suggest a minimum of 10 feet (3.05 m) between cages with normal rates of stocking, in water with abundant dissolved oxygen.

Species for cage culture

Of the fish species discussed in this book, the following have been successfully grown in cages: bluegill (*Lepomis macrochirus*), redear (*Lepomis microlophus*) and hybrid sunfishes, channel catfish, blue catfish (*Ictalurus furcatus*), yellow bullheads, common carp (*Cyprinus carpio*), rainbow trout (*Salmo gairdneri*), grass carp (*Ctenopharyngodon idellus*) and several species of tilapia. Based on our experience at New Alchemy, the yellow bullhead appears particularly promising for homestead use, but most other fishes should also be amenable to cage culture. For some reason, little or nothing has been done with cage culture of buffalofish. Yet these fish may be particularly suited for cage life since, as plankton feeders, they would obtain more natural food in cages than other fishes.

Stocking rates for fish in cages are much, much higher than most people would expect. The capacity of a cage can be expressed in total weight of fish, instead of numbers of individual fish. Commercial channel catfish farmers have produced 1,000 pounds (455 kg) of marketable fish in a growing season in a 64 ft³ (1.8 m³) cage (15.6 lb of fish per cubic foot or 250.7 kg per cubic meter). Since cultured channel catfish are normally marketed at a weight of about one pound (0.45 kg), that came to about 1,000 fish per cage, but it could as easily have been 2,000 half-pound (0.22 kg) fish. The maximum cage production figures so far reported for rainbow trout and common carp happen to be almost identical with the catfish figure. Under good conditions it should be possible to surpass this.

Two factors determine the capacity of a cage: the quantity and quality of feed being supplied, and the cage's water quality. This water quality will depend not only on the quality of the water in which the cage is placed, but also on the rate at which it can be replaced. Almost all cage culture done so far in North America has been done in lakes or ponds, but the water replacement would be more rapid (and fish production theoretically greater), in flowing water. Most natural streams do not maintain a stable enough flow rate to be convenient for cage culture, but irrigation canals and other artificial or controlled flowing environments seem to present a real potential. Such waters might be particularly suited for cage culture of trout, which have a high need for dissolved oxygen.

The lower limit of feasibility for channel catfish is about 2.3 *individuals* per cubic foot (81 per cubic meter); at lower densities fighting begins to occur. This figure may vary widely with species; at present we do not know. Although commercial cage culturists try to stock cages at or near the maximum safe rate, I recommend that inexperienced subsistence growers stock at the lowest rate that will prevent fighting and supply the desired weight of fish.

Stocking of cages is generally done about two weeks before the anticipated beginning of the growing season (as determined by water temperature). Where the growing season is year-round, cages should be stocked in the cool part of the year. The reason for this is that fish are less active at lower temperatures, and less apt to become excited and injure themselves during the period of acclimation to a new and restricted environment.

A certain amount of natural food enters floating cages: aerial insects, emerging aquatic insects, and various organisms which swim or drift in through the mesh. But, except in extremely fertile, flowing waters, the amount of such food entering a well stocked cage is insignificant. (This may not be true with plankton-feeding fish. In one trial at New Alchemy very young hybrid bluegills, which are generally considered to be plankton feeders, grew without supplementary feeding when stocked in cages in a fertile natural pond. Growth of these fish ceased when they reached a weight of 3.5 grams each—about the size when bluegills switch to prey larger than zooplankton.) Certainly, feeding is even more critical in cage culture than in other types of fish culture. Some commercial fish feed manufacturers, recognizing the contribution made by natural foods in conventional fish culture, make a special cage culture feed, richer in protein and vitamins than their standard product.

Using floating feeds is particularly worthwhile in cage culture. When feeding natural foods that sink, they must either be fed a little at a time, or some provision made to keep most of the food from being lost through the bottom of the cage. (An example is the earthworm feeder shown in Figure III–5–1.)

An as yet untested method for improving the nutrition of caged predatory fish at little or no cost grew out of our experience at New Alchemy. During 1978, Grassy Pond, where our cage culture experiments have been conducted, supported a large, wild population of golden shiners (*Notemigonus crysoleucas*). When our caged fish were fed with commercial dry feed, shiners and other small fishes would gather around the cages to pick up feed scraps which were spilled or had drifted out of the cages. John Todd suggested we harvest these fish periodically, perhaps by means of an umbrella net (see Chapter VI–1), to be placed in the cages as food for yellow bullheads. (Bullheads are much better at capturing small fish within an enclosure like a cage than in the natural environment.)

Cleaning

Almost invariably, algae will grow on the sides of cages. While algae provide some oxygen, they will also impede water flow through the cage. So algae, and any other attached organisms, must be removed regularly. Though most written materials on cage culture recommend some sort of chemical control, I emphatically disagree, particularly with regard to small-scale operations. It is a fairly simple matter to scrub off algae and most other attached organisms as necessary, using a toilet brush or janitor's broom. The only really stubborn creatures I have encountered growing on cages are colonial bryozoans; for them something on the order of a putty knife works best.

It is also possible to use fish to reduce algal growth on cages. One widely distributed native fish which will do this job is the golden shiner. As golden shiners are of no interest as food fish, they should be stocked in the smallest numbers that will still control the algae. Bear in mind that for algae-eating fish to be completely effective, there must be a substantial population outside the cage in addition to the few inside, in order to clean the outsides of the cages.

Disease treatment

I have mentioned that cage culture is conducive to the spread of contagious disease. Fortunately, cages also make the detection of disease easier. In most cases, the fish are visible at all times, and it is easy to net out a few for inspection and diagnosis. (See Chapter VIII–3.)

Inspection for disease may be easy, but treatment is not, since medicines in solution are almost immediately diluted in the larger body of water. Antibiotics and some other drugs may be applied to the feed. Commercial growers sometimes enclose the entire cage in a plastic bag prior to applying medication, but this technique requires mechanical aeration within the cage. In one instance, I "cured" an unidentified fungus disease by anchoring 10 pounds (4.5 kg) of quicklime in a burlap bag upwind of a cage. (At least, I like to think I cured the disease—it may have just gone away.) Such a method is, to say the least, imprecise and should only be used as a last resort.

There is some danger of disease being transmitted from caged fish to wild fish. With cage culture in public waters, you should always use either fish from the same body of water, or certified disease-free fish from a reputable source.

Losing feed to wild fish

A problem of cage culture in some bodies of water is that certain types of wild fish can suck food through cage walls or learn to create currents that force food out of the cage. At New Alchemy we experienced this with brown bullheads and in one Arkansas study it was estimated that carp and blue catfish caused 25 percent loss of feed. Where it is technically and legally feasible to harvest such fish, the problem may be turned into an asset. Cage-like traps of the type illustrated in Figure II–2–1 (page 44) are an excellent tool. At New Alchemy we were able to harvest enough of these bullheads with hook and line to justify the cost of cage culture even if no fish had been produced in the cages.

Overwintering

While the common practice is to put out cages in the early spring and harvest all the fish in the fall, it is usually possible to leave cages in all winter. I know of one instance in which rainbow trout were overwintered under 2 feet (0.6 m) of ice, with air temperatures which reached −50°F (−10°C). It was an interesting demonstration of the fishes' capability to escape freezing by providing their own water circulation, but it is impractical to keep fish in cages under such extreme conditions. Even cold water fish like trout realize no growth at near-freezing temperatures, and the convenience of leaving cages out does not justify the risk of loss of fish or damage to the cages. In less extreme climates it may be practical to leave cages out. I have heard claims of caged hybrid bluegills continuing to grow all winter in the Midwest, and certainly some species of fish would grow year-round in the deep South. In recent years, trout have successfully been raised from 5.9 inch (15 cm) fingerlings to commercial size fish (mean weight 0.7 lb or 318 g in one experiment) in 4½ months in cages during the winter in northern Alabama and Georgia. The same ponds, and in some cases the same cages, were used to grow catfish during the warm months.

Harvesting

Harvesting may be partial or total. Most commercial fish culture operations are based on total harvest at the end of a growing period, but partial harvesting is suitable for some types of commercial operations and particularly so for the subsistence grower. There is no method of fish farming better adapted to partial harvesting than cage culture. (Methods for partial and total harvesting of cages are described in Chapter VI–1.)

An ingenious combination of cage culture and trapping was developed in Japan. Rainbow trout placed in cages in the ocean were trained to associate the sound of a buzzer with feeding. When the fish were well trained, the cages were opened, allowing them access to the sea. Daily feedings at the sound of the buzzer continued. At harvest time the buzzer was sounded and the cages closed with the fish inside. In this way the trout received a combination of artificial and natural feed, and reached weights of 2.6 pounds (1.2 kg) in one year and 4.4–5.5 pounds (2–2.5 kg) in two years.

A similar experiment was carried out in Arkansas with channel catfish, which learned to respond to a variety of sounds and smells. Training for 57 days with feeding at the sound of a 512 Hz tone from an electric speaker permitted an 87 percent recapture; after 153 days of training 99 percent of the stocked fish could be recaptured. This technique would be practical only in waters closed to public fishing, and with abundant natural food.

Pond as hatchery for cage culture

I cannot leave the subject of cage culture without offering one more idea. One of the main limitations in culturing some species of fish is the difficulty of spawning them. The fish farmer must either set up a hatchery and learn how to operate it, or purchase stock from a commercial hatchery. Several species present the opposite problem. With bluegills, bullheads and tilapia, for instance, the trick is either to limit natural reproduction, or to harvest hard enough to compensate for it. (See Chapter V-1 for a detailed discussion of reproduction and breeding in small-scale aquaculture.) The many stunted populations of bluegills in the United States and tilapia in the tropics are examples of failure to accomplish either of these things.

It seems to me that many overpopulated American farm ponds which are presently almost useless from the standpoint of fish production, could become integral parts of food producing systems based on cage culture. Assuming recreational fishing is not an important factor, it should be possible to let bluegills, or other highly reproductive fish, breed as freely as they wish in a pond provided with floating cages. Periodically, young fish would be harvested from the pond proper and placed in the cages for intensive culture. In this way the pond would function as a completely unmanaged "natural hatchery." Since bluegills, bullheads, tilapia and many other fishes need access to the pond bottom to reproduce naturally, no breeding would occur in the cages.

Combining a "natural hatchery" with cages has not been tested in North America, but the Colombian aquaculturist Anibal Patiño has used such a system to grow *Tilapia rendalli* with excellent results. (Another form of the "natural hatchery" idea involves the use of station feeders and is discussed in Chapter III–7. It should be considered as an alternative strategy by any farmer considering cage culture as a solution to the problem of overpopulation.)

Polyculture

Patiño's work raises the question of why more filter feeding species have not been tested in cage culture, since they are the only fishes capable of obtaining significant amounts of natural food in cages. Possibly caged filter feeders could be added to conventional fish culture systems. The potential of this approach was shown by experiments at Auburn University in which blue tilapia were cage cultured in ponds being used for commercial type channel catfish culture or largemouth bass-bluegill "farm pond" polyculture. Results are shown in Table IV–3–1.

Table IV–3–1.   Results of cage culture of
blue tilapia in ponds used for culture of other fishes.

Type of Fish Culture	Average Weight of Fish Produced per 8.8 cubic foot (0.25 cubic meter) Cage			
	Type of Feed and Feeding Rate of Tilapia			
	Purina Trout Chow ® At 3% of Standing Crop Daily	Auburn No. 2 Feed – 3%	Auburn No. 2 Feed – 1.5%	No Feeding
Channel catfish monoculture (fed and fertilized)	30.17 lb (13.71 kg)	22.17 lb (10.08 kg)	20.81 lb (9.46 kg)	23.46 lb (10.66 kg)
Largemouth bass-bluegill polyculture (fertilized but not fed)	21.67 lb (9.85 kg)	11.06 lb (5.03 kg)	10.32 lb (4.69 kg)	8.90 lb (4.05 kg)

Tilapia production was greater in the heavily fed catfish ponds (normally characterized by a dense plankton population) than in the less fertile bass-bluegill ponds, and while the high protein trout feed had some effect in all cases, the less expensive feed produced no increase in growth in the catfish ponds. Production of catfish, bass and bluegills was not reported, but it is reasonable to assume that caged tilapia would confer at least some of the beneficial effects observed in polycultures of channel catfish with uncaged tilapia. (See Chapter IV–1.)

CHAPTER IV-4

Closed Systems

Closed systems
defined

All of the methods of aquaculture discussed so far are "open system" methods. That is, the water used by the farmer is directly linked to the natural hydrological cycle. Water enters the system by rainfall, seepage from ground water, and surface run-off. Water leaves by run-off, seepage and evaporation. While the cycle of precipitation and evaporation is the ultimate source of all water, it is possible to create cycles within that cycle. This is what occurs in closed system aquaculture, where the same water is routed through the system again and again. (Since there is always some loss of water to evaporation, requiring inputs of outside water, there are actually only open and "*semi*-closed" systems. For convenience, I shall refer to "closed" systems.)

Inappropriate
technology

If effluent from a habitat densely populated with aquatic animals is to be reused by those animals without benefit of lengthy natural purification processes, it must be restored by other means. This gives rise to a sizable technology. I suspect Western society's perverse fascination with technology is largely responsible for some of the more elaborate closed aquaculture systems. The few outstanding technological monstrosities to which aquaculture has so far given birth are all closed systems. I refer to such aberrations as giant fiberglass "silos" where trout are raised in water circulating vertically, and carp culture systems where fish are stocked so densely that they literally cannot turn around.

The mentality behind some of these systems is straight out of the *Guiness Book of World Records*. Yields of such systems have been measured, by extrapolation, in the millions of pounds per acre. Proponents of one vertical circulation scheme claim to have produced more protein per unit area than any other food growing method—to which I can only offer a resounding "So what?!"

Anyone who wants to challenge the protein per acre record need only apply the same technology in a taller tank (it is customary to measure food production per unit of surface *area*, so the taller the tank the better) and/or get a bigger and faster pump. Another approach could be to force feed one or two fish in a very small enclosure, and extrapolate from that. It is all equally meaningless. Obviously, the technology involved in such systems is inappropriate; it is not going to help solve any of the world's food problems.

Even when evaluated within their own economic frame of reference, these "supersystems" have been unsuccessful, and the few business ventures founded on them have been financial fiascoes. Though some slightly more modest versions of the same concept may be profitable, in the long run the technology is still inappropriate. In any case, energy-intensive, closed aquaculture systems on a commercial scale are probably beyond the financial means of most readers; the minimum initial investment for a closed trout culture system is estimated to be $250,000. Such systems are certainly outside the interest of this author.

Need for closed systems

Nevertheless, there is a place for closed systems in aquaculture. Although research dollars for and publicity about closed systems have gone largely to giant operations, it seems to me that no approach to aquaculture so clearly lends itself to the "small is beautiful" concept as closed recirculating water systems.

If you have one or more outdoor ponds which can be controlled to a sufficient degree, it will usually prove easier, cheaper, and less risky to practice open system aquaculture on an appropriate scale. Cage culture is an alternative for some people combining some of the advantages of closed and open systems. But neither open system growing nor cage culture meets the needs of many people who neither own nor have access to bodies of water, or sites on which to construct them. If aquaculture is to be made available to such people and to realize its full production potential in North America, what is needed is a productive system capable of operating within the confines of a backyard, a greenhouse, or even indoors.

Limitations of small aquaculture systems

You can raise 100 pounds (45 kg) of fish or more in a floating cage measuring 4 feet (1.2 m) on all sides by merely stocking and feeding. It is not possible to raise any appreciable quantity of fish in a backyard pool of that size without making some provision for replacement or purification of the water. Obviously, in a small chamber containing dense populations of fish, it would be necessary to periodically remove organic wastes simply to prevent pollution and deoxygenation. But, while such simple hygiene might permit *maintenance* of large amounts of fish, it will not do anything to promote growth. To understand why, we must see just what limits the growth of fish.

Overpopulation

In natural waters, a principal limiting factor on the growth of fish is scarcity of food; fish farming begins by doing away with this limitation. But, at high population densities, even with an abundance of high quality food and high concentrations of dissolved oxygen, fish may not grow. They will eat, perhaps reproduce and be perfectly healthy, but fail to gain weight. Why? Because they are physiologically overpopulated.

Fish perceive their own population density in two ways. Territorial species, at least, have some notion of how many of their species are present in a given area. This phenomenon is of little concern in closed systems. (See Chapter IV–3 for an explanation of why this is so.) But fish also perceive their own populations in terms of biomass. Unlike humans, they recognize when they are in danger of overexploiting their resources, and they do something about it. They could and may stop reproducing but, being cold-blooded animals, they also have another option, and they often "choose" it: they stop growing. In this way, the number of individuals may increase, but the total weight of fish does not.

You can induce such fish to resume growth by thinning the population or expanding their living quarters, but such tactics are not compatible with the goal of high production in a limited space. You can also pass water through the system continually, as is usual in raceway culture. (See Chapter VII–2 for design of raceways.) But often this will be wasteful and costly and, in a backyard, it might be hard to dispose of the run-off. (Also, we would no longer be dealing with a closed system.) So we employ bio-technology to fool the fish into perceiving a lower biomass than is really present.

Fish do not estimate their own biomass within a system by weighing each other. Instead they respond to the concentration of ammonia (and possibly other compounds) in their water which they themselves produce. When certain levels, characteristic of very high population density, are reached, growth ceases. If we can remove or break down these metabolites, then we can persuade the fish to continue growing at high population densities.

Fish limit their own growth

The techniques for deceiving fish in this manner were originally developed for use in public aquariums and government trout and salmon hatcheries. They are the basis for all closed aquaculture systems, large and small. While much more effort has been devoted to large closed systems, the potential of small-scale, closed system aquaculture has not gone entirely unnoticed.

Unfortunately, most attempts in this area have been to scale down and transfer the technology of large systems to the backyard and the basement. While avoiding the monstrous nature of some of the larger operations, these systems, to my knowledge, have failed to meet the needs of families and communities for productive, inexpensive and relatively risk-free fish production. And while the small prototypes which have been tested have, individually, very little measurable environmental effect, large numbers of small, energy and technology-intensive aquaculture installations would ultimately raise the same problems of energy use, technological expense, pollution, and rising operational costs that the "supersystems" have had to face.

We at the New Alchemy Institute were among the first, if not the first, to explore alternative closed systems designed, in the words of John Todd, "to replace hardware with information," so that the closed system concept becomes appropriate for the small grower. Specific goals for New Alchemy closed aquaculture systems were the following:

Goals of New Alchemy closed systems

- Low capital cost.
- Power to be derived primarily from the sun and wind; no fossil fuel-based heating or cooling.
- Modular design units so that capacity may be increased by simply adding sub-units.
- Fishes low on the food chain, especially phytoplankton feeders, to be emphasized.
- The bulk of the feed to be produced within or adjacent to the fish-raising facilities.
- Use of "waste" materials as feeds to be maximized, and costly ingredients, such as fish meal and grains, to be minimized.
- Polyculture to be developed, so as to optimize space utilization and maximize production.
- Fish culture system to be linked to adjacent agricultural systems, not only by

using agricultural products and wastes as feeds, but also by irrigating and fertilizing crops with fish pond residues.

We have by no means exhausted the range of approaches which might be taken to achieve these goals, let alone "perfected" our systems with respect to any of them. The systems mentioned in this chapter are discussed in the hope that they will be modified and improved by readers to meet individual needs.

Blue tilapia in closed systems

Before describing any systems, I should say something about the fish we chose for our initial experiments, the blue tilapia (*Tilapia aurea*). It is by no means the only fish which can be grown in New Alchemy-type closed systems. Over the years we have moved toward polyculture, incorporating other species of tilapia, mirror carp (*Cyprinus carpio*, var. *specularis*), chain pickerel (*Esox niger*), yellow perch (*Perca flavescens*), bullheads, channel catfish (*Ictalurus punctatus*), largemouth bass (*Micropterus salmoides*), rainbow trout (*Salmo gairdneri*), sunfishes, three species of Chinese carp, fathead minnows (*Pimephales promelas*), fresh water clams, crayfish and a variety of tropical fishes with varying degrees of success. But blue tilapia remains our principal crop, and it behooves us to defend our choice. We chose blue tilapia for the following reasons:

• It is capable of obtaining the bulk of its nourishment by filter feeding on phytoplankton, normally the cheapest of all fish foods to produce. According to ecosystem theory, the most productive fish pond would be one in which there is a large population of phytoplankton and an appropriate number of filter-feeding fishes to utilize it, and centuries of experience by Chinese fish farmers support this theory. (Among edible phytoplankton feeders are certain other species of tilapia, young common carp, *Cyprinus carpio;* bigmouth buffalo; *Ictiobus cyprinellus;* silver carp, *Hypophthalmichthys molitrix;* clams and a variety of exotic fishes not usually available in North America.)

• Tilapias are hardy fishes, among the most disease-resistant of all fish, and capable of tolerating rough handling, crowding, temporary oxygen depletion, fright, thermal shock and other traumas. These characteristics were important to us because, from the start, we wished to encourage people without prior fish-keeping experience to undertake aquaculture. We needed a fish which could withstand the mistakes and initial ineptness which were inevitable.

• Since (except in the few principal fish farming regions of North America), the availability of young food fish for stocking is undependable, to say the least, we needed a fish which could be reproduced by the inexperienced culturist. Not only are tilapia easy to spawn, they are among the most interesting of fish to breed. (See Chapter V–1.) Unlike most food fishes, they are not reluctant to breed in the close confines of a closed culture system.

• In most North American environments tilapia, should they escape into the wild, would not be able to survive the winter. Thus there is little danger of upsetting native ecosystems by establishing an unwanted exotic species. (The extreme southern parts of the United States are an exception; in a few places there tilapia are already established in nature. Elsewhere in the region we do not advocate the culture of tilapia without stringent measures to prevent their escape or release.)

• Blue tilapia is a very fine food fish. It appears similar enough to such popular North American food fish as perch, bass and sunfish, that it is not likely to be dismissed on the basis of food prejudice.

The Back Yard Fish Farm

Our first tilapia-based fish culture system, constructed in 1971, was dubbed "The Back Yard Fish Farm," and has evolved to the facility shown on page 241. The 25

foot (7.6 m) geodesic dome greenhouse which covers the pool is made of Kalwall®
fiberglass (see Appendix IV–4), a relatively inexpensive material which may be cut,
shaped, and fastened using ordinary woodworking tools. The frame is made of
wood. Two of the triangles of the dome are hinged to open outward, serving as
cooling vents; one doubles as a door.

*Interior of the "Back Yard Fish Farm," an 18 foot (5.5 m) diameter pond under a
geodesic dome greenhouse at the New Alchemy Institute, Hatchville, Massachusetts.
Plant at lower right is comfrey, a principal food for the blue tilapia cultured in the
pond. Comfrey leaves can be seen suspended in the water from the platform at the left.
Other plants are vegetables and ornamentals. Tank barely visible at right houses a
biological filter in which water cress is raised. Water is pumped into the filter through
the tube at the right and can be seen reentering the pond through a trough at the
center of the photo. (Photo by Ron Zweig, New Alchemy Institute.)*

Other types of greenhouses could be used as well as domes. We built our first
dome simply because we wanted the experience of building one; later we discovered that it is among the more effective shapes in terms of trapping heat and
admitting light at all daylight hours. On Cape Cod, we are able to keep tilapia six
months a year in this structure; further south, it might be possible to maintain
them in it year-round.

The pool under the dome is 5 feet (1.5 m) deep and 18 feet (5.5 m) in diameter, and
is lined with Hypalon® plastic. When fertilized once a year with about 5 pounds (2.3
kg) of horse manure it supports a thick "bloom" of green algae which serve multiple
functions. Not only are they the primary food of the tilapia, they oxygenate the
water, aid in processing fish wastes and toxins and increase the heat-retaining
capacity of the system by acting as a "black body."

When we constructed the system we did not know that some types of phytoplankton are able to process natural fish toxins, so we incorporated a bacterial filter. At present we cannot tell whether it is supplemental or superfluous, but such filters are necessary in high density fish cultures where phytoplankton either is not desired or is not abundant. I have already mentioned the growth-inhibiting chemicals naturally produced by fish. Ammonia and other possibly detrimental nitrogen compounds can be broken down to benign compounds by the action of de-nitrifying bacteria. Such bacteria will establish themselves near the water surface whenever two conditions can be met: (1) high dissolved oxygen concentration, and (2) a calcium carbonate substrate. We provided these conditions by setting up a series of tanks made from discarded refrigerator liners. Water from the bottom of the pool is pumped into the first tank by means of a small electric pump; this tank serves as a settling chamber for solid wastes. The water then passes into subsequent tanks by gravity feed. The calcium carbonate substrate consists of pieces of clam shells, a material readily available on Cape Cod. In inland areas it would be necessary to use some form of limestone, or possibly egg shells. The filter also continually buffers pH.

A tropical fish-raising friend, Cal Hollis of Houston, Delaware, saved us one mistake by pointing out that, while salmon hatcheries and public aquariums use oyster chips, with an average particle size of about ⅛ inch (32 mm) in their filters, we would need at least ½ inch (128 mm) pieces. Use of smaller particles would result in mechanical removal of algal cells. (The small holes between small particles trap smaller items. Thus algal cells will pass freely through a gravel filter but sand will remove them from water.)

Since the denitrifying bacteria are largely restricted to the upper few inches of the calcium carbonate substrate, the critical dimension of the filter bed is its surface area. The reports available on bacterial filters suggest that the filter surface area be at least half that of the fish culture tank; some compensation can be achieved by increasing the rate of pumping. Our filter beds have a combined area of 40½ square feet (3.76 square m), while that of the pool is 277 square feet (25.74 square m). Since neither a larger filter nor a more powerful pump is feasible for us, we are certainly running at sub-optimal levels of filtration; presumably the algae make up for it. The problem is greatly simplified when a cement pool is used. In this case the lime in the cement acts as a bacterial substrate so that, assuming some circulation, the pool walls are the filter.

Our biological filters of course also do a certain amount of physical filtration, and eventually they clog. We have found yearly removal and washing of the clam shell pieces necessary. A more frequent problem is clogging by filamentous algae, which flourish in the filters. These may be removed by hand as necessary and fed to the tilapia. Further advantage can be made of the filters by periodically stirring the shells to flush midge larvae, snails, leeches, daphnia and other small animals which take up residence there into the fish pool, by using them as "pots" to cultivate aquatic plants, and by using the bottom sludge as a fertilizer. Since the denitrifying bacteria and the various animals which inhabit the filters are aerobic, it is necessary to reoxygenate the water by splashing it back into the pool.

A variation on the New Alchemy filter has been developed by Steve Serfling and Dominick Mendola of Solar Aquafarms, Inc. in Encinitas, California. The filter in their Solar Aquadome is confined within a circular fiberglass wall in the center of the dome-shaped culture pool. An electric pump continually draws water from the filter chamber and jets it back into the pool at the perimeter, setting up a circular current which returns wastes to the center. The filter substrate, located within the fiberglass cylinder is a high surface area plastic mesh. (Why they do not find calcium carbonate necessary is unclear to me.) This mesh provides habitat for denitrifying bacteria and detritivores such as snails. Zooplankton and floating plants are also stocked in the center section to help process waste. The pump, an air lift system, and occasional manual removal of plants all serve to cycle food from the filter chamber to the fish.

Our Back Yard Fish Farm has proven virtually problem-free; routine labor inputs include only supplemental feeding and daily checking of the pump and filters. Failure to perform either chore can lead to reduced growth rates, but not to death of fish. Tilapia overpopulation is prevented by netting out the young at an early age, when they gather on the surface.

In 1976, we used the Back Yard Fish Farm as a combination breeding and rearing pond. It yielded 55 pounds (25 kg) of edible size tilapia, plus several thousand young fish which we used elsewhere at New Alchemy. We also took advantage of the greenhouse by growing vegetables in the space bordering the pond and on hanging structures.

The Back Yard Fish Farm fails to meet one of our eight criteria, as it is not modular nor easily adaptable to scaling up, though one could conceive of a larger greenhouse with several sunken pools. It has, however, shown itself to be a workable family fish production system.

A second New Alchemy closed system is modeled after a tropical river, as in nature such aquatic environments are among the most productive. The "Solar River" has been through many configurations; its present design consists of four components: a cement reservoir, 15 feet (4.6 m) by 15 feet (4.6 m) by 6 feet (1.8 m) deep, housed in a greenhouse roofed with double glazed acrylic plastic with an insulation value equal to thermopane glass; two smaller outdoor pools (15 ft by 5½ ft by 18 in [4.6 m × 1.7 m × 0.5 m] and 15 ft by 6½ ft by 18 in [4.6 m × 2.0 m × 0.5 m], respectively); and a water pumping sailwing windmill. As in the Back Yard Fish Farm, the greenhouse doubles as a vegetable growing area. In earlier versions of the Solar River we incorporated a 128 ft² (11.9 m²) solar heater and small greenhouses over the smaller pools, and we may reconstruct these to extend the growing season.

Formerly the main pool was used as the fish culture chamber, but currently it is maintained without fish, serving to produce phytoplankton and zooplankton, which are then pumped uphill into smaller pools where fish are grown. The water proceeds by gravity flow from the uppermost pool to the second one and back to the reservoir.

At present we have next to no control over the plankton population in the reservoir beyond knowing that fertilization will bring an increase in total plankton. But we have obtained some impressive preliminary growth data. Three hundred brown bullheads (*Ictalurus nebulosus*) stocked at a mean weight of 1.1 ounce (31.9 g) gained 54 percent in weight in a month, when provided with commercial trout feed at a rate amounting to 0.275 pounds (125 g) per day. The weight of five hundred young blue tilapia was increased 16 times in two weeks, with no feed other than the plankton from the reservoir and insects provided by a U.V. bug light feeder (most of which were too large for the tilapia to consume).

The Solar River

Though this present system cannot be used to produce substantial quantities of food fish, it makes an excellent nursery for young fish. We do believe that by scaling up and perhaps altering the relative size of the pools, we could construct a production scale Solar River. At present, the greatest stumbling block is the need for a larger, yet cost-effective windplant.

Aeration

Another approach to small, low technology closed systems substitutes aeration for circulation and filtration. We know that providing dissolved oxygen beyond the amount necessary for survival has a positive effect on fishes' growth. Proponents of aeration claim that production of small-scale fish culture systems can be tripled with no modification other than installation of a suitable aerator.

There are a number of types of mechanical aerators for fish ponds on the market. The main drawbacks to their use by small-scale fish farmers are the initial cost, their constant demand for electrical energy, and the necessity of an automated back-up system. You might think that maintaining population levels in the pond at a low enough level so that fish would not die for lack of oxygen if the aerator quit would eliminate the necessity for a back-up system. However, it has been shown that in order to pay back the cost of purchasing and operating the aerator you will usually have to produce fish in excess of what a pond could support without aeration. Hence it is crucial to have an automated back-up system ready at all times.

Rodale's aerator

In recent years, Rodale Research of Emmaus, Pennsylvania, have become interested in aeration; they now manufacture and sell a ½ horsepower aerator which is said to meet the criteria of low initial cost, low power demand and reliable operation. At their New Organic Gardening Experimental Farm, these aerators are used in outdoor "fish garden" ponds 12 feet (3.7 m) in diameter and 3 feet (0.9 m) deep, where channel catfish, carp and buffalofish are raised experimentally. In 600 gallon (2,544 liter) indoor tanks they found it necessary to augment aerators with bacterial filters similar to those described earlier in this chapter. But they also found that providing aeration and flow through the filter at ratios sufficient to produce significant quantities of fish required the use of pump-aerator combinations drawing power in amounts that made costs prohibitive for the small-scale grower. Rodale researchers are presently working on reducing this cost by combining the pumping and aeration functions. (See Chapter X–2 for a more detailed discussion of aeration.)

Solar-algae ponds

New Alchemy's newest type of aquaculture facility, the solar-algae pond (see photos on page 246) represents a biotechnical approach to aeration. Its advantages include in-system feed production, thermal characteristics, biological processing of pollutants and adaptability to various types of sites. The 5 foot (1.5 m) by 5 foot (1.5 m) diameter tubes (capacity approximately 750 gal or 3,180 *l*) are fabricated of 0.06 in (15 mm) Kalwall® fiberglass (discussed earlier in this chapter). Our solar-algae ponds are put together at the Kalwall® plant with a cement of unknown composition. However, one New Alchemy Associate Member has succeeded in making a solar-algae pond using silicone sealant, pop rivets, and an outside coating of epoxy cement. (Caution: some epoxies are toxic, and should under no circumstances be allowed to come in contact with the water.)

Water chemistry

With this simple device it is possible to expose the full water column to solar energy for photosynthesis and heating to occur (rather than just the upper few inches as in a sunken or opaque pool). One advantage of this arrangement is in providing food for algae-feeding fish. Further advantages come from the vast amount of photosynthetic activity which goes on in a solar-algae pond. During daylight hours supersaturation with dissolved oxygen is the norm. (See Chapter X–2 for a discussion of oxygen chemistry.) For example, at 89.6°F (32°C) water is saturated with dissolved oxygen at 7.4 ppm, though a biologist would consider 6 ppm a satisfactory level in a natural pond at that temperature. At the same temperature we have recorded dissolved oxygen levels over 20 ppm in solar-algae ponds stocked as densely as one fingerling tilapia per 3 gallons of water (1 fish per 12.7 liters). Though mechanical aeration is necessary at night, the amount needed is reduced by the high initial oxygen levels. The only other activity needed to maintain satisfactory water chemistry is periodic removal of bottom sediments.

Extending the growing season

Due to the dark green color of the water in solar-algae ponds, they absorb heat during the day and radiate it at night. As a result they can reduce heating costs and buffer temperatures in greenhouses. It is very easy to modify both their heat-absorbing and radiating capacity. If you want to extend the growing season for warm-water fishes, you can equip solar-algae ponds with an extra Kalwall® skin, with an air space between the layers, lids, reflectors, and/or polystyrene foam pads underneath.

During the extreme winter of 1976–1977 two solar-algae ponds with all the above devices were maintained outside at the New Alchemy farm. Though outside temperatures fell to 1°F (−17.2°C), the lowest water temperature reached was 37°F (17.8°C). The thermal flexibility of solar-algae ponds is well illustrated in New Alchemy's "Ark" on Prince Edward Island, Canada. There we have maintained two banks of indoor solar-algae ponds. The upper bank, uninsulated, has been used to house rainbow trout (optimum temperature range 55–69°F, or 12.8°–20.6°C, upper lethal temperature 82°F or 27.8°C). The lower bank, with double skins and styrofoam pads, has overwintered blue tilapia (optimum temperature range 80–90°F, or 26.7–32.2°C lower lethal temperature 48°F or 8.9°C).

When setting up solar-algae ponds, we fill them with tap water, let it stand for a day to leach out possible toxic chemicals in the fiberglass, drain, fill with tap water again (approximately 700 gallons), let stand for another day, and fertilize with a single small dose of human male urine. The fish and sunlight do the rest; in a few days an opaque green bloom is produced, rarely requiring refertilization, if ever.

Types of algae

Usually, solar-algae ponds become virtual algal monocultures, but we have yet to learn to control or predict the species of algae which will predominate in a tank. Since some algae are nutritionally and chemically superior, this is an important factor. Luckily for us, our solar-algae ponds are often dominated by *Scenedesmus*, which takes up ammonia directly during photosynthesis. Another common type is *Golenkinia*, about which little is known.

Importance of light

Evidence to date indicates that light is limiting in solar-algae ponds. So we place some of ours in "solar courtyards," i.e., in front of a reflective Mylar® screen, or a white painted wall oriented so as to take maximum advantage of the sun. The distance between the solar-algae pond and the reflective screen is important; a distance of about 2 feet (0.6 m) worked best for us. In one instance we solved an overheating problem by planting scarlet runner beans in front of the reflective screen in the spring. In the summer, given strings to climb, the bean plants shaded the screen with dense foliage. In fall, they were cut down.

Harvesting circular enclosures like the solar-algae ponds presents special problems. (See Chapter VI–1 for details of harvesting.)

Eight 700 gallon solar-algae ponds in a solar courtyard at the New Alchemy Institute, Hatchville, Massachusetts. The curved white walls (left) and white gravel on the ground reflect light and solar heat back to the ponds. (Photo by Ron Zweig, New Alchemy Institute.)

The same solar-algae ponds in mid-winter (below). With tight fitting lids, a double fiberglass wall with an air space between layers and polystyrene foam pads underneath, they do not freeze. The dark color indicates a vigorous growth of planktonic algae. (Photo by John Todd, New Alchemy Institute.)

You can guess at the potential productivity of solar-algae ponds from the results of our first polyculture experiment. A combination of blue tilapia (95.2% by weight), mirror carp (4.8%), and a very few silver carp and bighead carp, yielded 15.5 pounds (6.8 kg) of fish in 98 days. That amounts to 4.3 pounds per cubic yard (2.5 kg/m³)—and this despite a nocturnal aeration failure which killed 72 fish.

The adaptability of solar-algae ponds is remarkable. Scaling an operation upward or downward is a simple matter of adding or subtracting modular units. We have used them effectively individually or coupled together, in greenhouses or outside. With the aid of artificial light we have used 66 gallon (277 liter) solar-algae ponds to maintain tilapia in dense phytoplankton blooms indoors over winter. On one occasion, we even managed to culture young rainbow trout in a dense algal bloom, though only at the low density of one fish per 20 gallons (84.7 liters). Elsewhere in the U.S. and Canada, solar-algae ponds are being used a variety of ways in marine fish culture experiments, including culturing algae for use as oyster feed.

We see these solar-algae ponds as particularly promising for rearing warm water fish in northern climates where there is a need to extend the growing season; in arid lands where conservation of water is of the utmost concern; and in the urban environment, where space is limiting. At New Alchemy, we maintain single solar-algae ponds, stocked with blue tilapia and other fishes, in the gardens where natural supplementary feeds are close at hand and the water can readily be used in irrigation and fertilization. The largest scale use of solar-algae ponds to date is in New Alchemy's two "Arks," which I described briefly in Chapter IV–2.

Pairing solar-algae ponds

We have only recently begun to experiment with solar-algae ponds in pairs, but there are a number of interesting possibilities. If water is circulated back and forth between two solar-algae ponds, one of them may serve any of a variety of filtering functions: trapping sediments, removing nutrients by aquatic plants or hydroponically grown vegetables, etc. Or one solar-algae pond may be used to produce food for fish in the other. During the day, with the two ponds unconnected to each other, a zooplankton population can be allowed to build up in one pond. Then at night, when aeration is necessary, water may be exchanged between the two tanks, supplying food to the fish. We are also beginning to use solar-algae ponds as components in "solar river" systems similar in design to that discussed earlier in this chapter.

Pumps

All of the systems discussed so far in this chapter, with the exception of isolated single solar-algae ponds, require some sort of pump to move water. The choice of pump is one of the most important economic determining factors in closed system aquaculture, and cannot be considered too carefully. Begin by asking three questions:

(1) How far, vertically and horizontally, must water be lifted? (In almost all cases, water in closed aquaculture systems will not need to be lifted nor displaced far. It is a common mistake to install a pump far more powerful than needed.)

(2) How much water is to be moved?

(3) How fast must the water be moved?

Catalog specifications for commercial pumps should tell you whether a particular pump is adequate, overpowered or well matched to your job in terms of these criteria. Bear in mind, though, that catalog specs may be slightly exaggerated, and that the effectiveness of any pump will be reduced with time. When in need of advice, the best place to turn is usually a plumbing supply house.

Airlifts   Perhaps the simplest way of moving water is with airlifts. In this method there is actually no "water pump" at all. Rather, an air pump moves air, which in turn moves the water while aerating it at the same time. The functioning of an airlift is diagrammed in Figure IV–4–1.

The only real advantage of airlift pumps is the combining of the aeration and water pumping functions, which may prove more economical than installing separate air and water pumps. Air lifts are less efficient than other types of water pumps and are generally limited in application to shallow, small volume systems. The depth limitation is because 65–75 percent of the total height of the vertical tube has to be submerged, while a greater amount of energy is required to pump air into deeper water. Volume is limited by the need to limit pipe diameter to 0.2 in (5 cm) or less.

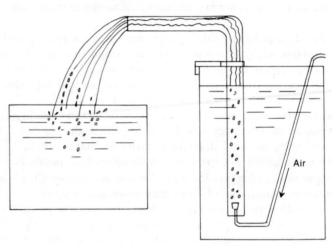

FIGURE IV–4–1. *Airlift.*

Airlifts are normally run by electric air pumps, though the Bowjon Co. manufactures a windmill designed to move water using the airlift principle. (See Appendix IV–3.) When selecting equipment for an airlift system remember that, assuming the volume of air is held equal, the smaller the bubbles produced the greater the lifting power—and the greater the rate of oxygenation. (See Chapter X–2 for air pumps, air stones, etc.)

Energy sources   Though large closed aquaculture systems must rely on electrical energy, the small-scale grower can certainly consider other sources, notably wind. The difference lies in the degree of brinkmanship practiced. For large closed systems to approach economic feasibility, they must be stocked at maximum capacity. Such systems cannot withstand a cessation, or even a reduction, of flow for more than a few minutes; water must circulate continuously. Under such circumstances wind or liquid fuel driven pumps are out of the question. (Though a gasoline or diesel pump is mandatory as a backup in case of power failure.)

Small-scale closed system operators can stock at similar densities if they are willing to take the risk, but cost factors do not force them to do so. If flow stops in a recirculating system stocked at less than carrying capacity, growth will be temporarily retarded and other malfunctions may occur, (including death of denitrifying bacteria in biological filters), but the fish are much less likely to die. This not only opens the door to the use of wind energy, it also permits the small-scale grower to consider doing without a backup pump.

Whether or not to include a backup pump in a moderately stocked system is a matter of economics. Will the improved growth rate which is almost certain to result pay for the cost of the pump? The answer to this question depends not only on the cost of the pump, other aspects of the aquaculture system, and the value of the fish produced, but also on the nature of the main pump and the prevailing climate and weather. *Backup pumps*

If the main pump is electrical, stoppage will occur either due to malfunction of the pump or power failure. In either case, the problem is likely to be repaired in a relatively short time or, in a real emergency, a second pump can usually be installed on short order. If wind is the main source of energy for moving water, the duration of stoppage will depend on the weather, and a significant amount of growing time may be lost. This, in turn, is related to the length of the growing season. For example, Cape Cod is generally a favorable location for windmills, but prolonged calm periods do occur, notably in July and August. With the short northern growing season, a few days lost in midsummer is more serious for us than for someone in the South. We, therefore, have found it necessary to install an electric backup pump in our wind-driven "Solar River" system. (It is a simple matter to rig a pressure switch, or some other device, to activate a backup pump automatically should the main pump stop.)

Of the energy sources which may be employed in closed systems, electricity is suitable for either main or backup pumps, liquid fuel due to its high cost is usually only appropriate for backup purposes, and wind is only appropriate to power the main pump, since if the wind fails a second windmill will be of no use.

Few technological gadgets are so diversely developed as the pump. The prospective purchaser of electrical, gasoline or diesel pumps will find a bewildering variety to choose from. Apart from emphasizing the importance of the three questions listed earlier, I can only advise that submersible pumps, though designed for higher lifts than needed by most small-scale aquaculturists, have the advantage of never losing their prime. Also, be aware that there may be great price differences in comparable pumps, and it will pay to shop.

Whether or not to consider wind as an energy source for moving water depends on three things: *Wind energy*

(1) Your local power rates compared to the costs and benefits of a windmill. (Be sure to think about future, as well as present rates.)

(2) The wind patterns at your location: If you have average winds of 7–8 mph (11.3–13.4 km/h) during the growing season, wind power is worth considering.

(3) Your own idealistic concern about energy sources: At New Alchemy we are dedicated to the maximum utilization of "clean" energy, as opposed to fossil fuel and nuclear energy, and so have opted for wind pumps whenever possible.

An often overlooked economic factor in wind energy is the saving resulting from reduced wear on an electrical backup pump as opposed to the same pump in full time use. Most electrical pumps have a life span of 6 to 10 years in full time use, following which you can count on a 60 percent replacement cost if the pump is to be repaired.

The use of wind pumps in aquaculture is very new and no manufacturer has yet produced a mill specifically for this purpose. Several manufacturers of quality water pumping mills are listed in Appendix IV–3. At present, it seems that if you are mechanically inclined, you can fabricate your own mill at a cost which will compare favorably with a commercial one. (Two papers on this subject are listed in Appendix IV–4.)

Biologically designed vs. engineered closed systems

I would like to end this discussion of closed system aquaculture with a graphic comparison of two very different systems (Table IV–4–1). The biological/structural approach to closed system aquaculture is seen in solar-algae pond polyculture set in a solar courtyard. The more conventional engineered/technological approach is shown by a salmonid culture system described in a 1974 report "Technology of Closed System Culture of Salmonids" by Thomas L. Meade, College of Resource Development, University of Rhode Island. Neither system is entirely "typical," and neither represents the highest development of its type, but I think they provide a fair basis for a general assessment of the two types of systems.

Table IV–4–1   Biologically Designed and
Engineered Closed System* Aquacultures.

Category	New Alchemy Solar-Algae Fish Culture Biological/Structural	Salmonid Culture in Recycling Silo System Engineered/Technological
Energy Source	The sun: supplemental nocturnal aeration.	Electricity
Energy use	Minimal: radiant, solar energy, plus supplemental compressed air.	Heavy electrical demand for re-circulating, heating, cooling, purifying, etc.
Design Emphasis	a. Passive: few moving parts. b. Internal, self-regulating and purifying ecosystems powered by a renewable energy source—the sun. c. Internal photosynthetically based.	a. Hardware based. b. Energy-intensive with rapid flow-through. c. External foods. d. Emphasis on technological regulation. e. Elimination of plants, animals other than cultured species. f. Sterilization.
Materials	a. Light transmitting .060 inch (1.5 cm) thick fiberglass cylinders. b. Placement in light-reflecting courtyards or walls. c. Auxiliary compressed air equipment.	a. Metal tanks, pumps, filters, water exchangers, heating-cooling units, sterilizers, automatic feeders, cleaners, back-flushing plumbing, co-lumnar gas exchangers, etc.
Structure	Simple: based on maximization of light absorption.	Complex: technologically.
Biotic Environments	Relatively complex: photo-synthetically based ecosystems.	Simple: Elimination of most organisms except for micro-organisms in bacterial filters used in water treatment.
Control	Primarily ecological and internal through interactions of micro-organisms, phytoplankton, zooplankton, detritus. Also through use of several fish species occupying individual niches, occasional water change and sediment siphoning.	Electrical, chemical and mechanical controls superimposed onto the water body in which fishes are housed.

Table IV-4-1 *Continued*

Category	New Alchemy Solar-Algae Fish Culture Biological/Structural	Salmonid Culture in Recycling Silo System Engineered/Technological
Likelihood of Contaminating Fish Flesh With Poisons	Slight: Fish feed predominantly low on food chain, particularly algae, in a protected environment. Dangerous chemicals *not used* for disease control and environmental management—less dependent upon contaminated commercial feeds. (Solar-algae ponds do, however, require thorough initial leaching.)	Highly likely: Fish fed exclusively upon commercial feed, contaminated with agricultural poisons. Also these feeds include fish meal from marine species which might contain poisons. Toxic contamination from disease and sterilizing chemicals, including algicides, and from stabilizers, etc.,in pumps, plastics, piping, etc.
Recirculation	Occasional: Low-flow, air lift pumps used to exchange water between two solar-algae ponds.	Continuous and rapid: Up to 1,500 l per minute (23,760 gal per hour) in a 4,888 gal (18,427 l) facility.
Thermal Regulation	a. Ponds are efficient year-round solar heaters: algae absorb solar energy thereby heating water. Summer venting and partial shading with plants prevents over-heating. b. Heat-tolerant species.	a. Refrigeration and heavy energy requirement for heating to maintain narrow thermal range throughout year. b. Heat-sensitive species.
Oxygen for Life Support	Generated internally by algae during photosynthesis. Supplemented at night with compressed air.	Continuously supplied from industrial oxygen cylinders.
Fish Species	Polyculture based on plankton-feeding fishes, particularly tilapia.	Monoculture, typically of carnivorous salmonids (though many types of fish could be used).
Primary Feeds	Internally grown foods: particularly attached and planktonic algae, also zooplankton. Bulk of feeds grown with fishes in the solar-algae ponds.	Processed commercial feeds incorporating fish meal, soy and grain meals, vitamins synthesized from petroleum, minerals, stabilizers, etc. Many ingredients potentially applicable directly to human nutrition.
Supplemental Feeds	Weeds, agricultural wastes, insects, earthworms, etc. Commercial feeds are used in moderate amounts to balance the diet and maximize growth, but their use is not essential.	Often animal by-products prepared from slaughter-house offal.
Disease Control	Tilapia, the principal fish, among the most disease-resistant of fishes. Mass mortalities from disease rare.	Salmonids sensitive to diseases. Control through water sterilization, chemical and antibiotic treatments and quarantine procedures. Mass mortalities from disease common.

Table IV–4–1   *Continued*

Category	New Alchemy Solar-Algae Fish Culture Biological/Structural	Salmonid Culture in Recycling Silo System Engineered/Technological
Biopurification	Toxins purified internally: algae and other organisms rapidly utilize them as nutrient sources.	Toxins purified through technological steps and tertiary water treatment, including bacterial filtration.
a. *Clarification*	None: particulates an important component of internal ecosystem.	Screening and steam cleaning to remove particulates.
b. *Sulfide Toxicity*	Not yet occurred.	Occurs when filters clog. Requires back-flushing to eliminate anaerobic layers in filters.
c. *Nitrite Toxicity*	Not yet observed: algae photosynthesis prevents nitrite build-up.	Occurs on occasion. Water change most effective treatment.
d. *Ammonia Toxicity*	Ammonia utilized by ecosystem. Green algae take up ammonium directly during photosynthesis. Nitrification takes place naturally in shallow sediments. If ammonia should ever build up, tilapia, carps and bullheads can withstand ammonia levels up to ten times greater than levels lethal to salmonids.	Ammonia toxicity major threat to system. Nitrification through chemical, physical and biological methods. Plastic media trickle filters and/or submerged aerobic bio-nitrification filters are used. Filter bacteria often pre-activated on synthetic growing media.
e. *Nitrate Toxicity*	Toxic levels not yet observed, as most nitrogen is incorporated by algae prior to nitrate stage in nitrification process.	Columnar denitrification requiring introduction of outside chemical sources of organic carbon, such as glucose and methanol—column filled with polypropylene flexi-rings.
Holding Capacity	Densities of three fish per gallon have been held for up to six months. Fish ranged from fingerling size to 0.55 lb (250 grams).	Salmon smolt densities of 8–10 fish per gallon, with continuous dilution by fresh well water at 5–10%.
Productivity	Projected fish production of 5.2/6.9 lb/gal (30–40 kg/m³) per year. First trials yielded 0.4 lb/gal (2.55 kg/m³) in 100 days (1.6 lb/gal or 9.31 kg/m³ per year). Equivalent to approximately 157,300 lb/acre (143,000 kg/ha) per year.	Projected fish production of 9.4 lb/gal (54 kg/m³) per year, or 2,450 lb (1,000 kg) of 1.2 lb (0.5 kg) salmon in a 4,907 gal (18,500*l*) system with flow rates up to 398 gal (1,500*l*)/minute. Average flow: 164 gal (620*l*)/minute.
Life Span	Pond material life span: 20–30 years.	Not known. Continual replacement of filters, pumps, sub-units, etc.
Management	Amenable to amateurs familiar with aquarium techniques. Novices can start with low densities.	Requiring high level engineering, therapeutic and handling skills.
Tolerance for Mistakes	Tolerant, except for too heavy supplemental feeding.	Intolerant.
Environmental Impact	Beneficial: pond water and by-products used to irrigate and fertilize gardens and orchards.	Varying from little impact to potentially harmful with discharge of treatment chemicals and back-flushed filter materials.

Table IV-4-1   *Continued*

Category	New Alchemy Solar-Algae Fish Culture Biological/Structural	Salmonid Culture in Recycling Silo System Engineered/Technological
Cost	Capital cost of 25–35 cents/gallon (6–9 cents/*l*) to build, depending on light reflecting courtyard for pond, and aeration system. Facilities with additional ponds will cost proportionately less.	Capital cost of scaled-up second and third generation facilities, each with an estimated capacity of 500,000 pounds (225,000 kg) of fish per year is expected to be one million dollars.** Estimated to be equivalent to 80 cents/gallon (21 cents/*l*) capacity. Smaller systems much more costly per gallon.
Running Costs	Low	High

*   Description of "Technology of Closed System Culture of Salmonids," based on 1974 report of Thomas L. Meade, College of Resource Development, University of Rhode Island.
** Personal communication with design engineer, 1977.

PART V

Breeding of Fish

CHAPTER V-1

Methods of Controlled Reproduction

*Reproductive
requirements*

The breeding of fish might seem to be of major concern for the fish farmer. Certainly aquaculture on a global or national basis would be severely limited without controlled reproduction of fish. (Although some important culture systems in Asia rely on annual collections of young fish from wild stocks.) However, many fish farmers find that they need not concern themselves with the techniques of spawning fish. To see why this is so, I will first break down the fishes we are considering into three categories based on their reproductive requirements.

(1) Fish which will reproduce naturally in ponds or other aquaculture enclosures, with no attention from the farmer. Of the fish being discussed, all the "farm pond" fishes (see Chapter II-1), the yellow perch (*Perca flavescens*), the bullheads, and the tilapias fall into this category.

(2) Fish which can be bred by the farmer, with some special care, but without extraordinary skill or technological aids, and under environmental conditions normal for the culture of the species in question. There is considerable overlap between this group and groups 1 and 3, but I would place in group 2 the channel, blue and white catfishes (*Ictalurus punctatus*, *I. furcatus* and *I. catus*), and the common carp (*Cyprinus carpio*). The buffalofishes probably also belong in this category, though there is comparatively little accumulated experience in their breeding.

(3) Fish with specialized reproductive needs, requiring skills, equipment and/or environmental conditions not found on all fish farms. In this group are the trouts and grass carp (*Ctenopharyngodon idellus*).

*Group 1: Fish which
may overpopulate*

The fishes in group 1 normally reproduce annually (or as often as four times annually, as do tilapia in the South) in any environment where they will survive. This frees the fish farmer from the need to obtain outside stock after the first season of operation, but it may introduce another problem. The sunfishes (excepting the redear sunfish [*Lepomis microlophus*] and some hybrids), the bullheads, and especially the tilapias are simply too cooperative when it comes to breeding. The farmer's problem is not to produce more fish, but to restrict their reproduction, or to cope with it in some way, so that the fish do not outstrip their food supply and become "stunted." Several types of measures may be taken against stunting due to excess reproduction including:

• In a few cases reproduction may largely be prevented by environmental manipulation. All of the native species included in group 1 are more or less seasonal spawners, usually spawning in the shallows. If your pond can be drained efficiently you can, when spawning activity is first observed in the spring, "pull the plug" and dry up the shallows. If the pond bottom slopes gently all the way to the middle, you may only cause the fish to move back. But if the profile of your pond is such that by adjusting the water level you can greatly reduce the expanse of shallow water, you may be able to limit spawning. Even tilapia are said to spawn only rarely in water over 39 inches (100 cm) deep. The usefulness of this method is limited by the fact that it may affect more than one species. For example, you might successfully control bluegill (*Lepomis macrochirus*) spawning, but at the same time cut down breeding of largemouth bass (*Micropterus salmoides*), which have a lower reproductive potential.

• You can destroy spawning nests and broods. Sunfish and bass "nests" consist of shallow excavations in the pond bottom. The parent fish continually clean debris off the nest so that, in moderately clear water, they may be seen as light-colored circles on the bottom. Bass nests can be distinguished from sunfish nests by their greater size.

For effective control, nest destruction must be carried out after spawning has taken place. You can sometimes see the eggs in the nest, but it is easier to detect the young fish immediately after hatching. The first impression is of a squirming or undulating mass in the middle of the nest. On closer inspection, you may be able to see the semi-transparent individual fish. A glass-bottom bucket or face mask will help. Either eggs or fry can be destroyed by burying the nest in sand or gravel.

Because bullhead nests are less conspicuous, they cannot be controlled in this manner. Tilapias' year-round spawning habits make nest destruction ineffectual. Even with sunfishes it is only effective and feasible in small, clear ponds.

• Predatory fish can be used to crop the young fish. This is the basic premise of largemouth bass—bluegill polyculture. (The application and management of predator-prey combinations is discussed in Chapters II–1 and IV–1.)

• Harvesting can be an important tool in managing fishes with high reproductive potentials, though very often such fish are underharvested. When edible size fish are harvested at the proper rate, more food and space are freed for smaller fish to grow. Sometimes it may also be necessary to harvest fish smaller than those normally eaten. (I hesitate to call these fish "inedible"; most of our notions of what constitutes an "edible" fish are based on cultural values.) If harvested fish are not to be eaten, you should consider possible alternative uses, such as for livestock feed or in compost. (Methods for harvesting fish of all sizes are discussed in Chapter VI–1.)

• Supplemental feeding and environmental improvement may improve the growth of fish, and allow a higher percentage of a spawn to reach an edible size. But aiming *only* to improve growth is not going to eliminate the problem of overpopulation, though it may delay stunting for a few years. Eventually, if the population is not controlled in some way, increased growth will result in intensified spawning, more young fish, and strained food resources.

• In bullheads, tilapia and probably numerous other fishes, reproductive behavior can be suppressed by stocking at extremely high densities (8,097 individuals per acre or 20,000 per hectare or more). But at such densities, more often than not, growth is also suppressed.

Cages • Small lots of fish to be reared to "harvestable" size can be kept in net cages, preventing reproduction of those fishes which require access to the bottom (including all in group 1 except the yellow perch).

Monosex culture • It is sometimes feasible to prevent reproduction by stocking fish of only one sex. This not only eliminates overpopulation, but may result in increased growth since the energy put into reproductive behavior is then reduced. Growth loss due to reproductive behavior is especially severe in tilapia. Normally one of the parent tilapia incubates the eggs and young in its mouth, and it is scarcely possible to eat with a mouthful of eggs or baby fish!

Sexing fish The most straightforward way to obtain a "monosex" population is by inspecting each fish before stocking. This labor-intensive method is feasible only in certain small-scale operations.

External sex indicators vary from species to species. With tilapia, you can achieve a fair degree of accuracy by inspecting the dorsal fin. If it tapers to a long *Tilapia* point at the rear, and if the breeding colors (crimson edging on the dorsal fin and tail in blue tilapia, different colors in other species) are brilliant, the fish is likely male. However, while males are ready to breed most of the time, they are not so *all* the time, and an occasional female with bright colors will come along. So, to achieve near one hundred percent accuracy, you will have to inspect the genital papilla. This is a small flap of flesh just behind the vent. (See Figure V–1–1.) If it is broad and rounded in shape, with a transverse slit about two-thirds of the way from the base to the tip, the fish is a female. If the genital papilla is more pointed, with a hole in the end, it is a male. My experience with this technique is almost entirely limited to blue tilapia and *Tilapia nilotica*, but it can be used with at least some other species.

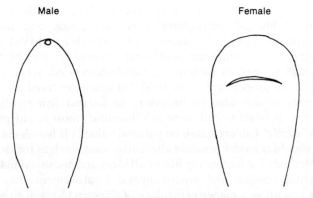

FIGURE V–1–1. *External sexual characteristics of tilapia.*

You cannot just look at Figure V–1–1, set down this book and go out and sex tilapia accurately. It takes practice. There is a trick to holding the fish at just the proper angle to the light, which I find impossible to describe. A magnifying glass may be helpful. Researchers in Puerto Rico have found that brushing India ink on the genital area makes the sexual characteristics much more apparent, particularly in small fish. Before attempting to sex tilapia for practical use, you should make a number of trials in the presence of an expert, or with fish which can be killed and dissected to check your accuracy.

If you dissect a fish you will find the gonads in the dorsal portion of the gut cavity, near the backbone. The ovaries are a pair of long, transparent or translucent bags. Except in fish which have recently spawned, you will nearly always find thousands of small, yellow-orange eggs in some state of development in the ovaries. The male gonad appears as a solid white mass, with a slightly angular quality; the surest distinguishing characteristic is the lack of eggs.

Dissecting to check accuracy

You may have some success visually sexing sunfishes, especially around the spawning period. At that time, male bluegills can be distinguished from females by the vivid red-orange breast and the darkening of the head and back. Similar intensification of bright colors and deepening of dark areas occurs in males of other *Lepomis* species. When handled, ripe males will usually spurt milt.

Sunfishes

J.L. Brauhn of The Bureau of Sport Fisheries and Wildlife's Fish-Pesticide Research Lab in Columbia, Missouri, has offered suggestions for sexing bluegills, which are outlined in Table V–1–1.

Table V–1–1.   External sexual characteristics of bluegills.

Males	Females
Opercular lobe squared, heavily pigmented.	Opercular lobe rounded, less pigmentation.
Gular area black.	Gular area yellow.
Overall body color dark.	Overall body color light.
Definite spot at posterior base of dorsal fin.	Spot less distinct.
Urogenital opening wedge shaped, darkly pigmented.	Urogenital opening rounder, with little or no pigmentation.

Using these characteristics, ninety-nine percent accuracy was achieved in separating one-year old males from the general population. With nine-month old fish, only eighty percent accuracy was achieved. I have not personally tested Brauhn's methods, but it seems that considerable practice would be required, and that regional variation might present problems.

A method proposed for sexing largemouth bass over 14 inches (35.6 cm) long is based on the shape of the scaleless area around the urogenital opening; it is said to be round in males, but pear-shaped or elliptical in females. Trials of this method produced ninety-two percent accuracy.

Largemouth bass

I have read about methods for sexing bullheads without dissection, but none have worked to my satisfaction.

You can also arrive at predominantly male populations of tilapia and sunfishes by means of hybridization. By far the most widely used monosex tilapia hybrid is ♂ *Tilapia hornorum* × ♀ *Tilapia mossambica*, though other hybrids have yielded 98 to 100 percent male offspring. At least 11 sunfish hybrids with a great predominance of males have been produced. The two which seem to have most potential for fish farming are the "hybrid bluegill" (♂ *Lepomis cyanellus* × ♀ *Lepomis macrochirus*) and the "hybrid redear" (♂ *Lepomis microlophus* × ♀

Monosex hybrids

Lepomis cyanellus). Both are sold under a confusing variety of trade names. These fish, unlike some hybrid tilapias, are not one hundred percent male. But they are so infertile that they are not ordinarily able to sustain populations in nature.

Monosex hybridization of bullheads has not been attempted, though the various species do hybridize in nature.

Regional breeding

Monosex tilapia and sunfish hybrids can be produced by any fish grower, but most farmers will probably prefer to devote all their waters to the production of food fish. The three hybrids mentioned are all available from U.S. dealers. At present they are quite expensive, but I can foresee the price dropping if they become popular. If a number of farmers in one region were growing hybrid fish, something similar to the traditional Chinese fish supply system might develop. One farmer, favored by skill, interest, location or water supply, could become a fry specialist, selling stock to other farmers in the vicinity. The same approach would be appropriate to the fish in groups 2 and 3. It is likely that more efficient production would result from a few individuals specializing in breeding and fry rearing. At present, breeders of most species are concentrated in a few localities. (Channel catfish are an exception.) If there were more small breeders serving local markets, it would be less costly and risky to ship fish. There would also be less transmission of fish diseases around the world, and more opportunity to develop appropriate regional fish varieties. Perhaps as aquaculture becomes more widespread and common in North America, this sort of situation will develop.

Breeding in solar algae ponds

Brief mention needs to be made of one special breeding problem. Blue tilapia and probably other species of tilapia as well, while naturally among the most enthusiastically reproductive of fishes, do not ordinarily reproduce in solar-algae ponds. (See Chapter IV–4 for a description of solar-algae ponds.) The reason is that male tilapia, when ready to construct a spawning nest, seek a dark corner. In a solar-algae pond, every time they back into a "corner" they come directly up against the translucent fiberglass wall. Conditions conducive to tilapia reproduction may be created in transparent tanks simply by wrapping the bottom portion with some opaque material.

Group 2: Fish fairly easy to breed

As for the fish in group 2, it is also possible to buy stock of these species. Channel catfish, in particular, are widely available. Many fish farmers, especially small-scale growers, will find it advantageous to buy, rather than breed stock. The decision involved is essentially an economic one; the time, labor and space required to breed fish must be weighed against the cost of stock. Possible income from the sale of young fish must also be considered. Methods for spawning catfish are well covered in the writings on catfish culture. (See Appendix IV–1.)

Catfish

Common carp

Carp breeding in the United States is usually done with the aid of pituitary injection, not because carp will not mature naturally, but in order to control the time of spawning. This is more important in temperate climates where carp are annual spawners than in subtropical and tropical regions where they will spawn year-round, since a sudden cold snap may cause naturally maturing carp to interrupt or postpone spawning.

Pituitary injection methods

Ripe female common carp are injected intraperitoneally at the axil of the pelvic fin with fresh or fresh frozen pituitary extract from another common carp of either sex and equal weight, or with 2 to 3 mg per kg body weight of acetone-dried carp pituitary in 1 cm^3 distilled water. It is very important to use only common carp pituitaries, for although most fish which have been tested with pituitary extracts respond to materials from unrelated species (even animals belonging to different classes), the common carp has so far been found to respond only to pituitary extracts of its own species. Males can be injected in the same way, but this is usually not necessary.

Whether using injected or naturally matured carp, the procedures for spawning are the same. You can spawn the fish "naturally" by placing equal numbers of females and males in well-oxygenated ponds at about 68°F (20°C). Spawning ponds can be as small as 269 square feet (25 square meters) if the breeders are small, and you should have some sort of collector for the eggs. Usually this is in the form of bundles of filamentous water plants, piles of brush, or more elaborate structures with a similar texture. One such device, known as a "kakaban" is used in Indonesia. (See Figure V–1–2.) Kakabans are made of the fibers of a plant known as "indjuk," which have previously been soaked in water for about five days. A thin layer of fibers 3.9–4.9 feet (1.2–1.5 m) long is pressed lengthwise between two bamboo lathes 1.6–2.0 inches (4–5 cm) wide. The margins are trimmed to even the end. The resulting structure, shaped like a two-sided comb with a width of 16–27 inches (40–70 cm) is laid crosswise on a long bamboo pole held in place between two pairs of shorter poles driven into the bottom at either end of the pond. They are spaced so that the fibers of adjacent kakabans just touch. The bamboo pole floats, and the whole structure moves freely with changes in water level, but the weight of the kakabans keeps it slightly submerged. The number of kakabans needed is calculated on the basis of 11 per 17.6 pounds (5 per 8 kg) of female spawners. Professional goldfish (*Carassius auratus*) growers in the United States use synthetic spawning "mops" as a substitute for natural materials, and these will work also with carp. As a rule of thumb, the total weight of fibrous material provided should be slightly greater than the total weight of spawners.

Spawning procedures

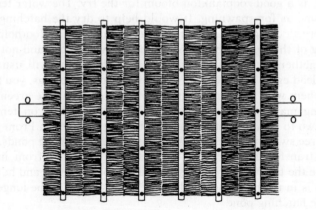

FIGURE V–1–2. *"Kakaban" used for collecting common carp* (Cyprinus carpio) *spawn in Indonesia.*

A slight current is beneficial, but not necessary. Spawning usually takes place at night. Egg collectors should be inspected each day during spawning and turned over if they become filled. When spawning is completed the eggs and parent fish must be separated. This can be done with less disturbance by transferring the collectors, with eggs attached, to another pond, than by removing the fish.

Injected fish can also be artificially spawned, but then the males must also be injected. Artificial spawning saves space by doing away with spawning ponds, and results in a higher percentage of fertilization while giving a high degree of genetic control.

Stripping of carp spawners may begin twelve to twenty hours after injection. (I will describe the technique of stripping later in this chapter.) One or more females are stripped of their eggs first, then milt from a male is stripped directly onto the eggs.

The eggs and milt should be mixed thoroughly with a nylon bristle paint brush in a plastic container, to which the eggs will not stick (as they do to glass or enamel). Take care to prevent any water from coming into contact with the eggs or milt during stripping, and it will help to wipe the fish dry with cheesecloth and work in dry gloves.

Shortly after the eggs and milt are mixed, small amounts of water are dripped onto them; fertilization occurs only if water is present. Once fertilized, the eggs will begin to clump and stick to each other. Before clumping progresses very far the eggs are lifted with the brush and shaken into a holding vessel containing about four to six inches (10–15 cm) water with a mat of Spanish moss, or some other egg collector. The eggs can also be poured onto the egg collector from the plastic container. In either case, the water in the holding vessel should be vigorously agitated while transferring the eggs to disperse them as carp would do in nature. Dispersion is important to prevent fungus—which inevitably forms on dead eggs—from smothering adjoining live eggs. The egg collectors with eggs attached are transferred from the holding vessel to ponds or hatchery troughs for hatching.

Hatching

The hatching pond should be no more than 31 inches (80 cm) deep, and fertile, so that there is a good zooplankton bloom for the fry. The water temperature should be the same as for spawning. It would help to dry the hatching pond for several days before stocking eggs, so that predators do not gain a foothold. You can prevent fungusing of the eggs by handling them with great care and not allowing them to bunch together. If fungus does occur, your first clue will usually be numerous opaque, dead eggs. If you can circulate water over the eggs, you may save them by flushing them with a 2 ppm solution of the dye, malachite green.

Hatching may take from two to six days depending on temperature, and in another two to six days the larvae will be free swimming. From the day after they become free swimming they can be transferred to growing ponds. The later in their life this transfer can be done, the less the mortality from handling. You can determine the time of transfer by the degree of crowding and how much zooplankton there is in the hatching pond. Three weeks is about the longest they should be left in the hatching pond.

Sexing spawners

Before you can successfully breed carp, you must know a little about sexing and selecting spawners. It is often not easy to determine the sex of common carp. When ripe, females usually show a fuller profile than males. Old males usually develop a few nuptial tubercles on the sides of the head and on the pectoral and ventral fins. The only sure way to sex young breeders is by pressing out the eggs and milt. To avoid having to examine each fish, carp culturists in India and elsewhere have developed the practice of spawning each female with two or more presumptive males of such a size that the total weight of the "males" approximately equals that of the female. Though a few immature males or females may, in this way, be included among the brood stock, there is very little chance that any female's eggs will not be fertilized.

Male and female carp to be used as spawners are usually kept separately from each other and from other stock. It is often recommended that ponds used for this purpose be in a sheltered spot, as exposure to a cold wind with resultant chilling of the water may retard spawning. Breeders should be given an especially good diet, but should not be overfed, as excess fat hampers gonad development.

Over the centuries culturists have developed and maintained a number of strains of carp considered to be especially desirable breeders. Fecundity of females is of course the main consideration, but to determine fecundity requires considerable expenditure of time and effort and the sacrifice of a number of fish. Consequently, external signs of fecundity have been sought, and the supposed characteristics of good breeders have been summarized as follows:

Desirable breeders

- Body moderately soft.
- Lower side of the belly broad and flattened so that the fish will stand on its belly.
- Relatively great body depth.
- Caudal peduncle relatively broad but supple.
- Small head and pointed snout.
- Rather large and regularly inserted scales.
- Genital opening nearer to the caudal peduncle than in the average carp.

In Asia "some farmers believe that the best mark of a good spawner relates to the insertion of the last scale before the genital opening; if a line is drawn from the head along the body to the center of the genital opening it should cross this scale and divide it into two equal parts." (Hora and Pillay 1962.)

There is also at least one behavioral sign used in selecting spawners: Females are considered poor brood stock if they release large numbers of eggs at one time, so that they are bunched on the collector.

Choosing spawners with the preceding characteristics of course amounts to selective breeding, a subject which I will cover in detail later.

Age and size of spawners are also factors to take into consideration. Age at maturity varies greatly with climate, as does growth. Usually in temperate countries males mature by their second or third year and no later than the fourth; females in their third or fourth year. In very cold climates, some individuals may not mature until the fifth or sixth year. In the tropics both sexes usually reach maturity within one year, sometimes in as little as six months.

Carp follow the general rule for all fish that the largest females produce the most eggs. Of course fecundity will vary with genetic and environmental factors but Table V-1-2 shows the general relationship between size and number of eggs. Take note, also, that the spawn of very old fish may be low in viability.

You might think it more efficient to spawn the largest and most productive females, but under the usual conditions of close confinement in the classical carp spawning methods it may be difficult or impossible to breed very large females. Small males are preferred for the same reason, and because they are more ardent courters. Most Asian culturists select females weighing 1–2 pounds (0.45–0.9 kg) and males of the same size or slightly smaller. If you induce spawning by pituitary injection and stripping of eggs and milt, it does make sense to take advantage of the high fecundity of large spawners.

Table V–1–2. Fecundity of female common carp.

Size in Inches (cm)	Number of Eggs
5.9–7.9 (15–20)	13,512
7.9–9.8 (20–25)	29,923
9.8–11.8 (25–30)	54,180
11.8–13.8 (30–35)	128,434
13.8–15.7 (35–40)	141,000
15.7–17.7 (40–45)	249,000
17.7–19.7 (45–50)	310,000
19.7–21.6 (50–55)	488,000
21.6–23.6 (55–60)	405,000
23.6–25.6 (60–65)	1,507,000
Over 25.6 (Over 65)	2,945,000

Spawning methods for buffalofish

During the brief period when buffalofish culture was popular in the South Central United States, they were spawned naturally. Experimental injection of buffalofish with pituitary hormones was moderately successful, but the "jar method," which requires constant circulation of water was used for hatching artificially fertilized eggs; this method is not generally available to the small-scale fish culturist.

The secret of successful buffalofish spawning is timing. In nature they spawn in the spring, when the water reaches 64–70°F (17.8–21.1°C). But if a cold snap occurs after that time, much of the spawn will be lost. To counteract this, farmers learned to take advantage of a natural trait of buffalofish to delay spawning until that danger was past. Buffalofish at high population densities give off a substance which inhibits their own reproductive behavior. Buffalofish farmers therefore crowded their broodstock in special wintering ponds. These ponds were about 0.7 acres (0.3 ha) in area with a mean depth of 3 feet (0.9 m), but with some water twice as deep. They were stocked with 264 to 1,320 pounds (120 to 600 kg) of adult buffalofish. Toward the end of the winter the breeders were conditioned by heavy feeding with commercial type feed.

Spawning ponds (identical to the breeding ponds), were kept dry until just before stocking, to prevent the establishment of predatory insects. Stocking with three to twenty pairs of spawners was done as the pond was filled with water from a source containing no buffalofish.

Sexing presents no problem; as spawning time approaches, the vent of the female becomes enlarged and inflamed and begins to protrude. Ripe males, if touched near the vent, will produce a small amount of milt. The males also develop nuptial tubercles, which make the body feel like sandpaper, while the female's body feels smooth.

The combined influence of freshly drawn water and greatly reduced population density should induce spawning within twenty-four hours of stocking, though it may be delayed if the fish are not fully mature or if a cold snap occurs. If spawning does not occur within a week, the pond can be drained nearly dry, and then refilled to trigger mating behavior.

Hatching

The fry hatch in five days at 64.4 to 69.8°F (18 to 21°C), and are free swimming in another two days. Newly hatched fry are very delicate and should be left in the spawning pond for some time. Meanwhile, the parents are removed to reduce the danger of transmission of diseases and parasites. If the fry are not to starve, an abundant supply of plankton must be available soon after they become free swimming. To this end, spawning ponds are fertilized after the eggs are laid. Commercial 8–8–2 fertilizer applied at 89 pounds per acre (100 kg/ha) was found to produce an adequate bloom.

Pituitary injection

Though the induction of spawning by means of pituitary injections may sound excessively technical, it is, in fact, a biotechnology which is becoming available to everyone. It allows the culturist to spawn fish out of season, and to breed fish such as grass carp, which normally have very specialized requirements for reproduction. Pituitary hormones are now available commercially, or hormone preparations can easily be made with a minimum of equipment and expertise.

Almost all of the fish I discuss here have been spawned by means of pituitary injection (or hypophysation as it is sometimes called), and there is every reason to believe that most other species could be spawned in the same way. Virtually any fish of either sex can be used as a hormone donor for any other fish. Many fish respond to hormones from other animals as well, even to human chorionic gonadotrophin, a hormone extracted from female urine. An exception to this rule is the common carp, which responds only to hormones from its own species. The common carp is the most widely used donor fish, due to its ready availability, the ease with which a small stock can be kept for use as donors, and the relative ease of locating the pituitary gland in the carp's large head.

Obtaining carp pituitary glands

Pituitary glands are obtained from carp by first "scalping" the skull so as to expose the brain. When you lift the brain out, you will find the pituitary either exposed on the floor of the brain case, or attached to the bottom of the brain. Figure V–1–3 may help you to locate the pituitary, but it is probably best to do the operation for the first time with an experienced person. Care must be taken not to damage the gland. The membrane covering the gland is removed as carefully as possible with a needle and forceps, and the gland picked up with the forceps. Practiced collectors can remove one hundred pituitaries per hour in this manner. Fresh fish give the highest percentage of success, but pituitaries removed from fish frozen for up to a year have also given satisfactory results.

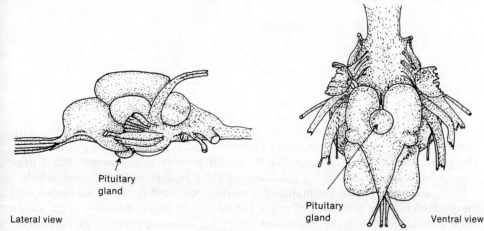

Pituitary gland

Lateral view

Pituitary gland   **Ventral view**

FIGURE V–1–3. *Two views of fish brain showing location of pituitary gland.*

Formerly fish were sometimes injected with a suspension of whole ground pituitary gland, but you can obtain better results by separating the hormone. If you use the pituitaries right away, they may be ground and made into a suspension in distilled water. Until recently, the next step in the process, separation of the hormone, required a laboratory centrifuge. However, a Peace Corps volunteer in the Philippines, Cody Best, developed a simple centrifuge, made from a hand drill, which will do the job. (See Figure V–1–4.) The pituitary suspension is placed in the tubes and the drill cranked for five minutes or longer, resulting in separation of the hormone.

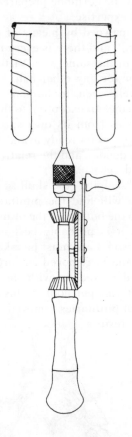

FIGURE V–1–4. *Centrifuge made from a hand drill for separating pituitary hormones from suspensions of ground pituitary gland.*

The optimal location, dosage and timing for pituitary injection varies with species, climate, size and ripeness of spawners and type and potency of hormone. The publications listed in Appendix IV–1 will provide some insight, but I urge anyone interested in practicing pituitary injection to first seek professional help.

There are fairly simple techniques for storing extracted pituitary hormones, and these extracts have remained effective for up to ten years. Probably most small-scale growers will only be interested either in immediate use or in purchasing hormones. In this case, simple refrigeration will suffice for storage. If you are interested in preparing and storing your own pituitary extracts, refer to Appendix IV–1.

I have several times referred to "stripping" fish as a step to induced breeding. Stripping requires a fair amount of skill to avoid injuring either fish or eggs. In some hatcheries injury to the spawners is reduced by anesthetizing them with MS-222. You can strip without anesthesia, as a one-person operation, but there is less fumbling and injury to the fish if it is done by a two-person team. One person grasps a ripe female by the caudal peduncle and pectoral fins and holds it over a pan, while the other gently extrudes the eggs by pressing the abdomen with thumb and forefinger, beginning just forward of the vent and proceeding forward to the pelvic fins. If you hold the fish tail down the eggs will flow naturally into the pan. Do not press forward of the pelvic fins, as this may damage the internal organs. It would be wise to wear wool gloves so that you can grip the fish firmly. Take care not to break any eggs, as the albumen from broken eggs will coat other eggs and inhibit fertilization. For this reason it is not wise to try to extract every egg from a female. Forcing out the last hundred or so eggs increases the chance of eggs being broken.

As soon as a female has been stripped, a ripe male is stripped into the pan using the same technique. The eggs and sperm are then gently mixed, to bring about fertilization by what is known as the "dry" method. At one time, a "wet" method was also popular, in which a small amount of water was added just before mixing the eggs and sperm, but the dry method is now almost universally preferred. If no eggs are broken it is customary to strip two or more pairs of fish before emptying the pan, in case one male is sterile.

Very few trout farmers carry their fish through the complete life cycle, including spawning, incubation, and fry rearing. If your water supply is dependable and rarely rises above 55.4°F (13°C), you may find that it is more profitable to sell fry or "eyed" eggs than to attempt to produce marketable size trout in water too cold for optimal growth. I would recommend, however, that anyone embarking on producing trout eggs or young on other than a hobby basis first acquire a year or two of experience in a public or commercial trout hatchery. (Those readers with a serious interest in trout breeding are referred to Lietritz. See Appendix IV–1.)

It is not clear whether grass carp will mature[1] naturally in North America. For the present, at least, it is necessary to use pituitary injections to produce young grass carp on fish farms. This in itself is not beyond the ordinary farmer's means, but large grass carp are extremely difficult to handle. Perhaps the legal attitude toward the fish is a greater obstacle to the aspiring grass carp breeder. (See Chapter II–8.) In few states other than Arkansas is there much likelihood of a private farmer obtaining permission to spawn grass carp.

---

[1] The term "maturity," in fishery biology does not carry exactly the same meaning as in other branches of zoology. A fish is "mature" when the gonads are in a "ripe" or potentially reproductive condition. We may speak of the same individual fish "maturing" annually, or even more often, as it enters the reproductive phase. A fish is "adult" if it has spawned once; it matures periodically.

CHAPTER V-2

Hybridization and Selective Breeding

Hybridization

Two aspects of fish breeding which interest many people are hybridization and selective breeding. Certain sunfish and tilapia hybrids and one buffalofish hybrid are discussed in this book. Except for the bullheads, most possible hybrids of the other species discussed here have been produced, but so far none of them have shown notable advantages for culture as food fish.

While the likelihood of truly new hybrids is very low, I would not rule out the possibility of the "rediscovery" of a hybrid which is in some way outstanding. Carl and Laura Hubbs developed the original sunfish crosses and noted their increased growth rates, skewed sex ratios and low fertility in the 1930's. (See Chapter V–1.) Their work may be regarded as a pioneer effort in the area of "monosex" hybridization. But it was thirty years before culturists gave serious attention to the concept of monosex culture and before anyone attempted to grow the hybrid sunfishes as food fish.

Racial differences

There may also be important racial differences between fish of the same species, which will affect the hybrids made with them in unpredictable ways. For example, Francis Bezdek of Aquatic Management, Inc. (among the first to experiment with hybrid sunfish as food fish), has said that in his opinion you cannot consider the hybrid ♂ green sunfish (*Lepomis cyanellus*) × ♀ bluegill (*Lepomis macrochirus*) from natural populations in Illinois to be the same fish as the offspring of the same cross made with Ohio stock.

This sort of genetic variability may also occur when mating different strains of the same species. In the early days of experimental tilapia culture at the Tropical Fish Culture Research Institute in Malacca, Malaysia, it was found that when female *Tilapia mossambica*, from stock acclimated for many generations in Malaysia, were mated with males of the same species from their native waters in Africa, the offspring produced were over ninety-eight percent male, rather than in their normal 50–50 sex ratio.

If you are intent on producing hybrids, you should recognize that, while sunfishes and tilapias will hybridize if deprived of the opposite sex of their own species, many fish have stronger natural barriers to hybridization. Artificial spawning, however, makes many hybrids possible. (See Chapter V–1.)

The value of selective breeding is illustrated by these two rainbow trout of the same age, fed on the same diet. The larger fish is of the selected Donaldson strain; the other of unselected wild stock. (Photo courtesy Lauren Donaldson, University of Washington.)

Selective breeding

As yet, selective breeding has played an important role in the culture of only two species of fresh water food fish: the common carp (*Cyprinus carpio*) and the rainbow trout (*Salmo gairdneri*). (See Chapter II–5 and II–6.) Some attempts have also been made with channel catfish (*Ictalurus punctatus*) and *Tilapia mossambica*. While indications are that these species lack the genetic variability of the common carp and rainbow trout, it is really too early to say. Development of improved strains of these and other fish will proceed in proportion to their economic importance, and there is very little the small grower can do about it. It is simply not feasible for anyone outside the large fish farms and experimental stations to maintain a breeding stock large enough for meaningful mass selection.

An aspect of selection which may concern the small grower is negative selection through inbreeding. If you keep a small stock of fish and do your own breeding, you would do well to bring in outside blood from time to time. It would be best to obtain fish from nearby sources, so as to minimize opportunities for long-distance spread of disease.

PART VI

Harvesting, Handling, and Marketing of Fish

CHAPTER VI-1

Harvesting

PART VI

Harvesting,
Handling and

*Total vs. partial
harvest*

You might think that in densely populated fish culture ponds it would be easy to harvest fish. In fact it is not always that simple, and in this chapter I will discuss various harvesting tools and techniques and explain which are best for various types of waters, species and degrees of harvest.

The very first point to consider in planning harvest strategies is whether you want a total or partial harvest. Commercial fish farmers often find it more profitable or at least more secure to contract for the sale of an entire crop at one time. The advantages of this sort of arrangement are increased when you have more than a few ponds.

If you are concerned only with home use or sale of small lots of fish, you will find it advantageous to make periodic harvests of whatever quantity of fish is desired. No fish is as good as fresh fish, and few tasks are as tedious as processing large quantities of fish with just a few people. At New Alchemy we are virtually forced to make an annual mass harvest by our use of a tropical fish (tilapia) and the short northern growing season. But if you are dealing with fish which can be over-wintered outdoors, and certainly if you live in the southern U.S., it is better to be set up to make periodic harvests of whatever size at will.

A further advantage of partial harvesting is that it tends to increase total production of the pond. The normal pattern of growth for fish is illustrated in Figure VI–1–1. While, as cold-blooded animals, fish never stop growing, growth is much more rapid early in life. If you harvest fish near the point where their growth rate begins to fall off, and replace them with smaller fish, greater production will result. The best way to achieve this is by intermittently removing a few of the largest individuals.

"Drawdown"

The most nearly sure way to get a one hundred percent harvest is by total drainage (drawdown), though in soft-bottomed ponds fish may be buried in mud and escape detection even after "drawdown." However, total drainage requires a rapidity in handling and processing fish that is virtually unattainable except in the smallest systems. This does not mean that drainage is not an important facet of harvesting cultured fish; partial drainage, particularly in properly designed ponds (see

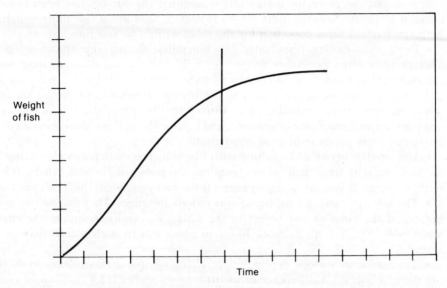

FIGURE VI-1-1. *Typical fish growth curve. Note that while growth never ceases altogether, it is much more rapid in the earlier part of life.*

Chapter VII–1) will be useful with most of the harvesting systems I am going to describe.

Except in emergencies, drawdowns should be made only when environmental conditions are good. If pH, ammonia concentration, temperature or D.O. (dissolved oxygen concentration) is already marginal, the added stress of a sudden change in water level may be lethal. Or, where there is a smaller volume of water, its lesser stability may speed up the rate of change of any of the factors just mentioned, and others as well.

Not only is it advisable to check pH, ammonia, temperature and D.O. before drawdown, they should be checked at all depths. The latter two are particularly likely to vary widely within a single body of water, and sudden loss of water from one portion may force fish into an unfavorable environment. It is usually better to drain water from the bottom, since bottom water is ordinarily lowest in D.O. and contains the most organic pollutants.

When using harvesting techniques which do not rely on draining it is always convenient and sometimes essential to know where the fish are. Often the presence and location of fish will be obvious, but not always. For instance, in the winter catfish feed little and tend to form schools in deep water. You can keep track of such groups by adapting an old angler's trick: capture one or two fish, attach floats to their dorsal fins with a few feet of line and release them.

Locating fish

The least intensive, least efficient method of harvesting fish is hook and line. However, it is quite suitable for harvesting small numbers of fish for table use. Subsistence growers may prefer angling to other harvesting methods because of the added recreational value. Among the species I have discussed, bluegills (*Lepomis macrochirus*) and bullheads are the most easily caught, but most species can be taken regularly on hook and line. Angling is not a dependable way of catching buffalofish, and angling for tilapia will generally select for the smaller fish.

Hook and line

This is not the place for a treatise on angling; the subject has been copiously covered in print. (See Appendix IV–1.) However, here are a few hints which may help you harvest farm pond fish for the table with hook and line.

Hints for angling

• Don't stop fishing (especially for bluegills), during the spawning period. Though it is often necessary to protect wild fish during the spawning season, remember your farm pond fish are likely to be too successful at reproducing. Bluegills and bass are particularly easy to catch with artificial lures during the spring when they are on the spawning beds. During that time you shook look for bluegill spawners in shallow water near shore, and fish for them. Bass spawners can also be harvested from ponds with good populations.

• Use smaller hooks when fishing with live bait. The main reason for using large hooks is so that small fish do not swallow the hook deeply and can be released without harm. If you are going to keep all the fish you catch, this does not matter.

• The key to choosing a fishing spot is to look for edges. They can be the sides or bottom of the pond or any object in the water, especially including the edges of weed beds. You are much more likely to catch fish in such places than in open water.

• Try fishing at various depths before changing spots. This is easy to do if you are using a bobber. A difference of as little as six inches (15.4 cm) can be crucial.

• During hot weather, fish early and late in the day, rather than in mid-day. Night fishing is effective for bass, bullheads and channel catfish (*Ictalurus punctatus*).

• Encourage kids to fish. A child who comes home with a dozen bluegills has made a real contribution to the family's nutrition, learned something about nature and had a good time in the bargain.

"Snag" or "snatch" hooking

A method of hook and line fishing which is considered unsportsmanlike and usually illegal in public waters is "snag" or "snatch" hooking. It is, however, used by fishermen in west central Florida to harvest the dense populations of blue tilapia (*Tilapia aurea*) in lakes there and could also be used in aquaculture situations where fish are heavily concentrated. Snag hooking involves a special type of treble hook or similar improvised devices. (Figure VI–1–2) The hook is cast beyond the fish, allowed to sink to an appropriate depth and retrieved by alternately making short, powerful jerks and taking up the slack line.

FIGURE VI–1–2. *Snag hook.*

Seining

More efficient harvesting can sometimes be accomplished by using nets. If you can drain your pond to any desired level and refill it relatively quickly you may want to periodically drain it down to a maximum depth of 3 feet (0.9 m) and net out a portion of the fish. Or, if your pond has extensive areas of open water, free from obstructions and sloping gradually from the bank toward deeper water you may be able to harvest significant amounts of fish by seining without manipulating the water level.

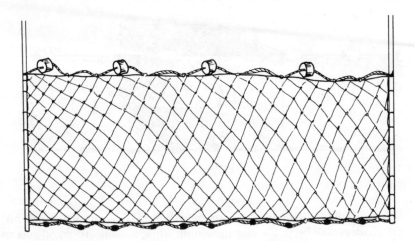

FIGURE VI–1–3. *Seine.*

The most commonly used type of net for harvesting fish from shallow waters is the seine. (Figure VI–1–3.) Most kinds of fish can be caught in seines, but adult centrarchids (black bass, sunfish, etc.) and tilapia are difficult to capture in this way. A seine is no more than a long piece of netting with a series of floats on one edge and lead weights on the other. Most commercial seines come with the lead line underweighted, so keep on hand some lead weights which can be hammered or clamped on the lead line between the original weights. The various net and twine companies (see Appendix IV–3) sell and sometimes rent virtually all lengths and depths of seines in mesh sizes from ⅛ inch (32 mm) to 3 inches (768 mm). The following will serve as rules of thumb in seine selection for small-scale fish growers. *The seine*

To maximize efficiency, seines should be at least one third longer than the widest place to be seined. However, fifty feet (15.2 m) is about the maximum length that can be comfortably pulled by two or three people. In large bodies of water, if additional labor or mechanization is not available, you may have to settle for more hauls and a lower percentage of harvest. In waters where there are numerous stumps, rocks, weeds, or other obstacles, short seines are more maneuverable. Long seines can be shortened in use by rolling, while short seines can be pieced together to make longer nets. *Length*

Commercial fishermen and fish farmers say that seines should be at least one third deeper than the deepest water to be seined. Readers of this book will probably find most use for seines four to eight feet (1.2–2.4 m) deep. *Depth*

Select the largest mesh size which will hold the fish you want to harvest without tangling. The smaller the mesh, the greater the likelihood of its becoming clogged with plants, unwanted small fish or debris, making the net very difficult to pull. Half inch (128 mm) or one inch (254 mm) mesh is a good choice for most harvesting jobs; smaller sizes are useful for sampling or transferring young fish. *Mesh size*

A variation on the standard seine design is the "bag" seine. (See Figure VI–1–4.) The bag increases the efficiency of capture in unobstructed waters, but bag seines are more difficult to pull than ordinary seines, and more prone to become caught on obstructions. *"Bag" seine*

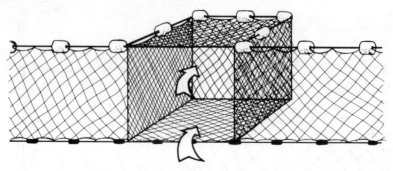

FIGURE VI-1-4.   *Bag seine.*

Seines can be pulled from boats or shore by ropes or with machines, but probably more readers will be interested in seining while wading. To do so you will need to attach two poles or "brailles" to the seine. (See Figure VI-1-3.) Brailles should be made of stout but light wood and be somewhat longer than the depth of the seine. Attach the bottom of the seine as near the bottom of the brailles as can be done without danger of its slipping off. The brailles should be notched at that point to prevent the seine from riding up.

Using the seine   In use the seine is drawn through the water so that the lead line rides on the bottom and the float line floats. It is extremely important that the lead line remains on the bottom, so that fish do not pass under it. To accomplish this, hold the brailles at approximately the angle shown in the top left photograph on page 277. You can experiment with speed. Sometimes very slow hauling is best, but it is often more effective to pull the seine as rapidly as is possible while still keeping the lead line down. Try to keep your feet and legs in front of the foot of the braille and make some disturbance to prevent fish from going around the end of the seine. At first you may tend to move closer to the person holding the other braille, but you should concentrate on keeping the lead line taut and straight by pulling away from the other operator.

It is best to seine into shallow water and right up onto the shore. In the final stages of the haul, you can drop the brailles and pull the seine in by hand, taking care to to keep the lead line down and stretched tight, and the float line up. (top right photograph on page 277.) The seine can then be picked up like a cradle and the fish carried to a convenient working place. Seines can be picked up in mid-water, when necessary, but this is less effective.

More often than not, it will take several passes of the seine to achieve anything approaching a complete harvest. You can greatly increase the efficiency of seining by not disturbing the area just before seining, and this will require proper setting of the seine. Figure VI-1-5 illustrates some methods of setting a seine.

Seining large bodies of water may call for power equipment and specialized techniques. Figure VI-1-6 illustrates a procedure for seining large ponds using a line hauler, a floating net barge and a set of blocks or pulleys. The lines are passed through the blocks and the seine paid out from the barge. The line is detached from first one and then the other set of blocks as hauling proceeds.

Umbrella nets   Another type of net, useful for partial harvests or periodic sampling of any fish which feed at the surface, is the umbrella net (see Figure VI-1-7, page 279). You can fish umbrella nets by means of a long pole, or from a boat, but one of the best

...oper seining technique demonstrated by ...embers of the New Alchemy Institute. The ...uthor (background) and Bill McNaughton ...e pulling the seine; note the angle of the ...ailles (poles). Linda Gusman is following ...hind the float line to free the net should it ...come snagged. (Photo by Ron Zweig, New ...chemy Institute.)

Close-up of seine in action. When the brailles are held at this angle, the float line rides slightly behind the lead line, creating a "bag." By pulling slightly away from each other, the two operators keep the lead line taut on the bottom, as shown. (Photo by Ron Zweig, New Alchemy Institute.)

This seine, on an Arkansas fish farm, has been bagged and lifted. Young channel catfish are being transferred from the seine to buckets by using a scap net. (Photo courtesy Maurice Moore, Aquaculture Magazine.*)*

A. To seine a small pool in its entirety. Can also be used in large bodies of water where entry can be made from deep water with a boat.

B. To seine a long, narrow body of water when it is not desired to seine the entire length.

C. To seine around obstacles and in restricted areas.

D. To seine in a large body of water

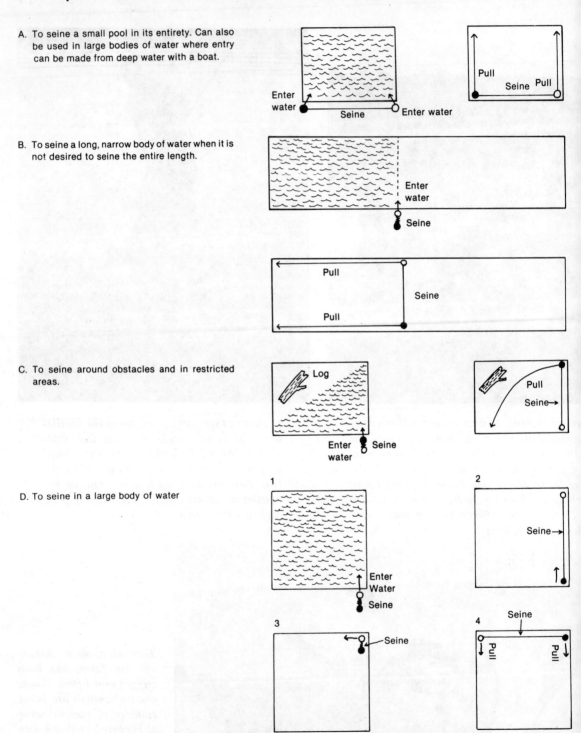

FIGURE VI-1-5   *Methods of setting and hauling seines.*

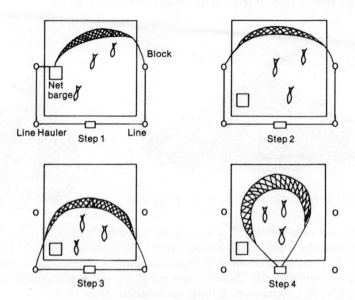

FIGURE VI–1–6. *Procedure for seining large ponds, using a mechanical seine.*

set-ups is to suspend the net with a pulley and winch. When you wish to make a harvest you lower the net until it is about 2 feet (0.6 m) under the surface. You then broadcast a little floating feed over the net, and when the fish are feeding vigorously, you crank the net up as rapidly as possible. Even fast moving fish like trout can be caught in this way. Most commercially available umbrella nets measure three to four feet (0.9–1.2 m) on a side, but you can also buy larger, battery powered nets, incorporating feeders, based on the same principle. You can apply the same concept in small pools by leaving a net, attached to a pulley and shaped to conform to the pool, permanently in place on the bottom, ready to be hauled up at any time.

Densely concentrated fish can be harvested with a cast net, or throw net (Figure VI–1–8). Traditionally cast nets are used primarily to catch schooling fish in salt water, but I have seen them used very successfully to catch tilapia in fresh water lakes in Florida. I don't know to what extent they can be used to catch other kinds of fish. The effective use of a cast net involves the development of certain skills, and as I do not possess these skills, I will not try to explain the technique of cast net fishing.

Cast nets

FIGURE VI–1–7. *Umbrella net.*

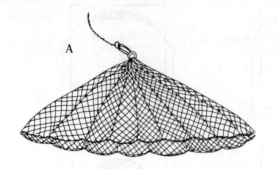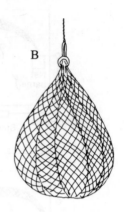

FIGURE VI–1–8. *Cast net: A. open; B. closed.*

Fish "traps"   The types of nets discussed so far are designed for "active" fishing. There are also nets, essentially traps, which fish passively. These are set, and, if the location is well chosen, fish swim into them and find it difficult or impossible to leave.

Gill nets   The most widely used "passive" fish net is the gill net. A gill net is not radically different from a seine: it is a long, rectangular piece of netting suspended in the water, with its short axis set perpendicular to the surface by means of floats and sinkers (Figure VI–1–9). But the gill net is not pulled. Instead, fish attempt to swim through it and find themselves unable to make it all the way through. When they try to back out, their gill covers are caught. Obviously, gill nets are quite size-selective. Fish below a certain size can pass through freely, while very large fish are unable to get enough of their bodies through to become caught.

In recent years considerable improvements have been made in gills nets with the introduction of nylon monofilament netting and camouflage colors. Still there are some fish, notably sunfish, bass and tilapia, which by and large refuse to enter gill nets.

Trammel nets   A deadlier variation on the gill net is the trammel net, which is built in layers of different sizes of mesh, with slack netting at the bottom (Figure VI–1–10) so that fish become thoroughly entangled. Trammel nets do not appear to have been used in aquaculture, so I will not presume to advise on sizes, etc. I would not bother with trammel nets unless and until gill nets are proven to be unsatisfactory in a given situation. The following comments on setting gill nets should apply to trammel nets as well.

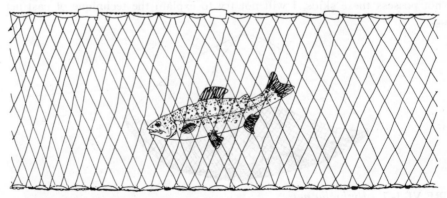

FIGURE VI–1–9. *Gill net in use.*

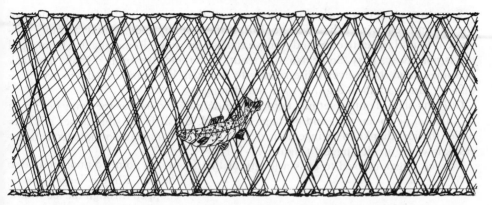

FIGURE VI–1–10.   *Trammel net in use.*

The first consideration in gill netting is what size mesh to get. You can determine this by attempting to pass various size fish of the species to be harvested through the netting, or by consulting with net and twine companies or commercial fishermen.

Fish caught in gill and trammel nets will struggle and injure themselves on the netting, which may detract from their appearance and reduce their market value. More important, gill netted fish eventually die in the net. They are also subject to predation, particularly by turtles. So, gill nets *must* be tended at least daily, even in foul weather.

Harvesting rainbow trout with a gill net on a Saskatchewan pothole. (Photo courtesy M.E Swanson, Saskatchewan Fisheries and Wildlife Branch.)

Sometimes you can work along the length of the net in a boat, lifting portions of the net and removing fish as you go. But, if the catch is good, it may be more convenient to pull the entire net and remove the fish on shore. Removing fish from a gill net is not always easy. Some fish, like trout, with their streamlined shape and lack of spines, lend themselves quite readily to removal, but spiny rayed fish are more difficult, and catfish in a gill net are maddening.

The ice-jigger

Gill nets should not be set just before an anticipated freeze, or they may be frozen in, impossible to harvest, and perhaps damaged. But it is possible to set gill nets under the ice, thanks to an ingenious device, used by Canadian pot hole trout farmers, known as an "ice jigger" (Figure VI–1–11).

The ice jigger is used to pass a length of line from one point to another under

Setting a gill net under ice with an ice jigger. (Photo courtesy M.E. Swanson, Saskatchewan Fisheries and Wildlife Branch.)

the ice. First a large hole is chopped through the ice where one end of the net is to be set. Then the wooden ice jigger, with line attached, is inserted. Being buoyant, it floats to the underside of the ice, and you can then propel it in the desired direction by jerking on the line. When it reaches the point where you want to set the other end of the net, you cut another hole to retrieve the jigger and use the line to pull the net between the two holes.

Thread size is also an important consideration. Commercial fishermen sometimes need stout gill nets to prevent breakage, but unless very large fish are known to be present, a culturist should use only the lighter sizes. The finer the thread, the more effective the net; size 210/2 is a good choice.

The next step is to find out at what points and depth your intended catch will be passing. The importance of having precise knowledge of this increases with the size of the body of water to be netted. Sometimes the only way to figure it out is by trial and error.

One end of the net can be tied to something on shore, or both ends can be anchored. In either case, a boat and two people are required for setting. The first step is to neatly fold the net in the boat so there are no twists or tangles. When the fishing ground is reached the net is paid out, lead line down, from the bow or off the side.

Table VI–1–1.   Recommended gill net mesh sizes
for harvesting cultured rainbow trout.

Fish length inches (cm)	Fish weight ounces (gm)	Recommended mesh size, stretched inches (cm)
7–9 (17.9–22.9)	3–8 (85–227)	2 (5.1)
9–11 (22.9–27.9)	8–14 (227–397)	2½ or 2¾ (6.4 or 7.0)
11–13 (27.9–33.0)	14–18 (397–511)	3 or 3¼ (7.6 or 8.3)

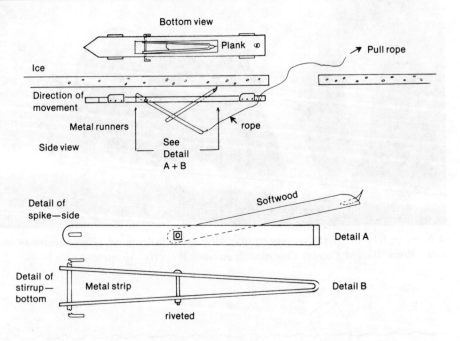

FIGURE VI–1–11.   *"Ice jigger" for setting gill nets under ice.*

It is best and safest to work with the boat running into the wind.

Make sure the net is pulled taut before anchoring the far end. A marker buoy with plenty of line is attached to the anchor as shown in Figure VI–1–12.

Because fish find it harder to see and avoid the net at night, the best schedule for running gill nets is to set in the late afternoon and harvest in the morning.

Probably the most important use of gill nets in aquaculture is to harvest large trout ponds, but they may also be used whenever deep water renders other harvesting methods impractical, or where the farmer wants to harvest fish in one size range without having to handle others.

Another type of trap used in commercial fisheries is the hoop net or fyke net. In small bodies of water, fyke nets can be set without the "lead" or "wings" illustrated in Figure VI–1–13, but usually these are included. Moving fish encounter these walls and are deflected into the series of hoops that constitute the body of the net and ultimately into the "pot" at the rear of the net. Ordinarily one wing, or the lead, if wings are not used, is extended very nearly to shore, to prevent fish from going around the net.

Hoop or fyke nets

A hoop or fyke net (without wings) in a Minnesota farm pond. (Photo courtesy Meredith Olson, Wright County Community Action, Waverly, Minnesota.)

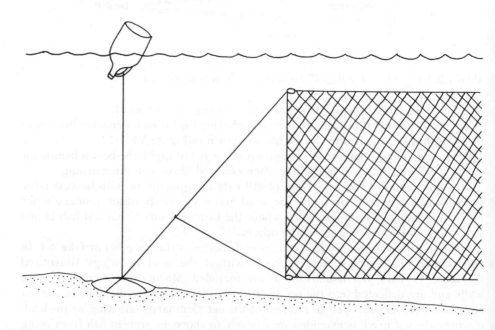

FIGURE VI–1–12.   *Set-up for anchoring and marking gill nets.*

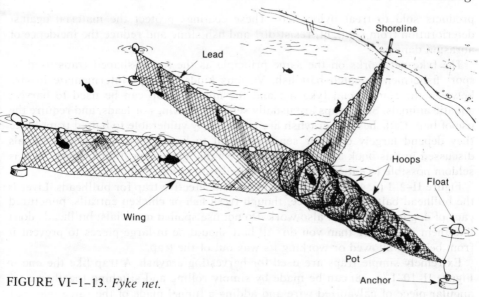

FIGURE VI-1-13. *Fyke net.*

Setting fyke nets requires two people and is fairly laborious. Though it can be done from a rowboat, an outboard motor is helpful. The first step is to organize the various components in the boat ready to be set out one at a time. When the fishing site is reached, the anchor is put over first and the anchor line paid out. The best way to do this and subsequent operations is to run the motor in reverse and pay line off the bow. Next the net itself is thrown out. Finally, the lead and wings are set and anchored. It is important in all phases of setting that the boat is run in as near a straight line as possible and that the anchor line, hoops, lead and wings are stretched perfectly taut. For convenience in hauling the net, and for safety if there is power boat traffic, the pot and the ends of wings and leads should be marked with conspicuous buoys.

Fyke nets can be harvested without removing them from the water, by pulling up the pot end, passing the boat under the net and emptying it into the boat. Some fyke nets are designed to be opened at the top; with others it is necessary to undo the end of the pot.

Fyke nets are very effective for harvesting bullheads, carp and buffalo. They are least reliable for taking centrarchids. While I have made great catches of bluegills and crappie at times with fyke nets, other times they have been almost a total failure with these fish. In corresponding with net companies and fishery biologists I have learned that this is a common experience—and that no one has an explanation for it.

An important point that is often overlooked is the care of nets. Nets should be *Care of nets* cleaned of debris after each use and spread to sun dry whenever possible. Inspect frequently for holes, so that small ones can be repaired before they become large. It will pay any net owner to learn net makers' knots and repair techniques. (See Appendix IV-1.)

Chemical treatment to protect nets may not be as critical in this day of nylon as formerly, but nylon is not indestructible, and it is often wise to use one of the products sold to treat nylon nets. These coatings protect the material against deterioration from sunlight, resist dirt and fish slime and reduce the incidence of abrasion damage.

Simple traps

The fyke net works on the same principle as the funnel-shaped traps used by sport fishermen to catch bait fish. Various kinds of traps intermediate in size between the commercial fyke net and the minnow trap can be used to harvest aquatic animals. Such traps are usually set without wings or leads, and require the use of bait. Catfishes and crayfish are particularly vulnerable to baited traps since they depend largely on their sense of smell to locate food. The other animals discussed in this book are visual feeders, benthic grazers or filter feeders, so it is seldom possible to entice them into a baited trap in any numbers.

Figure II–2–1 (p. 44) illustrates a simple and effective trap for bullheads. Liver is the bullhead bait par excellence, though fresh fish or chicken entrails, punctured cans of dog food, etc., will also work. Do not use spoiled materials; bullheads don't like them any better than you do. All bait should be in large pieces to prevent it from being swallowed or working its way out of the trap.

Extremely simple traps are used for harvesting crayfish. A trap like the one in Figure II–10–1 (p. 78) can be made by simply rolling and stitching together a triangular piece of galvanized wire and adding a funnel made of the same material. It is difficult to give advice on setting and baiting crayfish traps, since there are so many different species and habitats involved. Successful baits have included fresh fish, dog food and potatoes. As with catfishes, fresh bait and nocturnal sets are most effective.

In contrast to the multi-chambered fyke net, fish will eventually find their way out of such simple traps. Because of this, traps are harvested from fifteen minutes to an hour after setting, depending on water temperature (the warmer the water, the faster the response) and the potency of the bait. Night time sets are best. In dense populations of bullheads it is sometimes possible to take several hundred per trap with a single set, but a dozen or two is the more usual quantity.

Channel catfish traps

Channel catfish are considered more difficult to trap than bullheads, but some success has been had with an adaptation of the funnel type trap which incorporates an automatic feeder. Figure VI–1–14 is a sketch of such a trap. Catfish attracted by the feed enter through the funnels on three sides. Attempting to leave, they meet with the reversed funnel on the fourth side and swim into a holding pen, from which they may be harvested or returned to the pond. In preliminary studies, 120,000 pounds (54,480 kg) of fish have been removed with ten such traps in one month from a 120 acre (49 ha) pond. The use of dry feed and the automatic feeder open the way to trapping visual feeders as well; a similar trap has been effective in catching crappie.

Cages

Perhaps the simplest of all aquaculture systems to harvest are cages. It is often convenient to simply pick up the cage, though with heavily loaded cages you may need a winch. When you want a partial harvest, you can use a net. The most effective is a shallow rectangular net similar to a bait dealer's scap net with which to pursue fish into corners of the cage. Take care to avoid passing fish in the net directly over open water.

Trained fish

In an earlier chapter I have mentioned the Japanese system of training caged fish to respond to a signal, then releasing them to supplement their diet with natural food, and finally harvesting by calling them back and closing the cage. It is possible

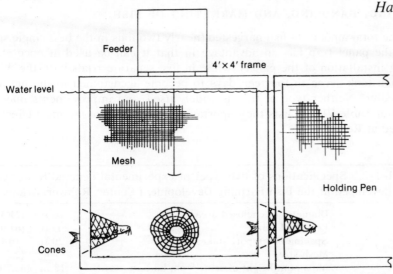

Feeder

4' x 4' frame

Water level

Mesh

Holding Pen

Cones

FIGURE VI–1–14. *Funnel trap for catfish.*

that uncaged fish could be similarly trained and harvested. In one experiment about ninety percent of the trout in a quarter acre (0.6 ha) pond were successfully trained in about forty-five trials with feeding and an electronically produced sound. In another experiment testing the feasibility of releasing and recapturing cage-reared channel catfish, a substantial portion of wild populations of channel catfish and green sunfish (*Lepomis cyanellus*) responded to feeding and were also captured in the cages. Grass carp (*Ctenopharyngodon idellus*) and goldfish (*Carassius auratus*) in the same pond were not successfully harvested though the grass carp, at least, responded to feeding.

Funnel traps are not always successful on catfish farms, and it seems impossible to predict when and where they will or will not work. Two other types of trap based on the use of a demand feeder were developed at the Fish Farming Development Center, Rohwer, Arkansas, and these may prove more dependable. The "drop-seine" consists of a seine set in a semi-circle around the feeder, with both ends secured on shore. Portions of the seine are lifted and hung on triggers so that the bottom of the net is clear of the pond bottom, and fish can easily enter the enclosure formed by the net to use the demand feeder. When fish are present and a harvest is desired the trigger mechanism is released, trapping the fish, which can then be removed by conventional seining.

"Drop-seine"

The "panel trap" operates on the same principle as the drop-seine, but incorporates a rigid wooden frame and chicken wire panels in place of the netting. While the drop seine must be installed and set on the triggers all at once, it is better to set the panel trap piecemeal. First the frame is installed, then one or two panels are added daily so that access to the feeding enclosure is gradually restricted to a single opening provided with a hinged, weighted gate. When a rod holding the gate open is removed, it drops, trapping the fish.

"Panel trap"

Both types of trap effectively captured channel catfish during the warm months, but in the winter, or at other times when catfish feed little, drop-seines, panel traps and funnel traps were all of no use. The drop-seine seemed the most effective in terms of the total amount of fish harvested (nearly two tons in the best single set). However, the panel trap has an advantage in that it can be used at once after completing installation of the panels, while in five separate trials with the drop-seine, catfish took seven to fourteen days to become accustomed to the net and resume feeding. Neither type of trap is widely used, and improvements may be expected, but Table VI–1–2 lists the important specifications of the most effective set-ups used at Rohwer.

Table VI–1–2.   Specifications of two types of experimental demand feeder cat-fish traps developed at the Fish Farming Development Center, Rohwer, Arkansas.

Drop-seine:			
	Diameter of enclosed area.	70 ft	(21.3 m)
	Length of seine.	200 ft	(61.0 m)
	Spacing of support stakes.	20 ft	(6.1 m)
	Mesh size.	1 in	(2.5 cm)
	Maximum height of seine from bottom.	18 in	(45.7 cm)
	Size of weights on lead line.	2 oz	(56.7 gm)
	Spacing of weights on lead line.	1 ft	(30.0 cm)
Panel trap:	Diameter of enclosed area.	45 ft	(13.7 m)
	Width of panels.	10 ft	(3.0 m)
	Height of panels.	5 ft	(1.5 m)
	Width of gate.	10 ft	(3.0 m)
	Height of gate	4 ft	(1.2 m)
	Mesh size.	1 in	(2.5 cm)

Gene Poirot's fish traps

Two ingeniously designed fish traps suitable for the small grower have been developed and manufactured by Gene Poirot of Golden City, Missouri. Poirot's traps are designed to be set on the bottom, in accordance with his observation that, since the natural enemies of fish come mainly from the top and sides, a trap set on the bottom eliminates the problem of fish becoming "trap shy."

The "Automatic Harvester" works as follows: A net is spread on the bottom of the pond in a place where fish are accustomed to being fed. In the center of the net is placed a buoyant plastic pan full of sinking fish feed. The pan contains an electrical switch which is open when the pan is full. As fish eat the feed, the pan begins to float, which closes the switch. This releases current from a six-volt battery, which in turn causes a solenoid to release a water-filled five gallon (22.7*l*) bucket. The bucket is attached to wires which, when relieved of its weight, pull up the sides of the net, trapping the fish.

Poirot's "inflatable seine" is set on the bottom, to surround the harvesting area. Fish are baited in with feed, and the float line of the seine can be automatically inflated at any time, in order to trap the fish. Then the harvest is completed as in conventional seining.

Harvesting solar-algae ponds

Small circular fish culture tanks, such as New Alchemy's solar-algae ponds, require a special harvesting device (shown in Figure VI–1–15). The circular basket, made of Vexar® nylon mesh, is lowered to the bottom in the half-closed position. This drives the fish to one side of the tank. Disturbance of the bottom and tapping on the sides of the tank will then cause the fish to rise. When all or most of the fish are above the raised lip of the harvester, the lines on that side are released, so that the basket comes to lie flat on the bottom. The harvester, containing the fish, is then

Harvesting tilapia from a solar-algae pond at the New Alchemy Institute, Hatchville, Massachusetts, using the harvester shown in Fig. VI–1–15. (Photo by Hilde Maingay, New Alchemy Institute.)

raised by one person on each side of the tank pulling down on the lines. If you have only one or two tanks, it may be more convenient to leave a similar, unhinged device permanently in place, to be hauled up as needed.

A newer tool for harvesting cylindrical tanks, invented by Carl Baum of New Alchemy, is a modification of an ordinary seine. The seine should be equal in depth to the depth of the tank to be harvested and in length to the circumference of the

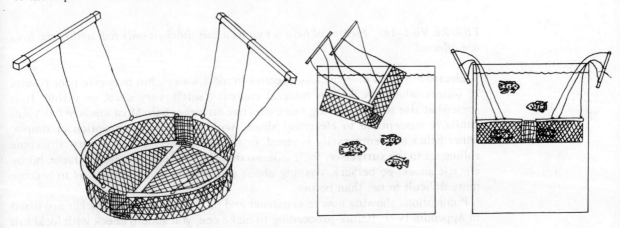

FIGURE VI–1–15.   *Harvester for solar-algae ponds.*

tank and is equipped with a drawstring on the lead line. The net is inserted at one side of the tank with the two brailles together. One person grasps each braille and brings it around the perimeter of the tank until they meet at a point directly opposite the starting point. Next the drawstring is pulled, pursing the bottom of the net. Finally the net is rolled on the brailles and lifted out with the fish inside.

Electro-fishing   A final method of harvesting which might be useful to small-scale fish growers is electro-fishing. Electro-fishing involves setting up an electric field in the water as shown in Figure VI–1–16. Fish entering the field are stunned and scooped up in a dip net, or collected in a seine drawn behind the shocker. Since workers are at all times behind the electrical field, accidents are rare, as is death or injury to fish or other aquatic animals.

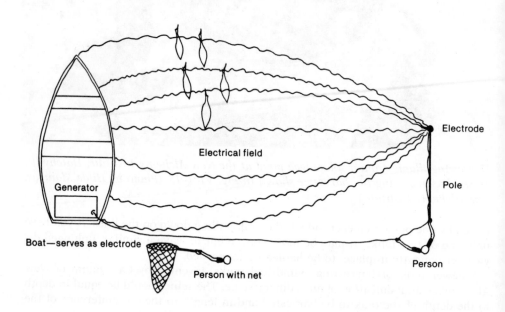

FIGURE VI–1–16.   *Electrical field set up by a fish shocker; only fish within the field are affected.*

Electro-fishing is more or less effective in most waters, but may give poor results in waters which are low in mineral content ("soft"), very deep, or turbid. It is somewhat size selective, being more effective on larger fish. Most kinds of fish seem similarly susceptible to electrical shock, with the possible exception of tilapia. Other fishes are temporarily stunned, or at least slowed down, and usually come rolling up to the surface on their side, so that they are easily netted. Tilapia, hit by electric shock, go berserk, dashing about and leaping so vigorously as to become more difficult to net than before.

Publications showing how to construct and use an electric fish shocker are listed in Appendix IV–1. Before proceeding to make one, you should check with local fish and game authorities. In some places it may be illegal for a private individual to even own such a device.

CHAPTER VI-2

Handling Fish as Food

Importance of freshness

A fish out of water, whether considered as a living animal or as meat, is a rather delicate entity. Compared to other food organisms, it is quite easy to unintentionally transform a live fish into a dead fish or a fresh one into a spoiled one. It is therefore to the aquaculturist's advantage to learn as much as possible about handling fish.

Whether fish are to be killed for consumption or transported live, the rule "less is better" applies to both the actual handling and the time factor. One reason North Americans don't eat more fish is the inordinate time and distance a fish must travel to reach most North American tables. No matter how advanced the technology or how careful the handlers, a fish that was put on ice several hundred miles off the New England coast, is not going to be a prime quality fish when it reaches Kansas City. More than a few of the people who "don't like fish" like it fine when it is truly fresh. But even fish that never leave the farm·will benefit from proper handling.

Killing fish

If fish are to be consumed or processed in the immediate vicinity, they should be killed as soon after harvest as is convenient. The best way to kill, or at least stun, large and medium sized fish is with a smart blow between the eyes with a hard, blunt object. Small fish are sometimes difficult to hit squarely without striking one's hand. You can achieve the same effect by throwing the fish forcefully against a hard surface, though this may bruise the flesh of delicate fishes. (Be sure to grip spiny finned fishes carefully so that the spines don't cut your hand.) Industrial-scale fish farmers kill whole lots of fish with a single powerful electric shock.

Frogs are best killed by severing the spinal cord at the base of the brain. The only way to kill crayfish and clams without tissue damage which would affect their edibility is by cooking.

Handling freshly killed fish

Once the animal is killed it should be preserved or prepared for the table as quickly as possible. In the interim, keep it moist (but *not* in water), shaded, and as cool as possible. Icing is helpful, but not ordinarily necessary.

If there is to be more than a few hours between capture and consumption or preservation, it is better to dress the fish at the harvest site. Otherwise the best site is the most convenient one. The importance of proper technique and equipment in dressing fish is often underestimated. Dressing is an unavoidable step in the process which begins with the cultivation of fish and ends in its consumption, and understanding it can contribute greatly to the enjoyment, efficiency and profitability of fish culture.

Dressing fish

In dressing fish it is very important that your knives are *sharp*. Cleaning fish will always be work, but it need not be the utter tedium many make of it, and nothing saves labor like sharp knives. You will gain speed, comfort and efficiency by providing a clean, flat, well-lighted work surface with water (preferably running water) at hand. Where large numbers of fish are to be skinned, hanging hooks will be useful, and clamps similar in operation to a clipboard will speed up scaling. Another convenience is a table with a slight slope, connected to a water tap, so that washing water and offal can be flushed from the dressing area and collected for use or disposal.

Cleaning trout, catfish and frogs

The simplest of all fish to dress are trout; they need only be gutted and washed. Catfishes are usually skinned (though for smoking the skin is better left on), while scaly fishes are usually scaled, not skinned.

Skinning and cleaning yellow bullheads. (Photo courtesy Meredith Olson, Wright County Community Action, Waverly, Minnesota.)

To skin a catfish take a very sharp knife or razor blade and cut through the skin in a circle completely around the fish just behind the head and pectoral fins. Be sure to cut deep enough that the muscle tissue is exposed all the way around, but try not to cut into the muscle. The sharper the blade, the easier this will be. Then, holding the fish firmly by the head, grasp the skin with a small pliers—a surgical hemostat is even better—and pull toward the tail. It may be helpful to first slide the point of a knife under the edge of the skin to loosen a portion for the pliers to grip. If you have cut into the muscle tissue, your attempt at skinning may result in tearing the head from the body, particularly if you are skinning a small bullhead. If you have avoided this mistake, the skin should come off like a glove, in one or two pulls. With a little practice, you can complete the entire process in a matter of seconds, with less labor than is required to scale a sunfish or perch of similar size.

The procedure for skinning frogs is the same as for catfish, except that the presence of legs prevents removing the skin in one piece. Scaly fishes are sometimes skinned to remove objectionable flavors. They are generally more difficult to skin than scaleless fish, but special fish skinning tools are available for the job.

Though you can scale with a knife, it is also an excellent way to dull a knife. *Scaling fish* Splendidly designed scalers are available for a few cents, or you can make one nearly as good using bottle caps. (See Figure VI–2–1.) To scale a fish with knife or scaler, fasten or hold the fish securely and scrape it from the tail toward the head. Use the point of the tool to get at the row of scales along the base of the dorsal fin, as well as the few scales on the back and underside directly forward of the tail and other hard-to-get-at places.

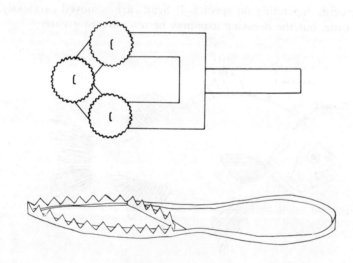

FIGURE VI–2–1. *Fish scaler made with bottle caps and commercial scaler.*

To gut a fish, slit it from the vent to the gills and remove everything that *Gutting* doesn't look like edible flesh. You can sometimes get away with a little sloppiness in this, but it is better to be fastidious for the sake of appearance, flavor and keeping qualities. Make sure you get rid of all the kidney (the blood-filled tissue which runs along the roof of the body cavity); your thumbnail is the natural tool for this chore.

People often remove and discard the head when dressing fish, but this can be wasteful. Most fish, particularly large ones, have much edible flesh along the top of the head in back of the brain, and a small but choice "cheek" piece on the gill covers. (See Figure VI–2–2.). To dress a fish leaving the head on, you must remove the gills. Continue the incision made in gutting all the way to the chin. Then, with a knife, cut across the point where the gill arches meet the roof of the oral cavity. Grasp the freed piece of tissue and pull the gills out whole. You may have to use a knife to sever the membranes which join the gills to the gill cover, or it may be possible to tear them out. In small fish, you can often simply tear the gills out.

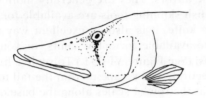

FIGURE VI–2–2. *Head of a chain pickerel* (Esox niger), *illustrating the "cheek" meat, located on the gill cover.*

If the head is to be removed, don't waste more meat than necessary. Take the trouble to cut as shown in Figure VI–2–3. Dressing loss on fish ordinarily runs forty to sixty percent, depending on species. If heads are removed carelessly, you may save some time, but the dressing loss may be ten percent greater.

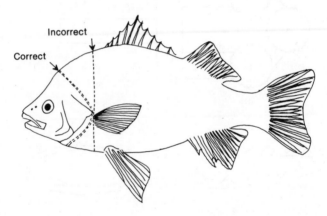

FIGURE VI–2–3. *Correct and incorrect ways of cutting off a fish head.*

Filleting

Dressing loss is, of course, much greater if fish are filleted. However, filleting improves the flavor of some fish, can result in a boneless product, is required for certain recipes, and may be commercially justifiable. In filleting, more than in any other type of fish dressing, a thin, sharp knife is of paramount importance to prevent waste. You can find instructions for filleting fish, steaking and other refinements of basic dressing methods in cookbooks. (See Appendix I.)

Use and storage of fish

If fish are filleted or the heads are removed, consider retaining the heads and carcasses for use in making soups, curries, etc. While fish culture production is conventionally measured in terms of pounds per acre (kg/ha) of live fish, the nutritional importance of fish culture (and fisheries) in North America could be enhanced if only our people would learn not to waste such a high proportion of the fish they use.

Once dressed, the sooner the fish is prepared for the table, the better. Refrigeration for up to twenty-four hours is acceptable, but if fish are to be served within twenty-four hours to a week of capture, simple freezing is preferable to refrigeration. Fish to be kept for longer periods of time should be frozen using fairly elaborate procedures, hard smoked, pickled or canned. These techniques of fish preservation have been well described in print. They are beyond the scope of this book, but I do wish to urge every reader who does not plan to consume each fish as soon as it is harvested to study some of the references listed in Appendix I. The result could be an immeasurable increase in the enjoyment of your cultured fish.

While probably nothing would contribute more to raising the quality of the fish consumed on this continent than making fresher fish available, it is a fair criticism of our culture to say that we are generally poor fish cooks. Though this is a large subject which cannot be covered here, Appendix I may help you get started. Its references have been selected from the myriad of cookbooks by New Alchemy's resident fish cooking expert, Susan Ervin, and myself, as among those likeliest to be useful in upgrading the product of the inexpert or inexperienced fish cook.

CHAPTER VI-3

Handling Fish as Live Animals

Handling fish in the hands

Even before the harvest is completed, the fish farmer must know how properly to handle fish which are to be retained live. When fish are harvested by total drainage, beach seining, or traps to be removed from the water, the first step in "handling" is just that—picking up the fish in your hands. Other times, handling begins when the fish are trapped in a small area, for instance near the spillway of a nearly drained pond. You might be able to perform an aquatic "greased pig chase" in such a situation, but it is more efficient and easier on the fish to use some sort of "dip" or "scap" net. Nets with a deep bag like a fisherman's landing net catch fish faster, but remember, you have to get the fish out of the net, too. Often, particularly when dealing with spiny fishes, it will prove faster to use a very shallow net even though you may drop a few fish.

Net or no, eventually the fish ends up in someone's hand. A firm, secure grip is essential, but so is gentleness. These two are not at odds. Generally, the most secure grip is the one that will result in the shortest handling time, therefore the best for fish and fish culturist alike. The old adage about wetting the hands before holding a fish, so as not to rub off the protective slime, is doubtful advice. Though the slime is important to the fish and some of it might be rubbed off by dry hands, wet hands make for a slippery grip. This may result in prolonged handling time, a dropped fish, or internal injuries caused by excessive squeezing. Dry hands or wet, bare hands or gloves, your choice should be made on the basis of security and efficiency.

No one grip is suited for all fish, and the situation is complicated by the various unpleasant things fish can do to your hands. Among the species I have discussed, bass, sunfishes and tilapia have multi-spined dorsal and anal fins which can puncture and cut; catfishes have single, but stouter spines in the pectoral and dorsal fins; pickerel and, to a lesser extent, trout and catfishes have teeth capable of drawing blood; crayfish nip with their pincers, and a sizable carp can apply a surprisingly strong pinch to your finger with its gill cover.

For fish over a few pounds you may need a two-fisted approach, but fortunately large fish are a good deal less delicate than small ones. Ideally, very small fish should not be handled at all. When you have to move such fish, or if they are inadvertently captured, it is best if they do not come into contact with hands at all. Use a soft, close-mesh net, and keep the fish out of the water for the shortest time possible.

Another aspect of fish farming which may call for handling your stock is checking the rate of growth. When dealing with individual fish, it is faster to measure length than it is to weigh. However, length is a less meaningful indicator of size. It is nearly impossible to accurately measure a live fish with a simple ruler or tape. You will need a special measuring board which can easily be made at home. (See Figure VI–3–1.)

Measuring and weighing

For more accurate assessment of size, or when dealing with volumes of fish, you will need to weigh. Precise weights are normally achieved only by killing the fish. The moisture which clings to the fish and the animals' general lack of cooperation make even the best field weighing an approximation. For consistency, shake or drip excess moisture off the fish before weighing. Generally it is best to weigh fish in a small amount of water, but if you use a hanging scale, large individual specimens may be placed "dry" in a sling made of soft netting.

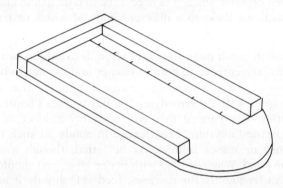

FIGURE VI–3–1. *Fish measuring board.*

Size grading

In some types of fish culture, notably intensive trout culture, it is customary to periodically grade young fish by size. Sorting and measuring hundreds or thousands of fish would be too time-consuming and could result in high mortality, so various types of graders are used. The simplest type consists of a slatted tray with handles at the four corners. (Figure VI–3–2.) Fish are loaded onto the grader and, with one person holding each end, it is shaken vigorously up and down at the surface of the water. Eventually all fish below a certain size pass between the slats, the larger ones being held for transfer to new quarters. More sophisticated graders make use of hand or power-driven rollers, but operate on the same principle.

"Artificial respiration"

If proper handling procedure is observed, there should be no problem in returning fish to the water they came from. If, despite all care, a few fish appear severely stressed and do not swim away immediately, it may help to give them "artificial respiration." Grasp the fish just behind the gills and move it fairly rapidly back and forth in the water. This produces an artificial flow of water over the gills and will often revive a sluggish fish.

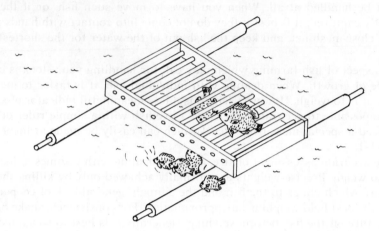

FIGURE VI-3-2. *Grader for sorting small fish by size.*

Problems often do arise when it is necessary to hold fish in close confinement for long periods, to move them to a different body of water or transport them long distances.

Fish held at high densities

When selling fish in small quantities it is often desirable to pen fish at high densities. It is a great convenience to both customer and farmer to be able to dip out a specified amount of fish at a moment's notice. Penning is also used when treating disease (see Chapter VIII-2), remedying "off-flavor" (see Chapter VIII-4), awaiting shipping facilities, etc. Penned fish are inevitably subject to physiological and social stresses beyond anything experienced in ponds. At such times, a short-term oxygen deficiency or minor injury may be lethal, though it might have passed unnoticed in the pond. When dealing with penned fish, you should be extra gentle in handling, inspect frequently for diseases, feed very lightly if at all, and keep the water well oxygenated. Artificial aeration will often be called for. (See Chapter X-2.)

Where feasible, cages or "live cars" floated in the larger body are much to be preferred to fully enclosed containers for impounding large numbers of fish. For reasons given in Chapter IV-3 on culturing fish in cages, such floating pens deal with excess feed, fish wastes and oxygen problems almost automatically. They are also most convenient when total harvest of the penned fish is desired.

One easy way to load live cars is in conjunction with traps like the one shown in Figure VI-1-14. Simply connect the live car to the outgoing funnel and harvest the fish.

Transferring fish

At New Alchemy we are frequently asked questions about transporting live fish from place to place. Often people are surprised to learn that live fish can be shipped anywhere in the world without great difficulty. Consider the fact that many of the tropical fishes kept by aquarists are shipped from Singapore and Hong Kong—and were so shipped in the days before air freight. Or, for a more mundane example, consider that thousands of pounds of catfish from the southern states are routinely trucked to Chicago and other northern cities and marketed live.

There is, in fact, more potential danger in the *transfer* of fish from one body of water to another—even where the distance between them is but a few feet—than in the act of transportation. So, I will consider transfer of fish before discussing transportation. Whether a fish has just arrived by plane from southeast Asia or is being moved from one aquarium to another in the same room, it is subject to a certain amount of shock. Though there is nothing you can do about the change from familiar to unfamiliar surroundings, there are ways to ameliorate other kinds of shock.

Fish may be shocked by sudden, dramatic changes in any of the physical or chemical factors affecting their physiology, but you usually need be concerned only with thermal shock. In transferring fish it is not enough to know that the temperature of the water into which they are to be placed is suitable for them. If the difference between the temperature of the transport water and that of the new environment is too great (even though neither temperature is extreme in terms of the fishes' tolerance), the fish can be killed outright. In less drastic cases, the shock of transfer from one temperature to another can predispose the fish to diseases and parasites which may not be detectable for days or weeks.

As a general rule, fish should not be subjected to sudden temperature changes greater than 2°F. But a temperature change which would be lethal if fish were just dumped, may be perfectly acceptable if adjusted gradually over a period of twenty minutes. The best way to achieve a gradual temperature transition is to "float in" the fish. First, determine the temperature of the transfer water and the surface temperature of the water the fish are to be placed in. If the difference is greater than 2°F, the fish should be floated in. (This temperature difference may be adjusted upward in the case of hardy fish like tilapia, or downward for fish which have experienced additional stress through long transport, disease, etc. Transferring fish to a lower temperature is generally riskier than moving them to a higher one.) By far the best container for transferring fish is a seamless plastic bag of appropriate size. Such bags may usually be obtained from aquarium shops. For purposes of transfer (*not* for transport) it does not matter if the bag has pinhole leaks. In the absence of plastic bags you can use plastic buckets or other such containers—the lighter the gauge plastic the better. Fish are often shipped in polystyrene foam containers. Do *not* use these for temperature acclimation, as the virtue of polystyrene foam is its effectiveness as an insulator to *prevent* temperature change.

Place the fish in the bag no more than half filled with water and tie or secure the neck. Float the bag in the water in which the fish will be placed, making sure that the bag has a rounded shape so that frightened fish do not pile up in a corner and suffocate. Check the temperature in and outside of the bag at five or ten minute intervals. Usually twenty minutes is enough for equilibration of temperatures. If more time is required, open the bag occasionally so that fresh air, loaded with oxygen, can enter. The process can be speeded up by occasionally exchanging small quantities of water between the bag and the outside environment. When the two temperatures are the same or nearly so, open and tilt the bag and let the fish gradually swim out. Do not dump them unless they refuse to leave.

Shock due to differences in pH, alkalinity, and other chemical or physical factors need not concern you unless there are known to be extreme differences between the transport water and the new water. In such cases, the technique of floating in will not do, and you must gradually mix the waters. Pour out a little of the transport

Thermal shock

Other types of environmental shock

water and replace it with the new water. Wait five minutes and repeat, and so on, until the waters are thoroughly mixed. This method also works for temperature equilibration when floating in is not possible.

The most important environmental factors in transport of aquatic animals are first, dissolved oxygen concentration and second, temperature. The second is easy to deal with. Insulate shipping containers well—plastic bags or bottles inside a polystyrene foam box are ideal. Also, avoid extremes of temperature. Do not, for example, put fish in the back window of a car on a sunny day or leave them outside for hours in the winter.

The maintenance of safe oxygen levels requires careful planning. One way to avoid the whole problem is to make use of the services of professional fish haulers, with aerated tank trucks, who advertise in the aquaculture trade magazines. (See Appendix IV–1.) The paragraphs which follow are for that majority of readers who, if involved in fish transport at all, will be moving small lots of fish by conventional freight systems, or on a do-it-yourself basis.

It is important to think about safe transport before the trip starts, when you have much more control over the conditions the fish will experience than you do en route. Unless absolutely necessary, only fish in the best of health should be shipped. Their water should be the best available, particularly with respect to dissolved oxygen, and it should be relatively low in temperature. Cool water holds more oxygen and, at the same time, the fishes' needs are less since their metabolism is slowed. Do not feed for a day or two before transport, as fish with full stomachs will excrete or regurgitate, and pollute their own water.

The best travel containers are plastic bags or bottles, preferably placed inside polystyrene foam boxes or otherwise insulated, though you can also use glass or metal vessels. If bags are used, the fish should be "double bagged" and each bag fastened separately with rubber bands. This is essential for sharp-spined fish like catfish, which can "unzip" plastic bags.

Never fill sealed containers for transporting fish more than half full with water. The reason for this is that air holds more oxygen than water, and as oxygen in the water is used by the fish, it will be replaced by oxygen from the air above the water.

The number of fish which may be shipped per container is larger than you might suppose, but when shipping only a few there is everything to gain by giving them more space than is absolutely necessary. The maximum rate at which fish can be packed depends on the species shipped and the temperature. The figures in Table VI–3–1, originally published by the pioneer German fish culturist Wilhelm Schäperclaus, will serve as a guide for fish with high and relatively low oxygen requirements (brown trout, *Salmo trutta*, and common carp, *Cyprinus carpio*, respectively), when transported without mechanical aeration. Of the fish I have discussed, brook trout (*Salvelinus fontinalis*) have the highest oxygen requirement. Most fall somewhere between trout and carp, though walking catfish (*Clarias batrachus*), tilapia, and bullheads are even more tolerant of low oxygen concentrations than common carp. The lower rates in the table are for higher water temperatures and vice versa. A rule of thumb which may also be helpful is that a fish twice as long needs twice as much oxygen per unit of body weight.

If additional oxygen can be provided, you can increase the density of fish. Many shippers inflate their shipping containers with compressed air or, better, pure

Table VI–3–1. Recommended loading rates
for transporting fish without artificial aeration.

Species	Size		Ratio of fish weight: water weight
Brown trout	Fingerlings		1:70–200
	2.0–2.7 in	(5–7 cm)	1:50–200
	3.9–5.1 in	(10–13 cm)	1:45– 90
Common carp	less than 2.0 in	(less than 5 cm)	1:38– 75
	3.5–4.7 in	(9–12 cm)	1:13– 23
	5.9–7.0 in	(15–18 cm)	1:9 – 13
	0.55 lb	(250 gm)	1:7 – 12
	1.65 lb	(750 gm)	1:5 – 10
	3.3 lb	(1.5 kg)	1:5 – 7

oxygen. (Never use your breath to inflate fish containers; you will be giving the fish carbon dioxide, not oxygen.)

You can also provide aeration en route, which is definitely recommended for shipments lasting longer than twelve hours. One way to aerate is with a product known as Otabs®. Otabs® are small metal canisters containing barium peroxide, detoxified with calcium sulfate, which when exposed to water gives off bubbles of oxygen. The manufacturer estimates that they are good for eight hours, but I have found that they last much longer. One professional shipper friend uses two Otabs® per small container. He opens one as instructed, but just punches a few small holes in the metal top of the other, to provide a small but more prolonged supply of oxygen. Otabs® seem to be marketed exclusively to the sport fishing trade, for use in minnow buckets. They should be more widely known by aquaculturists and aquarists.

Another device formerly advertised for fishermen, which I have not seen in recent years, is a small windmill designed to clamp on the outside of a car. As the car goes along, the windmill operates a small compressor which bubbles air into the water.

For moderately large shipments, the best rig is a truck with one or more shallow 60 to 125 gallon (254–530*l*) tanks, each equipped with a twelve-volt agitator run off the alternator. (Always carry a spare, because a mechanical failure can destroy an entire shipment.) When using such equipment, Stan Hudson, a columnist for *Aquaculture Magazine* (previously *The Commercial Fish Farmer*), recommends the rates listed in Table VI–3–2 for golden shiners (*Notemigonus crysoleucas*), a species with an intermediate oxygen requirement.

Table VI–3–2. Maximum loading rates for transporting
golden shiners in tanks equipped with a single 12–volt agitator.

Size of fish in lb (kg) per thousand	Loading rates in lb per gal (g/*l*)	
	high temperature	low temperature
6 or less (2.7 or less)	¾ (75)	1 (100)
10–12 (4.5–5.5)	1 (100)	1¼ (125)
20 (9.1)	1½ (150)	1½ (150)

Using compressed air, pure oxygen or constant aeration permits shipments of two days or more. If you use transparent containers you can easily check on the fish in transit. Where you think there may be a problem, you can reoxygenate the water by opening the bags, exposing the water surface to the air. Or you can make a partial change of water if water of the right temperature can be found. Remember that splashing, pouring, stirring—any process which breaks up or circulates the water—aerates it. In the days before aeration, hatchery truck drivers, when faced with a possible loss of a load of fish, used to drive like Hell, not only to get there faster, but also to bounce their load and splash water more.

Another technique of prolonging safe shipping time is to partially anesthetize or tranquilize the fish. This is particularly useful for trout and other high-strung fish. Table VI-3-3 lists concentrations of some drugs used for this purpose. Bear in mind that just what constitutes a safe dosage varies with many factors, including temperature, pH, size of fish, etc. If possible seek professional advice or, better yet, allow enough time or space so that drugs are not necessary.

Table VI–3–3.   Concentrations of some drugs used to tranquilize fish for transport

Drug	Concentration
sodium bicarbonate	642 mg/l
tertiary amyl alcohol	2 ml/gallon
methylparafynol (Dormison)	1–2 ml gallon
chloral hydrate	3.0–3.5 g/gallon
sodium barbital	6.7–7.7 micrograms/l
urethane	1–4 g/l
methane tricainesulfonate (MS-222)	40–150 mg/l
quinaldine	2–20 mg/l
mixture of MS-222 and quinaldine	20–30 mg/l. MS-222 and 5 mg/l of quinaldine

Shipping fish

When entrusting fish to commercial freight carriers be sure to seal all containers tightly and mark each box LIVE FISH and THIS SIDE UP as conspicuously as you can. Whether you are the shipper or the receiver, assume that no one but you will do anything right. Allow for twice as many hours as the trip should take, and be sure to insure the shipment for live arrival. When going to the airport or depot to pick up the fish, always have the waybill number and plenty of cash in hand, because those extra few minutes could be critical. Be sure to open each box while you are still in the office. In this way, if there is any mortality it will be the shipper's responsibility to replace the fish. If oxygen gas under pressure is used, you should ordinarily not unseal the bags until you are ready to stock the fish in their new home. In case of obvious stress, or when fish are packed with air instead of oxygen, you can open the bags to replenish the oxygen supply before moving the fish.

CHAPTER VI-4

Economics and Marketing

The cost of growing an aquatic crop is of interest to every aquaculturist, commercial or not. Though the rather alarming rate of change in this society makes specific cost figures fairly meaningless, I have tried, in the chapters on individual animals, techniques and problems, to give some feeling of the relative costs of various approaches. Here it seems appropriate to review the major costs likely to be incurred by any aquaculturist, commenting on them as they apply to the small-scale grower.

Costs of aquaculture

Some aquaculturists will be able to reproduce their own stock, while others will have to buy young fish, usually annually. When buying, the costs of purchase and shipping will be obvious. If you breed your own stock it may be necessary to economically analyze breeding as a separate operation. Perhaps you will be able to secure "free" stock from natural populations, but even then you must take into account the cost of capturing and transporting the fish.

Stock

One of the ironies of present day aquaculture is that it often costs less to obtain stock of those fish which are among the more costly to raise. For instance, it can safely be said that yellow bullheads (*Ictalurus natalis*) can be more cheaply raised to pan size than channel catfish (*Ictalurus punctatus*). But, while it is almost impossible to locate yellow bullhead fingerlings for sale, and those who do have them will often ask a premium price, channel catfish fingerlings are usually available at competitive prices. The reason is simply that there is an established commercial channel catfish culture industry which requires a constant supply of stock, while bullheads remain a specialty item.

When computing the cost of stock, don't forget mortality. Almost inevitably, a few fish will die during or soon after shipment. In the absence of sound advice or personal experience, ten percent is a safe figure to anticipate, though disease, mishandling, or accidents can cause much higher mortality. In my own experience, most commercial suppliers offset mortality by packing considerably more fish than were ordered.

Feed

Feed accounts for over fifty percent of the cost of North American commercial catfish and trout culture, and will probably be as expensive for any non-commercial grower who relies on concentrates. As the costs of the major ingredients of commercial feeds (fish meal, grains, synthetic vitamins) rise, so will the percentage allotted to feed in aquaculture budgets, though some compensation will certainly be achieved through feed industry research. Since the only fish for which commercial concentrated diets have been developed (catfish, trout and salmon) have extremely high protein requirements, the available concentrates are even more expensive than feeds which could be developed for less demanding fishes. While the convenience factor of packaged feeds is economically advantageous, all readers are advised to consider the cost and effectiveness of some of the alternative feeds and feeding strategies described in Part III.

Supplies and equipment

Supplies and equipment purchased specifically for aquaculture may or may not be major items, depending on the type of aquaculture. In either case, the costs of purchase are obvious.

Land and construction

Land and construction costs may range from prohibitive to almost none where there is a previously existing, unused pond. If a pond has to be constructed the cost will ordinarily be determined by a contractor, largely on the basis of the volume of earth which has to be moved. (See Chapter VII–1.) If there is any other use of the particular piece of land which might be contemplated, to this cost should be added a land cost (based on local land values).

Water costs

Water is most expensive if it has to be pumped into the aquaculture system. Water costs should also be calculated if the operation causes a net loss of water through evaporation or lowering of the water table. On the other hand, ponds supplied by rainfall, run-off, or diversion may be water conservers, and water costs become negative.

Depreciation

Depreciation on capital items can be calculated as in any business.

Loans

As well as calculating the interest on loans, it should be pointed out that if money is withdrawn from savings it should be treated as a loan, with the interest the savings would have drawn charged against the aquaculture operation.

Permits

Licenses and permits will normally be a very small item for the non-commercial aquaculturist, but may be considerable for the commercial grower. I will discuss this aspect of aquaculture further in Chapter VIII–5.

Labor

The cost of labor is straightforward if you are hiring others, but less so if you, or members of your family, do the work. The traditional practice is to value one's own labor as equivalent to labor which would be hired to do the same job, but this leaves a number of questions unanswered. Are there other, and possibly more lucrative, uses of the same time? If you do the job more or less efficiently than a hired laborer, should the cost of that labor be based on the time put in on the job, or on the basis of the time it would take a hired worker to do the job? If one person does a job out of necessity, while another enjoys doing the same job, should this be reflected in economic calculations?

Energy

The main sources of energy costs are vehicles, pumps and electrical apparatus. In these times, I scarcely need to urge anyone to think about saving money by cutting down on energy use. Aeration is one facet of aquaculture which lends itself particularly well to energy conservation through proper selection and utilization of equipment. (See Chapter X–2.)

Taxes

Property taxes affect all aquaculturists. Taxes on sales are of concern to all commercial growers, but some small producers do not report all their sales. If some of these were more scrupulous, they might not be able to remain in business.

All aquaculturists should be insured against accidents occurring on the property and potential damage to neighboring property. Commercial growers should also have insurance to cover possible claims against their products.

Miscellaneous expenses might include travel, advertising, publications, postage and telephone costs.

Of the thirteen types of costs I have named, six (stock, land and construction, depreciation, interest, licenses and insurance) can be calculated quite precisely before the first growing season has begun. The costs of supplies and equipment, labor, energy and miscellaneous items are more subject to change within a growing season, but these too can be fairly well estimated, particularly if you are in contact with experienced growers. Water costs are more difficult to estimate when pumping is necessary, since the need to pump will largely depend on the weather. Taxes on sales and feed costs are particularly difficult to estimate, since they depend on the size of the harvest.

Feed is the single most costly item in many aquaculture ventures, so you should attempt to estimate it at the outset. If you are raising catfish or trout, using an established brand of commercial feed, you should be able to predict feed costs and production fairly accurately, using information from the feed company and other growers.

For example, suppose a catfish farmer in Arkansas stocks a ten acre (4 ha) pond in March with three inch (7.6 cm) fingerlings, which normally run ten per pound (22/kg). From the experience of other growers in the area channel catfish fingerlings can be raised to harvestable size (approximately one pound or 0.45 kg) during warm weather in twenty-five weeks, using a commercial dry feed with an advertised conversion ratio of 1.2:1. Other local growers aim for a harvest of 2,500 pounds per acre (2,750 kg/ha). So, the total harvest should be 2,500 pounds per acre × 10 acres or 25,000 pounds (11,250 kg). Then to achieve the desired average size of one pound, 25,000 fingerlings should be stocked, plus a ten percent allowance for mortality during the season, for a total of 27,500 fingerlings. Multiplying the number of fingerlings by the known average weight we get 27,500 × 0.1 lb = 2,750 pounds (1,237.5 kg) of fish stocked in March. Subtracting this weight from the 25,000 pound (11,250 kg) harvest, we estimate a total production of 22,250 pounds (10,012.5 kg). This figure, times 1.2 (the known conversion ratio of the feed), gives the amount of feed (26,700 pounds or 12,015 kg) which will need to be purchased.

Assuming a constant feeding rate of three percent of body weight daily, six days a week, the precise amount of feed needed the first week of grow-out is 0.03 × 2,750 pounds × 6 days = 495 pounds (223 kg). If all goes according to plan 0.03 × 25,000 lb × 6 days = 4,500 pounds (2,025 kg) of feed will be needed during the last week. Using these figures, and assuming somewhat more rapid growth in the early weeks, the amount and cost of feed needed in any portion of the growing period can be estimated.

You can see that the preceding example is tidier than anything that is likely to happen in real life, but it is about as cut-and-dried a situation as can be imagined. At the opposite extreme, suppose an inexperienced aquaculturist is attempting polyculture of several species, incorporating fertilization, cultured earthworms and bug lights as food sources, in a region where there are no other fish farmers. Here, the best advice that can be given is to devote the first season to pilot studies on a small scale, and to err on the high side in estimating costs and on the low side for production.

For the time being, the experience of most readers will probably be closer to the second example. Nevertheless, try to estimate the crop as accurately as possible in advance, and remember to share your experiences with other aquaculturists, so we can all do a better job.

*Cooperating to
cut costs*

Part of the solution to the economic problems of small-scale aquaculture lies in the development of a greater degree of cooperation. As an example, take Humphreys County, Mississippi, with one of the densest concentrations of fish farms in North America. The farmers of Humphreys County realize some of the lowest catfish feed costs in the U.S. by owning and operating their own feed mill. While this is possible only with a concentration of large-scale fish farmers, the same approach could allow a group of small farmers to develop earthworms as an alternative feed, where it would not be feasible for a single grower. Another area where cooperative effort makes sense is in the sharing of expensive equipment which does not require constant use, such as vehicles, harvesters, and processing equipment. Cooperation among farmers within a region may eventually prove to be the key to making stock of such presently expensive fish as bullheads, hybrid sunfishes, monosex tilapia, etc., available at reasonable prices. Small-scale commercial growers may want to look into cooperative marketing arrangements. These are by no means the only opportunities to save money by cooperation and sharing. To find out what alternatives there are, all aquaculturists should get to know any other producers within a day's travel, and explore how they can help each other.

*Cost of
non-commercial
aquaculture vs.
other protein sources*

Simply knowing how much of a crop can be produced, its dollar value and production costs, will not be enough economic information for most would-be aquaculturists. The non-commercial grower will want to compare the crop to other protein foods he or she might produce or purchase. Despite the theoretical advantages of fish, many forms of aquaculture, given the present state of the art, are a less economical home protein source than raising poultry or even purchasing fish. If you have a half acre (0.2 ha) to five acre (2 ha) warm-water pond, it is very likely that, using the kind of low intensity methods described in Chapter II–1, you will be able to put home-grown fish on your table more cheaply than any other form of animal protein. Though other forms of aquaculture may work as well in a given situation, they need careful economic analysis before you make a commitment.

It is difficult to say much more to the non-commercial grower, since subjective factors play such an important role. If you like fish better than chicken, or bass better than bullheads, or vice versa, you just do, and it is up to you to weigh your preference together with the economic factors.

*Cost of small scale
commercial
aquaculture vs.
fisheries*

On the other hand, it is fairly clear what the small-scale commercial grower is competing against: other fish, whether taken by capture fisheries or produced by aquabusiness. In some cases, there is competition only with other sources of a particular species, but some North Americans still consider "fish" a food category and will buy whatever reasonably attractive fish is cheapest. Most consumer attitudes are probably somewhere between these extremes. For instance, the customer used to buying channel catfish will often accept South American catfish as a substitute, but not buffalofish.

At present commercial aquaculture production costs do not give the aquaculturist a notable advantage over the commercial fisherman unless there is a shortage of a particular fish. Of the three major North American commercial fresh water aquaculture crops, it is worth noting that cultured catfish became important only as catfish fisheries declined, the sale of wild trout is illegal everywhere, and crayfish farmers must count on selling their crop before the fishing season. However, the prognosis for fresh water aquaculture is brighter than for oceanic fisheries. The prospect for fisheries is one of increasing costs as the environment deteriorates, as commercial species become scarcer, and the costs of fishing, especially fuel costs, storage, transportation and other petroleum-dependent items increase. However, the aquaculturist employing controlled reproduction is less at the mercy of environmental deterioration, less dependent on fuels, usually closer to markets, and stands to benefit from the increased demand for fish and reduced supply of fishery products.

Before going on to discuss the economics of small- versus large-scale commercial aquaculture and offer some suggestions to the small-scale grower, I want to make it clear that I approach this subject with some trepidation. Though I certainly do not want to impede the development of small-scale aquaculture, neither do I want to lead even one reader down the garden path to financial fiasco. If I were writing on agriculture, I could draw on the accumulated experience of small farmers to assess the possibilities, while cautioning the reader against some of the pitfalls of a road often travelled. But in aquaculture there is precious little accumulated experience. The road ahead is both enticing and dangerous, and it is difficult to say around which bend the rewards and perils are to be encountered. Whatever else you read in this chapter, remember there is no substitute for reliable local information, if it exists.

Suggestions for small-scale commercial aquaculturists

Not only is there a scarcity of practical experience in small-scale commercial aquaculture, but such experience is also often not divulged. I have mentioned that much of the propaganda in North American aquaculture proposes that only a large-scale approach will yield a profit. Some of the blame for this inaccuracy lies with the big producers, the media which report on aquaculture, and economists who seem unable to take small enterprise seriously. But it is also true that many small-scale producers are reticent about divulging information. Friends in the major fish farming regions tell me that a fair percentage of small growers wish to avoid publicity for reasons related to taxes or other aspects of government regulation. A few growers simply refuse to provide information when contacted by surveys, while others claim ignorance of production and sales figures. Admitting that we are operating in an information vacuum, it is still clear that the small-scale grower should not try to beat the aquabusiness producers at their own game. Instead, the small producer can try to offer special services or products. Following are some suggestions.

• Try to identify local markets which you can supply with fresher fish than they can ever get from more distant sources. Restaurants are a good possibility, but only if you can provide a constant supply. Live or fresh, dressed fish at pondside justify a premium price, and most large-scale growers are not willing to deal with the small volumes involved in individual sales.

Local markets

• Offer a superior quality product. For example, trout fed on live foods draw higher prices than trout fed on commercial feeds. Or the superior product may be a different kind of fish. For example, the rainbow trout (*Salmo gairdneri*) is undoubtedly the most efficient trout for mass culture in most places, but many people

Quality

consider the brook trout (*Salvelinus fontinalis*) a superior food fish.

• Evaluate the economics of processing. In those states and provinces where sunfishes enter the commercial fishery, they bring from fifteen to forty cents a pound (33 to 88 cents/kg). But in Ohio, bluegill (*Lepomis macrochirus*) fillets, representing perhaps twenty percent of the fish's total weight, can be sold for four dollars a pound ($8.82/kg). Experienced operators can produce a pound of fillets from whole quarter pound (0.1 kg) bluegills in six or seven minutes. Allowing for dressing loss and costing the labor of filleting at four dollars per hour, the price received for bluegill fillets comes out near the upper end of the range of prices paid for whole, fishery-caught bluegills.

• Consider specialty products. Smoked fish is the most obvious possibility, but various pickled and preserved products are also marketable. As an example of the potential, consider that fresh fishery-caught sturgeon in Iowa bring forty-five cents a pound (99 cents/kg), but smoked sturgeon sells for $2.50 a pound ($5.51/kg).

• Identify ethnic or localized markets. The use by ethnic minorities of foods which are unknown to or not accepted by most North Americans is well known. Perhaps less known are such regional markets as the one for goldeye (*Hiodon alosoides*), which is eaten only in the vicinity of Winnipeg, Manitoba.

• There may be potential in the "health," "natural," or "organic" food market. Given the concern over uncontaminated foods, the known advantages of fish over other sources of animal protein, and the contamination of wild fish from the oceans and large inland waters, it might be worthwhile for some health food outlets to deal with the storage problem posed by fish.

• Consider diversifying by combining sale of food fish with non-food aquaculture in the same body of water. An example is the experience of Bill Gressard of Trail Lake Fisheries, Kent, Ohio. Over a period of thirteen months, he sold nearly two thousand dollars' worth of bluegill fillets harvested from less than forty-five acres (18.2 ha) of water, which was also used for fee fishing and to produce breeder largemouth bass (*Micropterus salmoides*), for sale.

Gressard's combination of food fish, sport fish, and fee fishing is only one of many possibilities. The major types of non-food aquaculture are discussed in Part IX; any of them might be used in combination with production of aquatic food for domestic use or for sale. Fee fishing, in particular, lends itself to this. Even if you cannot make your entire living in this way, it could be used to underwrite the cost of domestic food production.

Keeping these suggestions in mind, together with the sobering fact that (according to the United States Department of Agriculture), ninety-five percent of those attempting commercial aquaculture in the U.S. drop out of the business for financial reasons, we will look at the economics of the three major North American fresh water food crops in relation to the small farmer.

Roughly 1,400 to 1,800 persons were producing catfish for sale as food in the United States in 1976. Although catfish farming is both the most common and most studied form of aquaculture in North America, data on the average size of operations and the relative success of large and small operators is difficult to come by and sometimes contradictory. Clearly, one cannot make sweeping statements about the feasibility of small-scale commercial catfish culture. The small entrepreneur's prospects will vary, not only with his or her aquacultural and business skill, but also with geographic location, type of market exploited, and the method and time of harvesting and handling the crop.

More problematical for the small farmer are annual variations in the catfish market price and production cost. These are entirely due to factors outside the farmer's control, including feed and other production costs, landings of wild catfish, imports of competitive fishes (mainly catfish from Brazil), and the overall production of the catfish culture industry. Of course such factors affect both large and small growers, but it seems that small producers are less able to withstand short-term adverse conditions. The economic history of catfish farming reflects this situation. High catfish prices in certain years cause leaps in production, followed by drops in price. When the price falls, the less efficient operators, mainly small growers, are forced out of the business.

Just what are the economies of scale which give the large growers an advantage? There are at least four factors involved:

• The larger growers are more highly capitalized, and in some cases enjoy corporate backing. They are thus in a financial position to withstand a poor season.

Economic advantages for large growers

• In the opinion of most commercial catfish growers, a full time manager is essential. This means the farm must generate enough money to pay the manager's salary. Small operations can be owner-managed, but it is a general (and wise) practice to diversify by combining catfish with other aquaculture and agriculture crops, which tends to give the owner-manager more than a full time job.

• Certain capital costs are more or less constant, regardless of the size of the farm. This is particularly true of harvesting equipment. As catfish farmer Henry Anthony of Lake Village, Arkansas, said at a recent convention of the Catfish Farmers of America "It takes just as much investment in equipment to harvest a forty acre (18.2 ha) pond as it does ten forty acre ponds."

• Marketing opportunities are limited for the small grower. Restaurants, one of the best paying markets, usually demand a constant supply of table-size fish, and this is difficult to provide without a number of ponds containing catfish in various stages of development. The small producer may also run into difficulty with processors refusing to buy crops under a certain minimum weight. Five tons (5,080 kg) is a typical minimum.

The advantages to the larger and/or more efficient grower are pronounced in the principal catfish producing areas, where price competition is most severe. In 1973 in Arkansas, the second ranking state in catfish production, farmers received an average of fifty-two cents a pound ($1.14/kg) for catfish, while in California, which has relatively few producers, the price was ninety-five cents a pound ($2.09/kg). But growers outside the main producing areas experience greater difficulty in obtaining stock, supplies and technical assistance, and suffer from the inaccessibility of major markets for catfish.

Suitability of various regions for beginning farmers

Most cultured and fishery-caught catfish marketed in the U.S. are sold in the South or to southern emigrants in mid-western urban centers. For various reasons, the industry began in the lower Mississippi Valley, and production is still concentrated there, though catfish culture has spread to other parts of the South. Catfish are grown in some northern states as well, but much of the catfish consumed in the midwest is brought in from southern farms and fisheries.

(1) The greatest opportunities are in portions of the South where catfish is a

traditional food and the growing season is long, but the catfish culture industry has not yet become established.

(2) Parts of the midwest, particularly near large urban centers, offer established markets which are not fully supplied. Here the advantage of closeness to the market may offset the limitations due to a short growing season.

(3) The lower Mississippi Valley, Alabama, east Texas, and other southern areas where catfish culture is well established, offer a long growing season, established markets and easy availability of stock, supplies and services. However, prices tend to be low and the competition may be too heavy for the new small producer.

(4) At present, the poorest opportunities exist in the Pacific coast states, the Rockies, the northern Great Plains, the Northeast and Canada, where the growing season is short and catfish is not a traditional food. Local exceptions to this situation may exist, and there is also the possibility that people in these regions will eventually become educated as to the merits of catfish.

The types of market

There are four types of market open to catfish farmers: processors, local retailers, fee fishing facilities and live haulers. The relative importance of these markets has been only roughly estimated, and the figures given here are based on a 1973 preliminary study conducted by C.A. Oravetz of the National Marine Fisheries Service. (No more recent study of catfish markets is available.)

Processors

Processors are a market usually available only in areas where catfish farming is concentrated. Fish sold to processors are resold as iced or frozen whole dressed fish, fillets, or steaks, for distribution to fish markets, restaurants, groceries, supermarkets, and fast food chains. Most farmers prefer to saturate the more profitable retail and fee fishing markets before calling on processors to buy their surplus, though you can contract an entire crop to a processor in advance of harvest. Despite low prices, processors purchased about a third of the 1973 catfish crop.

Retail

About twenty-four percent of the 1973 crop was retailed locally. This may be in sales to local restaurants, fish markets, etc., or it may involve individual on-farm sales of live or processed fish. Amounts received by farmers through retail sales vary greatly, from the lowest prices paid by any processor on up to three or four times these prices when fish are sold directly to the consumer during the off season.

Fee fishing

Fee fishing facilities represent an important market in all catfish growing areas, particularly for over-size fish. Nationally, they only accounted for an estimated thirteen percent of sales in 1973, but in some areas the majority of cultured catfish are marketed in this way. You can exploit this market by harvesting fish for sale to outside fee fishing operators, maintaining a separate fee fishing facility on the fish farm, or by allowing paying fishermen access to culture ponds at certain times. Though the amount paid per fish caught varies widely, it is always more than the price paid by processors or live haulers. If you own the fee fishing facility, you can make additional profits from admissions, growing other types of fish, and services such as cleaning fish, selling bait, etc. (See Chapter IX–1 for more information on fee fishing.)

Live hauling

While live haulers buy thirty-one percent of the catfish crop, they are the least understood of the catfish markets. Fish markets in southern and midwestern cities are the destination of many live-hauled fish, but live haulers also sell to fee-fishing operators and individual farm pond owners. Based on known sales by live haulers, Oravetz's estimate of the distribution of farm-raised catfish in the United States was adjusted by Brown (1977): processors, 32%; retail outlets, 26%; fee fishing, 33%; and farm pond owners, who use the fish for home consumption and recreation, 10%.

Some of the discouraging advice often given about the feasibility of small-scale commercial catfish farming may be caused by overemphasizing sales to processors. Since small growers seldom have surpluses large enough to interest processors, this major market is rarely advantageous—or even available—to the small farmer. But, live haulers often buy small crops, sometimes at better prices than processors. The advantages of fee fishing to the small grower are obvious from the preceding discussion. As for retail marketing, the suggestions made earlier in this chapter apply to catfish farmers but methods and timing of catfish harvesting bear special mention here.

The main method of harvesting farm-raised catfish is mechanical or manual haul seining, which can result in capturing ninety percent of a pond's population in a single haul. With this sort of harvest goes the need for rapid, efficient transfer, or processing the captured fish to prevent mortality or spoilage. When labor and equipment can be provided, mass harvesting with haul seines is the best method of harvesting catfish for sale to processors or live haulers. However, the various trapping methods described in Chapter VI–1 are less costly. Equally important to the small grower, traps can be used to capture part of a population without disturbing or endangering the rest. You can achieve still greater flexibility by providing cages, live cars or other facilities for holding live fish. (See Chapter VI–3.) The catfish farmer, large or small, who can provide a given quantity or size of fish on short notice is in a position to charge a premium price.

Meeting market demands

Another way to realize a better price is to harvest when catfish supplies are low and prices high. The greatest number of mass harvests are made in January, February and March, when prices are lowest. Harvests are fewest and prices highest in midsummer. The reasons for winter harvests include the availability of farm labor at that time, and the greater danger of mortality when fish are handled during hot weather.

Not only are producer prices higher in the summer, but the fish are actually worth more. In the South, catfish kept in ponds from the end of the feeding season in the fall until the start of the next feeding season in the spring lose about nine percent of their weight. (Winter weight loss is even more severe in the North.) If the farmer can sell in the fall when fish are "fat" he has an advantage. The small-scale grower, using gradual or partial harvesting methods which do not endanger the fish in hot weather, and who does not have to deal with processors or live haulers, is better able to take advantage of this opportunity than the farmer with hundreds of acres of water to manage.

The trout culture industry is much less dominated by large producers than the catfish industry. The size of the usual trout farm is limited by the modest flow of suitable water which can be maintained in most locations. (Some hold the opinion that recent technological developments in closed system aquaculture have overcome this limitation, but I do not agree. See Chapter IV–4.) An exception exists in southern Idaho, where there are enormous isothermal springs ideal for trout culture. Here trout farms tend to be large, some maintaining flows as high as 14,583 gallons per minute or 918.7 liters per second. The region accounts for an estimated

Economics of trout culture

Water flow rates

ninety percent of the trout sold to processors in the United States. Elsewhere, commercial trout farmers operate on less than 1,000 gallons per minute (63.0 l/second) and the average is probably less than 300 gallons per minute (18.9 l/second). This means that the average trout farm at any one time can have on hand little more than a ton (907 kg) of trout of all sizes.

There are no U.S. official state or federal production statistics for trout, but it is certain that a substantial amount of the profit in trout culture is made by hatcheries specializing in egg and/or fingerling production. (See Chapter IX–1.) There are three groups of small-scale food trout producers:

Types of producers
- Hatchery-farm operators who breed and raise their own fish.
- Farmers who annually purchase eyed eggs or fingerlings for grow-out. This group accounts for the majority of trout farmers.
- Grow-out specialists who raise fish on consignment for another producer, never taking title to the fish, and paid on the basis of weight gain. This may be an attractive option for small operators located near major trout farms.

Trout prices
Ultimately, the trout produced by all three groups find their way to the same types of outlets as farm-raised catfish. But processors are believed to take about seventy-four percent of the available trout, according to the best available guess. Processors who supply iced and frozen dressed, boned and filleted trout to restaurants and retail stores offer the lowest prices and usually deal exclusively with the large-scale growers. The great majority of trout sold by small-scale growers are sold to fee fishing operators (who usually do not grow their own fish, but buy "catchables" of a half pound [0.23 kg] or more), or to live haulers, who in turn sell to fee fishing operators. Prices obtained in various markets during 1976 by Idaho trout farmers are listed in Table VI–4–1.

According to Brown, large-scale trout farmers in Idaho indicated in 1976 that production costs were nearly equal to prices. The small-scale trout farmers I interviewed in the East in 1977 and 1978 reported production costs of one dollar to one dollar fifty per pound ($2.20 to $3.30/kg), but prices in at least some states were double those cited in Table VI–4–1 for Idaho.

Trout marketing in Canada
Marketing trout in Canada mainly involves selling dressed fish to local retailers. The only commercial processor in Canada is the Fresh Water Fish Marketing Corporation, which buys some trout produced in "potholes" in the prairie provinces. Fee fishing has scarcely begun. A profitable specialty product from Quebec is large fresh or smoked trout, exported at a premium price to the New York market.

Economics of crayfish culture
Commercial crayfish culture for food is almost entirely confined to Louisiana, though I know of one bait grower in Missouri who sells his largest crayfish as food. The industry is mainly made up of small farmers, though there is one large corporate farm. Crayfish farmers sell their catch live or as peeled tails, to restaurants, retailers, individuals, or one of the forty or so processing plants in southern Louisiana. The trick in crayfish marketing in Louisiana is to get your crop to market ahead of the wild fishery catch. In early spring, before the wild fishery comes in, cultured crayfish may bring seventy cents a pound ($1.54/kg), but in May when wild crayfish are readily available, the price may drop to fifteen cents a pound ($0.33/kg). As production costs are low, crayfish farming in Louisiana continues to be profitable.

Table VI-4-1. Prices paid for cultured trout in Idaho in 1976.

Size of fish	Form	Market	Price per lb ($U.S.)	Indexed price¹ per lb ($U.S.)	Price per kg ($U.S.)	Indexed price¹ per kg ($U.S.)
12 in (30 cm)	Live or "round"	FOB Hatchery	1.19	1.19	2.61	2.61
12 in (30 cm)	Frozen, dressed	FOB Hatchery (wholesalers)	1.39	1.11	3.05	2.44
12 in (30 cm)	Frozen, boned	FOB Hatchery (wholesalers)	1.44	0.98	3.16	2.17
12 in (30 cm)	Frozen, dressed	Restaurants and retail stores	1.80	1.44	3.96	3.16
12 in (30 cm)	Frozen, boned	Restaurants and retail stores	1.85	1.26	4.07	2.79
12 in (30 cm)	Frozen, dressed	Retail to individuals	2.29	1.83	5.03	4.02
12 in (30 cm)	Frozen, boned	Retail to individuals	2.34	1.60	5.14	3.53
½ lb (0.2 kg) or more	Live	Fee fishing operators	1.00	1.00	2.20	2.20
½ lb (0.2 kg) or more	Live or "round"	Patrons of fee fishing facilities	$0.10 per inch of length, which amounts to approximately $1.50 per lb ($3.30 per kg)			

Table VI-4-2. Prices paid for cultured trout in Ontario in 1979.

Form	Market	Price per lb ($ Canadian)	Indexed price¹ per lb ($ Canadian)	Price per kg ($ Canadian)	Indexed price¹ per kg ($ Canadian)
Live	Other Farms	1.20–2.70	1.20–2.70	2.64–5.94	2.64–5.94
Live	Wholesalers	1.28–2.80	1.28–2.80	2.82–6.16	2.82–6.16
Live	To be stocked for recreational fishing	1.88–3.00	1.88–3.00	3.04–6.60	3.04–6.60
In the round	Retailers	2.52–2.63	2.52–2.63	5.54–5.79	5.54–5.79
In the round	Retail to individuals	1.45–2.90	1.45–2.90	3.19–6.38	3.19–6.38
Whole, dressed	Retailers	1.50–3.90	1.80–4.68	3.30–8.58	3.96–10.30
Whole, dressed	Retail to individuals	1.84–3.50	2.21–4.20	4.05–7.70	4.86–9.24
Dressed, smoked	Retail to individuals	5.49	6.58	12.08	14.48

¹ "Indexed" price is price as compared to the weight of whole fish, allowing 20% loss for dressed fish and 31.3% loss for boned fish. The aquaculturist must then evaluate the labor cost of processing.

Channel catfish, trout and crayfish are not the only fresh water creatures sold as food in North America. Table VI–4–3 lists prices received in 1978 for fresh water fishes and invertebrates in the various states and provinces. Except where noted, the prices are for fishery-caught animals, and are intended to help you judge the feasibility of growing a particular species. Note the great differences in price for a single species from state to state. For instance, in Nebraska fishery-caught bullheads bring only five cents a pound ($0.11/kg) while in Minnesota and Alabama they bring thirty-eight cents a pound ($0.84/kg), and in New York the price varies from twenty to seventy cents a pound ($0.44 to $1.54/kg).

Table VI–4–3.   1978 wholesale prices of fresh water food animals in the United States and Canada.*(All prices are for fishery catches unless otherwise indicated.)

Animal[1]	Form	State or Province[2]	Price ($/lb)	Price ($/kg)
Black bass	whole	Utah	price unknown	
Sunfishes	whole	Florida	0.25–0.40	0.55–0.88
(mainly		Maryland[7]	0.09	0.20
bluegill and		New York	0.30–0.35	0.66–0.77
redear)		Ontario	0.22	0.48
Bluegill (cultured)	fillets (80% dressing loss)	Ohio	4.00	8.80
Bullheads	whole	Alabama	0.38	0.84
		Delaware[7]	0.15	0.33
		Michigan	0.17	0.37
		Minnesota	0.38	0.84
		Missouri	0.26	0.57
		Nebraska	0.05	0.11
		New Jersey[7]	0.18	0.40
		New York	0.20–0.70	0.44–1.54
		Ohio	0.14	0.31
		South Carolina	price not available	
		South Dakota	0.10	0.22
		Wisconsin	0.13	0.29
		Ontario	0.28	0.62
Bullheads	dressed (45% dressing loss)	Florida	0.69–0.95	1.52–2.09
		Manitoba	0.16	0.35
		Saskatchewan	0.16	0.35
Channel[3] catfish (fisheries)	whole	Alabama	0.38	0.84
		Arkansas	0.53–0.67	1.17–1.47
		Florida	0.33–0.37	0.73–0.81
		Georgia[7,8]	0.41	0.90
		Indiana	0.23	0.51
		Iowa	0.65	1.43
		Kansas[7,8]	0.50	1.10
		Kentucky[7,8]	0.45	0.99
		Louisiana	0.40–0.60	0.88–1.32
		Maryland[7,8]	0.13	0.29
		Michigan	0.36	0.79
		Mississippi[7,8]	0.40	0.88
		Missouri	0.50	1.10
Channel[3] catfish (fisheries)	whole	Nebraska	0.80	1.76
		New York	1.00	2.20
		North Carolina	0.35	0.77
		North Dakota	0.35	0.77

*   These figures are cited primarily for comparative purposes; in most cases it is impossible to obtain up-to-date prices for fishery products on a state-or-province-wide basis.

Table VI-4-2. *Continued.*

Animal[1]	Form	State or Province[2]	Price ($/lb)	Price ($/kg)
Channel[3] catfish (fisheries)	whole	Ohio	0.44	0.97
		Pennsylvania	0.18	0.40
		South Carolina	prices not available	
		South Dakota	0.40	0.88
		Tennessee[7,8]	0.45	0.99
		Texas[7,8]	0.52	1.14
		Virginia[7,8]	0.22	0.48
		West Virginia	0.40–0.80	0.88–1.76
		Wisconsin	0.21	0.46
		Ontario	0.37	0.81
Channel[3] catfish (fisheries)	dressed (42–44% dressing loss)	North Carolina	1.30	2.86
		Manitoba	0.17	0.37
		Saskatchewan	0.17	0.37
Channel[3] catfish (cultured)	whole	Alabama	0.38	0.84
		Arkansas	0.50–0.60	1.10–1.32
		California	0.95	2.09
		Florida	0.33–0.37	0.73–0.81
		Georgia	price unknown	
		Hawaii	1.50–1.75	3.30–3.85
		Idaho	price unknown	
		Illinois	price unknown	
		Indiana	0.23	0.51
		Iowa	0.65	1.43
		Kansas	price unknown	
		Kentucky	prices not available	
		Louisiana	0.60–0.75	1.32–1.65
		Minnesota	price unknown	
		Mississippi	price unknown	
		Missouri	0.50	1.10
		Nebraska	0.80	1.96
		Ohio	0.44	0.97
		Oklahoma	prices not available	
		South Carolina[7]	0.21–0.24	0.46–0.53
		Tennessee	price unknown	
		Texas	price unknown	
		Utah	price unknown	
		West Virginia	0.85–1.30	1.87–2.86
Channel[3] catfish (cultured)	dressed (42–44% dressing loss)	California	1.89	4.16
		Florida	0.65–0.95	1.43–2.09
Blue catfish (fisheries)	whole	Missouri	0.50	1.10
Blue catfish (cultured)	whole	Arkansas	0.73	1.61
		Missouri	0.50	1.10
White catfish	dressed (45% dressing loss)	Florida	0.69–0.95	1.52–2.09
Flathead catfish	whole	Missouri	0.50	1.10
		Oklahoma	0.65	1.43
Buffalofish (fisheries)	whole	Alabama	0.12	0.26
		Arizona	0.33	0.73
		Arkansas	0.30–0.36	0.66–0.79
		Indiana	price unknown	
		Iowa	0.20	0.44
		Kansas[7]	0.25	0.55
		Kentucky[7]	0.21	0.46

Table VI–4–2. *Continued.*

Animal[1]	Form	State or Province[2]	Price ($/lb)	Price ($/kg)
Buffalofish (fisheries)	whole	Louisiana	0.15–0.40	0.33–0.88
		Michigan	0.17	0.37
		Mississippi[7]	0.24	0.53
		Missouri	0.23	0.51
		Montana	price unknown	
		Nebraska	0.40	0.88
		North Dakota	0.17	0.37
		Oklahoma	0.25	0.55
		South Dakota	0.17	0.37
		Tennessee[7]	0.19	0.42
		Texas[7]	0.22	0.48
		Wisconsin[7]	0.03	0.06
Buffalofish (cultured)	whole	Arkansas	0.26–0.28	0.57–0.62
Carp	whole	Alabama	0.02	0.04
		Arkansas	0.06–0.20	0.13–0.44
		California	0.34	0.75
		Delaware[7]	0.14	0.31
		Idaho	0.20	0.44
		Indiana	price unknown	
		Iowa	0.05	0.11
		Kansas[7]	0.25	0.55
		Kentucky[7]	0.09	0.20
		Maryland[7]	0.02	0.04
		Michigan	0.06	0.13
		Minnesota	0.04	0.09
		Mississippi[7]	0.07	0.15
		Missouri	0.09	0.20
		Nebraska	0.40	0.88
		New Jersey[7]	0.08	0.18
		New York	0.05–0.10	0.11–0.22
		North Dakota	0.05	0.11
		Ohio	0.07	0.15
		Oklahoma	0.15	0.33
		South Carolina[7]	0.10–0.14	0.22–0.30
		South Dakota	0.06	0.13
		Tennessee[7]	0.05	0.11
		Texas[7]	0.07	0.15
		Virginia[7]	0.05	0.11
		Washington[7]	0.01	0.02
		Wisconsin	0.04	0.09
		British Columbia	price unknown	
		Ontario	0.19	0.42
Carp	dressed	Manitoba	0.065	0.14
		Saskatchewan	0.065	0.14
Trout[4] (cultured)	whole	Alabama	price unknown	
		Arizona	1.50–2.00	3.30–4.40
		Arkansas	0.90–1.50	1.98–3.30
		Colorado	price unknown	
		Connecticut	2.00	4.40
		Georgia	price unknown	
		Idaho	0.65–0.80	1.43–1.76
		Maine	prices not available	
		Massachusetts	2.00	4.40
		Michigan	prices not available	
		Minnesota	price unknown	
		Missouri	price unknown	

Table VI–4–2. *Continued.*

Animal[1]	Form	State or Province[2]	Price ($/lb)	Price ($/kg)
Trout[4]	whole	Montana	price unknown	
(cultured)		Nebraska	price unknown	
		Nevada	price unknown	
		New Hampshire	(0.10 per in)	(0.04 per cm)
		New Mexico	price unknown	
		New York	price unknown	
		North Carolina	price unknown	
		North Dakota	0.60	1.32
		Ohio	price unknown	
		Oregon	price unknown	
		Pennsylvania	price unknown	
		South Carolina	prices not available	
		South Dakota	price unknown	
		Tennessee	price unknown	
		Utah	price unknown	
		Vermont	price unknown	
		Virginia	price unknown	
		Washington	price unknown	
		West Virginia	1.50–2.00	1.98–4.40
		Wisconsin	price unknown	
		Wyoming	price unknown	
		Alberta	0.50–1.60	1.10–3.52
		British Columbia	0.85–1.60	1.87–3.52
		Manitoba	0.50–1.25	1.10–2.75
		Ontario	0.85–1.60	1.87–3.52
		Yukon	price unknown	
Trout[4]	dressed	California	1.89	4.16
(cultured)	(20% dressing loss)	Manitoba	0.90–2.00	1.98–4.40
Lake trout	whole	Illinois	0.96	2.11
		Indiana	0.50	1.10
		Michigan	0.57	1.25
		Minnesota	0.92	2.02
		Pennsylvania	0.65	1.43
		Wisconsin	0.52	1.14
		Alberta	price unknown	
		British Columbia	price unknown	
		Ontario	0.67	1.47
		Yukon	1.25	2.75
Lake trout	dressed	Manitoba	0.32–0.37	0.70–0.81
	(20% dressing loss)	Saskatchewan	0.32–0.37	0.70–0.81
Yellow perch	whole	Illinois	0.46	1.01
		Indiana	0.65	1.43
		Maryland[7]	0.17	0.37
		Michigan	0.39	0.86
		Minnesota	0.18	0.40
		New York	0.80–0.85	1.76–1.87
		Ohio	0.37	0.81
		Pennsylvania	0.55	1.21
		Virginia[7]	0.18	0.40
		Wisconsin	0.39	0.86
		Manitoba	0.32	0.70
		Ontario	0.47	0.97
		Saskatchewan	0.32	0.70
Grass carp	whole	Arkansas	0.27–0.32	0.59–0.70
(fisheries)		Missouri	0.25	0.55
Grass carp	whole	Hawaii	2.50–3.50	5.50–7.70
(cultured)				

Table VI-4-2. *Continued.*

Animal[1]	Form	State or Province[2]	Price ($/lb)	Price ($/kg)
Tilapia	whole	Florida	0.15–0.30	0.33–0.66
Whitefish	whole	Illinois	0.77	1.69
		Indiana[7]	0.32	0.70
		Michigan	0.70	1.54
		Minnesota	0.36	0.79
		Pennsylvania	0.65	1.43
		Wisconsin	0.80	1.76
		Alberta	price unknown	
		British Columbia	price unknown	
		Manitoba	0.10–0.50	0.22–1.10
		Ontario	0.27–0.51	0.59–1.12
		Saskatchewan	0.10–0.50	0.22–1.10
		Yukon	0.90	1.98
Whitefish	dressed	Manitoba	0.19–0.57	0.42–1.25
		Saskatchewan	0.19–0.57	0.42–1.25
Ciscoes	whole	Illinois	0.36	0.79
		Indiana	0.36	0.79
		Michigan	0.26–0.57	0.57–1.25
		Minnesota	0.28–0.29	0.62–0.64
		New York	0.33	0.73
		Wisconsin	0.29–0.54	0.64–1.19
		Alberta	price unknown	
		Ontario	0.19–0.59	0.42–1.30
Ciscoes	dressed	Manitoba	0.17–0.25	0.37–0.55
		Saskatchewan	0.17–0.25	0.37–0.55
Goldeye	whole	North Dakota	0.46	1.01
		South Dakota	0.63	1.39
		Wisconsin	price unknown	
		Alberta	price unknown	
		Ontario	0.20	0.44
Goldeye	dressed	Manitoba	0.25	0.55
		Saskatchewan	0.25	0.55
Mooneye	whole	Wisconsin	price unknown	
Pike	whole	Minnesota	0.23	0.51
		New York	0.02	0.04
		Wisconsin	0.18	0.40
		Alberta	price unknown	
		British Columbia	price unknown	
		Ontario	0.26	0.57
Pike	dressed (20% dressing loss)	Manitoba	0.15–0.20	0.33–0.44
		Saskatchewan	0.15–0.20	0.33–0.44
Pickerel	whole	New York	0.02	0.04
Sacramento blackfish	whole	California	0.16	0.35
Hardhead (*Mylopharodon conocephalus*)	whole	California	0.20	0.44
Hitch (*Lavinia exilicauda*)	whole	California	0.20	0.44
Chubs[5]	whole	Idaho	0.20	0.44
Suckers[6]	whole	California	0.10	0.22
		Georgia[7]	0.14	0.31
		Idaho	0.20	0.44
		Indiana	0.30	0.66

Table VI–4–2. *Continued.*

Animal[1]	Form	State or Province[2]	Price ($/lb)	Price ($/kg)
Suckers[6]	whole	Michigan	0.03	0.07
		Minnesota	0.06	0.14
		Mississippi[7]	0.13	0.29
		Missouri	0.08	0.17
		Nebraska	0.05	0.11
		New York	0.06–0.10	0.13–0.22
		North Dakota	0.10	0.22
		Ohio	0.05	0.11
		Wisconsin	0.03	0.07
		Alberta	price unknown	
		Ontario	0.04	0.09
Suckers[6]	dressed	Manitoba	0.08	0.18
		Saskatchewan	0.08	0.18
White sucker[6]	whole	Montana	price unknown	
Carpsuckers[6]	whole	Arkansas	0.10–0.23	0.22–0.51
		Missouri	0.08	0.18
		Montana	price unknown	
		Nebraska	0.05	0.11
		North Dakota	0.07	0.15
		Oklahoma	0.25	0.55
		South Dakota	0.06	0.13
Quillback[6]	whole	Kentucky[7]	0.17	0.37
		Michigan	0.11	0.24
		Mississippi[7]	0.05	0.11
		Ohio	0.04	0.09
		Tennessee[7]	0.08	0.18
		Wisconsin[7]	0.05	0.10
Eels	whole	Delaware[7]	0.36	0.79
		Georgia[7]	0.29	0.64
		Maryland[7]	0.34	0.75
		Missouri	0.14	0.31
		New Jersey[7]	0.03	0.07
		New York	0.56	1.23
		South Carolina[7]	0.25	0.55
		Virginia[7]	0.32	0.70
		Ontario	0.53	1.17
White bass	whole	Michigan	0.22	0.49
		New York	0.25	0.55
		Ohio	0.26	0.57
		Oklahoma	0.25	0.55
		Pennsylvania	0.20	0.44
		South Dakota	0.35	0.77
		Wisconsin	0.20	0.44
		Ontario	0.27	0.59
White bass	dressed	Manitoba	0.03	0.07
Yellow bass	whole	Tennessee[7]	0.30	0.66
White perch	whole	Connecticut	0.50–0.60	1.10–1.32
		Maryland[7]	0.10	0.22
		New Jersey[7]	0.24	0.53
		New York	0.20–0.30	0.44–0.66
		Virginia[7]	0.21	0.46
		Ontario	0.12	0.26
Crappie	whole	Florida	0.25–0.40	0.55–0.88
		Michigan	0.58	1.28
		New York	0.18	0.40

Table VI–4–2. *Continued.*

Animal[1]	Form	State or Province[2]	Price ($/lb)	Price ($/kg)
Crappie	whole	Ontario	0.32	0.70
Rock bass	whole	Michigan	0.20	0.44
		New York	0.18	0.40
		Ontario	0.32	0.70
Walleye	whole	Minnesota	0.55	1.21
		New York	0.85–0.95	1.87–2.09
		Pennsylvania	0.70	1.54
		Wisconsin	0.70	1.54
		Manitoba	0.62	1.36
		Ontario	0.71	1.56
Walleye	dressed	Manitoba	0.73–0.85	1.61–1.87
Sauger	whole	Minnesota	0.20	0.44
		Ontario	0.33	0.73
Fresh water drum	whole	Alabama	0.03	0.07
		Arkansas	0.17–0.27	0.37–0.59
		Iowa	0.35	0.77
		Kentucky[7]	0.11	0.24
		Louisiana	0.15–0.30	0.33–0.66
		Michigan	0.09	0.20
		Mississippi[7]	0.15	0.33
		Missouri	0.14	0.31
		Nebraska	0.05	0.11
		New York	0.05	0.11
		North Dakota	0.06	0.13
		Ohio	0.04	0.09
		Oklahoma	0.20	0.44
		South Dakota	0.03	0.07
		Tennessee[7]	0.06	0.13
		Texas[7]	0.16	0.35
		Wisconsin	0.08	0.18
		Ontario	0.04	0.09
Paddlefish	whole	Alabama	0.07	0.15
		Arkansas	0.20–0.29	0.44–0.64
		Missouri	0.19	0.42
		Oklahoma	0.35	0.77
Diadromous goby	whole	Hawaii	3.50–4.50	7.70–9.90
Frogs	dressed	Florida	2.00–2.25	4.40–4.95
Crayfish (fisheries)	whole	Louisiana	0.30–0.60	0.66–1.32
		Wisconsin	2.24–3.00 per hundred	
		Ontario	price unknown	
Crayfish (fisheries)	peeled tails (90% dressing loss)	Louisiana	3.50–6.00	7.70–13.20
Crayfish (cultured)	whole	Louisiana	0.15–0.70	0.33–1.54
Crayfish (cultured)	peeled tails (90% dressing loss)	Louisiana	3.50–6.00	7.70–13.20
Clams	shell	Arkansas	176.00–274.40 per ton	
		Tennessee[7]	0.15	0.33
Fresh water shrimp (cultured)	whole	Hawaii	3.50–3.75	7.70–8.25

Table VI–4–2. *Continued.*

Animal[1]	Form	State or Province[2]	Price ($/lb)	Price ($/kg)
Turtles	whole	Arkansas	0.25–0.26	0.55–0.57
		Maryland[7]	0.30	0.66
		New Jersey[7]	0.21	0.46
		Tennessee[7]	0.25	0.55
		Texas[7]	0.33	0.73
		Virginia[7]	0.39	0.86
		Wisconsin	0.20	0.44

[1] Only animals likely to be of interest to small-scale aquaculturists are mentioned.

[2] Information was obtained largely from a survey of state and provincial fishery authorities and/or from "Fishery Statistics of the U.S." Some information was also supplied by private correspondents. Where no price information was obtained, animals known to be marketed are listed without prices. The following states and provinces did not return their questionnaires: Georgia, Illinois, Minnesota, Mississsippi, Tennessee, Texas, Vermont, Virginia, New Brunswick, Northwest Territories, Nova Scotia, Prince Edward Island.

[3] Most of the states keep data for "catfish," with no distinction made between channel, blue and white catfish. (Bullheads and flathead catfish are usually marketed separately.) Since the channel catfish is dominant in both commercial fisheries and aquaculture, data for "catfish" are listed under that species.

[4] Virtually all trout marketed are rainbow trout. The few brook trout sold as food may bring slightly higher prices.

[5] The term "chub," applied by Great Lakes commercial fishermen to ciscoes, here refers to a number of cyprinids of the genera *Gila* and *Hybopsis*.

[6] Commercially, the term "sucker" usually includes all members of the family Catostomidae except the buffalofishes. In certain states, the white sucker, the carpsuckers and/or the quillback are marketed separately.

[7] 1975 data.

[8] No distinction made between "catfish" and bullheads.

Marketing of cultivated fresh water food products other than the "Big 3" in North America is very limited and unorganized. As information is scarce and scattered, the following paragraphs are intended to serve more as a springboard for ideas than as information which can be put directly to use.

Marketing of miscellaneous crops

A handful of growers in the South are experiencing some success with tilapia. Marketing tests conducted by Auburn University showed that, at competitive prices, Alabama consumers were as willing to buy tilapia as the types of fish customarily available in local markets. Commercial fishermen in Florida profitably sell blue tilapia (*Tilapia aurea*), taken from phosphate pits, at fifteen to thirty cents a pound ($0.33 to $0.66/kg) wholesale, and it seems likely that aquaculturists taking advantage of fertile phosphate waters could produce tilapia competitively. Probably the greatest obstacle to commercial tilapia culture in North America is that it is illegal in most places.

Tilapia

At least some of the eleven or so farmers currently growing fresh water shrimp in Hawaii are wholesaling profitably at three dollars fifty to three dollars seventy-five a pound ($7.72 to $8.27/kg).

Shrimp

A few buffalofish grown in polyculture with catfish are sold in the lower Mississippi Valley. However, at the average price of fifty cents a pound ($1.10/kg), brought by buffalofish it would not be profitable to grow them as the main crop.

Buffalofish

Sunfishes may represent an untapped commercial opportunity. At present the only successful commercial food sunfish culturist I know of is Bill Gressard, whose work I discussed earlier.

Sunfishes

Bullheads

There is a market for bullheads in some regions. Culturing them commercially was certainly H.S. Swingle's aim in the experiments described in Chapter II–2. Then channel catfish came along and stole the spotlight just as Swingle was perfecting his bullhead culture system. When more attention is focused on the small-scale commercial aquaculturist, we may find that bullheads can be raised and sold at a profit.

Common carp and crappie

There is some incidental marketing of common carp (*Cyprinus carpio*) and crappie (*Pomoxis* spp.) from low intensity polyculture systems in the South. But these unmanaged systems differ from wild stock capture fisheries only in that they are deliberately stocked and public access restricted.

Importance of flexibility

Whatever animal or culture system the small-scale aquaculturist chooses, flexibility is of prime importance for economic success. One of the most successful trout farmers I know (with only three small ponds and no employees outside his immediate family), offers his customers the following options:
- Fee fishing, customer pays for fish "in the round."
- Fee fishing, with fish dressed on the spot.
- Fish in the round, size to order, net harvested.
- Fresh, dressed fish, size to order.
- Iced, dressed fish, delivered, in small or large lots.
- Customer-caught or net harvested trout cooked and served at pond side.
- Catered trout dinners.

Prognosis

Finally, remember it is best not to believe all you read. When panelists at a convention of The Catfish Farmers of America say that in order to make money in the catfish business you need at least three hundred acres (121 ha) under water, what they are really saying is "If you want to compete with the major growers, make your entire living from catfish farming, and service the processors and other mass markets, then our experience strongly suggests you should put at least three hundred acres (121 ha) under water." The established large-scale catfish farmers are, probably unintentionally, proposing a single economic model, which I assume does not meet the needs of most of my readers.

I am not an economist, and most of the economists I have met seem to want to pretend that, in developed countries like the United States and Canada, small producers do not exist. Those few economists concerned with small producers have not yet turned their hands to aquatic farming. So the best advice I can offer to the would-be small scale grower of aquatic crops is to be resourceful, and try to profit by the experience of other aquaculturists, large and small. Even though you may decide that aquaculture would be an unprofitable or excessively risky operation for you at this time, don't lose faith. There already are more successful small-scale aquaculturists in North America than we are led to believe. Also, the increasing demand for fish, along with economic difficulties faced by fisheries and much of "aquabusiness," suggests that eventually local and regional producers of cultured aquatic products will become dominant in the food fish business.

PART VII

Siting, Design, and Construction of Ponds, Tanks, and Raceways

CHAPTER VII-1

Ponds

PART VII

Siting Design,
and Construction

Introduction	Most small-scale aquaculture is carried out in some sort of pond. As the most difficult part of a pond culture system to change is the pond itself, its siting, design, and construction should be thoroughly studied before you make any commitment to a particular system. As an example of the sort of problems which can be prevented, in Florida in 1975 the cost of deepening a lake ran from two thousand to five thousand dollars per acre ($4,940–$12,350 per ha).

Though aquaculture ponds can be as diverse as the creatures which make and inhabit them, we may distinguish three categories on the basis of the way they are used.

Types of ponds

"Wild" ponds

The "wild" pond, though artificially constructed, is built primarily with esthetics and conservation in mind. (See Figures VII-1-24, 25 and 26, pp. 358–359.) Even though aquaculture is a secondary purpose here, it should be considered during the planning stages. Appropriate forms of aquaculture for the wild pond include low-intensity polyculture (to provide an occasional meal), intensive culture in floating cages, or perhaps use of the pond as a "natural hatchery" producing small fish to be grown out elsewhere.

While you can build an attractive pond for wildlife purposes which is only a few feet deep, if you are planning aquaculture it will be necessary in most climates to make at least one third of the pond five to ten feet (1.5 to 3.3 m) deep. (Depth, shape and other aspects of pond design are discussed in detail below; I refer to them here only in order to distinguish the three types of pond.) In other respects, the wild pond which doubles in fish production is radically different from other ponds. Among the features usually incorporated in wild ponds, but avoided in more intensive aquaculture systems, are an irregular and wooded shoreline, extensive areas of shallow water and abundant aquatic vegetation.

"Multiple use" ponds

The "multiple use" pond, more artificial in appearance than the wild pond, is designed with homestead fish production and/or recreation uppermost in mind, though esthetic and wildlife values are often considered as well. The design criteria are those of the classic "farm pond." (See Chapter II–1.) Fee fishing ponds are similarly designed. (See Chapter IX–1.) Harvesting is normally by hook and line, so there is less vegetation on the shore line than there is on the wild pond. A steep profile is best, to discourage growth of aquatic plants. There are few or no obstructions such as rocks, fallen trees, etc., which add so much to the appeal of the wild pond. A very high percentage of the multiple use pond is in deep water. If uses such as swimming, boating, stock watering, etc., are planned, they need consideration in the design stages, to minimize interference with fish production and fishing.

Intensive culture ponds

Intensive culture ponds are specifically designed for efficient production and harvest of a particular fish or fishes (usually channel catfish, *Ictalurus punctatus*, in North America), and other values are incidental. Such ponds are very regular in shape, usually rectangular. An efficient drainage system is almost always needed. A "harvest basin" is often incorporated, and provision for access by truck is desirable.

Sources of information

Before going on to discuss some of the specifics of pond design and construction, I want to add some words of warning. As I am not an engineer, and my practical experience in pond construction is limited, the purpose of this chapter is to help the beginner ask the right questions and avoid some common errors. Before proceeding with construction, be sure to consult other sources of information. In the United States, the first place to turn is the Soil Conservation Service, which offers a variety of publications, plus free consultation if you agree to follow a conservation plan on your property, (which I imagine most readers will want to do). SCS advice is only as good as your local agent, and so may vary from insipid to brilliant, but at least it won't cost you anything. In Canada, there is no government agency equivalent to the U.S. Soil Conservation Service.

Other agencies which may be helpful are university extension services, county agents and state or, in Canada, provincial conservation officers. It may also help to get a different viewpoint by consulting a building contractor. Don't neglect to talk with other pond owners you may know; even just looking at existing ponds may provide useful ideas and help you avoid mistakes. Finally, be sure to look at some of the publications listed in Appendix IV–4, which I have chosen as being the clearest and most detailed of the readily available guides to pond design and construction.

Do-it-yourself construction

Probably, most readers will employ a contractor or bulldozer operator in making a pond, but this chapter would not be complete without mention of the experience of Elmer Hedlund and his bulldozer, Alice. Elmer had no experience with heavy equipment when he bought Alice, and his instructions consisted of being shown how to start and stop the engine and raise and lower the blade. Armed with that information, he began by constructing tiny potholes for duck habitat, and advanced gradually to his masterpiece-in-progress, a thirty acre (12 ha) "natural" lake. He demonstrates just how easy it all is by virtually forcing visitors to take a turn with Alice.

Most North American writings make the assumption that it is impossible to construct a pond of any size without machinery. Having participated in the construction of sizable ponds in Latin America in places where machinery was out of the question, I can refute that assumption. Anyone with a lack of money or machinery (or a love of back-breaking labor) who wishes to construct ponds by hand, can refer to Marilyn Chakroff's Peace Corps/VITA manual for the real nitty-gritty. (See Appendix IV–4.)

Site selection

In constructing an aquaculture pond the first task is to select the site. A site which is suitable for one type of pond may be unsuitable for another, but the same three factors must be considered in all cases. They are, in order of importance, water supply, soil type and topography.

Water supply

Both the quality and quantity of water must be considered in planning a pond. Some aspects of water quality were discussed earlier in chapters dealing with specific organisms and fertilization, and will be covered in detail in Part X. In this chapter I will discuss sources of water, how to get water into the pond, and how to keep it there.

Sources of water for ponds include springs, wells, the water table, streams, rainfall and surface runoff.

A flowing spring is among nature's most precious resources, and whatever its use, first thought should be for its conservation. Keep it shaded to retard evaporation and heating. Guard against contamination by not allowing livestock direct access to the spring, protecting the watershed area above the spring and, if necessary, constructing a ditch to carry contaminated runoff around the spring. With the future of the spring protected, the next step is to determine its potential usefulness by measuring its flow rate. This should be done at whatever time of year the flow is least. Always observe a spring for a full year before tapping it for aquaculture, and then make some allowance for exceptionally dry years.

Measuring flow rate

To measure the flow rate of a spring, simply build a temporary dam to trap the flow. Insert a pipe through the dam, allow the outflow to fill a container whose capacity you know, and measure the amount of time required to fill it. You can convert the figure obtained with simple math to arrive at cubic feet per second (cfs) or some other conventional unit of flow. (There are 7.48 ft³ in a gallon.)

You can pump well water to supply an aquaculture pond, but this introduces a technology cost. There is also the danger of overdoing it and running out of water at a critical time. Table VII-1-1 may be helpful in evaluating the potential of a spring or well to supply ponds. Also consult well logs for your locality, available from the Soil Conservation Service or from commercial water drilling firms.

Table VII-1-1. Approximate pumping time for one acre-foot of water.

Pumping rate gallons per minute (liters per minute)	Number of hours required to pump one acre-foot of water
100 (378)	54.3
200 (766)	27.2
300 (1,134)	18.1
400 (1,515)	13.6
500 (1,890)	10.9
1,000 (3,780)	5.4
1,500 (5,670)	3.6
2,000 (7,560)	2.7

Excavating below water table

In some regions it is possible to excavate a pond below the top of the water table, so that it fills naturally, though this may involve the difficult removal of semi-liquid mud. Usually, this type of construction will result in pond banks which are

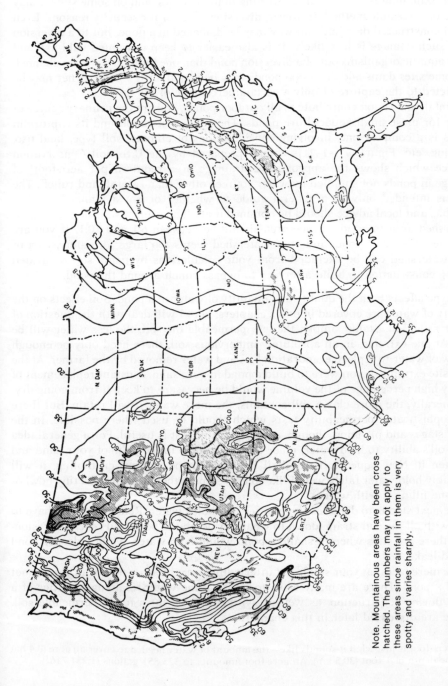

Note. Mountainous areas have been cross-hatched. The numbers may not apply to these areas since rainfall in them is very spotty and varies sharply.

FIGURE VII-1-1. A guide for estimating the approximate drainage area (in acres) required for each acre-foot of storage in ponds supplied solely by rainfall and runoff.

inconveniently high. Except in extremely humid locations, ponds built into the water table should always have an auxiliary water supply for emergency use.

You can construct ponds by damming or diverting flowing surface waters. Even normally dry channels can be dammed and a pond made by trapping seasonal runoff. Damming is simpler than diverting in many cases, and on some sites it may be the only feasible method. However, diversion is better for security reasons. Even a well constructed dam and spillway may be damaged in a flood, but to a diversion pond such damage is less likely. It is also easier to keep unwanted wild fish and other aquatic organisms out of a diversion pond than one built on a stream channel. On some sites dams and diversion ponds may be illegal, or the pond owner may be restricted to the capture of only a certain percentage of the flow.

Rainfall and runoff contribute to all outdoor ponds, and are the primary source of water for most small aquaculture ponds. The subject of runoff and its capture in ponds is a complex one, involving considerations of climate, soil type, land use, ditching, etc. Figure VII–1–1 is a map prepared by the U.S. Soil Conservation Service which shows the approximate drainage area required per acre-foot[1] of storage in ponds not supplied by water sources other than rainfall and runoff. The map is intended only as a general guide to whether or not a planned pond is feasible, and local advice should be obtained if at all possible. As a rule the ratio of watershed area to pond surface area should be between 6:1 and 10:1. If you are thinking of a small pond, and your watershed seems too large for safety, part of that watershed can be excluded from your calculations by using soil excavated during construction to build a terrace to bypass runoff around the pond.

Soil type

Soil type affects both the quality and quantity of water in a pond. Soil effects on the quality of water are covered in other chapters; here I will deal with the question of water retention. Stated simply, the less permeable the soil, the less water will be lost. At one extreme, in an absolutely impervious soil, there need only be enough inflow of water to replace evaporation loss and water removed by the farmer. At the opposite extreme, if you were to build a pond in pure sand, constant replacement of a very high percentage of the volume would be necessary to keep it from going dry.

Generally, the more clay a soil contains, the more water it will retain. But there are sophisticated permeability tests which should be used where possible. In the early stages and planning, though, a simple, crude test is useful. For a general idea of a soil's ability to hold water, place a small sample in the palm of your hand and moisten it. Now squeeze. If the soil holds its shape without crumbling, it will probably hold water fairly well. I like to go a step further by forming a tiny bowl of soil and filling it with water.

The next step is to dig a test hole to see if it holds water. Remember, you have to deal with all the soil strata, down to whatever depth the pond will reach. Occasionally, there are sites where a shallow pond would hold water, while a deeper pond would leak.

The mere fact that your soil is not utterly impermeable does not mean you cannot build a pond. There are many ways to improve the water-retaining qualities of a pond during construction, as well as synthetic substitutes for impermeable soil. These are discussed later in this chapter.

---

[1] An acre-foot is just what it sounds like—the amount of water needed to cover an acre (0.4 ha) to the depth of 1 foot (30.5 cm). An acre-foot amounts to 325,851 gallons (1,231,716 *l*).

A first consideration of topography is elevation. Many will assume that the "natural" site for a pond is the low point of a given piece of land. But though it may be natural, the lowest spot is usually not optimal for aquaculture. It is always more convenient to have a pond surrounded by higher ground on three sides and lower ground on one, so that drainage is aided by gravity and the danger of flooding reduced.

Topography is also important because of its effect on shading. A steep, narrow valley may be a good site for trout ponds, which depend on cold water. But in most other types of aquaculture it is best to expose the water to the sun, both for warmth and to encourage the photosynthetic activity necessary for a productive food chain. So it is usually better to choose a flatter, more open site.

The most important topographic factor in pond siting is slope. The degree of slope of a piece of land can be expressed as a ratio or a percentage. For example, a slope of 20:1 means that for every twenty units of length, there is a change in elevation of one unit. The same slope expressed as a percentage is 1/20 or five percent. Perfectly flat land is said to have a slope of zero.

Any piece of land has a longitudinal and a horizontal slope. The longitudinal slope, measured along the main axis of water flow, is important in controlling drainage and/or flow rate. A longitudinal slope of 0.2 to 1.0 percent is preferred for most pond construction.

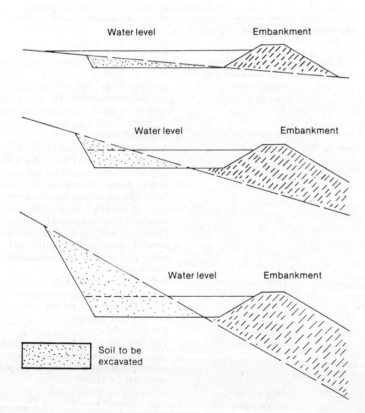

FIGURE VII–1–2. *Diagram showing the amount of earth which must be excavated to construct a dugout pond of the same size on a steep and a shallow slope.*

Though it is possible to construct virtually any sort of pond on any slope, as Figure VII–1–2 shows, the amount of excavation is greatly increased where an inappropriate slope is chosen. In this respect horizontal slope is as important as longitudinal slope. Both factors may be compensated for to a considerable degree by choosing an appropriate type of construction.

Type of construction

Impoundment, dugout, and levee ponds

All artificial ponds may be classed into three types on the basis of construction. Ponds of one type, called impoundments, dam or barrage ponds, depend on a dam at one end to back up water which would normally run off. Another type, called dugout, contour or excavation ponds consists, in fact, of holes excavated in the ground. When earth is banked up to make a high wall, or levee, around a pond, the pond is referred to as a levee pond. Levee ponds are a hybrid type, combining some of the characteristics of impoundments and dugouts.

Factors determining type of pond

Which type of pond will be best in a given place is based on a number of factors, including water supply, legal and ecological considerations, the nature of deep soil strata and the type of use. Here I will point out the importance of topographic factors, especially slope, in choosing the most efficient type of construction. Table VII–1–2 summarizes the topographic considerations in choosing a dugout, an impoundment, or a levee pond.

Table VII–1–2. Relation of longitudinal and horizontal slope to choice of pond type (impoundment, dugout, or levee pond).

Longitudinal slope	Horizontal slope	Type of pond preferred	Reasons for choice
Steep	Steep	None (or small impoundment if other sites not available)	While it is possible to build a pond in a steep, narrow valley, construction would be difficult and engineering safety doubtful. The most that should be done is to construct one or more small flowing-water impoundments.
Steep	Low or flat	Dugout	Any dam constructed on such terrain would have to be both high and wide, and would be subject to severe stress at times of high flow. With this combination of slopes it is easy to supply a pond with water diverted from a natural channel. However, less excavation will be needed on a lower longitudinal slope and such a site is usually preferred.
Low	Steep	Impoundment	A small dam will work well in such a site, while a large amount of excavation would be necessary to construct a dugout.
Low	Low	Dugout or levee pond	Effective dam construction is virtually impossible on this type of site. Such sites often have large watersheds suitable for large ponds supplied by runoff.
Flat	Flat	Dugout or levee pond	A dugout or levee pond is the only type of pond possible on a perfectly flat site.

A number of factors apart from the critical ones of water supply, soil type and topography may be important in choosing a pond site. One is competing land use. Often, the best pond site is also the best agricultural land. What to do then becomes a question of personal priorities. If on a given farm there is a location which is of little or no value agriculturally, it should be considered in selecting a pond site.

Another important factor is access, and the type of access will depend on the use you have in mind. A fee fishing pond should be easily accessible by car, preferably near a highway, with space for a parking lot. Commercial fish production ponds should also be accessible by vehicle, but unauthorized fishing and other nuisances are reduced by choosing a site away from roads, though visible from the farm house. If wildlife, esthetic values or recreational fishing are important to you, a secluded site may be more enjoyable, but don't make a pond so remote that you will not feel like fishing or caring for it. A pond several miles back in the woods may be beautiful, but it will usually be a neglected pond.

Access

Be aware of safety in siting impoundments. Any dam can break, no matter how well designed and constructed. For this reason impoundments should never be built directly above habitations, roads, other valuable structures, or crop land.

Dam safety

Finally, in choosing your site, consider the other resources you may want to integrate with your aquaculture, such as feeds, fertilizers, livestock, household wastes, etc., as well as the auxiliary uses you may want to make of the pond for stock watering, irrigation, etc. Chapter IV–2 can be helpful here.

Ponds integrated with resources

Once you have decided on a site, you can proceed to design your pond or ponds. The main things to think about are type of construction (impoundment, dugout or levee pond), surface area, depth, shape, planting and landscaping of banks and surroundings, and inlet and outlet structures.

Choosing pond type and design

Whether you build an impoundment, dugout or levee pond will in many instances be determined by topography or legal considerations. But, for readers for whom those factors will not completely restrict the kind of pond to be built, I have listed the various advantages of the three types in Table VII–1–3. Usually, the dugout is the best type of pond for non-commercial aquaculture, while impoundments or levee ponds are preferred by commercial fish culturists.

Table VII–1–3. Advantages of impoundments, dugouts and levee ponds (other than those determined by topography or legal restrictions).

Impoundment	Dugout	Levee pond
Usually cheapest to build.	No danger of dam or banks giving away.	Greatest depth from least excavation.
Least subject to drought.	Easier to exclude unwanted aquatic organisms.	Easier to exclude unwanted aquatic organisms.
Easiest to obtain a a constant flow of water, if desired.	Outlet structures least expensive. (An emergency spillway will usually suffice.)	Warms more readily.
Easiest to maintain cool water, if desired.	Warms more readily.	Less loss of feeds and fertilizers via outflow.
Natural fertility most likely to be high.	Less loss of feeds and fertilizer via outflow.	Can achieve greater depth per area.
	Can achieve greater depth per area.	

Surface area   The surface area of a pond depends partly on topography, but you will usually have some choice in the matter, particularly on fairly flat ground. Sometimes the choice will be whether to make one large pond or several small ones. I can give no general rule, but you should consider the advantages of small versus large ponds (Table VII-1-4).

Table VII-1-4.   Advantages of small versus large ponds.

Small ponds	Large ponds
Easier and quicker to harvest.	Less construction cost per area, since less soil must be moved to achieve equal surface area.
Can be drained and refilled more quickly.	
Easier to treat disease, apply fertilizer, etc.	Take up less space per area of water surface.
If stock is lost, there is less financial loss.	More subject to aeration by wind.
Less subject to wind erosion of banks.	Biologically more stable in that removal or addition of fish represents less of an oscillation in population.
Several small ponds offer a safety factor if disease strikes one pond.	
Several small ponds permit segregation of breeders, fish of different sizes, etc.	
Several small ponds permit more simultaneous experimentation.	

Other considerations in choosing the size to make a pond are based on the owner's personal needs, and the species and type of aquaculture to be undertaken. All other things being equal a large pond is more manageable than a small pond. Ponds under half an acre (0.2 ha) in surface area are very likely to develop population imbalance.

Depth   You will have to consider a number of depth factors in designing a pond, including maximum, average and minimum depths, and the depth profile. Where the water supply and drainage system are adequate, you can maintain each of these constant throughout the year. However, some ponds are subject to considerable fluctuation. For convenience, I will give depths for late spring or early summer— the beginning of the time of greatest evaporation loss.

Whatever a pond's maximum depth, it should be maintained over at least ten percent of the pond's surface area in regions where surface water freezes in winter, to prevent "winterkill." (See Chapter X-2 for the biological nature of winterkill, along with possible preventive measures after the pond is built.) Here I will mention the first line of defense, which is to make ponds deep enough to provide a refuge for fish in the coldest weather. Suitable maximum depth can vary from five feet (1.5 m) to fifteen feet (4.6 m) depending on the severity of the climate, so be sure to seek local advice on this.

Where freezing does not occur, the biological considerations are different. In truly tropical regions, assuming there is an adequate supply of water, it is a waste of energy and space to make a pond more than three feet (0.9 m) deep unless a constant flow is provided. Since the air temperature hardly changes, water temperature changes even less, so there is little exchange of water between top and bottom. The result is that bottom waters in deep ponds may be without oxygen.

Where temperatures become cool but without freezing, there is more exchange between surface and bottom waters so that, up to a point, deepening a pond enhances its productivity by increasing its volume. In such climates depth should be prescribed by water supply and convenience.

By "average depth" I really mean the depth existing over most of the pond. In southerly regions, where there is no danger of winterkill, average depth may be the same as maximum depth. Elsewhere, the most suitable average depth will depend on the rate of loss of water. In some locations there is enough water for constant replacement. But more often the rate at which water can be replaced is lowest exactly when loss is highest—mid and late summer. This must be compensated for by extra water collected during times of high precipitation and low evaporation.

The natural rate of water loss, aside from human use, is dependent on evaporation and seepage. Because evaporation loss occurs only at the surface, it is less important in a deep pond than in a shallow one of the same size. How much water is lost by seepage has nothing directly to do with depth, but clearly a deep pond can withstand a longer period of constant seepage loss than can a shallow pond. Nothing can substitute for sound local advice in designing a pond to withstand evaporation and seepage loss. Figure VII–1–3 gives a general idea of the average depth of water to be maintained in ponds in the United States, assuming seepage loss does not exceed 3 inches (7.6 cm) per month. Allow extra depth when constructing ponds in very porous soil.

The minimum depth usually given for fish culture ponds is two feet (0.6 m). In practice, this minimum is rarely achieved, as soil materials require an easier slope than the perfectly vertical bank which would allow the two foot minimum. But to maximize production you should try to come as close as possible to this ideal. Doing so will ease weed problems (see Chapter VIII–1), slow down siltation, and prevent the creation of a refuge for small fish which may then become too abundant. From a naturalist's point of view, a certain amount of very shallow water enriches the biological diversity of a pond, so you may choose to include shallows in a pond design, while understanding the trade-off involved.

The term "depth profile" refers to the contours of the bottom. If you plan any sort of intensive aquaculture or periodic complete drainage, the bottom should be smooth, with no irregularities. On the other hand, with an irregular profile you increase the diversity of habitats and add interest to recreational fishing in wild or multiple use ponds. Rocks, logs, standing deadwood and brush piles are all permissible.

"Depth profile"

Some aquaculture ponds are built with perfectly flat bottoms, but a slope of between 1,000:1 and 1,000:6 toward the outlet is better. Occasionally, shallow trenches are made in the bottom to facilitate drainage.

A common feature in commercial catfish ponds is the harvest basin. (See Figure VII–1–36 and 37, p. 366.) The portion immediately around the outlet, which may be from one to ten percent of the pond total area, is made one to one and one half feet (0.3 to 0.5 m) deeper than the rest. Then, when the pond is drained, the fish are concentrated in the harvest basin where they can be captured without too much

Harvest basin

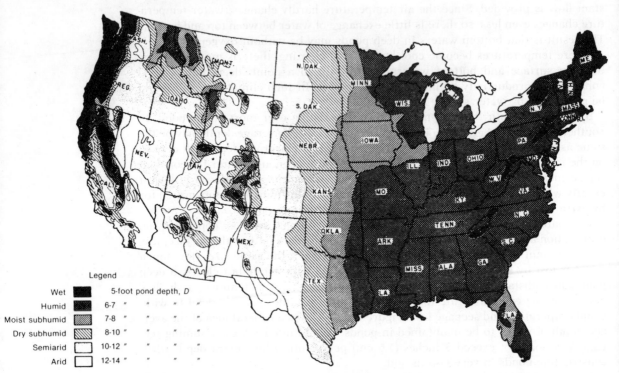

Legend

Wet	5-foot pond depth, *D*
Humid	6-7 ″ ″ ″ ″ ″
Moist subhumid	7-8 ″ ″ ″ ″
Dry subhumid	8-10 ″ ″ ″ ″
Semiarid	10-12 ″ ″ ″ ″
Arid	12-14 ″ ″ ″

FIGURE VII–1–3. *Recommended average depth (in feet) of ponds in the United States to provide adequate water for aquaculture. (It is assumed that seepage loss does not exceed 3 in, or 7.6 cm, per month.)*

exposure to air and sun.

Shape   The shape of ponds, particularly impoundments, is often dictated by topography. Where the pond builder has a choice, a regular shape is better for intensive operations, though irregularities in the shoreline will increase the pond's value to the owner who values a natural appearance.

If any sorts of nets are to be used, and particularly if complete drainage will not always be possible, a rectangular shape is far superior to any other, as rounded corners make capture of fish difficult. Small ponds may be square, since a square shape requires less construction to achieve a certain surface area. As the size of a planned pond increases, there is more reason to consider an oblong shape. Harvesting with nets and uniform distribution of food and fertilizer are easier in an oblong pond, and drainage is usually more efficient. An oblong shape is particularly helpful if hook and line is to be the main harvest method and boats are not to be used. With a square pond, it may be impossible to fish the center of the pond.

Oblong ponds are usually laid out to conform with natural contours and gradients, but on flat ground you may have the option of siting them perpendicular or parallel to the prevailing winds. If a pond is large, it is better to construct it with the long axis perpendicular to the wind, to reduce bank erosion. If it is small, the long axis should be parallel to prevailing summer winds to take advantage of wind aeration.

This pond, on a catfish farm in Arkansas, has a properly designed harvest basin. While the rest of the pond has been drained completely dry, enough water remains in the harvest basin to keep the fish alive during harvest. (Photo courtesy Maurice Moore, Aquaculture Magazine.)

Pond banks

Almost all pond banks and surroundings need some sort of planting, to prevent soil erosion, if for no other reason. Other possible purposes for pondside planting include appearance, providing wildlife habitat, shade, and provision of food for the fish. Before discussing what should be planted around a pond, I will mention two things which should not be planted adjacent to a pond: lawns and row crops. These provide only fair to poor protection from erosion and may be sources of unwanted nutrients.

Planting

If you feel you must have a lawn right down to the edge of your pond, recommended grasses are carpet, bermuda and St. Augustine grass. Especially avoid *Panicum* grasses, which will grow out into the water. Include a swale to trap nutrients or eroded soil. (Figure VII–1–4.) Don't use lawn fertilizers, especially those which are not recommended for food plants, as they may bring heavy metals and other pollutants into the pond's food chain. And try not to be obsessive about mowing. Frequently and closely mowed grass develops a shorter root system than grass which is allowed to grow, offering less protection against erosion.

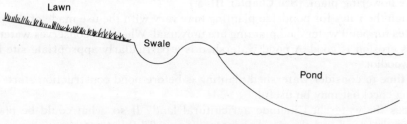

FIGURE VII–1–4. *Swale to intercept runoff from a lawn.*

If you must have row crops adjacent to the pond, be zealous about conservation measures such as contouring, cover cropping and mulching. Try to avoid pesticides, or at the very least don't apply them on windy or rainy days. Leave at least forty feet (12.2 m) of grass, bushes or natural vegetation, perhaps including a swale, between field and pond. If possible, install drainage tiles on adjacent agricultural land. In this way soil and agricultural fertilizers will be kept on the land where they belong.

The banks of intensive aquaculture ponds are planted only to prevent erosion. Anything taller or stouter than grasses may interfere with management operations. It is best to get local advice on what grasses or other plants should be seeded.

The wild pond is at the opposite extreme from the intensive aquaculture pond, and can be planted to suit your fancy. Though you may prefer a completely forested shoreline, alternating trees, shrubs and grassy areas will provide diversity and facilitate fishing and access to the pond. You may choose flowering plants to attract insects which serve as food for fish and frogs. It is beyond the scope of this book to recommend specific plants on the basis of insect attraction, esthetic value, and hardiness. Such advice may be obtained from state or provincial conservation departments or county agents.

Plantings for the multiple use pond fall between the intensive aquaculture pond and the wild pond. Access for vehicles or machinery, except perhaps at one point, is usually not important, but too many trees or overhanging bushes will interfere with fishing. If you prefer dense planting, a boat is one option. Or you can provide small fishing platforms or "docks" every few feet along the water's edge.

All dams and levees, regardless of use, should be planted only with grass or other low-growing plants like vetch, and should be kept mowed more closely than other pond banks. Trees, shrubs or tall grass may mask the efforts of muskrats, wood-chucks and other burrowers which can weaken dams and levees. Plants with long, vigorous roots may be as damaging as burrowing animals.

Though shade is a pleasant addition to the wild or multiple use pond, it generally has a negative effect on fish production by reducing photosynthetic activity and lowering temperatures. (Trout ponds in marginal locations are an exception, since cultured trout are dependent on cold water and artificial feed.) If you want shade, large trees planted well back from the bank are preferable to smaller plants which give shade only if planted close to the water's edge. Smaller plants may actually add to the problem they were intended to solve by causing water loss through transpiration. Try to choose trees which do not drop excessive amounts of branches and leaves.

An aspect of pond-side planting for which there is still not much North American experience is raising fodder plants for herbivorous fishes. Two plants are worth considering: dasheen (*Colocasia esculenta*), which thrives in moist pond-side environments, and hairy vetch (*Vicia villosa*) which is a good erosion controller and attractive flowering plant. (See Chapter III–5.)

Watershed planting

Though the rules for pondside planting may vary with the use made of the pond, the rules for pond watershed planting are universal. Whatever conserves water and retards erosion is good. A pond watershed is a particularly appropriate site for a farm woodlot.

The time to consider watershed planting is before pond construction starts. The follwing checklist may be useful:

• Does the watershed include agricultural land? If so, what could be planted between that land and the pond to intercept runoff? What conservation practices might slow down runoff from agricultural land?

• Are there ditches or gullies in the watershed? The pond is just one more reason to plant them.

• Are there natural streams entering the pond? If so, could the stability of their banks be improved by well planned planting?

• Does the watershed have roads passing through it? If so, what about planting banks and cuts? The highway department may be cooperative if you explain your concern.

• Gravel pits, parking lots and other areas without vegetation are particularly bad sources of eroded materials. Are these areas essential where they are in the watershed? If not, is there some undemanding plant which can help you start patching them up?

• Were you thinking of any sort of tree planting project? If so, why not start planting, in the watershed, before you begin pond construction?

• Do you control the entire watershed? If not, is there a neighbor you should talk with? You might offer to share fish, extend fishing rights, or cooperate on a conservation project which will benefit you both.

No inlet structures are needed in ponds supplied entirely by rain, surface run-off, or springs which emerge within the pond, or in impoundments built directly on natural watercourses. Ponds supplied partly or totally by wells, by springs outside the pond or by diverted watercourses will all need inlet structures of varying complexity. The main purposes of inlet structures are: *Inlet structures*

• To prevent water evaporating or seeping away between source and pond. You can retard evaporation by shading or covering the inlet, and you can prevent seepage by lining the inlet channel with cement, plastic or any other impermeable material. You can do both by passing the inflowing water through a pipe or culvert, but be sure to use non-toxic materials. Most cements and cement-like materials are safe. Among plastics, polyethylene and pvc are acceptable. Few metals are completely safe and copper is absolutely out of the question. Size of pipe for inflow structures is determined by flow rate. (See Table VII–1–5.) *Preventing water loss*

• To prevent erosion. Any of the methods used to prevent seepage will work, as will lining the inlet with stones. The most endangered place is where the water actually falls into the pond. Entering water should never dribble over a cut bank, but should always fall clear—over a rock, board or cemented area. *Preventing erosion*

Table VII–1–5. Relation of inlet pipe diameter to rate of inflow
(where there is a free outlet and a minimum pipe slope of 1%, with the
water level 0.5 foot, or 0.15 m, above the top of the pipe at the upstream end).

Pipe diameter inches (cm)	Rate of inflow in cubic ft/sec
15 (38.1)	0– 6
18 (45.7)	6– 9
21 (53.3)	9– 13
24 (61.0)	13– 18
30 (76.2)	18– 30
36 (91.4)	30– 46
42 (106.7)	46– 67
48 (121.9)	67– 92
54 (137.2)	92–122
60 (152.4)	122–158

Oxygenation

• To oxygenate incoming water, particularly well water and some spring water, both of which are usually low in dissolved oxygen. Oxygenation occurs whenever water is broken up, exposing greater surface area to the air, as when it is splashed off a flat surface from a height or passed through a perforated container. Never allow water to enter a pond from below if you can possibly prevent it. (See Chapter X–2 for detailed discussion of dissolved oxygen.)

Raising water temperature

• To warm extremely cold water. You can accomplish this by passing the water first into a broad, shallow pool exposed to the sun, then into the pond. The trade-off, of course, will be a certain amount of evaporation loss.

Purifying water

• To settle or filter out unwanted organisms and particles. You can also use the warming pool just described as a settling tank for silty water or water with an excessively high iron content (revealed by rust-colored sediment). These materials may also be filtered out by passing the water through a gravel filter. Filters are also used to prevent unwanted wild fish, fish eggs and other organisms from entering the pond, and to prevent fish from leaving. Three of the many possible types of filters are shown in Figures VII–1–5, 6, and 7.

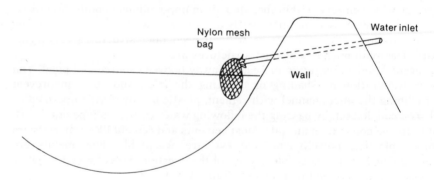

FIGURE VII–1–5. *Saran or nylon "sock" filter.*

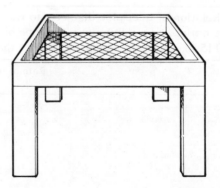

FIGURE VII–1–6. *Box filter.*

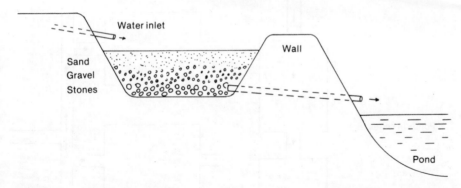

FIGURE VII–1–7. *Sand and gravel filter.*

Large ponds may require a sluice gate at the entrance, into which you can build one or more wire screens to filter out fish and other organisms. Figure VII–1–8 shows a sluice with screen in place. The rate of inflow is regulated by removing and adding dam boards.

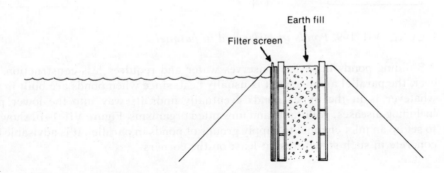

FIGURE VII–1–8. *Sluice with filter screen.*

Pond filter material should have approximately the same size mesh as mosquito netting, because anything larger will admit larval fish. This of course implies daily to weekly cleaning, depending on the amount of debris which accumulates. A good material for pond filters is Saran screen, manufactured by National Filter Media Corporation. (See Appendix IV–3 for address.)

It is often desirable to supply two or more ponds from the same source. There are two ways of arranging such groups of ponds. If ponds are constructed "in series" (sometimes called "rosary" ponds), water passes from one pond, through a channel or pipe, into the next pond. If they are constructed in parallel, the water is divided before it reaches the ponds. Figure VII–1–9 illustrates the two types of arrangement.

Ponds in series or in parallel

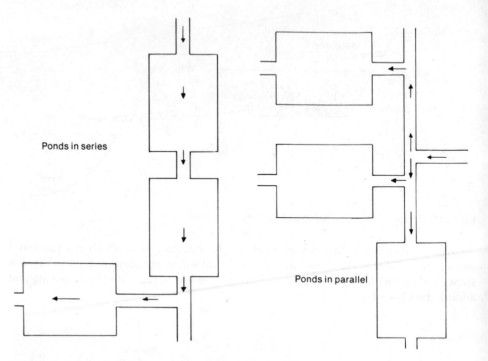

FIGURE VII–1–9. *Ponds in series and in parallel*

Building ponds in series conserves water and requires less construction. However, the parallel arrangement is usually best, since when ponds are built in series whatever is in the upper ponds eventually finds its way into the lower ponds, including diseases, pollutants and unwanted organisms. Figure VII–1–10 shows how to set up an inlet channel to supply groups of ponds in parallel. It is advisable to use concrete in such structures, at least on the corners.

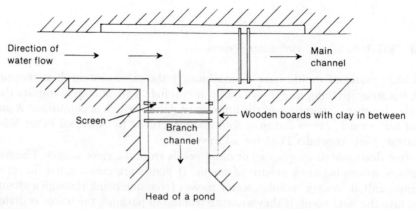

FIGURE VII–1–10. *Sluice system to regulate the flow of water from a main inlet channel to ponds set up in parallel.*

All ponds need an outlet structure or structures. At the very least they need an emergency spillway to save the pond in case of an exceptionally severe flood. There is no more to an emergency spillway than a channel about eight inches (20.3 cm) deep, built into the dam or levee, or preferably to one side of it. A rule of thumb for deciding the width of an emergency spillway is to divide the number of acres in the pond watershed by two. The result, plus a small margin of error, is the spillway width in feet. The slope of emergency spillways should be at least two percent, to prevent puddles forming. If the soil is at all loose, pave the spillway with stones, bricks, cement or boards to prevent erosion. In firm soil a good growth of grass gives enough protection. On the rare occasions when an emergency spillway is in use, some fish may be lost, but that is a whole lot better than losing the pond.

Emergency spillways function only during floods – or, ideally, never. Other structures serve as overflows, to keep the water level below a certain maximum during "normal" weather and/or to totally or partially drain the pond for harvesting or other management tasks. Both these purposes are served by a sluice. You can get better water retention when installing a sluice at the pond outlet by using two sets of dam boards with earth fill packed tightly between them. Since replacing the earth fill is laborious, I don't recommend this system for ponds which must be regulated frequently.

A wooden, metal or concrete structure known as a monk amounts to no more than a sluice which is not built into the pond wall. (See Figure VII–1–11.) Water leaves by a rigid pipe passed through the pond wall so that the lower end is twelve to fifteen inches (30–38 cm) below the pond bottom. Well built monks are stronger than sluices and can be opened and closed more easily. Their disadvantage is that drainpipes may become clogged.

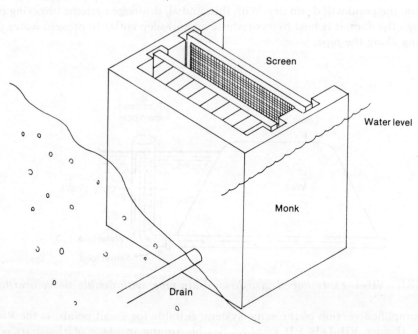

FIGURE VII–1–11. *Monk.*

It is often best to remove water from the bottom of a pond, where water quality is lowest. This can be done with a modified monk known as the Herrguth monk. (Figure VII–1–12.)

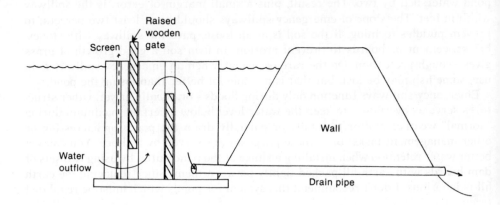

FIGURE VII–1–12. *Herrguth monk.*

Turn down pipe

You can also use a variety of simpler devices, made with only pipe and tubing, to serve as overflows and to drain ponds. The L-shaped turndown drain pipe shown in Figure VII–1–13 is one frequently used. By regulating the level of the pipe, you can automatically draw the pond down to any level. If the pipe is placed on the bottom, the pond will drain dry. With this, and all drainage systems involving pipes through the dam, it is best to incorporate an anti-seep collar to prevent water from seeping along the pipe.

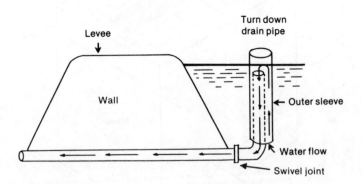

FIGURE   VII–1–13. *L-shaped turn-down drain pipe, with double sleeve overflow.*

The Rivaldi valve

A simplified version of the same system, suitable for small ponds, is the Rivaldi valve. (Figure VII–1–14.) It employs flexible tubing in place of that part of the L-shaped pipe which is not buried.

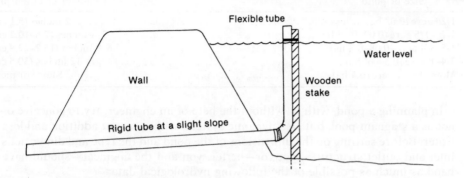

FIGURE VII-1-14. *Rivaldi valve.*

You can adapt either the L-shaped drain pipe or the Rivaldi valve to drain water from the bottom of the pond without draining it below a certain level. This is done by using the double sleeve overflow, shown in Figure VII-1-13.

Perhaps the simplest, though least adaptable, system of all involves two rigid pipes, as shown in Figure VII-1-15. You can also use siphons and pumps. Their respective disadvantages are low speed and energy costs.

Simplest drainage system

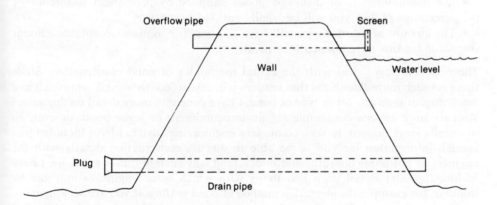

FIGURE VII-1-15. *Simple pond drainage system involving two rigid pipes.*

All drains and overflows should be screened to prevent clogging and escape of fish. It is also helpful to spread gravel on the portion of the pond bottom near the outlet wherever water is to be drawn off the bottom.

You may have noticed that I have not mentioned dimensions in the preceding section on inlets and outlets. If at all possible, try to short-circuit the question by getting advice from an engineer. However, the table of pipe sizes for outlet structures below may be helpful.

Table VII–1–6. Size of outlet pipes for various sizes of pond.

Size of pond	Diameter of outlet pipe
1/20 acre (0.02 ha) or less	2 inches (5.1 cm)
1/20–1/5 acre (0.02–0.08 ha)	3–4 inches (7.6–10.2 cm)
1/5–3/4 acre (0.08–0.3 ha)	6–9 inches (15.2–22.9 cm)
3/4–1 acre (0.3–0.4 ha)	12 inches (30.5 cm)
More than 1 acre (0.4 ha)	Sluice or monk

Summary of hydrological considerations

In planning a pond, with or without the help of an engineer, try to conceive of it, not as a stagnant pool, but as a dynamic system, with constant addition and loss of water. Before settling on the dimensions of the pond and the type and dimensions of inlet and outlet structures, you, or—better, you and the engineer—should have at hand as much as possible of the following hydrological data:

• Average, maximum and minimum precipitation, and its seasonal distribution.
• Size of the pond watershed.
• Rate of runoff (i.e., what percentage of the precipitation that falls runs off.) Don't forget to take snowmelt into consideration.
• Average, maximum and minimum flow of any watercourses entering the pond.
• Average annual evaporation loss.
• Normal seepage loss of ponds in your region, taking into account any variation from the norm in your soil, and anything you may be able to do to improve its water-retention.
• An approximation of the amount of water you might withdraw for such uses as irrigation or livestock watering.
• For impoundments or diversion ponds supplied by permanent watercourses the percentage of flow you will be legally entitled to capture.
• The amount of dilution you will have to allow for to meet acceptable effluent standards for intensive aquaculture ponds.

Mechanics of pond construction

The next few pages deal with the actual mechanics of pond construction. Since there is much more likelihood that readers will take a "do-it-yourself" approach to a small dugout than any other type of pond, I have gone into more detail on dugouts. I strongly urge anyone contemplating an impoundment or levee pond, or even an unusually large dugout, to seek competent engineering advice. I have included just enough information for you to be able to discuss construction details with the engineer or bulldozer operator and understand why certain things are done. I have included greater detail on a few items with which some engineers may not be familiar, for example the gleization method of pond sealing. If you are seeking more detailed technical information, the single most readable source is U.S. Soil Conservation Service Agriculture Handbook No. 387. (See Appendix IV–4.)

The first step in constructing a dugout, assuming the watershed is in good shape, is to stake out the surface dimensions of the pond. Try to leave room for later enlargement, in case your assessment of the rate of inflow of water is incorrect.

The next step is to determine the angle of the side slopes. For many aquaculture purposes, the more nearly vertical the slope, the better, but this must be tempered by knowledge of the soil characteristics. Few soils will suppport a perfectly vertical slope for long. Usually the steepest feasible slope is 1:1, and 2:1 or 3:1 is more typical. Soils saturated with water at the time of excavation or soft sandy soils may require even gentler slopes.

With the correct side slope in mind, you can roughly calculate the amount of excavation necessary by using the formula:

$$V = \left(\frac{A + 4B + C}{6}\right)\left(\frac{D}{27}\right)$$

where

V = volume to be excavated in cubic yards.
A = surface area of the pond in square feet.
B = area of the excavation midway between the surface and bottom, in square feet.
C = area of the pond bottom, in square feet.
D = depth of the pond in feet. (The formula assumes a fairly flat bottom slope; a strongly sloping or irregular bottom would require more complicated calculations.)

As an example, take a pond to be made 200 feet (60 m) long, 100 feet (30 m) wide at the surface, and 10 feet (3 m) in depth, with 2:1 side slopes. Then $A = 200 \times 100 = 20,000$ square feet (1,820 m²). If the side slope is 2:1, that is, if for every 2 feet (61 cm) horizontally out into the water there is a 1 foot (30.5 cm) increase in depth, then the horizontal distance from the shoreline to the bottom of an excavation 10 feet (3 m) deep will be $10 \times 2 = 20$ feet (6 m). Doubling this figure, to allow for two sides (or ends) of the pond, we get 40 feet (12 m), which is the difference between top and bottom length or width. Thus the area of the bottom, $C = (200 - 40) \times (100 - 40)$, or $160 \times 60 = 9,600$ square feet (883 m²). You can compute B by averaging the top and bottom dimensions:

$$\frac{20,000 + 9,600}{2} = 14,800 \text{ square feet } (1,332 \text{ m}^2)$$

After plugging these numbers into the formula,

$$V = \frac{20,000 + (4)(14,800) + 9,600}{6} \times \frac{10}{27} = 4,810 \text{ cubic yards of soil to be excavated.}$$

This figure can be used to calculate the number of machine hours required to do the job, or as a basis for taking bids from contractors.

The next step is to remove all trees, stumps and other obstructions from the site, followed by removal of the topsoil, which should be stockpiled for later use. Normally a bulldozer or scraper will be the best machine for this and following excavation jobs, but in wet soils a dragline may be necessary. You can finish off and smooth the pond contours with hand tools.

With the topsoil safely set aside, you can begin excavation of the subsoil and deeper materials. In making a dugout, where no embankment construction will be done, it is often most convenient to remove the spoils from the site completely, though it may be cheaper not to do so. In any event, consider the possible uses of the material before you haul it away and then wish you still had it.

The two possibilities are either to stack or to spread the spoils. In either case you do not want soil tumbling or eroding back into the pond, so certain rules must be observed. When stacking spoils, a space or "berm" should be left between the edge of the pond and the toe of the stack. The berm should be at least as wide as the pond is deep and never less than twelve feet (3.6 m) wide. The material should be stacked with slopes no steeper than the natural angle of repose, or 3:1, whichever is less. The stacks need not merely be unsightly heaps, but can be made to double as a windbreak or snowfence. Spoils may also be used inventively in landscaping.

It is perhaps easier to dispose of spoils by spreading them, in which case no berm is necessary. The disadvantage of spreading is that the material covers a greater area, and may conflict with adjacent land use. Figure VII–1–16 illustrates the proper dimensions for stacked and spread spoils.

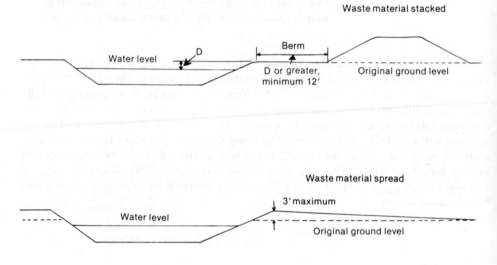

FIGURE VII–1–16. *Cross sections of a pond showing correct methods of stacking and spreading spoils.*

The topsoil set aside earlier in the process should be used to cover the stacked or spread spoils. You are more likely to get complete coverage if you stack the spoils. On the other hand, you can more easily establish vegetation on the gentler slope of spread spoils.

Sealing porous soils

Meanwhile, the hope is that the newly excavated hole will hold water. Whether or not it will is partly a function of site selection, but there are several processes you can use to seal even quite porous soils.

Compaction

Compaction is the simplest and cheapest method, and will usually work where the soil contains a variety of particle sizes, with at least ten percent clay. For each soil there is a certain moisture content for optimum compaction, so the weather plays a role in the success of compaction. A good rule of thumb is that a soil compacts best when it is too wet to plow but not wet enough to discharge water in the process of compaction. The first step in compaction is to disk or rototill the soil to a depth of 8 to 10 inches (20.3 to 25.4 cm). The best tool for the actual task of compaction is a sheepsfoot roller, but the wheels of a heavy tractor will do the job. In small ponds, a lawnroller is a feasible but laborious substitute.

Porous bottom soils can sometimes simply be blanketed with well-graded course grained soil containing at least twenty percent clay and compacted as described. Spread the blanket material at least 1 foot (30.5 cm) deep when the water is to be 10 feet (3 m) deep or less. Increase this thickness by 2 inches (5.1 cm) for each additional foot (30.5 cm) of water. Below the high water level, if the bottom will be exposed to freezing and thawing, the blanket material should be covered with a 12 to 18 inch (30.5 to 45.7 cm) layer of gravel.

Bentonite clay

Perhaps the best known pond bottom sealant is bentonite clay, which absorbs large amounts of water and expands to 8 to 20 times its original volume, plugging up the pores in the soil. It is applied, preferably to the dry pond bottom, at 1 to 3 pounds per square foot (5 to 15 kg/m²) spread, mixed and compacted as described above. (You will need a laboratory soil analysis to determine precise dosage.) Bentonite is readily available and not expensive, but immense quantities may be needed, so that, if you cannot find a nearby source, shipping costs may be prohibitive. An added caution is the occasional unexplained failures that occur with bentonite. Some say that it is the low montmorillonite content of some bentonite which is responsible, but I cannot confirm that.

Chemical sealants

Certain chemical sealants, collectively called dispersing agents, may be used to reduce the permeability of soils containing at least fifty percent fine grained material and at least fifteen percent clay. Among these materials are sodium chloride (which, contrary to what you might expect, does not increase the salinity of the pond water if properly applied to the dry bottom soil), tetrasodium polyphosphate, sodium tripolyphosphate and soda ash (99 to 100% sodium carbonate). These salts are used in quantities much smaller than bentonite, but get technical help if you decide to use them.

Sealing with membrane liners

A newer, and increasingly popular method of pond sealing involves lining the pond with a thin membrane of polyethylene, vinyl or butyl rubber. The obvious disadvantages of this method are cost (approximately $10,000 per acre or $24,000 per ha in Florida in 1975), and the fact that whatever natural fertility is present in the bottom soil will be lost. In time, however, plastic or rubber lined ponds will develop their own fertile organic mud. Methods for installation of pond liners are available from the manufacturers and distributors.

It has always seemed incredible to me that, even with careful preparation and cautious use after installation, a plastic film as thin as 0.2 mm can remain intact under a pond. One theory is that the liners' impermeability is quite temporary, and that their proven long-term effectiveness is better explained in biochemical terms.

Gleization

Which brings me to a final method of pond sealing developed (or perhaps rediscovered?) in the Soviet Union, called gleization. To use the gleization method, proceed as follows:

- Cover the pond bottom and sides completely with moist animal manure.
- Cover the manure layer with green vegetation, preferably broad-leafed sorts. Paper or cardboard may also be used.
- Cover that with a third layer of soil.
- Moisten and compact, as described above.
- After several weeks, add water.

The process set in motion by the manure-green leaves-soil compaction process involves creation of anaerobic conditions under which certain bacteria form a "biological plastic" called "gley." It is likely that the same sort of process goes on at a slightly slower rate under plastic or rubber liners, giving a fast, temporary seal together with a slower but permanent one.

Publication of an article on gleization in *The Journal of the New Alchemists* (see Appendix IV-1) led to a number of responses describing variations on the method. In some cases, the animal manure was omitted, resulting in a slower, but eventually effective reaction. Perhaps the most interesting response came from Lee Schilling of The Folk Life Center of the Great Smokies, Cosby, Tennessee. Lee reminded us that farmers in that region used to let their hogs do the work for them. The pigs were simply allowed access to, and perhaps fed in, a newly excavated pond. After a year or so, the resulting combination of fertilization and compaction brought about a natural seal.

Similar natural organic processes are often involved where ponds are leaky in their first years, but gradually become impervious. In ponds sealing is accelerated in direct proportion to the amount of fish, feeds and fertilizers placed in the pond.

Plugging leaks

You can plug small localized leaks by sprinkling bentonite on the water surface directly above the leaking spot. Cinders passed through a 1 inch (2.5 cm) screen can also be used to plug such leaks.

Protecting the shoreline

The area around the pond's waterline should be protected against wave erosion and damage by people and animals. If the slope of your pond is gentle and wave action and human traffic light, planting should give adequate protection. However, many pond owners prefer the greater protection of a rip-rap (described later in this chapter). A good fence offers further protection from humans and livestock. If a pond is to be used for stock watering, it is better to construct a separate pool for that purpose. (See p. 361 Figure VII-1-29.) This is usually not feasible with a dugout pond, since gravity flow cannot be used to fill the pool. The next best thing to do is to fence stock out of all but one small portion of the pond and provide that portion with a gravel, concrete or asphalt apron. (See Figure VII-1-17.)

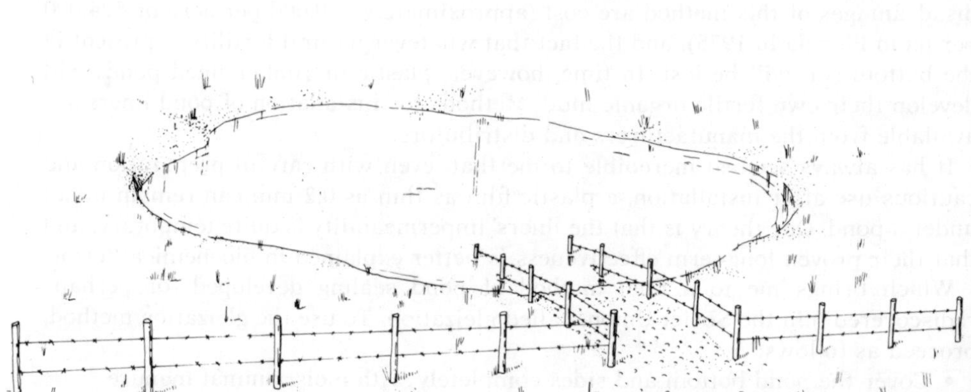

FIGURE VII-1-17. *Aquaculture pond used for stock watering, with fencing to restrict access by animals.*

The most important specialized aspects of constructing an impoundment or levee pond have to do with dam and levee construction. Dams and levees need to be made as impermeable as possible, just as do the bottom and sides of a dugout. But in addition to retarding the seepage of water, they must resist its pressure as well. An inadequately built dugout will leak, and can often be repaired at relatively low cost. An inadequate embankment on an impoundment or levee pond may "blow out," causing costly damage downstream and requiring repairs which may cost as much as the original cost of constructing the pond. For this reason, and because impoundments and levee ponds are often larger than dugouts, I discourage amateur efforts unless they are on a very small scale.

The following paragraphs cover some of the main considerations in embankment construction. The essential difference between a dam and a levee is that a dam lies perpendicular to the main axis of flow in the valley to be impounded. If there is even a small stream flowing through the valley it can create erosion problems during dam construction. If there is a time during which the stream is dry, schedule construction for this period. Though you may be able to complete the dam on both sides of the stream first, and then close it with an all-out effort, there are two other preferable possibilities. Either you can divert the stream around the dam before beginning construction, or you can lay a culvert which will be blocked up as the last stage in construction.

If you find that you have to import special soil for construction purposes, you should question the feasibility of the site. Normally, soils selected and stockpiled during excavation can be used for embankment construction. Where there is a choice, save the most impervious materials for the "cores" or "keys" (described below). The most pervious materials can be used on outside slopes of embankments, while intermediate materials can go on the inside walls. All exposed surfaces of dams and levees should consist of a mixture of clay and more pervious materials. Pure clay becomes gooey when wet, sand is loose and unstable when dry, but the mixture remains firm and stable in all weather, as well as being better suited to planting.

Levees and dams are usually trapezoidal in cross section. (See Figure VII–1–18.) The important dimensions are height at the crown, freeboard, width at the crown, thickness or width at the base, inside slope and back slope.

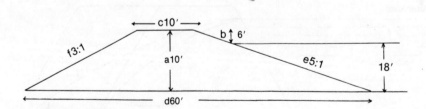

FIGURE VII–1–18. *Cross section of a dam or levee, showing:*
   a) height at the crown
   b) freeboard
   c) width at the crown
   d) thickness or width at the base
   e) inside slope
   f) back slope

The height will be determined by the desired depth of the pond and the degree to which the water level fluctuates. The height of an embankment, then, is equivalent to the depth of the pond at the outlet, plus a safety factor, minus any natural topographic features which can be incorporated in it.

Freeboard is the difference between the water level and the top of the dam or levee. Most North American authorities cite a minimum freeboard of two feet (0.6 m) at normal summer water levels.

Dams and levees should never be triangular in cross-section, so that a peak is reached at the top. Beyond that, width at the crown will depend largely on the use to be made of the pond. The following table lists minimum crown widths for various size embankments, assuming vehicles are to be driven on them.

Table VII–1–7. Minimum crown width,
in feet (m) as related to height of pond embankments.

Height of dam in feet (m)	Minimum crown width in feet (m)
Under 10 (3.3)	8 (2.4)
10–15 (3.3–4.6)	10 (3.3)
15–20 (4.6–6.6)	12 (3.7)
20–25 (6.6–7.6)	14 (4.3)

If you do not plan to use vehicles, a crown as narrow as 3 feet (1 m) is permissible on a pond 3 to 4.5 feet (1–1.5 m) deep; deeper ponds should have wider embankments. Before designing a narrow-crowned embankment, think about who will be walking on it and what equipment they might carry.

Thickness of dams and levees is determined primarily by the need for structural strength. Two guidelines offered for estimating width at the base are that it should be 6 to 7 times the height, or equal to the width of the crown plus 2.5 to 5 times the height. The looser the soil the thicker the embankment. Generally, if the inside and back slopes and the height are carefully chosen the thickness will automatically be adequate, but it is best to make sure.

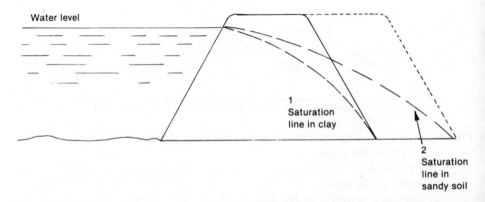

FIGURE VII–1–19. *Saturation curve in embankments made of highly and slightly permeable soil.*

Where it is impossible to construct a perfectly impervious embankment, water retention becomes a factor in determining thickness. The degree of permeability of a heaped-up soil and the force of gravity combine to produce a characteristic "saturation curve." (See Figure VII-1-19.) By determining this curve (in consultation with an engineer or Soil Conservation Service agent) one can calculate the width necessary to prevent excessive leakage.

The ideal inside slope for a particular pond may not be the most practical from a construction point of view. A slope ratio that is nearly vertical is theoretically ideal for many forms of intensive aquaculture. For esthetic reasons, some may prefer an unusually gentle slope. I have seen inside slopes from 1.5:1 to 4:1 recommended for dams and levees, but 3:1 seems to be the most commonly cited figure for small ponds and 4:1 for ponds of over 10 acres (4 ha).

The outside, or back slope, can often be steeper than the inside slope, which will reduce the construction cost. However, the range of figures cited is almost the same as for inside slope. Both, of course, depend on the nature of the soil. If clay soil is scarce, save it for the inside slope, where permeability is important. In that event, the use of looser soils on the back slope may dictate a more gentle slope than would otherwise be necessary.

You can improve water retention in impoundments and levee ponds by the same measures described for dugouts: compaction, lining or other improvements of the bottom soil. The selection of materials will also be a factor, and, as we have just seen, the dimensions of embankments made of somewhat pervious soils. However, there are some special construction techniques which usually must be employed to ensure impermeability of dams and levees. In most locations, it will be possible to excavate down to an impervious soil stratum. If this is not possible, the entirety of the pond bottom and sides will have to be sealed using the methods described above for dugouts.

Assuming you can reach an impervious stratum, proceed to place the base of the dam at that level. Where the levees are to go, a trench 3.0–6.1 feet (1–2 m) wide is dug to the same level. (See Figure VII-1-20.) As construction proceeds, fill the trench with layers of impervious soil and compact repeatedly to form an impervious band extending from the top of the natural impervious stratum to the top of the embankment. (On small ponds, where impervious soil is scarce, concrete may be used.) The resulting structure resembles the cross-section in Figure VII-1-21. Usually the core or key is placed at the center of the embankment, but an alternative approach is shown in Figure VII-1-22.

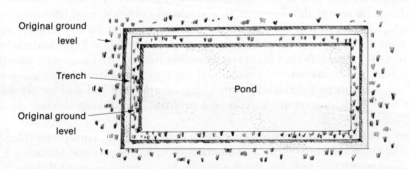

FIGURE VII-1-20. *Excavation of a core trench.*

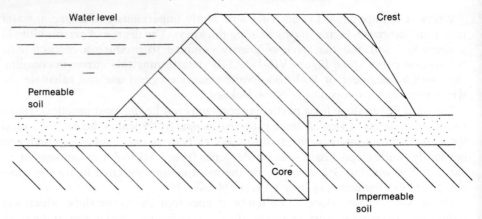

FIGURE VII–1–21. *Cross-section of an embankment showing impermeable core or key.*

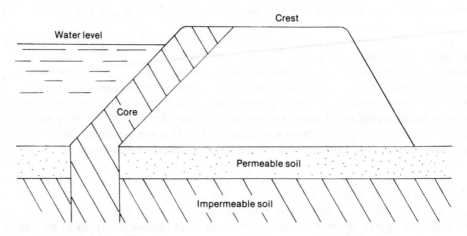

FIGURE VII–1–22. *Embankment built with the impermeable key outside the embankment proper and incorporating a clay blanket on the inside slope.*

Preventing embankment erosion

The finishing touches on dams and levees are to prevent erosion, so undertake them immediately when construction is finished. Planting with perennial grasses and/or legumes is the first line of defense. Reed canary grass (*Phalaris arundinacea*) is a good choice. A recent development is a strain of common reed (*Phragmites australis*), selected specifically for control of erosion on pond banks. Known as "Shoreline" reed, this plant is available from the Soil Conservation Plant Materials Center in Knox City, Texas. Do not plant trees or other plants with long roots; water can follow a root and break an embankment. If the weather is dry, do not put off planting. Your investment in irrigation, mulching or applying sod will be repaid many times over. Mulching on slopes can be a problem; try staking chicken wire over the mulch.

Apart from planting, erosion protection can include graveling, riprapping (placing large rocks along the embankment), construction of log booms and including a berm. Graveling is suitable only for the flat crown, but is recommended if there is to be vehicular traffic.

Rock riprap may extend to the top of dams, but on levees it is usually confined to a band three feet (0.9 m) on either side of the anticipated low water level. For best results place riprap by hand. You can prevent erosion under the riprap by spreading a bed of gravel or crushed stone at least ten inches (25 cm) thick before placing the rocks.

Log booms are a second-best substitute for riprap. They offer considerable hindrance to fishing and other activities, but may be less expensive for some pond owners. Booms consist simply of end-to-end rows of floating logs ten to twelve inches (25–30 cm) in diameter. They are not placed on the bank, but are anchored about six feet (1.8 m) offshore, allowing some slack for fluctuations in the water level.

If a very stable water level can be guaranteed, a berm eight to ten feet (2.4–3.0 m) wide between the pond edge and the toe of a levee can be left at or barely above the water level. However, miscalculation of the water level or unforeseen conditions can turn such a berm into a muddy morass and/or haven for aquatic weeds.

Table VII–1–8. Sample of table to be constructed in estimating volume of earthfill needed for a dam.

Station	Ground elevation	Fill height[1] ft	End area[2] sq ft	Sum of end areas sq ft	Distance ft	Double volume[3] cu ft
0 + 50	35.0	0	0			
				44	18	792
+ 68	32.7	2.3	44			
				401	32	12,832
1 + 00	25.9	9.1	357			
				1,066	37	39,442
+ 37	21.5	13.5	709			
				1,564	16	25,024
+ 53	20.0	15.0	855			
				1,730	22	38,060
+ 75	19.8	15.2	875			
				1,781	25	44,525
2 + 00	19.5	15.5	906			
				1,730	19	32,870
+ 19	20.3	14.7	824			
				1,648	13	21,424
+ 32	20.3	14.7	824			
				1,805	4	7,220
+ 36	18.8	16.2	981			
				2,030	4	8,120
+ 40	18.2	16.8	1,049			
				2,064	3	6,192
+ 43	18.5	16.5	1,015			
				1,911	3	5,733
+ 46	19.6	15.4	896			
				1,771	13	23,023
+ 59	19.8	15.2	875			
				1,650	41	67,650
3 + 00	20.8	14.2	775			
				1,023	35	35,805
+ 35	27.7	7.3	248			
				324	25	8,100
+ 60	31.6	3.4	76			
				76	36	2,736
3 + 96	35.0	0.0	0			
					Total[3]	379,548

[1] Elevation of top of dam without allowance for settlement.
[2] End areas based on 12-foot top width and 3:1 slopes on both sides.
[3] Divide double volume in ft[3] by 54 to obtain volume in yd[3], e.g.,

$$\frac{379,548}{54} = 7,029 \text{ yd}^3.$$

Allowance for settlement (10%) =   703 yd[3]
Total volume = 7,732 yd[3]

A final and important aspect of pond construction is the cost. With a little thought, anyone can come to grips with labor and materials costs and decide whether he or she is about to be ripped off. But the primary factor used by engineers to set a price is the volume of earthfill required. So, you may find it useful to make your own estimate of the volume of a dam, levee, or excavation. The simplest method of doing this is called the "sum of end area" method. As an example, take a dam 346 feet (105.5 m) long and 12 feet (3.7 m) wide at the crown with identical back and inside slopes of 3:1, set across a natural valley. To estimate the volume of any as yet unbuilt dam, construct a table like Table VII-1-8 and follow these steps:

• Sketch the profile of the dam and determine the elevation (second column in Table VII-1-8) at a number of points along the centerline of the space to be filled, as shown in Figure VII-1-23. If there are points where the elevation changes abruptly, these are where measurements should be made. Otherwise, measure every 5 to 20 feet (1.5-6.0 m), depending on the steepness of the site. If you cannot obtain the actual elevation above sea level, you can designate the low point on the profile as zero elevation and further measurements can be made as distances above that point.

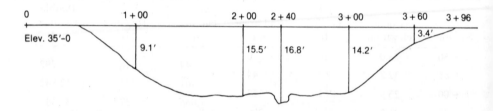

FIGURE VII-1-23. *Profile of dam indicated in Table VII-1-8, showing height measurements.*

• Obtain the height of fill needed at each point by subtracting the ground elevation from the top elevation of the dam (third column in Table VII-1-8).

• Using Table VII-1-9 and the known slopes and top width of the dam, find the end area of each measured section (fourth column in Table VII-1-8), as follows: To determine the end area at a point where the fill is 15 feet (5 m) high, first look in Table VII-1-9 under the appropriate fill height and combination of slopes. You will find the number 675. Then look under the same fill height and the given top width. You will find the number 180. Adding these two numbers gives the end area at that point, or 855 square feet (79 m²).

• To find the total volume, in cubic feet, of fill between two points for which the end areas have been computed, sum the two end areas (fifth column in Table VII-1-8) and multiply by the distance between the two points in feet to obtain the figures shown in the final column of Table VII-1-8. For example, the volume of fill required between Station 1 + 53 and the next station, 1 + 75, is computed as follows: (855 square feet + 875 square feet) (22 feet) = 38,060 cubic feet (1,077 m³).

• Add the volumes in the last column of Table VII-1-8 and divide by 54 to obtain the volume in cubic yards, the unit customarily used by contractors.

• Add a percentage for settling of the dam. If you do not have an engineer's estimate based on the type of soil in the dam and its foundation, 10% is a good approximation.

Table VII-1-9. End areas, in ft², of embankment sections for different side slopes and crown widths.[1]

Fill Height (ft)	Side slopes 2.5:1 2.5:1 2:1 3:1	2.5:1 3:1 2:1 3.5:1	3:1 3:1 2.5:1 3.5:1	3.5:1 3.5:1 3:1 4:1	4:1 4:1 3:1 5:1	Crown width (ft) 8	10	12	14	16
1.0	3	3	3	4	4	8	10	12	14	16
1.2	4	4	4	5	6	10	12	14	17	19
1.4	5	5	6	7	8	11	14	17	20	22
1.6	6	7	8	9	10	13	16	19	22	26
1.8	8	9	10	11	13	14	18	22	25	29
2.0	10	11	12	14	16	16	20	24	28	32
2.2	12	13	15	17	19	18	22	27	31	35
2.4	14	16	17	20	23	19	24	29	34	39
2.6	17	19	20	24	27	21	26	31	36	42
2.8	20	22	23	27	31	22	28	34	39	45
3.0	22	25	27	32	36	24	30	36	42	48
3.2	26	28	31	36	41	26	32	38	45	51
3.4	29	32	35	40	46	27	34	41	47	55
3.6	32	36	39	45	52	29	36	43	50	58
3.8	36	40	43	50	58	30	38	46	53	61
4.0	40	44	48	56	64	32	40	48	56	64
4.2	44	49	53	62	71	34	42	50	59	67
4.4	48	53	58	68	77	35	44	53	61	71
4.6	53	58	63	74	85	37	46	55	64	74
4.8	57	63	69	81	92	38	48	57	67	77
5.0	62	69	75	87	100	40	50	60	70	80
5.2	67	74	81	94	108	42	52	62	73	83
5.4	73	80	87	102	117	43	54	65	75	87
5.6	78	86	94	110	125	45	56	67	78	90
5.8	84	93	101	118	135	46	58	69	81	93
6.0	90	99	108	126	144	48	60	72	84	96
6.2	96	106	115	135	154	50	62	74	87	99
6.4	102	113	123	143	164	51	64	77	89	103
6.6	109	120	131	152	174	53	66	79	92	106
6.8	116	128	139	162	185	54	68	81	95	109
7.0	123	135	147	172	196	56	70	84	98	112
7.2	130	143	156	182	207	58	72	86	101	115
7.4	138	152	165	193	219	59	74	89	103	119
7.6	145	159	174	203	231	61	76	91	106	122
7.8	153	168	183	214	243	62	78	93	109	125
8.0	160	176	192	224	256	64	80	96	112	128
8.2	169	185	202	235	269	66	82	98	115	131
8.4	177	194	212	247	282	67	84	101	117	135
8.6	186	204	222	259	296	69	86	103	120	138
8.8	194	213	232	271	310	70	88	105	123	141
9.0	203	223	243	283	324	72	90	108	126	144
9.2	212	233	254	296	339	74	92	110	129	147

Table VII–1–9. *Continued.*

Fill Height (ft)	Side slopes					Crown width (ft)				
	2.5:1 2.5:1 2:1 3:1	2.5:1 3:1 2:1 3.5:1	3:1 3:1 2.5:1 3.5:1	3.5:1 3.5:1 3:1 4:1	4:1 4:1 3:1 5:1	8	10	12	14	16
9.4	222	244	266	310	353	75	94	113	131	151
9.6	231	254	277	323	369	77	96	115	134	154
9.8	241	265	289	337	384	78	98	117	137	157
10.0	250	275	300	350	400	80	100	120	140	160
10.2	260	286	313	364	416		102	122	143	163
10.4	271	298	325	379	433		104	125	145	167
10.6	281	309	338	394	449		106	127	148	170
10.8	292	321	350	409	467		108	129	151	173
11.0	302	333	363	424	484		110	132	154	176
11.2	313	344	376	440	502		112	134	157	179
11.4	325	357	390	456	520		114	137	159	183
11.6	336	370	404	472	538		116	139	162	186
11.8	348	383	418	488	557		118	141	165	189
12.0	360	396	432	504	576		120	144	168	192
12.2	372	409	447	522	595		122	146	171	195
12.4	385	424	462	539	615		124	149	173	199
12.6	397	437	477	557	635		126	151	176	202
12.8	410	451	492	574	655		128	153	179	205
13.0	422	465	507	592	676		130	156	182	208
13.2	436	479	523	610	697		132	158	185	211
13.4	449	494	539	629	718		134	161	187	215
13.6	463	509	555	648	740		136	163	190	218
13.8	476	523	571	667	762		138	166	193	221
14.0	490	539	588	686	784		140	168	196	224
14.2	505	555	605	706	807		142	170	199	227
14.4	519	570	622	726	829		144	173	202	230
14.6	534	586	639	746	853		146	175	204	234
14.8	548	602	657	767	876		148	178	207	237
15.0	563	619	675	788	900		150	180	210	240
15.2	578	635	693	809	924		152	182	213	243
15.4	594	653	711	830	949		154	185	216	246
15.6	609	669	730	852	973		156	187	218	250
15.8	625	687	749	874	999		158	190	221	253
16.0	640	704	768	896	1,024		160	192	224	256
16.2	656	722	787	919	1,050			194	227	259
16.4	673	740	807	942	1,076			197	230	262
16.6	689	758	827	965	1,102			199	232	266
16.8	706	776	847	988	1,129			202	235	269
17.0	723	795	867	1,012	1,156			204	238	272
17.2	740	814	888	1,036	1,183			206	241	275
17.4	757	833	909	1,060	1,211			209	244	278
17.6	774	852	930	1,084	1,239			211	246	282
17.8	792	871	951	1,109	1,267			214	249	285
18.0	810	891	972	1,134	1,296			216	252	288
18.2	828	911	994	1,160	1,325			218	255	291
18.4	846	931	1,016	1,186	1,354			221	258	294

Table VII–1–9. *Continued.*

	Side slopes						Crown width (ft)				
Fill Height (ft)	2.5:1 2.5:1 2:1 3:1	2.5:1 3:1 2:1 3.5:1	3:1 3:1 2.5:1 3.5:1	3.5:1 3.5:1 3:1 4:1	4:1 4:1 3:1 5:1	8	10	12	14	16
18.6	865	951	1,038	1,212	1,384			223	260	298
18.8	884	972	1,060	1,238	1,414			226	263	301
19.0	903	993	1,083	1,264	1,444			228	266	304
19.2	922	1,014	1,106	1,291	1,475			230	269	307
19.4	941	1,035	1,129	1,318	1,505			233	272	310
19.6	960	1,056	1,152	1,345	1,537			235	274	314
19.8	980	1,078	1,176	1,372	1,568			238	277	317
20.0	1,000	1,100	1,200	1,400	1,600			240	280	320
20.2	1,020	1,122	1,224	1,428	1,632			242	283	323
20.4	1,040	1,144	1,248	1,457	1,665			245	286	326
20.6	1,061	1,167	1,273	1,486	1,697			247	288	330
20.8	1,082	1,190	1,298	1,515	1,731			250	291	333
21.0	1,103	1,213	1,323	1,544	1,764			252	294	336
21.2	1,124	1,236	1,348	1,574	1,798			254	297	339
21.4	1,145	1,254	1,374	1,604	1,832			257	300	342
21.6	1,166	1,283	1,400	1,634	1,866			259	302	346
21.8	1,188	1,307	1,426	1,664	1,901			262	305	349
22.0	1,210	1,331	1,452	1,694	1,936			264	308	352
22.2	1,232	1,356	1,479	1,725	1,971			266	311	355
22.4	1,254	1,380	1,506	1,756	2,007			269	314	358
22.6	1,277	1,405	1,533	1,788	2,043			271	316	362
22.8	1,300	1,430	1,560	1,820	2,079			274	319	365
23.0	1,323	1,455	1,587	1,852	2,116			276	322	368

[1] To find the end area for any fill height, add ft² given under staked side slopes to that under the top width for total section. Example: 6.4–foot fill, 3:1 front and back slopes, 14–foot top width—123 plus 89 or 212 ft² for the section. Any combination of slopes that adds to 5, 6, or 7 may be used. A combination of 3.5:1 front and 2 5:1 back gives the same results as 3:1 front and back.

In our example, the only earthfill structure is the dam. If you plan to include levees or other elevated structures, their volume must be calculated and added in before the settling allowance is made. It is simpler to estimate the volume of levees than of dams, since the height of levees tends to be uniform from one end to the other.

If you plan to include a core trench, ditches or dugout construction, their volumes must be added in after making the settlement allowance. If such excavations are trapezoidal in cross-section, estimate the volume by using the formula given earlier in this chapter, under dugout construction. If they are rectangular, a simple arithmetical calculation of volume is all that is necessary.

The concluding portion of this chapter is devoted to diagrams of five aquaculture ponds, with explanations of some of the design and construction features discussed above as they apply to these ponds, plus drawings of each pond. It is unlikely that you will find any of these diagrams useful as a "blueprint." Instead they are intended to further clarify the principles of pond design and encourage the type of

Examples of aquaculture ponds

thinking which leads to successful pond construction. The five ponds are:

(1) A "wild" pond which produces fish for the family table as a by-product.

(2) A "multiple use" pond, more productive than the wild pond. This pond is based on the farm pond model (see Chapter II–1), with slight modifications to increase its efficiency as a source of fish for home consumption.

(3) A fee fishing pond operated as a family commercial enterprise.

(4) A small-scale commercial crayfish pond, such as you might find in Louisiana.

(5) An intensive fish culture pond, using design criteria developed by large-scale commercial catfish culturists in the southern United States.

FIGURE VII–1–24.   *"Wild" pond in surrounding environment.*

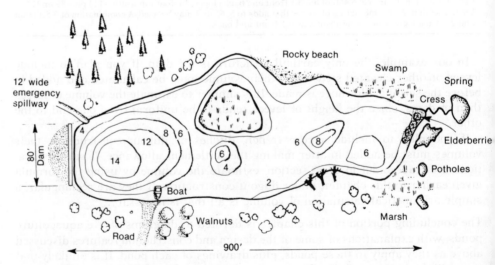

FIGURE VII–1–25.   *Map of "wild" pond.*

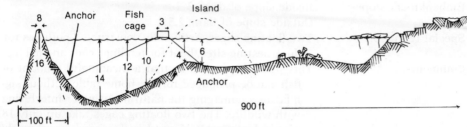

FIGURE VII–1–26. *Longitudinal cross-section of "wild" pond.*

"Wild" Pond (Figures VII–1–24, 25 and 26)   *"Wild" Pond*

Purposes — principal:	Esthetics, wildlife conservation.
secondary:	Fish for home consumption, recreational fishing, soil conservation.
Production:	200 lb (90 kg) per year in cages; 50 lb (22.5 kg) per year to hook and line fishing.
Species:	Mixed game and pan fish based on largemouth bass (*Micropterus salmoides*)-bluegill (*Lepomis macrochirus*) model, bullheads in cages, forage fishes, frogs.
Water supply:	Primarily surface runoff, also small spring.
Soil type:	20% clay with no coarse materials.
Sealing:	Natural, compaction of dam.
Topography:	Varied and rolling, pond set in a narrow valley with overall 2% natural longitudinal slope.
Location:	Northeast United States.
Type of construction:	Impoundment, with excavation to create dam.
Disposition of spoils:	All used in dam.
Access:	Private road, secluded.
Use of surrounding land:	Primarily natural; some food plants which do not require intensive cultivation.
Surface area:	Approximately 2 acres (0.8 ha).
Depth — maximum:	14 feet (4.3 m).
average:	4 feet (1.2 m).
minimum:	Slopes naturally up to shoreline.
Shape:	Very irregular, dictated by natural land forms.
Planting of banks:	Mostly natural forest, a few food and flowering plants. Dam is incorporated into a "natural" meadow, with abundant wildflowers.
Bank erosion control devices:	Plants only.
Aquatic vegetation:	Water lilies in several shallow areas.
Shoreline features:	Fallen trees, brush, combination of steep and gentle slopes.
Inlet structure:	Spring has been dug out and a series of rock dams incorporated in the natural stream which is its outlet to oxygenate water. This accounts for perhaps 10% of the water supply.
Outlet structure:	Emergency spillway only, planted to grass and 12 feet (3.7 m) wide.

Embankment slopes: Inside slope of dam: 3:1
                                Outside slope of dam: 2.5:1.

Special features: Two floating cages, artificially constructed potholes for duck nesting sites, boat for tending cages and fishing.

Comments: This pond demonstrates that a meaningful amount of fish can be produced for the home table without significantly damaging the natural setting or interfering with wildlife. The two floating cages take up only 18 square feet (2 m²) of pond surface and are scarcely visible to the casual observer. The most unnatural aspect of the pond, the dam, is made to blend in by incorporating it in a larger rolling meadow and planting it with wildflowers.

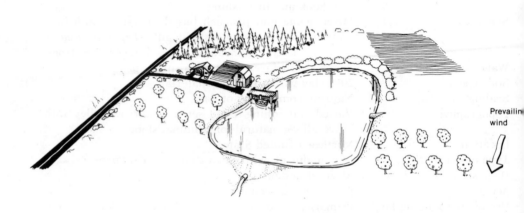

FIGURE VII–1–27. *Multiple use pond in surrounding environment.*

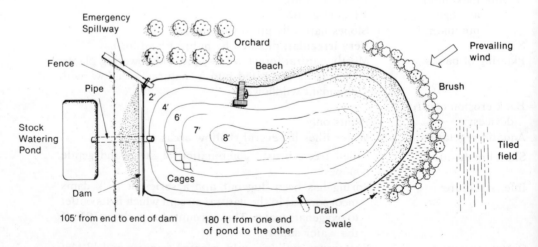

FIGURE VII–1–28. *Map of multiple use pond.*

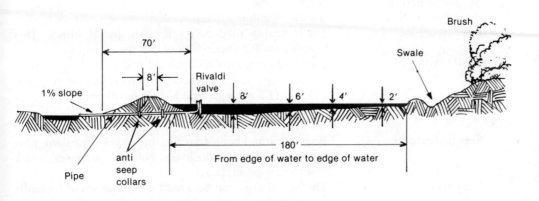

FIGURE VII–1–29. *Longitudinal cross-section of multiple use pond.*

Multiple use pond (Figures VII—1–27, 28 and 29)

Purposes—principal:	Fish for home consumption, recreational fishing, swimming, stock watering.
secondary:	Esthetics, irrigation for orchard, soil conservation, water conservation, wildlife conservation.
Production:	400 lb (180 kg) per year in cages, 50 lb (22.5 kg) per year to hook and line fishing.
Species:	Mixed game and pan fish based on largemouth bass-bluegill model, bullheads in cages, a few channel catfish, forage fishes. A similar design could be used to grow trout where cold enough water is available.
Water supply:	Surface runoff, including that collected from tilled field.
Soil type:	20% clay with no coarse materials.
Sealing:	Compaction.
Topography:	Rolling, set in a wide valley with slight natural longitudinal slope. Pond oriented with the long axis parallel to the prevailing winds.
Location:	Central Great Plains.
Type of construction:	Principally dugout, with dam at lower end.
Disposition of spoils:	Some used in dam, rest stacked to form brushy area separating tilled field from pond.
Acess:	Through yard.
Use of surrounding land:	Mixed agricultural.
Surface area:	½ acre (0.2 ha).
Depth—maximum:	8 feet (2.4 m).
average:	6 feet (1.8 m).
minimum:	2 feet (0.6 m).
Shape:	Regular, but in conformity to natural land forms.
Planting of banks:	Sod around edges, brush starting about 40 feet (12.2 m) back from the water's edge.
Bank erosion control devices:	Plants only.

Aquatic vegetation:	Almost none.
Shoreline features:	Swale. (Shallow ditch.)
Inlet structure:	Ditch drains tiled field, through gravel filters. This accounts for about 50% of the water supply.
Outlet structure:	8 inch (20 cm) pipe with Rivaldi valve, leading to stock watering pond. Emergency spillway, planted to grass, 4 feet (1.2 m) wide and 8 inches (20 cm) deep.
Embankment slopes:	Inside slope of dam: 3:1. Outside slope of dam: 3:1.
Special features:	Boat dock and boat, four floating cages near dam, plus one attached to dock for holding purposes, stock watering pond, beach.
Comments:	The beach area can be used for seining small bluegills to remove excess or for stocking in cages.

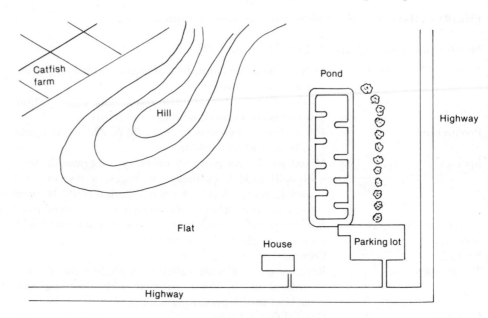

FIGURE VII–1–30. *Fee fishing pond in surrounding environment.*

Fee fishing pond

Fee fishing pond (Figures VII–1–30, 31, 32 and 33)

Purposes—primary:	Income through fee fishing.
secondary:	Disposal of excess stock from commercial catfish culture operation.
Production:	(Production is not a meaningful statistic for a fee fishing pond, since fish are stocked at catchable size, though they may grow while in the pond.)
Species:	Principally channel catfish, but other species as available.
Water supply:	Well.
Soil type:	20% clay, with no coarse materials.
Sealing:	Compaction.

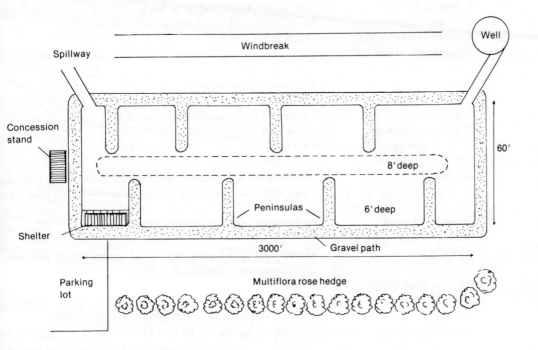

FIGURE VII–1–31. *Map of fee fishing pond.*

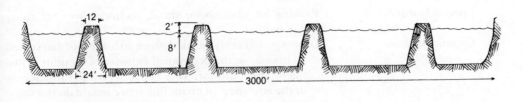

(Note: walls are not in scale)

FIGURE VII–1–32. *Longitudinal cross-section of fee fishing pond.*

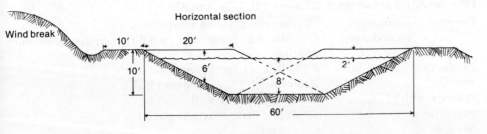

FIGURE VII–1–33. *Horizontal cross-section of fee fishing pond.*

Topography:	Flat at pond site, ridge in back.
Location:	South Central United States.
Type of construction:	Levee pond.
Disposition of spoils:	Used in dam, windbreak on one side of pond, and peninsulas.
Access:	Via two public highways.
Use of surrounding land:	Woodlot, commercial along highways.
Surface area:	5 acres (2 ha), counting peninsulas.
Depth—maximum:	8 feet (2.4 m).
average:	6 feet (1.8 m).
minimum:	2 feet (0.6 m).
Shape:	Oblong and rectangular, with eight peninsulas jutting out into the water to increase accessibility for anglers.
Planting of banks:	Permanent grass and hairy vetch.
Bank erosion control devices:	Riprap 2 feet (61 cm) above and below water surface, path on top of levees graveled.
Aquatic vegetation:	None.
Shoreline features:	Peninsulas.
Inlet structure:	8 inch (20 cm) pipe, with splash box, from well.
Outlet structure:	Emergency spillway, 10 feet (3.3 m) wide, 8 inches (20 cm) deep, paved with cement. Occasional pumping down.
Embankment slopes:	2:1 on both sides of all embankments, including peninsulas.
Special features:	Parking lot, concession stand, sheltered area for rainy day fishing.
Comments:	Drainage is rarely needed since fish are not harvested en masse, so the occasional expense of a pump is justified. The multiflora rose hedge serves to limit access to the entrance. (A chain link fence would do the same job less esthetically.) This pond is located adjacent to a commercial catfish farm so that there is a constant source of stock.

Commercial crayfish pond

Commercial crayfish pond (Figures VII–1–34 and 35)

Purposes—primary:	Commercial crayfish production.
secondary:	Crayfish for home use, wildlife conservation, duck hunting.
Production:	2,800 lb/yr (1,260 kg/yr).
Species:	Red crayfish (*Procambarus clarki*).
Water supply:	Ditch, emergency well.
Soil type:	10% clay, little coarse material.
Sealing:	Clay blanket.
Topography:	Flat and swampy.
Location:	Southern Louisiana.
Type of construction:	Levee pond.

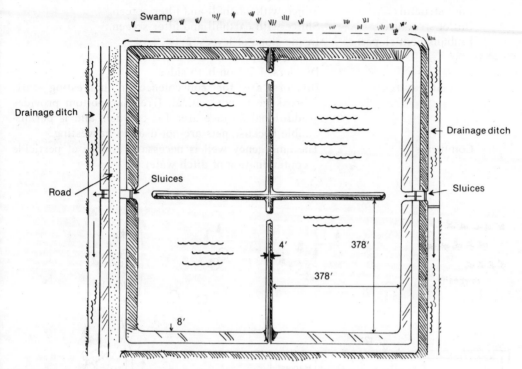

FIGURE VII-1-34. *Four interconnected commercial crayfish ponds.*

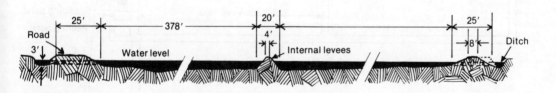

FIGURE VII-1-35. *Longitudinal cross-section of commercial crayfish ponds.*

Disposition of spoils:	All used in levees and dividers.
Surface area:	4 acres (1.6 ha); part of a unit of four ponds totalling 16 acres (6.4 ha).
Depth—maximum:	3 feet (1.2 m).
average:	2 feet (0.8 m).
minimum:	8 inches (20 cm).
Shape:	Rectangular.
Planting of banks:	Grass.
Bank erosion control devices:	Plants only.
Aquatic vegetation:	Entirety of pond planted to alligator grass.
Shoreline features:	None.

Inlet structure:	Sluice with earthfill and filter screen.
Outlet structure:	Sluice with earthfill and filter screen.
Embankment slopes:	Inside slope of levees: 2:1
	Outside slope of levees: 2:1.
	Dividers: 1.5:1 on both sides.
Special features:	Dividers are for convenience in harvesting only, breakage is not critical. Irregular bottom provides additional surface area for crayfish, and is permissible because nets are not used in harvesting.
Comments:	The emergency well is necessary in case of pesticide contamination of ditch water.

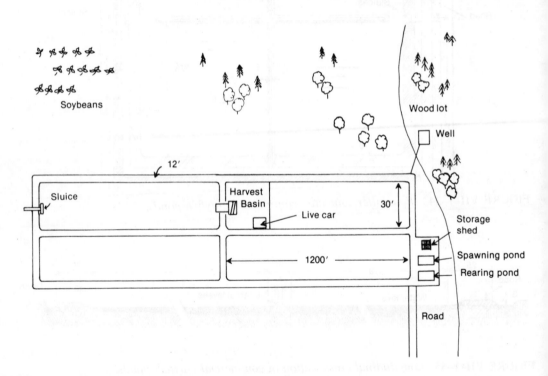

FIGURE VII-1-36. *Set of four commercial catfish ponds, with spawning and rearing ponds.*

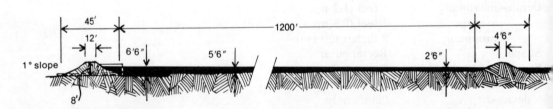

FIGURE VII-1-37. *Longitudinal cross-section of commercial catfish pond.*

Commercial catfish pond (Figures VII–1–36 and 37)

Purpose:	Commercial catfish production, rice and soybeans in rotation.
Production:	25,000 lb/yr (11,250 kg/yr).
Species:	Channel catfish.
Water supply:	Well.
Soil type:	20% clay with no coarse materials.
Sealing:	Compaction, clay core in dam and levees.
Topography:	Flat, long axis perpendicular to prevailing wind.
Location:	South Central United States.
Type of construction:	Levee pond.
Disposition of spoils:	Levees and windbreak.
Access:	Private road.
Use of surrounding land:	Agricultural, woodlot.
Surface area:	10 acres (4 ha).
Depth—maximum:	6½ feet (2.0 m).
average:	3½ feet (1.1 m).
minimum:	2½ feet (0.75 m).
Shape:	Rectangular and oblong.
Planting of banks:	Grass.
Bank erosion control devices:	Riprap 3 feet (0.9 m) above and below water level.
Aquatic vegetation:	None.
Shoreline features:	None.
Inlet structure:	Water pumped via 24 inch (61 cm) pipe from well.
Outlet structure:	Concrete monk.
Embankment slopes:	Inside levee slope: 2:1.
	Outside levee slope: 2:1.
Special features:	Harvest basin, live car to be used in conjunction with trap, smaller ponds for breeding and fry rearing, storage shed.
Comments:	This pond is included, not because it will be of direct interest to small-scale aquaculturists, but because it illustrates design features which could be incorporated in other ponds. For example, a scaled-down version of this pond, with a cement floored pen on the levee, could be used in an integrated fish-livestock system, such as described in Chapter IV–2.

CHAPTER VII-2

Raceways, Very Small Tanks, and Pools

Some forms of aquaculture are carried out in enclosures very much smaller than what would normally be called a pond. In this group are included floating cages (Chapter IV–3) and solar-algae ponds (Chapter IV–4). Remaining to be discussed are raceways and various sorts of small tanks or pools, all of them simpler in design and construction than ponds.

Raceways The name raceway suggests a very rapid rate of flow, but here it will be defined as any system where there is a perceptible current throughout all the pools or enclosures at all times during normal operation. Though raceways are traditional in intensive trout culture, they are considered a radical departure in the production of other fishes. One reason for this situation is that the raceway environment resembles that commonly associated with wild trout. But it is also true that, in most parts of North America, spring water is usually too cold for culture of warm water fishes, and one rarely encounters warm surface water of adequate quality and quantity for intensive trout culture. (A well which can economically be used to supply raceways is also very rare.) I will confine this discussion, then, to design criteria developed in trout culture.

Advantages The advantages of raceways over standing water are the following:

• Production per unit of area is many times greater, since polluted water is constantly being replaced by clean water. Stocking density of course varies with many factors other than flow rate, but it is not unusual to stock raceways at thirty to forty times the rates normal for standing water of similar quality.

• If you build occasional drops into the system, it will be easier to add oxygen to a flowing water system than to a pond. (This is of course offset by greater stocking and feeding rates.)

• It is often difficult to raise cold water fish in ponds, which tend to heat up. Raceways supplied by cold spring water largely eliminate this problem.

• Raceways are preferred in sport fish hatcheries, since fish reared in flowing water exercise constantly, even at high population densities, becoming more vigorous and better adapted to survive in the wild. However, current aquaculture theory holds that maximum weight gain is achieved when exercise is minimized. (The effect of exercise on the quality of flesh of fish has not been studied.)

Many readers will rule out raceways simply after inspecting the available construction sites. Some sort of pond can be built in most places, but few sites have both suitable topography and an adequate supply of water for raceways. If you think raceways are a possibility for you, be sure to get engineering help in determining the rate and volume of flow that can be maintained, cost of construction as compared to ponds, and the amount of fish that might be produced. (If you are not planning to grow trout, you will be operating largely in the dark on the last item, but try to make a conservative estimate.) Construction is sure to cost more per area of water than for ponds, since raceways make much less efficient use of space and must be built to withstand erosion. Another disadvantage of densely stocked and heavily fed raceways is the production of a considerable volume of polluted effluent. Disposing of this water may confront you with legal and ethical problems which would not arise in pond culture.

Siting raceways

It is possible to pump water into or out of raceway systems, or even to recirculate it through filters. However, as we saw in Chapter IV-4, the energy required for such methods tends to render them uneconomic. Though in theory it is possible to construct a raceway based on gravity flow in any terrain, costs will be least if you select a site which has a natural slope of 1 to 3 percent.

Raceways may be rectangular or circular. The main advantage of rectangular raceways is that they make by far the most efficient use of space. There may be a marginal advantage in the greater ease of harvesting in a rectangular enclosure, but this is not nearly as important in such small, readily drainable enclosures as raceways as it is in ponds.

Raceways' shape and dimensions

The main disadvantage of the rectangular shape is that "dead spots," with low flow, occur in the corners, where waste may be accumulated and dissolved oxygen depleted. You can compensate for this to some extent by rounding the corners.

Rectangular raceways used in trout culture are 6.6 to 13.1 feet (2–4 m) wide and up to 109 yards (100 m) long. In use they may be horizontally or vertically subdivided by temporary barriers. Maximum depth is 6.6 feet (2 m), but most are 3.3 feet (1 m) deep or less. The less the depth, the less the variation in current velocity. There is always some velocity reduction due to "drag" exerted by the sides and bottom, and if this is excessive, it may cause settling of waste materials and deoxygenation. Larger, slow-flowing, "semi-raceways," up to 50 feet (15 m) wide and 1,200 feet (366 m) long have been used in catfish culture, but so far have not been proven to be practical. Rectangular raceways are usually drained by sluices (See Figure VII-1-8) or simple stand pipes.

A circular shape avoids dead spots, particularly if the drain is placed at the center of the raceway. A good design for this purpose is the Venturi drain (Figure VII-2-1). Bottom water enters at the base of the standpipe system and exits over the top of the inner standpipe, which you can adjust up or down to regulate water level in the raceway. Wastes are trapped between the pipes and may be flushed by periodically removing the inner pipe briefly. Conventional standpipes or monks (see

Figure VII–1–11) may also be used to drain circular raceways.

The maximum practical diameter for a circular raceway is probably about 66 feet (20 m). Depth is about the same as, or slightly deeper than, rectangular raceways. A large circular raceway with a Venturi drain would require an inner standpipe with a diameter of 3.9 inches (10 cm) and an outer standpipe of 11.8 inches (30 cm).

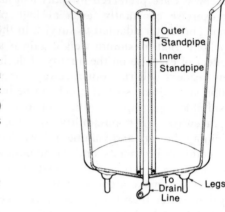

FIGURE VII–2–1. *Venturi drain in circular raceway tank.*

Raceways in parallel or in series

Raceways may be set up in parallel or in series (see Figure VII–1–9). Biologically, the parallel system is better, because it minimizes the transmission of waste materials, pollutants, escaped and wild fish, and (especially) diseases from one pool to another. At the same time, this system facilitates disease treatment or other operations requiring isolation or drainage of a single raceway. Further, temperature and dissolved oxygen of water entering each of several parallel raceways will be identical.

Yet there are some practical advantages of raceways in series. Since water passes through more than one raceway before being discarded, a lesser volume of inflowing water is required to support an equal volume of water used in culture. Also, except on perfectly flat ground, by constructing raceways in series you can save on construction costs by taking better advantage of gravity. In practice, many raceway culture facilities consist of small parallel groups of raceways in series, perhaps with a variety of species to take advantage of the different environments within a series. For instance, some trout hatcheries rear brook trout (*Salvelinus fontinalis*), with their very exacting water quality requirements, in the head raceways, and the less demanding brown trout (*Salmo trutta*) or rainbow trout (*Salmo gairdneri*) in lower raceways. Some commercial trout culturists can grow warm water fish in their tailwaters.

Water flow rates

Optimal flow rates for trout vary greatly with many factors. These include species and size of trout, stocking density, water quality, feeding rate, volume of water available, etc. For fish other than trout little is known about these factors. Commercial flow regulators are available and advisable, so that the beginning culturist can experiment in order to learn what flow rates are suitable. One flow parameter which has been measured is that a minimum velocity of 0.8 to 1.0 feet per second (0.24–0.30 m/sec) is necessary for self-cleaning.

In circular raceways, the rate of flow in highest nearest the center. If small fish are being sucked in, you can protect them by placing a tubular screen around the outer standpipe. Or you can reduce the flow rate by cutting out portions of the outer pipe and replacing them with screening.

Supplemental aeration

Supplemental aeration is necessary in raceways in series. While you can achieve this mechanically (see Chapter X–2), it is cheaper and more reliable to take advantage of gravity. The minimum vertical drop between raceways in series is 1.0 to 3.3 feet (0.3–1.0 m) depending also on the factors just cited for optimum flow rate. Regardless of how you achieve aeration, the maximum length of a single series of trout raceways is considered to range between 218 and 327 yards (200–300 m).

There are two types of very small aquaculture enclosures: excavated pools and tanks. It is very much simpler to construct a small sunken pool than a pond since it may be done by one person with a shovel. If possible, site the pool on or near a slope to permit drainage without pumping. Size and depth are up to you, though there is seldom any advantage in building a pool deeper than 5 feet (1.5 m).

Sunken pools

Round pools are more common than rectangular ones for a number of reasons. Not only does the round shape provide better circulation, it is easier to dig, easier to line and makes better use of space inside such structures as New Alchemy's geodesic dome greenhouse. Since very small systems can easily be drained, the convenience of harvesting rectangular enclosures is of little importance. Still, a rectangular shape is worth considering if you are working with such hard-to-catch fish as tilapia.

Due to the high ratio of vertical to horizontal surface, the chance of success with such "natural" sealing methods as compaction and gleization is less in very small pools than in ponds. But the cost of lining such pools is quite reasonable, and given the high density of population of fish which must be maintained for meaningful production, the sterility of an artificial liner may be advantageous. At New Alchemy we have successfully used a number of types of plastic liners, but the most durable substance we have worked with is Hypalon® rubber.

Sealing round pools

To install a plastic or rubber liner in a sunken, round pool, first run a measuring tape from one lip of the pool down the side, across the bottom on the diameter, and up the other side, then add eight feet (2.4 m). A square of liner material of this width will usually be adequate to leave a little for anchoring the edges.

So that you can anchor the edges of the liner, dig a trench eight to ten inches (20–25 cm) deep and twelve inches (30 cm) wide around the perimeter of the pond before putting it in place. Tuck the edges of the liner into this trench, bury them with earth, and trim the excess. Take care to locate the trench far enough from the edge of the pool that the bank does not crumble and fall down behind the liner. You can add soil or sand to the bottom, but do not place gravel or sharp objects on the liner, or they may scrape and seriously damage the liner if the pool is walked in.

Though tanks may be set in the ground, they are usually built above-ground. I have seen a unique sunken tank adapted from the base of a silo.

Tanks

You can make aquaculture tanks of almost any material which is impermeable to water, durable, non-toxic to fish, smooth on the inside (to prevent injury to fish and snagging of gear), and easily cleaned. The most practical materials for home-made tanks are concrete and wood.

Concrete tanks have one disadvantage: they are non-portable. Also, note that concrete is high in calcium carbonate. This may be advantageous in establishing populations of denitrifying bacteria, but it may also raise pH, particularly in new tanks. Whether or not this is beneficial will depend on the the original pH of the culture water.

Wooden tanks must be coated inside and out with a non-toxic epoxy. The effects of water pressure on the points where wooden tanks are joined can be quite complicated. To learn more about this I suggest you seek professional help.

Ready-made tanks   A great variety of ready-made fish tanks may be purchased, and some of them are quite specialized. I suggest that the small-scale culturist stick to one of the two basic types I will describe. (The reader interested in tank design is referred to Wheaton's "Aquacultural Engineering." See Appendix IV–1.)

The best commercially available tanks are circular and molded of opaque fiberglass. The optimum design would have a Venturi drain. But you can also use simpler systems involving a single standpipe, a removable plug, a faucet or even tanks without outlets, which must be pumped or siphoned down. A variation in fiberglass tanks is the transparent "solar-algae pond" described in Chapter IV–4.

Cheaper, but serviceable tanks can be adapted from small swimming pools with separate vinyl liners. Some such pools incorporate drain plugs, but they are not designed to facilitate hooking them up to any kind of a flow system.

You may have access to other kinds of tanks. It will be easy to determine whether a material meets most of the criteria I have listed, but it may not be so easy to find out which are toxic. In general, fiberglass and concrete are safe. Among the metals, aluminum and stainless steel are safe. Iron is non-toxic, but rust may interfere with biological processes. Galvanized materials and anything containing copper should be avoided. Plastics are something of an unknown quantity, with the exception of polyethylene, PVC, acrylics and most vinyls, which are safe so far as is known. Ordinarily, the manufacturer of a plastic container will know whether it contains toxic materials which might affect aquatic organisms. Wooden tanks should not be chanced unless they are well coated and the type of coating can be determined. Some epoxys and almost all wood preservatives are toxic.

Leaching   Even the safest materials should be leached several times before aquatic animals are introduced so as to safeguard against possible contamination. You can do this by repeatedly filling the tank, letting the water stand for a day or so, and draining.

PART VIII

Special Problems
in Aquaculture

CHAPTER VIII-1

Weeds

PART VIII

Special Problems

in Aquaculture

"Weeds" defined

A friend of mine, a fancier of wild foods, is fond of saying "A 'weed' is just a plant that's unloved." I will devote this chapter to explaining why you shouldn't love certain plants in certain situations, and how you might most effectively express your feelings.

It would not be fair to completely pass over the beneficial effects of plants in aquatic ecosystems. They provide oxygen to all aquatic animals and food and habitat for many. Often, aquatic plants are beautiful, and they may curb the effects of excess nutrients which would otherwise become pollutants. Some may even provide food or other useful products to humans.

In the very early days of pond fish culture in North America, farmers were encouraged to imitate nature by stocking their ponds with selected plants. This practice soon fell into disfavor, and today the orthodox view is that any and all plants (other than planktonic algae) are "weeds." Among the weed effects of aquatic plants all of the following concern aquaculturists.

Detrimental effects of aquatic plants

• Dense plant growth may result in oxygen depletion rather than in supplementation. While all green plants produce oxygen during the daylight hours, at night they respire, giving off carbon dioxide and using up oxygen. Depending on environmental conditions, this may more than offset their contribution in oxygen production.

Another unfortunate aspect of the relationship between aquatic plants and dissolved oxygen has to do with the tendency of entire masses of plants to die off, especially following pronounced weather changes. The sudden entry into the system of large amounts of decomposing organic material can have catastrophic effects on D.O.

Floating plants provide hardly any oxygen to the water. In fact, they may deplete it by forming dense mats which close the water surface to atmospheric exchange and shade out oxygenating plants.

• Though plants may be important as habitat for aquatic animals, too much cover can lead to over-survival of forage organisms leading to population imbalance. This can occur, for example, with young bluegills (*Lepomis macrochirus*) in farm ponds.

• Large aquatic plants (as opposed to phytoplankton) can short-circuit nutrients and so lessen fish production. If a given amount of nutrient material is incorporated by natural processes into planktonic algae, it has entered a food chain that is reasonably likely to lead to human food: the phytoplankton is eaten by zooplankton, which is eaten by small fish, and so on. But if the same nutrients go into something like water lilies or cattails, it may be many months before they are returned to the cycle. And they may re-enter in large amounts which are difficult for the system to assimilate, or at a time of year when fish do not grow rapidly.

• Plants with leaves above the surface lose water by transpiration. In extreme cases, this can actually result in drying up a small body of water.

• Dense growth of plants can trap silt particles and contribute to the gradual filling in of a body of water. Mud is also formed on the bottom through the decay of plants which die.

• Weeds offer tremendous mechanical resistance to harvesting. Hook-and-line fishermen are sometimes helped in locating fish by noting weed beds, but I know of no other way in which plants are an asset to any kind of fish harvesting.

• Aquatic plants are sometimes connected with undesirable flavors in fish. A causal relationship has been established only for certain types of blue-green algae, but my own experience strongly suggests that other types of plants are also implicated in producing "weedy" flavor.

• There may be toxic effects (lethal to fish or mammals in extreme cases) from blue-green algae. The release of tannins in the decay of emergent plants may also produce minor toxicity.

• Aquatic weeds can interfere with the esthetic quality or recreational uses of farm ponds.

• Floating plants can have a cooling effect, which may or may not be desired.

In intensive farming of non-herbivorous fish in North America, weeds are considered a curse, while in Europe controlled plant growth is welcomed. Perhaps, as North American fish culture becomes more sophisticated we will come to take a more balanced view of aquatic plants. The orthodoxy may swing back to judicious stocking and management of limited amounts of plants in farm ponds and other small-scale fish culture systems. Even now, for esthetic or ecological reasons, many food fish growers will opt to maintain some aquatic vegetation, albeit at the cost of lower fish production.

To clarify one term, I will begin by saying that most of the problematic plants we shall be considering are *not* algae. Algae is a term used to distinguish a great number of aquatic plants which consist of a single type of cell, or even just a single cell, and which do not have the specialized structures (roots, leaves, flowers, etc.) which we ordinarily associate with plants. Most algae, as individual plants, are very small or even microscopic, though they may aggregate in impressive quantities. Most aquatic weeds are called "macrophytes" or "higher plants" and have identifiable parts similar to those of terrestrial plants.

Types of aquatic plants

Aquatic plants may be classed as four types on the basis of their growth habits:

• Planktonic plants (phytoplankton) are unattached and found drifting passively at all levels of the water column where light penetrates. There are no planktonic macrophytes, and the usually microscopic phytoplankton probably pose the fewest "weed" problems of the four types. Rather than eradicated, they are more often cultivated as part of the food web. Where problems do occur with phytoplankton they are most often esthetic or recreational ones. At high densities, however, phytoplankton can interfere with the normal behavior of visual feeding fish and may create the danger of sudden die-off.

• Submerged plants grow mainly or entirely beneath the surface and are rooted. ("Attached" is a better term than "rooted" for submerged algae.)

• Emergent plants are rooted underwater (or on the shore near the water's edge) but a major portion of the plant is above water. Emergent plants as a rule are hard and tough, as opposed to other types of aquatic plants which are soft-bodied.

• Floating plants have their roots trailing in the water, rather than penetrating the substrate.

Some plants seem to straddle the categories. For example, water lilies display some of the characteristics of submerged, emergent and floating plants. Others adopt different life habits according to circumstance. For example, *Ceratopteris* spp. will float in deep water, but may take root in shallow water. Table VIII–1–1 lists some of the more common and widely distributed plants which cause problems in aquaculture ponds with brief comments. These plants are illustrated in Figures VIII–1–1 through VIII–1–15.

Table VIII–1–1. Common weeds of aquaculture ponds.

Common name	Scientific name	Growth form	Comments
Alligatorweed or Alligator grass	*Alternanthera philoxeroides*	Emergent, submerged, floating	Confined to the South, sometimes grown as feed for crayfish, biological control being developed.
Bladderwort	*Utricularia* spp.	Floating	
Cattail	*Typha* spp.	Emergent	Commonest emergent weed, resistant to chemicals, controlled by muskrats, some parts edible by humans.
Duckweed	*Lemna* spp., *Spirodela* spp.	Floating	Important food for ducks, good nutrient remover, prodigious reproductive rate, may survive on moist soil, easily introduced accidentally.
Elodea, Anacharis, or ditch moss	*Elodea* or *Anacharis* spp.	Submerged	Fragments can take root, confined to North, prefers hard water, popular aquarium plant, very rapid grower.
Eurasian water milfoil	*Myriophyllum spicatum*	Submerged	Introduced from Europe, a major pest, much more noxious than native *Myriophyllum* spp. or the similar coontail (*Ceratophyllum* spp.). Avoid introduction of this plant.
Filamentous algae, pond scum or frog spit	*Spirogyra, Cladophora, Zygnema,* etc.	Submerged	Usually indicative of overfertility, eaten by tilapia.
Hydrilla or Florida elodea	*Hydrilla verticillata*	Submerged	Confined to the South, where it is a major problem, spreads from fragments.

Table VIII–1–1. *Continued.*

Common name	Scientific name	Growth form	Comments
Musk grass	*Chara* spp.	Submerged	Actually an alga, despite appearance, resistant to chemicals, prefers hard water, not always a pest. Short forms of *Chara* may inhibit other weeds.
Pickerel weed	*Pontederia cordata*	Emergent	Showy blue flowers, rapid grower, can grow on moist soil, common breeding site for mosquitoes.
Pondweed	*Potamogeton* spp.	Submerged, emergent	Many species with widely differing leaf shape, some species good food for ducks, some species resistant to chemicals. Species illustrated (*P. crispus*) is an early summer plant which may retard other plants.
Smartweed or knotweed	*Polygonum* spp.	Emergent, submerged	Pinkish flowers, can grow in moist soil, one of the worst plants for interfering with boat traffic due to its tough horizontal stems. Edible for humans.
Spatterdock, yellow cow lily or bullhead lily	*Nuphar* spp.	Submerged, with floating leaves	Prefers muck or silt bottom, this and the showier water lilies (*Nymphaea* spp.) are prized as ornamentals, good habitat for some kinds of fish. Usually should be controlled, not eradicated.
Water buttercup	*Ranunculus*	Submerged, with emergent flower	Yellow flower with star-shaped float is very distinctive.
Water hyacinth	*Eichornia crassipes*	Floating	Beautiful showy flowers, good nutrient remover which can be used in filters, native to Venezuela, overwinters only in the South, perhaps the world's most economically significantly plant pest. Causes deoxygenation, excessive transpiration, interference with fishing and navigation, and flooding. Biological control being developed.

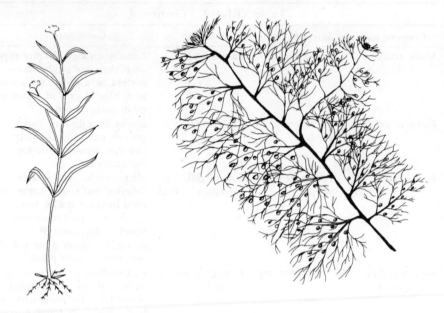

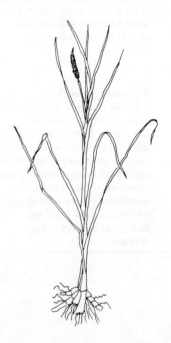

FIGURE VIII-1-1. *Alligatorweed or alligator grass*, Alternanthera philoxeroides.

FIGURE VIII-1-2. *Bladderwort*, Utricularia *sp*.

Top view

Side view

FIGURE VIII-1-3. *Common cattail*, Typha latifolia.

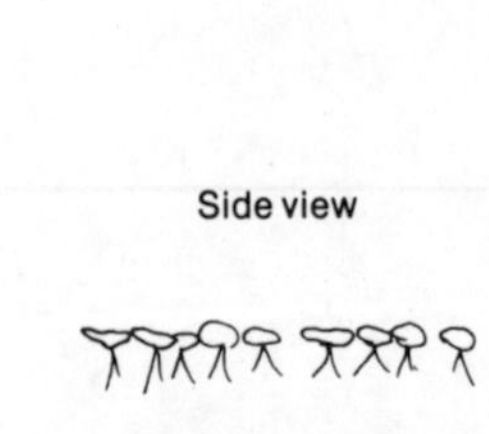

FIGURE VIII-1-4. *Common duckweed*, Lemna minor.

FIGURE VIII–1–5. *Elodea or anacharis,* Elodea (Anacharis) canadensis.

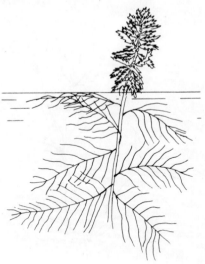

FIGURE VIII–1–6. *Eurasian water milfoil,* Myriophyllum spicatum.

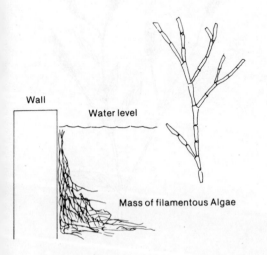

FIGURE VIII–1–7. *Filamentous algae,* Spirogyra *sp.*

FIGURE VIII–1–8. *Hydrilla or Florida elodea,* Hydrilla verticillata.

FIGURE VIII–1–9. *Musk grass,* Chara *sp.*

FIGURE VIII–1–10. *Pickerel weed,* Pontederia cordata.

FIGURE VIII–1–11. *Curlyleaf pondweed,* Potamogeton crispus.

FIGURE VIII–1–12. *Water smartweed,* Polygonum *sp.*

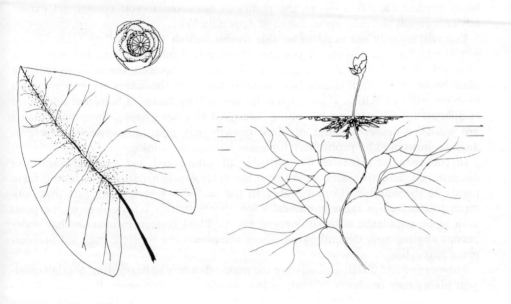

FIGURE VIII–1–13. *Spatterdock or bullhead lily,* Nuphar *sp.*

FIGURE VIII–1–14. *Water buttercup,* Ranunculus *sp.*

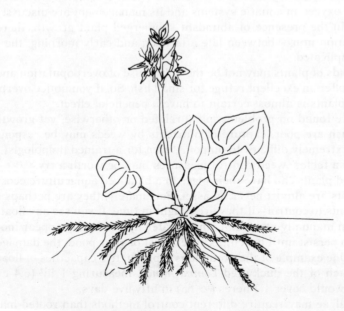

FIGURE VIII–1–15. *Water hyacinth,* Eichornia crassipes.

In choosing your weed control strategy you may, in some cases, have to make a fairly precise identification of the plants in question. If you cannot get expert advice, consult the references listed in Appendix IV–1.

You will usually not need to be able to distinguish *Myriophyllum* from *Ceratophyllum* to determine whether you have an excess of aquatic plants. In some cases, as where recreational value or esthetic criteria are involved, the point at which a plant becomes a "weed" is largely a matter of taste. In other cases, the existence of a problem will be obvious. If you plan to harvest fish by means of hook and line, but it is difficult to cast without becoming entangled in plant growth, then those plants are weeds. A lesser amount of plant growth may offer no serious obstacle to angling, but could be prohibitive for mass harvest by seining.

In conventional intensive fish culture, all submerged and floating plants are discouraged, and the same holds true for extensive fish culture, including farm ponds, in the southern United States. In the northern states, it appears that optimum production in farm ponds calls for approximately ten percent of the pond area to be populated with submerged plants. Plant concentrations on that order permit angling and, depending on where the plants are located, may not prohibit mass harvesting.

Submerged and floating plants are the main offenders in harvesting, though emergent plants may render it difficult or impossible to beach a seine. Emergent plants most frequently contribute to siltation problems and share the blame with floating plants for water loss through transpiration. Both effects are most severe in small bodies of water, which can least afford to lose depth. A thin band of emergent plants may serve as bank stabilizers, but should be closely watched. If from year to year the band grows wider, it is time to adopt control measures.

A cover of floating plants may directly deoxygenate water, but the role of submerged plants in deoxygenation is usually as one of a complex of factors. The role of dissolved oxygen in aquatic systems and its management are discussed in Chapter X–2. If in the presence of abundant submerged plant growth, there is a dramatic reduction in D.O. between late afternoon and early morning, the plants are probably implicated.

Dense stands of plants may not be the only cause of overpopulation and stunting, but they do offer an excellent refuge for small fish. So, if you notice overpopulation, thinning of plants is almost certain to have a beneficial effect.

If you have found no problems, plant-related or otherwise, yet growth and production of fish are poor, capture of nutrients by weeds may be responsible. This problem is extremely difficult to pinpoint, even for a trained limnologist, but if you suspect it is a factor, weed control measures may be worth a try.

Submerged plants can have their place in a balanced aquaculture ecosystem, but floating plants are almost never an asset. Fortunately, they are perhaps the easiest group of plants to control—if they are caught in time. If you see any floating plants, remove them manually at once. A single water hyacinth may appear inocuous, but if allowed to persist until hyacinths cover most of your pond, the damage will have been done. One example of the prodigious reproductive capacities of floating plants is that a patch of the duckweed *Lemna minor*, measuring 1 in² (6.4 cm²), if left unchecked, would cover 1.2 acres (½ ha) in fifty-five days.

Attached algae may require different control methods than rooted macrophytes, but the criteria for assessing their desirability are the same. A word about the Cyanophyta, or blue-green algae (which are usually, but not necessarily, blue-green in color): concentrations of blue-greens dense enough to be considered weeds usually indicate a state of eutrophication not permitted by even the untidiest fish farmer. It is possible, however, for a dense growth of benthic blue-greens to develop in an otherwise healthy pond if there is poor circulation in the bottom layer of water. Under such conditions fish may acquire a musty flavor. The remedy is to continually turn over the bottom sediments, either through aeration, or, still better, by stocking suitable benthic animals.

Before discussing any additional methods of controlling aquatic plants, a word on behalf of the hoary adage about an ounce of prevention. The most important environmental factors in prevention of aquatic weed problems are depth, watershed stability and fertility.

Preventing plant problems

Submerged and emergent plants cannot obtain enough light for photosynthesis below a certain depth. (The precise depth varies with the clarity of the water; in very clear water it is about ten feet or three meters.) This is one of the main reasons for building fish ponds with regular shorelines, and banks as nearly vertical as possible—to keep at a minimum any water shallow enough for weeds to get started.

Often, over the years, depth of water is reduced by siltation from the surrounding watershed and weed problems usually follow. Prevent this as you would prevent erosion. Proper use and planting of all areas draining into a body of water (particularly stream banks, steep hillsides and the water's edge) is essential to maintaining a stable depth structure. A swale may be constructed around the pond. (See Figure VII–1–4.) When planting pond edges, take care to select plants which will not spread out into the water, collecting soil around their bases, and eventually having an effect opposite to the one for which you planted them.

Livestock is often overlooked as an agent of bank erosion. If you plan to let livestock use your pond, restrict it to one small area where the bank's slope is relatively gentle. (See Figure VII–1–17.) It will be helpful to pave or gravel the bottom at this point. (Detailed advice on erosion and siltation problems is usually available through local offices of the United States Soil Conservation Service.)

I will discuss how you may use fertilization of a body of water as a control method for weeds, but it may also stimulate weed growth. The key word in all alterations of natural fertility is "control." Uncontrolled fertilization should always be avoided. (See Chapter VIII–4 for sources of, and control measures for unwanted fertilization.)

Lastly, do not introduce aquatic plants or allow them to be introduced. The great majority of aquatic plants which become weeds are not native to the waters where they create problems. While some are brought in as ornamentals, a possible source of unintentional introductions is with fish. When stocking fish take care that no bits of vegetation come with them, because many of the most noxious weeds can regenerate from a tiny piece.

If it is already too late for prevention there are two critical times for undertaking weed control measures. The first is the early spring. If you can anticipate weed problems and set weeds back before they are fairly started, both weed damage and control costs will be minimized. The second critical period is late summer. With cool weather beginning, or sometimes before, massive plant die-offs may occur, with predictable effects to aquatic animals. Even with no die-off, the nocturnal respiration rate of plants reaches a peak when the plants are most abundant. It is advisable, then, to remove the biomass of plants in mid-to-late summer.

Timing of weed control

Methods of controlling aquatic weeds can be broken down into three categories: chemical, physical and biological. Since most development of North American aquaculture has coincided with the chemicalization of agriculture, it is not surprising that chemical control has been "pushed." A number of synthetic herbicides and algaecides have been developed or adapted for aquatic use. They range from wide-spectrum types to compounds so specific that they will kill one species of plant while not affecting another in the same genus.

In the last few years, the trend toward increased use of synthetic chemicals in aquaculture has met with setbacks. As the industry has come of age, fish farmers are finding it difficult to get government clearance to use chemicals in their ponds. Chemical manufacturers complain that it is more difficult to get a compound registered for aquaculture use than for human medicinal use. Though I am not an admirer of federal red tape, I believe that this is as it should be, for we simply know too little of the behavior of herbicides, especially chlorinated hydrocarbons, in aquatic ecosystems.

Aquatic herbicides' harmful effects

The most prevalent harmful effect of aquatic herbicides we do know about, is not directly related to their chemical character: it is the depletion of dissolved oxygen which may follow herbicide treatment, as large quantities of dead weeds decay in the water. Attempting to remove this material would amount to adding the costs and problems of physical weed control to those of chemical treatment.

An equally serious problem which has occurred is the poisoning of fish and other aquatic animals. Damage to terrestrial plants via windborne spray or irrigation water has also been reported. Weed infestation may even be increased by decaying plants fertilizing the bottom soil. And herbicides may pave the way for herbicide-resistant species to supplant the original problem with a worse one.

Little as we are able to predict the short-term effects of herbicides, we have even less knowledge of their long-term fate and consequences. We have learned the hard way how synthetic hydrocarbons, which are not readily broken down in nature, may build up in terrestrial food chains, with ultimate toxic effects. Entry of such chemicals into aquatic ecosystems is easier, since animals may absorb them by respiration, as well as by feeding. There are also known to be microbial and photolytic processes which increase the toxicity of some hydrocarbons in water.

Chlorinated hydrocarbons

In 1970, the chemist George Harvey, one of the leading authorities on chlorinated hydrocarbons in aquatic systems, characterized the use of these compounds in aquatic food production as "insane." We know a little more about them now than we did then, but there is nothing which would incline me to soften Dr. Harvey's choice of words. I urge all non-commercial fish culturists to voluntarily abide by the restrictions imposed on commercial fish farmers and refrain from using any chemicals which do not have a long history of safe use in aquatic systems.

Copper sulfate and other non-hydrocarbon chemicals

One chemical which does have a long history of use in controlling aquatic vegetation is copper sulfate, used at least since the 1920's to combat algae and fungi. The problems with copper sulfate stem from its tendency, in alkaline waters, to precipitate as copper carbonate. This not only makes it ineffective below a depth of two feet (0.6 m), but results in a build-up of toxic copper on the bottom. Though algae in the upper portion of the water column may be killed, they are soon replaced, requiring further treatment—perhaps as often as weekly. When copper sulfate is effective against algae, the decaying algae may cause temporary dissolved oxygen depletion. The standard remedy for this is treatment with potassium permanganate. (See Chapter X–2.) But this does not always work and, if the copper sulfate has not killed all the algae, the permanganate may also act as an algaecide, leading to a further reduction of dissolved oxygen.

Apart from the expense and the possibility of a fish kill, use of copper sulfate will almost certainly result in toxicity to benthic organisms. All the scientific studies which used copper sulfate in ponds reduced fish production, and probably for this reason. (The same applies to those few synthetic chemical algaecides which have been federally approved.)

A chelated copper sulfate algaecide is marketed by Applied Biochemists, Inc. of Mequon, Wisconsin, under the name Cutrine-Plus. Cutrine-Plus does not form a precipitate with carbonate ions. As a result, the activity of the copper is spread out over a longer period of time, and the algaecide is effective at depths up to ten feet (3 m). Since no precipitate is formed, toxicity of Cutrine-Plus to fish and benthic organisms is much less likely than with ordinary copper sulfate. And because of the smaller amounts of Cutrine-Plus needed, there is still less toxicity. (See Table VIII–1–2.)

Cutrine-Plus has even been approved for use in drinking water supplies. Nevertheless, copper is toxic to animals and plants other than target algae species, and unwanted lethal effects can occur. You should be able to avoid fish kills with careful attention to instructions and dosages, but the long-range effects of chelated copper compounds on pond productivity remain unknown.

Several other non-hydrocarbon chemicals are suggested by Marcel Huet (University of Louvain, Belgium) and are included in Table VIII–1–2. It is not at all certain that these treatments, most of which are based on experience in Europe, would be suitable for North American waters. (Since I am unable to recommend synthetic chlorinated hydrocarbon herbicides under any circumstances, I have not included them in Table VIII–1–2.) Chemical control of aquatic plants, even with simple compounds such as copper sulfate, is complicated by the great chemical differences between individual bodies of water. A chemical may behave in a completely different manner in each of two bodies of water. Often the only expert consultation you can obtain comes from a chemical manufacturer or dealer, whom you cannot expect to be objective. In all, beware of "cookbook" approaches to chemical weed control, be sure to explore alternative solutions, and proceed only with extreme caution.

Table VIII–1–2.   Some non-hydrocarbon chemical herbicides with recommended dosages for aquatic use.

Compound	Target plants	Treatment	Comments
Copper sulfate	Algae (all types)	0.5–1.0 g/m³ (acid water); 1.5 g/m³ (alkaline water), as often as necessary	To avoid toxicity to fish treat only ⅓ to ½ of a pond at a time. Apply in solution or by towing a sack of crystals from a boat. (See discussion in text.)
Cutrine-Plus (copper sulfate chelated with triethanolamine)	Algae (all types)	0.6 gallons per acre-foot (liquid form) or 60 lb/acre (66 kg/ha) (granular form), 2 or 3 times a year	Dilute liquid form at least 9 to to 1. (See discussion in text.)

Table VIII-1-2.   *Continued.*

Compound	Target plants	Treatment	Comments
Copper oxychloride	Filamentous algae (except *Chara*)	0.9 lb/acre (1 kg/ha)	
Malachite green	" "	0.3 g/yd³	
Superphosphate (powdered)	" "	535 lb/acre (600 kg/ha)	
Quicklime (powdered)	Floating algal masses	as necessary to temporarily raise the pH to 10.0–10.2	Caution required; danger of toxicity to fish.
Sodium chlorate	Macrophytes	267 lb/acre (300 kg/ha)	Applied to dry pond bottom as a preventive.
Calcium cyanamide	"	668 lb/acre (750 kg/ha)	Applied to ponds filled with water but not stocked. Ponds so treated must be drained and refilled before stocking.
Sodium arsenite	"	_____	To control weeds in stocked ponds.
Ammonium nitrogen	*Elodea*	200 lb/acre (225 kg/ha)	Applied to dry pond bottom as a preventive.
Ammonium sulfate	*Fontinalis*	200 lb/acre (225 kg/ha)	Applied to dry pond bottom as a preventive.

Removing weeds physically

The physical removal of weeds is a much simpler approach, and successfully eliminates the problem of oxygen depletion by decaying weeds. The oldest form of physical control is, of course, hand removal. The limitations are obvious, but a little digging, cutting or pulling at the very beginning of a problem may be all the control needed for some plants, particularly emergents.

As an indication of how deeply indoctrinated we are with chemical pest control propaganda, I was once asked by a commercial catfish farmer which chemical I would recommend to kill cattails. Looking at his pond, I saw small patches of cattails in two of the corners, and estimated that three hours of hard labor would suffice to eradicate them!

If you have weed problems covering more than a few square feet, and especially for submerged plants, removal by hand or with conventional hand tools may not be adequate. It does not follow, however, that you should then buy or rent a machine costing $40,000 or more and weighing several tons, like those sometimes used on recreational lakes. In Europe, though not in North America, you could choose from an abundance of intermediate weed control technology, ranging from specially designed scythes through cutters used with a small boat and outboard motor, to amphibious weedcutters built like tractors. These tools and machines are described and illustrated in Huet's *Treatise on Fish Culture*, and as I have no experience with them, I will recommend that text to the reader.

No matter how you harvest aquatic weeds, you not only eliminate the immediate problem, but, where the underlying problem is excess nutrients entering the water, you may reduce the rate of eutrophication by removing large amounts of nutrients. As a result, reinfestation by weeds may be less severe.

The idea of weed harvesting would be more attractive if there were economic uses for the plants. Two uses which have been proposed are as compost materials and livestock feed. Development of such uses has been hindered by the low nutritive value of aquatic plants as compared to terrestrial ones and by their extremely high water content, which multiplies handling and processing costs. Among the few exceptions are the duckweeds. The relative ease of harvesting these tiny plants, their high nitrogen and protein content, favorable amino acid balance, and their ability to remove large quantities of nutrients from water, all suggest that they might be a valuable and feasible livestock feed or fertilizer for the small farmer.

Uses of aquatic weeds

Even if you are not using herbicides, the act of draining a pond, or at least drawing it down to a level which exposes all the normally shallow areas may contribute to discouraging weed growth. Plowing and/or planting the dry bottom to rye or other cover crops may be helpful. In temperate regions, drawdown or draining can be done at any time. In Florida, two annual drawdowns—one during February or March and the other during August or September—have been found best for discouraging weed growth in recreational lakes. In the North a drawdown every five to ten years may be enough. Drawdown should be carried out only with knowledge of the likely effect on the fishes, particularly with respect to spawning.

Drawdown

Another simple traditional method of weed control is through shading. Planting shade trees can be effective where shallow water is confined to a zone near shore. However, shade trees may take several years to grow to a useful size and can interfere with use of a body of water. Further, shading also has a cooling effect which may or may not be desirable.

Shading

A different approach to shading is the use of dyes to suppress plant growth. Dyes are of course ineffective against plants at or near the surface, and their long-range chemical effects are not known. (Many dyes are toxic. In fact, they are mainly used in aquaculture as fungicides.) However, the biggest problem associated with shading by dyes is their impermanence.

The principle of shading is also applied in the most common form of "biological" weed control in North America. The idea is to fertilize in such a way that a phytoplankton bloom is produced, suppressing photosynthesis by macrophytes. This method is not suited to ponds used for swimming or where the concern over weeds is in part an esthetic one, since the end result is replacement of macrophytes with an opaque cloud of phytoplankton. Sometimes, such a phytoplankton bloom may overdevelop, die-off and stimulate further weed growth by releasing nutrients.

Biological control

Fertilization for weed control

The standard treatment for controlling weeds by fertilization is 100 pounds per acre (110 kg/ha) of synthetic 10–5–5 fertilizer. Another technique employs hay bales as a natural fertilizer. Ten bales per acre (25 bales/ha) weighted down in shallow water, should be sufficient. Application of baled hay or other fertilizers is more likely to be effective in the southern U.S.A. than in the northern states or Canada, in soft water than in hard, and more effective against macrophytes than attached forms of algae.

Herbivorous fish   Economically, the best form of biological weed control is to use an edible herbivore to harvest the weeds and turn a nuisance into an additional crop. As it happens, the North American continent has evolved few large herbivorous fishes, so most of the fish considered in this context are exotics. Easily the most widely discussed is the grass carp (*Ctenopharyngodon idellus*), but it is illegal in more places than not. (For a detailed discussion of grass carp as a weed control agent and cultured food fish, see Chapter II–8.)

Next to the grass carp, the fishes most widely used to control aquatic weeds are the tilapias, particularly two relatively small species, *Tilapia zilli* and *Tilapia rendalli* (sometimes called Congo tilapia and, mistakenly, *Tilapia melanopleura*). These fish have been effective in controlling both macrophytes and filamentous algae when stocked at 1,012 to 2,025 individuals per acre (2,500–5,000/ha). Their small size and their tendency to overpopulate, which is high even for a tilapia, limits their value as food fish. The larger and more commonly cultured blue tilapia (*Tilapia aurea*), Nile tilapia (*T. nilotica*) and Java tilapia (*T. mossambica*) may control filamentous algae effectively — but not macrophytes.

Though other exotic fishes may be found to effectively control weeds in North America, I will limit this discussion to species native, established, or generally available commercially here. Among the native or established fishes treated in this book, the common carp (*Cyprinus carpio*) is perhaps the only one effective in control of macrophytes. Huet has suggested using 162 carp, weighing at least nine ounces (250 g), per acre (400/ha). A problem with the common carp is that it generally uproots plants rather than consuming them. In the process, it may create so much turbidity as to offset its positive value. You may obtain better results with the mirror or Israeli strain of common carp, which will eat considerable quantities of filamentous algae.

Other fishes which have been observed to eat filamentous algae and might find use in controlling such plants include channel catfish (*Ictalurus punctatus*), goldfish (*Carassius auratus*), golden shiners (*Notemigonus crysoleucas*), chubsuckers (*Erimyzon* spp.) and the little known flagfish (*Jordanella floridae*), which rarely exceeds two inches (5.1 cm) and is largely confined to the state of Florida.

Herbivorous crayfish   Another group of aquatic food animals which will eat aquatic plants are the crayfish. Over a period of years the introduced crayfish *Orconectes causeyi* caused marked reduction in infestation of New Mexico lakes by pondweed, and *Orconectes nais* has reportedly been used to control weeds in Kansas. However, crayfish are likely to function as a crop which makes use of aquatic weeds, rather than as an effective control. An important obstacle to their use in fish culture situations is the fondness of most predatory fishes for crayfish as food. Also, they are useless against hard-stemmed emergent plants, and are usually unable to reach floating weeds.

Waterfowl   Among the most promising biological controllers of weeds in small ponds are waterfowl. Ducks, as you might guess, are particularly effective against duckweeds. Muscovy ducks stocked at two to three per acre (5–8/ha) will control this pest.

Geese are likely to be more effective in controlling other weeds. In a trial in Hawaii, sixty-five Chinese white goslings were placed in a 2.47 acre (1 ha) pond which was completely choked by para grass (*Brachiara mutica*) and cattails. Though mechanical and chemical methods had failed in this pond, the geese eliminated the weeds in two and a half years. Geese (and swans) are among the few animals which not only eat aquatic plants, but uproot them to get at the roots. They are most effective in shallow water, and usually prefer narrow-leaved grass-like weeds to the broad-leaved types.

Before introducing waterfowl you should consider other possible effects in addition to weed control and food production. These may include fertilization of the water, competition for fish food, predation on fish or damage to banks. (See Chapter IV–2 for a discussion of waterfowl in aquaculture.) It will also help to learn as much as possible about the nutritive value of the available weeds and other forage. In some cases you may have to supplement the diet of the fowl with grain in order to prevent malnutrition. But if you provide too much grain, they may refuse to eat weeds.

Muskrats are a natural control for some aquatic plants, including cattails.

Biological control by insects

Where a specific weed is the cause of a problem, it may be best to use the strategy usually taken in terrestrial biological control and introduce an invertebrate or pathogen specific to the undesired plant. The control of water weeds by insects and other invertebrates has been researched for some time, while the use of pathogens is only beginning to be studied. At present neither method is commercially available. However, two specific weed controls involving introduced invertebrates are far enough along that they may be available in the not-too-distant future. These and other experimental biological controls for water weeds are discussed in a special issue of the journal "Economic Botany." (See Appendix IV–1.)

Alligator grass, although used as a fodder crop by crayfish farmers in Louisiana, is a major pest in many waters of the southern United States. It is a native of South America, as are its control agents, the alligatorweed beetle (*Agasicles hygrophila*) and the alligatorweed stem borer (*Vogtia malloi*). In field trials in South Carolina, release of these insects resulted in dramatic reduction of alligatorweed.

Water hyacinth has become practically a world-wide pest in warm climates, though it is relatively inocuous in its native Venezuela. This is probably due at least in part to the presence there of two species of weevil, *Neochetina eichorniae* and *Neochetina bruchi*, and a leaf-boring mite, *Orthogalumna terebrantis*, which feed on water hyacinth. Field trials with these animals in Florida have been very promising. (As in any introduction of exotic organisms, there is the risk that the introduced organism will have harmful, as well as beneficial, effects in the new environment. Academic and government agencies, in their enthusiasm for new ideas sometimes gloss over this aspect, and the individual farmer should look into this carefully before introducing exotics to his or her own waters.)

Pond owners with alligatorweed or water hyacinth problems could contact their nearest agricultural extension agent, or conservation officer. In some cases, it may already be possible to make use of the biological controls just described.

Competitor plants

An underdeveloped method of biological control involves the use of competitor plants. For example, establishment of fresh water eel grass (*Vallisneria* spp.), which is unlikely to get out of hand, may prevent invasion of virtually unmanageable submerged weeds like hydrilla and Eurasian water milfoil. Slender spikerush (*Eleocharis tenuis*) is a good and controllable shore stabilizer which will resist encroachment by other emergent plants.

Whatever weed problems you may experience, you should maintain an ecological perspective. In the natural course of things every pond eventually evolves through the stages of eutrophication from deep pond to shallow pond to swamp to bog to dry land. To maintain an aquaculture pond will require, in part, the stemming of this process. But the problem is complicated by the fact that many human activities, including aquaculture, tend to accelerate eutrophication so that what would normally take thousands of years occurs within one person's lifetime. No matter what steps you take to eradicate symptoms of eutrophication (such as weeds) in a given season, the problem will reappear the next season as long as you emphasize control instead of prevention. Even where it is not clear whether the cause of a weed infestation is siltation, over-fertility, poor pond design or some other factor, it will pay to research the situation. Your only alternative is to repeatedly face the costs of weed control.

CHAPTER VIII-2

Diseases and Parasites: Prevention and Control

Introduction

Virtually all wild fishes host a certain number of parasites, but only rarely are they present in such numbers as to harm the host. Disease organisms occur everywhere in nature, but it is unusual to encounter a fish with disease symptoms. When a diseased fish is found, it is usually an isolated individual which has been injured or otherwise stressed; natural epizootics are rare.

In aquaculture, however, diseases and parasites are sometimes a major problem, and when they occur, a substantial portion of the fish population is likely to be affected. This is so because aquaculture begins by increasing what would normally be the natural population density. The more closely any animals are crowded together, the greater the opportunity for contagion, and water, as compared to earth and air, is a superior medium for disease transmission.

Symptoms of stress

Most people sense intuitively if another person, or a warm-blooded animal, is feeling well or not, but if you are unused to dealing with live fish you may overlook symptoms of stress in fish. If any of the following signs are seen, fish are not feeling well:

- Dead fish. (An occasional dead fish is no cause for concern, but if numbers are seen, either collectively or consecutively, something is wrong.)
- Loss of interest in feed.
- A "hollow-bellied" profile.
- Sluggish swimming, or periods of resting on the bottom which are longer than normal for the species.
- Resting on the side, upside down, head up, or head down.
- Scratching by repeatedly rubbing the body against a solid object.
- Frantic or erratic swimming.
- Loss of equilibrium.
- Loss of ability to adjust buoyancy, shown by uncontrolled floating or sinking.
- Fins constantly folded.

- Blood in the fins.
- Gasping for air at the surface.
- Wounds on the body.
- Any growth or swelling on the body, including visible parasitic organisms.
- Splotches, spots (other than natural pigmentation) or discoloration.

Diagnosis

The presence of one or more of these symptoms is likely to mean one of two things: either there is a disease or parasite problem, or there is an environmental problem which may lead to disease. Though reams have been written on the biology and control of fish diseases and parasites, much remains to be learned. Some of the major fish disease works are referenced in Appendix IV–3. It would be of little help to you if I were to go into detail on disease organisms and their treatment here. Not only would I be duplicating others' work, but to make much use of the fish pathology publications often requires a microscope and some familiarity with laboratory techniques.

Some of the common diseases and parasites of fish, once seen, are readily identifiable by anyone but many are not. Should you encounter a disease you cannot diagnose, seek professional help. In the United States, try Agricultural Extension Services or the Division of Fish Hatcheries of the U.S. Fish and Wildlife Service. In Canada, begin with the Fisheries Research Board of Canada in Nanaimo, British Columbia, or Halifax, Nova Scotia. Some large-scale fish farms have their own laboratories and pathologists, but beware the "expert" who diagnoses from the pond bank, without handling the fish.

Therapy

Many diseases and parasites respond to fairly simple and inexpensive chemical therapy. If you are growing aquatic animals only for your own use it may be easy to take advantage of this form of treatment. But if you are raising fish for sale in the United States, you may meet another obstacle. As of 1982, the only chemicals definitely cleared by the U.S. Food and Drug Admininstration (FDA) and the U.S. Environmental Protection Agency (EPA) for use in controlling diseases of food fishes were sodium chloride, terramycin or oxytetracycline (an antibiotic), sulfamerazine (an antibacterial), formalin and the various forms of lime. This leaves the vast majority of successful chemical disease treatments legally unavailable to U.S. food fish growers. In Canada, the situation is quite different; use of drugs on aquatic crops is almost unregulated.

Even without the problem of regulation, it is better on all counts to prevent disease and parasite outbreaks than waiting until it is necessary to attempt to control them. The remainder of this chapter deals with this aspect of the problem.

Preventive hygiene

Pond design

You can begin preventive hygiene in the design stages. Given the nature of the aquatic environment, disease usually spreads throughout any infected body of water, so that it is safer to have two or more small ponds than one large one. Wherever possible, ponds, raceways or tanks should have separate water sources, or at least be set up in parallel rather than in series, minimizing opportunities for the spread of disease. Before construction begins, location of quarantine and disease treatment facilities should be carefully planned.

Other things being equal, disease-resistant animals are the best choice. Of the animals discussed in this text, the tilapias are, in my experience, almost disease-free. Trout, the larger catfishes, buffalofish and frogs are relatively delicate, with the other animals occupying intermediate positions. Since few aquaculture animals have been selectively bred, claims for disease-resistant strains are doubtful.

Probably more important than choice of species or strain of stock is the source of supply. You can assume that all wild fish harbor some pathogens, which may erupt when the animals are stressed by relocation. The best you can do when harvesting wild stock is to select an apparently healthy environment, and minimize stress in handling.

Commercially available stock varies in health from superb to abysmal. If at all possible, visit prospective suppliers or obtain an endorsement from previous customers. A few suppliers offer certified disease-free stock, which is to be preferred when available.

Proper care during shipping is extremely important. Aspects of shipping over which you can exercise control are covered in Chapter VI–3. In evaluating the quality of shipping by commercial suppliers, consider the following points:

Shipping and stocking

• Does the supplier take time to help you work out the shortest and fastest possible route?

• Are animals packed in insulated containers?

• Are bags filled with pure oxygen, or just with air?

• Does the supplier avoid crowding, even though it may increase costs for both parties?

• Will the supplier treat the shipping water with methylene blue? This dye treatment, while not cleared for use on fish to be sold as food, is a good preventive for fungus. It is safe if fish are not to be eaten immediately.

• Does the shipper guarantee live delivery? This indicates confidence in the health and handling of stock.

All other things being equal, it is best to minimize stress by selecting the supplier closest to you. You may then be able to pick up the fish yourself, rather than risking the stress of long-distance shipping and possible careless handling.

In the long run, it is very important to the entire aquaculture industry to cut down on long-distance shipping. While it is good to take advantage of selectively bred strains or certain exotic species, and while a certain amount of exchange is helpful in avoiding inbreeding, a number of diseases have been introduced to new environments with shipments of aquatic animals. Further, indiscriminate shipping of "superior" stock tends to reduce genetic diversity, with the possible loss of localized disease-resistant strains.

The time of shipping is also important, since aquatic animals are temperature-sensitive. Suppliers in the South often refuse to ship during the summer, and with good reason.

The concept of "stress" has already been mentioned several times, and will recur often in this chapter. Any physiological stress lowers aquatic animals' resistance to everpresent disease organisms. One of the times when forethought can do most to reduce stress is in the period between receiving stock and placing it in the culture enclosure. The time spent by animals in the confines of a small shipping container is always stressful, so proceed as rapidly from receipt to release as you can while being consistent with the procedures outlined in Chapter VI–3. Do not leave animals in the hot sun or outside in the winter cold. Avoid rough handling, and do not handle animals unnecessarily.

A precaution which is usually advisable, but not always feasible, is to quarantine new stock for a week or two. Ideally, you will receive the stock on time, in good condition, and will have on hand a suitable tank in which to place the fish for observation. In practice, many culturists prefer to forego the expense of maintaining an observation tank, which may require aeration and/or filtration, and take their chances on introducing pathogens. Also, under certain conditions, quarantine may do more harm than good. For example, if fish arrive visibly stressed, but without apparent disease symptoms, it is probably better to get them into the growing enclosure as quickly as careful handling will allow, than to subject them to the additional stress of another crowded environment and another move.

Nutrition

Though minimizing stress is of first importance after your culture facilities are stocked, nutrition is also extremely important. One of the leading authorities on fish nutrition, Tom Lovell of Auburn University, has pointed out that nutritional disease is almost unknown in wild fish. For while a natural diet may be quantitatively inadequate, it is usually balanced, even where fish are underfed. But, since aquaculturists aim for unnaturally rapid growth, they may encounter nutritional disease. This is particularly true in the more "artificial" culture environments (cages, raceways, cement or plastic tanks, etc.) or in extremely densely stocked systems. For this reason, use the best feeds possible, and provide a variety of feeds, preferably with at least one live or fresh item. If you discover a pathological condition which you cannot immediately diagnose, it should be checked against published lists of known nutritional deficiency syndromes. Many of the publications on catfish or trout farming contain lists for these species. A more complete treatment of the subject is presented by Laurence M. Ashley in the text edited by Halver (1972). Or you can consult fish pathology texts.

Daily preventive measures

Water quality

Probably the single most important approach in preventing diseases and parasites is avoidance of stress by maintaining water quality. Here, the three main environmental factors are temperature, dissolved oxygen concentration and ammonia content. Fish may be stressed by water that is too hot or cold, while sudden temperature changes are even more dangerous. Hygiene requires dissolved oxygen to be maintained at the highest concentration possible. Ammonia is of most concern in closed systems. I shall discuss these and other environmental parameters in detail in Part X.

Professional aquarists suggest that a fairly deep layer of gravel on the bottom of an aquatic enclosure is valuable in preventing disease. The idea is that sediments, including fecal matter, sink into the interstices, so that they are not stirred up by the fish, in this way preventing constant re-exposure to pathogens from feces. This might be a wise precaution in small tanks such as are often used in closed systems.

Handling

Handling is a source of stress which should be avoided when possible. When fish are to be handled, give forethought to how to do it gently and efficiently. Hot weather is particularly dangerous, so during the summer months try to restrict handling to early and late hours, or to cool or cloudy days. There are numerous minor sources of stress, including any sort of disturbance in the water, visual presence on the shore and loud noises. Use common sense in restricting these.

Population density

As population density increases contagion becomes more of a possibility, stress related to water quality becomes more likely, and nutritional deficiencies may appear. Therefore, from the point of view of hygiene, less intensive culture systems are preferable though other considerations may override this. The non-commercial culturist can use an ounce of prevention by stocking a system at the lowest density which will provide the desired harvest, but the commercial grower is economically pressed to seek the maximum feasible density.

A health precaution many aquaculturists neglect is disinfecting equipment to be used in more than one body of water. Disinfection is mandatory if a disease or parasite outbreak occurs in one, but not all, culture chambers. Nets may be the most critical item. Minimal hygiene requires nets to be cleaned and sun-dried after use. The most practical net disinfectants are chlorine bleach, lime water, and dilute formalin. You may either immerse or spray dry nets, depending on their size. For items such as feed and fish containers, water quality instruments, boats, and the hands and clothing of workers, washing with soap and water is generally adequate. During disease outbreaks, keep equipment used in infected ponds completely separate from other gear. Though this would be a desirable practice at all times, my experience is that it is inconvenient, difficult to enforce, and often uneconomic. It is also difficult to prevent the transport of disease organisms on the bodies of water birds.

Equally as important as preventive hygiene is inspection. The aquaculturist who does not visit each culture enclosure at least daily is simply not doing the job. Visual inspection of the water and, when possible, the animals themselves, is most conveniently done together with feeding in the early morning, when stress symptoms are most likely to be apparent. You can often monitor water quality at the same time.

When you know that stress has occurred inspection is particularly important. No matter how careful you are, stress is not completely avoidable, and the wise culturist knows when to expect it. Some stress is inevitable whenever fish are handled (stocking, sampling, harvesting), following any management procedures involving prolonged disturbance of the environment, during extended hot weather or unseasonable cold snaps, after any sudden change in temperature, water level, water quality, etc., or following mistakes such as accidental overfeeding.

A more recent form of prevention is through the use of vaccines derived from immunized individuals. So far fish vaccines are commercially available only for salmonids. They should probably be used only when a specific disease is known to be prevalent in a particular region or stock of fish.

Knowing that the most careful culturist may encounter disease or parasite problems, it is well to be prepared. Keep the telephone number of your county agent, nearest fish hatchery or whomever you might wish to consult by your telephone as you do your doctor's. Plan ahead by incorporating treatment facilities in your overall design. These might include small, aerated tanks separate from the main culture facility, giant plastic bags for enclosing cages, or simply a mechanism for cutting water flow to and from ponds or raceways so that treatment chemicals are not lost before they have done the fish any good. If, due to past experience or the presence of a disease or parasite on a neighboring farm, you are expecting problems, you will want to have medication ready in advance. The time lost in purchasing or ordering medicine could be critical.

North American aquaculture, with the exception of trout culture, has, to date, been relatively little troubled with diseases and parasites. It would be unrealistic not to expect some deterioration as more and more people become involved, as some facilities become more intensive and/or artificial, and as stock is exchanged more widely. In time the practice of preventive hygiene will become more important, both to the individual aquaculturist, and in terms of his or her responsibility to other growers.

CHAPTER VIII-3

Animal Pests and Predators

Humans as pests

The most troublesome pest and predator species for aquaculturists is *Homo sapiens*. Theft, vandalism, and unintentional blunders by humans probably cause greater losses than all other non-pathogenic animals combined. The concept of live fish as property seems to be a tough one for Western society. The novice should be cautioned that many individuals who wouldn't steal your car, your money or your chickens will steal your fish. Some of this theft is simple poaching by hook-and-line fishermen looking for sport or dinner. But serious theft occurs, too. Cages are a particularly easy target: in 1979 we lost seventy-five percent of our caged bullheads at New Alchemy when the growing season had hardly begun. In major aquaculture centers, organized fish rustlers equipped with mechanized haul seines and tank trucks are not unknown.

Vandalism runs the gamut from yahoos bent on wholesale wanton destruction down to kids who yank dam boards out of sluice gates just to see what will happen, and the lunatic who for some reason sank a fire extinguisher in one of New Alchemy's pools. The prevention and control measures for both theft and vandalism —fences, locks, guards, threats, community relations, etc.—are familiar to all readers.

Extraneous fish

The second most important group of predators and pests is fish, often including the cultured species themselves. While fish of all ages are essential to a healthy natural population, the presence of a few old, oversize, hard-to-catch trout, bass or catfish in an aquaculture pond can lead to cannibalism and loss of production. The best prevention is to design all systems to allow for occasional total drainage and harvest.

Extraneous fish species can interfere with aquaculture in the following ways:

• Predation on crop species. This is a particularly severe problem in spawning and fry rearing ponds, or where the crop consists of small animals such as crayfish, bait minnows or ornamental fishes.

• Predation on eggs. You should suspect almost any species of catfish, cyprinid, sucker or perch, as well as numerous other species in this regard.

- Competition for natural or introduced feed.
- Competition for spawning sites. Centrarchids and cichlids are the worst offenders.
- Creation of turbid water by fish such as common carp (*Cyprinus carpio*), goldfish (*Carassius auratus*) and bullheads, which root in the bottom. Such behavior may have the beneficial effects of making feed available to other fishes, controlling rooted weeds or blue-green algae (see Chapter VIII–4) or by facilitating chemical processes involving the substrate. But turbidity may also limit phytoplankton blooms, interfere with the normal behavior of visual feeders, reduce hook-and-line fishing effectiveness, and restrict visual inspection. Turbid water may also be considered unattractive. (See Chapter X–1 for a general discussion of turbidity problems.)
- Interference with harvesting. Where harvesting is by hook and line, certain aggressive pest species may make it almost impossible to get the hook to your quarry. Where nets or traps are used, time is lost in sorting out extraneous fish.
- Unwanted hybridization. Under natural or semi-natural conditions most species rarely hybridize. However carp and goldfish cross frequently, producing a hybrid without their parents' desirable traits. Sunfishes, tilapias, and pickerels also hybridize fairly frequently in nature. Though such hybrids are sometimes superior to the parents, more often they are not. Hybrid fish for aquaculture should be produced only under controlled conditions.

Problems with pest fish occur most often when aquaculture facilities are directly connected to natural bodies of water, or where different species are cultured very near each other. The best control is to provide screens and filters on inlets and outlets and to design systems to prevent flooding. But some species, notably green sunfish (*Lepomis cyanellus*), bullheads and eels, are uncanny gate crashers. One possible source of invasions by these fishes is the carrying of eggs or fry on the feet of wading birds. However, among fishery biologists, the expression "birdsfoot theory" is often used to mean "unexplained presence," suggesting that human error is the most likely explanation.

You can sometimes control large predatory fish by netting or diligent hook-and-line fishing, but the small species causing the bulk of pest problems are often controllable only by drainage or poisoning. Drainage has the disadvantage that it usually involves losing a season's production. Though methods are self explanatory, be aware that some fish may survive for a while in wet mud. When you drain, drain dry and leave the pond dry for at least several days.

The only poisons recommended for fish pond management in North America are rotenone and quicklime. Pure rotenone is a natural organic product (powdered derris root) and perfectly safe when used according to instructions. You can take fish killed with rotenone from the water and safely prepare them for the table immediately. However, rotenone has not been cleared for use in commercial aquaculture by the Food and Drug Administration and Environmental Protection Agency. The reason is that most commercial rotenone is not pure, but contains a synthetic chemical, piperonyl butoxide, which acts as a synergist and compensates for the variation in strength which makes pure rotenone unreliable. Since piperonyl butoxide has not been proven non-carcinogenic, it cannot be used in a commercial food production facility. If you do not plan to engage in commercial aquaculture you can make up your own mind about the risk of contracting cancer from piperonyl

Fish poisons

Rotenone

butoxide in rotenone used to control "weed" fish.

Selective poisoning

Though you can use rotenone for a near-total kill of aquatic life, with a little care you can also use it to eliminate only unwanted small fish. Undertake such treatment only in summer, when water temperatures are over 60°F (15.6°C) and large fish are likely to be in deep water. Choose a clear, windless day with little prospect of a weather change. Calculate the dosage of rotenone according to the length of shoreline to be treated. (Do not treat deep water unless you plan a total kill.) For each three hundred feet (90 m) of shoreline to be treated, allow one pound (0.45 kg) of five percent rotenone or its equivalent, i.e., five pounds (2.3 kg) of one percent rotenone, etc. Mix powdered rotenone with just enough water to form a stiff paste, or mix emulsified rotenone with about twice its volume of water. Starting ten to fifteen feet (3–4.5 m) out from the bank, depending on the location of a convenient depth, apply a line of the pasty mixture just under the water. Wading gives better control than working from a boat. Do not treat more than 600 feet of shoreline per acre (450 m/ha) of water at a time. Repeat at intervals of about a week until you achieve the desired results. Follow up each treatment by removing as many of the dead fish as possible.

If you see too many dead fish, or if mortality is occurring in parts of the pond where it was not intended, there is some hope of neutralizing the rotenone with potassium permanganate. Allow two pounds of potassium permanganate for each pound of rotenone applied. Dissolve the permanganate in water and spray or scatter the solution on the affected area. As this method is slow, expensive, and uncertain, it is best to be conservative when applying rotenone in the first place. You can add more rotenone much more easily and cheaply than you can compensate for an overdose.

Quicklime

So far, no methods for partial elimination of pest fish with quicklime have been developed. For a total kill, apply it at 140 pounds per acre (157 kg/ha). (See Chapter III–2 for a general discussion of quicklime and its use.)

The effects of poisons on the lower levels of the aquatic food chain have hardly been studied. Though rotenone seems to have no effect on crayfish, it may kill smaller crustaceans. Quicklime is most effective against soft-bodied animals such as fish. In considering poisons, assume that some damage will be done, and then assess the severity of the problem which caused you to consider poisoning in the first place.

Other predators

Most aquaculturists imagine far more predators—apart from fish and humans—than really exist. Among the animals which are seldom if ever serious predators are crayfish, turtles, frogs, geese and gulls. Water snakes do eat small fish. The best way to control them is by restricting habitat. A clean pond bank with no brush, logs, rocks, etc., at the water's edge will greatly reduce the number of snakes. Alligators may be serious predators under some circumstances, and may even break into fish cages. Since harvest of alligators is illegal in some states and strictly regulated in others, you should check with local conservation authorities before beginning an alligator control campaign.

Mammals

Raccoons, cats, and possibly other carnivores prey on frogs, and muskrats will eat clams, but the only mammalian predators which need concern fish culturists are mink and otters. Mink might better be termed murderers than predators; on one occasion I was the victim of a mink raid on a cage which resulted in about two hundred dead fish, with few or none consumed. Otters, which confine their damage to adult fish, are among a very few animals capable of capturing even such a fast and elusive fish as a healthy adult trout.

Both mink and otters must be excluded from intensive fish culture systems. As both are valuable furbearers, you may have an added incentive to trap them, but always check with local conservation authorities before proceeding. Mink and otters are nowhere over-abundant, and trapping is governed by seasons and other regulations, or it may be prohibited altogether. You might try to contract professionals to livetrap the offending animals and take them where they will be an asset.

Birds

Predatory birds are often more significant and difficult to deal with than mammals. Kingfishers and herons[1] cause the most problems, though ospreys, diving ducks, and cormorants may also prey on cultured fish. In small ponds, herons and probably ospreys can be stopped by criss-crossing the pond with a grid of strings a few feet apart. A similar, more closely spaced, grid will inhibit kingfishers. Such devices are impractical in large ponds and may cause more problems than they solve in small ones. Other means of control, such as shooting, traps and poison are of course available but nearly always illegal.

If you have a severe bird predation problem, contact local conservation authorities. But first try to make sure it is severe. The occasional presence of a piscivorous bird does not mean your stock is being significantly depleted. Birds are as lazy as any other predator, and will take slower moving species when they are available. What this sometimes means to the aquaculturist is that the birds are actually pest control agents. But it is clearly necessary to exclude predatory birds from shallow, densely populated monocultures—for example, fry rearing pools in a trout hatchery.

Insects

Even insects can be significant predators of small fish. The main insects involved are predaceous diving beetles, giant water bugs, and dragonfly larvae. (See Figures VIII–3–1, 2, and 3.) Insects may eat small fry whole, but they also suck the blood or internal juices, which means they need not be larger than the prey.

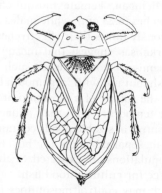
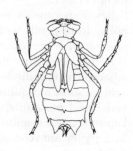

FIGURE VIII–3–1.
Predaceous diving beetle
(Dytiscus *sp.*) *larva.*

FIGURE VIII–3–2. *Giant water bug* (Belostoma flumineum), *adult.*

FIGURE VIII–3–3.
Dragonfly (Epicordulia *sp.*) *larva.*

[1]Not all long-legged wading birds are herons. A number of such birds, including cranes, dowitchers, oystercatchers, etc., are not herons and do not eat fish. Consult any good field guide if in doubt.

Most of the treatments suggested for predaceous insects (chemicals, oiling the water surface) are ecologically unacceptable, but you can use prevention. The first step is to control aquatic weeds which protect the insects from large fish. All the insects mentioned can fly as adults, so to prevent their spread try also to eliminate nearby marshy areas, shallow ditches and temporary pools, or at least control weeds in them. If you will hold small fish in separate ponds or pools, keep these dry until at least two weeks before stocking, to prevent build-up of predatory insect populations.

Mosquitoes

Mosquitoes are a pest which must be mentioned though they may have no effect on fish production, and may even be beneficial as a source of fish food. Some mosquito production is likely to occur wherever there is standing water, but it need not become a plague. The first line of defense is eliminating all areas of extremely shallow water—another reason for building pond banks as nearly vertical as your soil structure allows. Also, try to keep aquatic weeds under control.

Even more important is the elimination of temporary pools unconnected to the pond. A rain-filled tin can or shaded mud puddle may provide a perfect mosquito habitat, where the larvae are protected from their natural predators. Drain, dump, or otherwise eliminate all such stagnant water.

In the Tennessee Valley Authority reservoirs mosquitoes are controlled by constant raising and lowering of the water level, so that their floating egg masses are left high and dry. As far as I know, this has never been attempted in small ponds, but it might be effective where there are drainage facilities and adequate water supplies for replacement.

I can recommend no chemical control systems for mosquitoes, but biological control of the larvae by fish is possible. For this, perhaps the most effective fish is the appropriately named mosquitofish (*Gambusia affinis*). This prolific fish is native as far north as southern Illinois and New Jersey, and has been acclimatized as far north as southeastern Michigan. Female mosquitofish only reach a maximum length of two and a half inches (6.25 cm), while males seldom exceed one and a half inches (3.8 cm). This obviously limits their effectiveness where there are predatory fishes. Experiments in Arkansas suggest that mosquitofish can be maintained in polyculture ponds with semi-piscivorous fishes such as channel catfish (*Ictalurus punctatus*) or green sunfish, but not with largemouth bass (*Micropterus salmoides*). However, remember that the source of a mosquito problem is more likely to be an adjacent marsh, ditch, or temporary pool than aquaculture ponds. You may even achieve some mosquito control in sizable ponds containing highly piscivorous fish if you breed mosquitofish in separate pools (or buy them) and stock heavily just before heavy mosquito populations are expected. Such mosquitofish may also serve as a significant food source for cultured food fish.

Other fish species which may control mosquitoes include any of the livebearer family (Poeciliidae), the golden shiner (*Notemigonus crysoleucas*), the bluegill (*Lepomis macrochirus*) and some of the tilapias, notably *Tilapia mossambica*. Some of these may fare better with predators than the mosquitofish, but few can match its ability to hunt down mosquito larvae in extremely shallow water, or in tangled mats of weeds and algae.

Damage to banks by animals

Next to predators, the most common type of non-fish pest problem in aquaculture is damage to pond banks and embankments. Domestic livestock and humans are among the most frequent offenders, but they can easily be fenced out. (There is also some hope of training the humans.) Domestic waterfowl often damage banks, and you may have to restrict their access, but usually there are not enough wild birds to cause damage. (Puddling ducks cause more damage than diving ducks; Muscovy ducks in particular spend less time in water and tend to be easiest on banks.) In most cases, frequent human presence will prevent wild waterfowl from accumulating in large numbers.

Burrowing animals have caused more than a few dams to break, and must be controlled on all earthen embankments. The muskrat is by far the commonest burrower, though other types of rat and certain species of crayfish may also cause damage. Muskrats, like mink and otters, are valuable furbearers, but they are nowhere endangered. So no aquaculturist should feel badly about legally harvesting and selling a few muskrats from his pond. (Muskrats are also edible.)

The best way to prevent damage from muskrats and other burrowing mammals is to keep dams and levees free of brush and closely mowed. In this way they will be less attractive to rats, and you will be able to spot any that do set up housekeeping.

Burrowing crayfish are most likely to be a problem in the South. The best time to deal with them is during pond construction. Your local soil conservation agent can tell you if there are burrowing crayfish in your area and suggest measures to prevent problems.

CHAPTER VIII-4

Pollution

Definition

Environmental pollution is a term which eludes precise definition. Webster defines "pollution" as "defilement" or "impurity." This definition is not very useful to aquaculturists; certainly no natural or artificial aquaculture environment contains water which remotely approaches purity. But if we attempt to stop short of the absurdity of referring to all water inhabited by living things as impure, and therefore polluted, we find ourselves in a maze of value judgements. If I dump pig manure into a pond, I may be "enriching" it from an aquaculturist's point of view. But to a swimmer, pig manure in a pond is certainly a pollutant.

An aquaculturist's working definition of "pollutant" might be: "Any unwanted substance in the aquatic environment." Such a definition would include weeds, pathogenic and parasitic organisms, and unwanted large animals including "weed" fish. These problems were dealt with in the three immediately preceding chapters. Excluding these, I will divide water pollutants into the following categories, based on their physical and chemical nature and mode of action:

Categories of pollutants

• "Organic" substances: These become pollutants only when present in excess, as their mode of action is to compete with the cultured animals for dissolved oxygen.

• Microorganisms and algae: These, while not directly harmful to the creatures being cultured, are harmful to the human consumer. They range from the potentially deadly dinoflagellates which sometimes infest marine shellfish to organisms which cause undesirable flavors in fish.

• Physical pollutants: These may interfere with the activities of the cultured organisms or the culturist. Silt is the only one of these which merits discussion.

• Chemical toxins: These poisons may cause fish kills, but often the main danger is presented in more subtle, long-term effects, not only on the cultured organisms, but on the consumer and/or the entire ecosystem. (Natural toxins, including the leaves or sap of certain trees, are known in the tropics, but I have yet to hear of a fish kill of that nature in North America.)

All these types of pollution are similar to disease in that it is easier to avoid or prevent than to cure them. And in the case of chemical toxins there usually is no "cure." Of course, aquaculture itself is a potential source of all four types of pollution, both to the culture environment and the outside world. In this discussion I take into account the moral responsibility of the aquaculturist to the total environment, but it is only fair to emphasize that the danger to aquaculture from contamination of larger ecosystems is far greater than the pollution which can be attributed to aquaculture.

Sources of water available for aquaculture may be classed as more or less "safe" in terms of the likelihood of pollution. The riskiest environment is the ocean, which is outside the scope of this book. However, if you propose, for example, to do cage culture in large lakes or rivers, you must confront similar risks. Even if a large body of water is known to be relatively uncontaminated at present, you will have little control over future pollution of a water body shared with a myriad of individuals, governments, and businesses. All aquaculture systems dependent on surface water share this problem to a degree, depending on the source's size and the number of entities exercising control over it. If you happen to control an entire watershed, surface water may be as safe as any other, but as soon as even one other person begins to share the water or the watershed, the risk increases. It is always essential that the aquaculturist knows what is happening upstream.

Rain water used to be considered among the purest sources, and in some places it still is, but as the phenomenon of acid rain (discussed later in this chapter) shows, you can no longer take for granted even the purity of rain water. Few, if any, rain water supplied systems rely solely on rain falling directly on the water surface, so you must also take into account any pollutants picked up by the water between the moment it strikes the earth and its entry into your aquaculture system.

Wells and springs are least likely to be polluted, but increasing contamination of ground water is a wide-spread phenomenon.

Finally, we come to municipal drinking water supplies and another value judgment as to what constitutes pollution. Chlorine, with which drinking water is treated to render it safe for our consumption, is harmful to fish. However, it can be removed safely, simply and cheaply. (See p. 410.) So, for very small aquaculture systems with a reasonable cost for filling with municipal water, the public drinking water supply may actually be the safest source.

Sources of organic pollutants include the following:

Organic pollution sources

- Metabolites of living aquatic animals.
- Uneaten feed.
- Organic fertilizers used in aquaculture.
- Decay of dead animals.
- Plants, through respiration of live and decay of dead ones.
- Organic matter eroded from cropland.
- Sewage.
- Organic industrial wastes.

Fish metabolites include various nitrogenous compounds. Some, like ammonia, are toxic. Others serve to inhibit growth or reproduction under crowded conditions. Still others decay in water and compete for dissolved oxygen with aquatic animals. Beginning aquaculturists sometimes fail to take into account the fact that the volume of these metabolites (and the BOD of the fish) increases proportionally to the

Fish metabolites

increase in animal biomass. This can lead to inhibition of growth, or even mortality, in small, very heavily stocked systems.

The best preventive strategy for fish metabolite pollution is thoughtful stocking. If you are a non-commercial culturist you can avoid problems by stocking at a rate adequate to supply your needs, yet below the feasible limit. And, however densely you stock, be sure to base your stocking rate on the total weight of fish at the end of the growing season—or be prepared to thin your stock. Do not forget to consider animals other than your crop. These may include "weed" or desired species. For example, you might fertilize hoping to produce a dense population of zooplankton and then be so successful that the resulting plankton overloads the pond with metabolites, competing for oxygen with the fish.

Controlling fish metabolite pollution includes thinning of the crop species, removal of unwanted organisms, and biological filters which chemically modify the metabolites themselves. (This last method, practical only in small closed systems, is described in Chapter IV–4.)

Pollution by feed

Overfeeding may be the commonest form of organic pollution from within on fish farms. Some difficulties may arise from feeding in quantities which though above the optimum, are consumed by the fish. Here, however, the actual pollutant is not feed but excess fish metabolites. True feed pollution occurs when food goes uneaten, usually because given in excessive amounts. Other causes of feed pollution are incorrect choice of feed, and feeding at the wrong time or place. Your first line of prevention is to carefully calculate feeding rates. The danger of overfeeding is also one of the best arguments for using floating feeds, which allow you to see whether feed is being consumed. If you can use live feeds which will survive for a while in the environment so much the better. You may be able to compensate for overfeeding at times by removing uneaten portions, but this is wasteful and inefficient at best. (See Chapter III–5, –6, and –7 on feed rates, floating and live feeds.)

Natural fertilizers as pollutants

Excessive pond fertilization can have the same effects as overpopulation or overfeeding. It may result in overproduction of plankton, but even more common is the simpler problem of oxygen depletion through decay of manure. The same cautions apply as to feeding. Always remember that an amount of fertilizer which is beneficial if added bit by bit can be lethal if dumped into a pond all at once.

Natural fertilizer may also accidentally find its way into an aquaculture system. Runoff from manured fields following heavy rain or flooding is the usual mechanism. Even though in a given instance this may prove beneficial, take every possible precaution to prevent it. I consider any unplanned fertilization to be pollution. The best practice is to site agricultural land and ponds so that the former cannot drain into the latter. If that is impossible, at least plan to leave a shelter belt and/or a swale between field and pond. Better yet is to install drain tiles, as most organic matter will filter out when passing through soil into a tile.

If the problem originates with a neighbor who fronts on the pond or is located upstream, begin discussion by pointing out that fertilizer which washes off into your pond is a loss to the neighbor as well as a detriment to your pond.

Drainage and polluted effluent

The types of organic pollutant discussed so far are most active in the bottom waters. It is a wise precaution, then, to design ponds so that polluted bottom water can be drained off first. (See Chapter VII–1.) If your water supply is adequate for replacement, this is the best control measure for all forms of organic pollution arising within the aquaculture system.

In some cases drainage of polluted bottom water may solve your problem by passing it on to a downstream neighbor's ecosystem. The seriousness of this will depend on the volume of water being discharged and its chemical and physical characteristics, the size and nature of the downstream ecosystem, and the distance between pond discharge and the natural system or neighboring property which receives the effluent. In some cases, you will be legally required to treat such effluent in some way. Even where this is not so, some effort should be made to recover any nutrients escaping from the pond and to improve the quality of the effluent. You can do this by using the effluent on agricultural land, by cultivation (without feeding or fertilization) of organisms low on the feed chain in the tailwaters, or by the use of settling pools, filters or aerators.

Where any organic pollution problem has reached the point that fish or other animals are killed, you must first of all deal with the original source of pollution. But do not forget that the decay of dead animals is a further source of pollution, which should be removed immediately. The same also applies to dead plants.

Dead animals and plants

The remaining organic pollutants from cropland, sewage, and industry originate outside the aquaculture system. Organic matter which erodes from land is a symptom of poor land use; prevention is the same as for fertilizers applied to crop land. Though sewage and organic industrial wastes (for example from feedlots, food processing plants, breweries, etc.) may be applied intentionally as fertilizers or feeds, where this is not the case they are pollutants. Each of these materials may enter via surface water, in which case you must pinpoint the source so that you may proceed with appropriate persuasion or legal action. Sewage may also reach ponds in residential areas from leaky cesspools and septic tanks. Always keep an eye on such systems when they are located near water, or better yet switch to a composting toilet.

Outside organic pollutants

The deadliest of the various toxic microorganisms which may be found in the flesh of aquatic animals is the dinoflagellate *Gonyaulax tamarensis* (the famous "red tide") which has caused more than a few human deaths. *Gonyaulax* occurs only in salt water, and no fresh water counterpart is known. However, there are bacterial and virus diseases (notably hepatitis) which can be contracted from eating aquatic food. While fin fish and arthropods, including crayfish and shrimp, may carry such diseases, the chance of contracting disease is greatest if you eat mollusks. This is because the mollusk's mode of feeding results in concentration of microorganisms in its flesh at levels many times those found in water. Both fin fish and shellfish may carry bacterial or virus diseases without being affected in any way; it is not safe to assume that because a fish looks healthy it is perfectly safe to eat.

Human toxins and disease organisms

All reported cases of virus or bacterial diseases associated with the consumption of *fresh* aquatic foods have been linked with some form of organic pollution. In the great majority of cases, sewage was the offending pollutant. If you have any doubt as to whether your water is polluted by sewage, contact your local health authorities, who can make a simple test for coliform count. This test, while it measures only the concentration of a single species of bacteria (*E. coli*) is a reliable index of sewage pollution.

Sewage is sometimes used as a fertilizer in large-scale culture of food fish in Europe and Asia, and small-scale use in North America would be technically feasible. However, I do not recommend it on a homestead scale until the health considerations are better understood. The use of animal manures is discussed in this book, and while the danger of disease transmission is less than with sewage, it is still a possibility. I recommend the following precautions to anyone using animal manures in aquaculture:

Precautions with manure

- Have the coliform count checked periodically.
- Keep tabs on the health of livestock used as a source of manure.
- If possible, compost or ferment the manure; this will kill most pathogens.
- Clean and cook or preserve aquatic animals soon after harvesting. Most bacterial disease associated with aquatic foods arises from proliferation of bacteria after an animal's death.
- Thoroughly cook foods from heavily fertilized water. Avoid raw foods. (One of the aquatic foods most often linked with human disease is raw water cress.)

"Off" flavor

The problem of undesirable flavors in fish, variously referred to as "off," "musty," "lakey," or "muddy" is still poorly understood. Though off flavor is not associated with any health hazard to the consumer or the fish, it can be financially ruinous to the culturist if undetected. I discuss it in this chapter because it is often associated with blue-green algae, which in turn are usually by-products of organic pollution. Therefore, to guard against off flavor, keep an eye out for blue-green algae and follow the precautions against organic pollution outlined above.

One further precaution is necessary, as sad experience at New Alchemy taught us. In a seemingly unpolluted aquaculture system the low dissolved oxygen concentrations conducive to blue-green algae growth may exist on the very bottom, leading to off flavor. So, in densely populated or heavily fed or fertilized systems, always keep the bottom sediments turning over by water flow, mechanical agitation, aeration, or by stocking animals which root about in the sediment.

Off flavor sometimes occurs even in apparently unpolluted water. Certain feeds, such as silkworm pupae, may be implicated. Or clams may acquire the muddy flavor of the sediments in which they live. Some Canadian pot hole trout farmers find that fish harvested in summer or early fall taste muddy, while fish harvested later, from colder water, do not. Always sample one or two fish before making mass harvests. If you detect off flavor, a waiting period may be all that is needed. Heavy aeration might also help, and feeding and fertilization should be cut back or discontinued. Hydrated lime is also said to help. (See Chapter III–2 for application rates.) But the surest control measure is to transfer the animals to a clean, sterile environment, such as a concrete tank, for a few days.

Silt

Silt consists of the finest soil particles (less than $1/16$ mm in diameter) whether in suspension or as sediments on the bottom. It is a product of erosion and so may be associated with certain types of organic pollution, but its pollutant effects are physical, not biochemical.

Silt affects aquaculture systems in the following ways:

Effects of silt

- By settling out on spawning nests and interrupting the reproductive act or smothering eggs.
- By settling on the gills of young fish (or adults of certain delicate species) and suffocating them.
- By lowering productivity through blanketing more desirable bottom materials such as plants, organic mud and gravel.
- By reducing light penetration, which lowers the rate of photosynthesis, interfering with both oxygenation and natural production of food.

- When suspended, by interfering with the natural behavior of visual feeders such as bass, sunfish and trout.
- In the same way, silt makes it difficult for the aquaculturist to see what is going on.
- By interfering with harvesting or the use of machinery such as pumps.
- Finally, unchecked siltation can fill in a body of water over a period of years.

There are control measures for suspended silt, and these are discussed below. (See "turbidity," Chapter X–1.) Here I wish to emphasize silt problem prevention. The only possible source of silt in a body of water is surface runoff. Where the aquaculturist's land is a potential source of silt, the preventive measures are simply good soil conservation practice. A minor source of siltation which may escape the notice of even a conscientious farmer is the point where water enters the pond. Be sure water does not enter by flowing over bare soil, and if necessary, cement the entry point or provide a tile. If surface water from higher in the watershed is the source of silt, you are faced with a larger conservation problem, in proportion to the size of the watershed. Often the problem of silt in surface water is seasonal or temporary, and if you can cut off the flow of surface water during periods of high turbidity, you may avert the worst. If all else has failed, passing surface water through a gravel filter (see Figure VII–1–7, p. 339) or settling pool may somewhat reduce the problem.

Preventing silt problems

Occasionally, a shallow pond, though not receiving silt at present, has been partially silted in in the past, and disturbance of the bottom may cause resuspension of this silt. If you can deepen the pond, or if the drain system is designed so that bottom water can be drawn off, carrying the silt with it, you may eliminate the problem. Otherwise, wading, swimming and access by livestock should be restricted, particularly during spawning or when eggs or young fish are present.

Chemical toxins are a newer problem and less well understood. Should you be concerned about any of the chemicals mentioned in the following paragraphs, the best first step is to call the United States Environmental Protection Agency, or Department of the Environment in Canada. If you find dead or suffering fish, preserve a few by wrapping them in aluminum foil and freezing. If there is no biological evidence of pollution do not take samples unless instructed. The EPA may be able to do the necessary tests for you, refer you to a commercial laboratory, or assure you that tests are not necessary. If there is loss of fish or if you are convinced there is a chemical pollutant presenting a serious danger to human health, the next step after identification of the chemical is to get an attorney to gather evidence on its toxicity from government agencies, including the EPA.

Chemical toxins

Keep in mind that some chemical pollutants are ubiquitous, and the mere presence of a foreign chemical in an aquatic crop will not justify a lawsuit. The purpose of this section is only to alert you to some possible dangers and to suggest possible courses of action.

Chlorinated hydrocarbon pesticides represent a multiple hazard. They may kill fish directly, disrupt reproduction, interfere with the aquatic food chain by killing desirable insects, forage fish, plankton, etc., or they may accumulate in the food chain with little or no short-term (acute) effect but potentially disastrous long-term (chronic) effects. Chronic effects are related to the tendency of the metabolites formed by chlorinated hydrocarbons as they break down to accumulate in animal tissues. Here they may be mutagenic, teratogenic, carcinogenic or adversely affect reproduction or social behavior.

Chlorinated hydrocarbon pesticides

Since these effects were discovered, a number of chlorinated hydrocarbon pesticides have been banned or restricted. The newer pesticides developed in their place, including the organophosphates, are said to break down more rapidly in the environment, without the cumulative effect. However, some of them are still more toxic in their original forms. And, as a cynical friend observed, "In our society we use a technology until it is shown to be unsafe, then replace it with a new technology, the side effects of which are completely unknown." Who knows what we may still learn about the newer, "safe" pesticides?

No chemical pesticides have received EPA and FDA clearance for use in commercial aquaculture in the United States, but some such compounds, particularly herbicides, are sold for use in aquaculture. With the very few exceptions I have mentioned, I do not advocate the use of any chemical pesticides in aquaculture. Chlorinated hydrocarbons should never be used.

Aquacultural use is not the only way in which pesticides find their way into aquaculture systems. Probably, agricultural use is responsible for the great majority of these compounds in aquatic food chains. Possible sources include:

Possible sources

• Aerial spraying: If aerial pesticide spraying is practiced near your aquaculture site, advise your neighbors of the risk to your crop and health and try to persuade them to use some other method. Failing that, at least alert pilots to the situation and try to get them to spray when the wind is away from the aquaculture system and water supply. Should a problem still occur, record the serial number of the plane, if possible, and contact your Agricultural Extension Agent.

• Runoff from agricultural lands: Dangerous pesticides should never be used on any farm where aquaculture is practiced. But following this advice will not protect you from polluted surface water or runoff from neighbors' land. Since it is difficult to get water analyzed for pesticide residues, you may have to make do with an educated guess as to what is entering the system, based on your knowledge of local agricultural practice. If large amounts of pesticides are used, you may want to avoid using surface water, even if it means extra expense for a well and pump.

• Residues in soil: An extreme example is the case of a man in Arkansas who tried to convert abandoned cotton land to a bait minnow farm. He failed because pesticide residues in the soil were so high minnows could not survive in his ponds.

• Accidents: One year Auburn University lost a large portion of its tilapia stock when someone spilled toxaphene into the water supply. This is another argument for ground water or on-site collection of rain water as water sources for aquaculture.

• Entry via the food chain: About fifteen years ago, chlorinated hydrocarbons were used to control gypsy moths in pine forests on Cape Cod, Massachusetts. Many of the poisoned moths fell into ponds, leading to virtual extinction of insectivorous fishes in small ponds. (There was no mortality of non-insectivorous species.) The fact that certain sources of brine shrimp eggs have been found unacceptable has been explained by accumulation of pesticides which, while not perceptibly harmful to the brine shrimp, are lethal to young fish consuming them. Contaminated grains, fish meal, etc., in prepared feeds may have similar effects.

Though the overall level of environmental contamination by chlorinated hydrocarbon metabolites in North America seems to be declining as a result of restriction, pesticides are still everywhere, and residues of DDT have even been found in Antarctic penguins. So, while you cannot totally escape these poisons, you can cultivate and eat aquatic foods and at the same time minimize their presence in your diet. In addition to the precautions implied above, you should remember the following:

• Mollusks accumulate chemicals just as they do organic pollutants, and are a greater risk than fish from the same source.

• Apart from mollusks, the higher on the food chain an animal is, the greater the amount of pesticides it is likely to accumulate. For this reason, it is safer to eat herbivorous or filter-feeding fishes than predators.

• As the effect of pesticide exposure is cumulative, small, young fish may be less contaminated than larger ones.

• Though pesticide metabolites are stored in virtually all types of tissue, they are concentrated in the fat. Dry fish (sunfishes, tilapia) are safer than oily ones (trout, carp).

• For the same reason, avoid fatty or oily feeds, insofar as is consistent with fishes' proper nutrition.

Adherents of "organic" agriculture tend to be as firmly opposed to synthetic fertilizers as to chemical pesticides. In water, however, synthetic fertilizers are in no way comparable to pesticides as toxins. In fact, they are less dangerous than natural "organic" fertilizers. Only when present in great excess, as might occur if a flood follows application of fertilizer to crop land, are such fertilizers apt to kill aquatic organisms. Where there is less severe erosion of synthetic crop or lawn fertilizer unwanted plankton blooms can result. Prevention is the same as for pollution by manures.

Antibiotics may also be introduced to the environment by the aquaculturist. Though there are times when nothing else will solve a disease problem, antibiotics should be used only as an emergency measure, realizing that their overall effect on the fish, the natural food chain or the human consumer is unpredictable. Antibiotics should never be used immediately preceding a harvest.

Among the most widespread pollutants are petroleum-based fuels and lubricants, and their metabolites. Some of the pollution problems engendered by these products are:

• Acute toxic effects. (Kills of adult fish occur only in cases of severe spillage, but eggs, larval forms and invertebrates can be killed by relatively low concentrations.)

• Chronic toxic effects which reduce productivity in the long run, though not causing observable mortality.

• Possible carcinogenic effects on the human consumer of contaminated aquatic foods.

• Tainted flavor.

• Formation of a surface slick which is unsightly, and interferes with exchange of gases, including oxygen, between air and water.

Sources of these pollutants include:

• Spills: These are much less prevalent in fresh water than in the oceans, but could present a problem if you depend on surface water. It is also conceivable that an accidental spill of enough gasoline, oil, etc. to affect a small system could occur during normal farm work.

• Outboard motors: Outboards leak some gas and oil during normal operation. If you need an outboard in your aquaculture pond, consider a non-polluting electric "trolling motor" which will do everything a gas outboard does, except go fast.

• Applying motor oil in order to control mosquitoes. This should never be done in aquaculture ponds, though some authorities have shortsightedly recommended this treatment for predatory aquatic insects. If surface water has been treated in this way there will be a visible film of oil.

- Oil used as an emulsifier in applying pesticides.
- Automobile exhausts: Try not to locate ponds near busy roads, and keep vehicular traffic on farm roads to a minimum. If a pond must be located near a road, a row of trees, or a high levee between road and pond will help.
- Pollution of ground water through dumping of used oil, leaky fuel storage tanks, etc. You can usually detect this by odor.
- Improperly designed or poorly cared for pumps: Though only a minor problem in large ponds, in very small systems lubricant leakage can be severe enough to cause fish kills. Explain your needs when purchasing a pump as many pumps are designed to avoid this problem.

PCB's

Polychlorinated biphenyls (PCB's) are a ubiquitous group of synthetic chemicals used in a myriad of industrial processes. Only quite recently has it been recognized that they are potent carcinogens and cause severe disruption of fishes' reproductive cycles. Their manufacture and use has been restricted, but the problem remains, since their half life in the environment is estimated in the thousands of years. The problem of PCB pollution is severe enough that, in some places, for instance the fresh waters of New York state, sport fishing licenses come with an advisory notice not to eat too many fresh water fish.

You cannot completely avoid PCB's, but you can minimize their presence in aquaculture systems and the human diet. Avoid using surface water which receives industrial pollutants, even if the fish in it appear healthy. Think twice when choosing plastics, paint, and electrical equipment which will be in contact with water. As for the diet, PCB's behave like chlorinated hydrocarbons so you can restrict PCB intake in the same ways.

Heavy metals

Heavy metal ions are among the most toxic substances known, and some of them may accumulate in fish flesh. Since the deaths of a number of people in Minamata, Japan, after they had eaten ocean fish and cultured shellfish, attention has been focused on mercury. But lead, copper, cadmium and some other metals are potentially as dangerous. Possible sources of heavy metal pollution are:

- Industrial contamination of surface waters. (This was the cause of the Minamata tragedy.)
- Fall-out from leaded gasoline.
- Contamination from substances in contact with water, including pipes, machinery, paints, etc.
- Processed sewage sludge used as a fertilizer. These fertilizers should ordinarily not be used in or near bodies of water.
- Copper compounds sometimes used to control aquatic weeds. (See Chapter VIII-1.)

Chlorine

Among the common additives to drinking water, only chlorine is a "pollutant" from the aquaculturist's point of view, but it can be lethal to fish. You can replace up to ten percent of the total volume of any system with chlorinated drinking water with no harm to fish. Perhaps twice that percentage can be added if the water is very warm, or if it is splashed vigorously as it enters. If you must add a greater volume of chlorinated water, as for example in filling a dry tank, simply let the water stand for twenty-four hours, and the chlorine will dissipate into the atmosphere. Aquarium shops sometimes sell tablets for more rapid dechlorination of water. These are nothing more than photographic "hypo" (sodium hypochlorite) which you can obtain much more cheaply at a photo store. However, even at photo store prices, it is not worth the expense to treat a container much larger than a large aquarium.

A special word of caution is in order for operators of closed systems or other very small aquaculture facilities. All that I have said about oil (as fuel or lubricants for pumps), PCB's (in plastics, paints and electrical equipment), or heavy metals (in paints, pipes and machinery) goes double in small systems. Pollution can go from harmless to lethal levels much more rapidly in such systems than in large or open systems. And in many small systems the very containers in which the fish live, such as plastic or metal tanks, are potentially toxic. (See Chapter VII–2 for advice on choosing equipment and leaching out possible toxins.)

Dangers to very small systems

Acid seepage is a danger to surface and ground waters wherever coal is mined. Its effects are simple and direct; only a small quantity is required to destroy almost all aquatic life through drastic lowering of pH. Your only defense is careful site selection. The same applies to the avoidance of contamination by radioactive wastes, a pollution hazard which affects us all, aquaculturists or not.

Mine and radioactive wastes

It has generally been assumed that the small pond owner is better able to control pollution than the aquaculturist who must rely on rivers, lakes or the oceans, which all too often function as public dumping grounds. However, a relatively new and spreading form of pollution has thrown that assumption into question. This is acid rain, a by-product of the burning of fossil fuels in homes, vehicles and industries, which already afflicts most of North America and Europe. While the conditions that lead to acid rain are created where there are heavy concentrations of population and industry, its effects are felt downwind of these areas. As an index of the severity of these problems, consider this: The Adirondack Mountains of New York are perhaps the least populated and industrialized region of the Northeast. Yet in the Adirondacks over half of all ponds and lakes above two thousand feet (600 m) in elevation are now devoid of fish, due to acid rain originating in the industrial cities of the Midwest.

Acid rain

The long-term effects of acid rain could eventually be disastrous to all waters—and soils—downwind of industrial centers, but at present lethal effects are far more likely to be felt in small ponds at high altitudes, and/or where soils are non-porous and poor in calcium. The immediate problem is not so much due to acid rain which falls directly on the water's surface or in the watershed, because the aquatic system can deal with this up to a point. But lethal effects are felt mostly by eggs and very young fish, and occur in spring, when the winter's accumulation of acid snow melts, dumping virtually an entire winter's accumulation of acid pollution into the water at once. So, the problem is most severe where the most snow falls, and more dangerous to small ponds because of their higher ratio of shoreline to area. The implications of acid rain in terms of pond siting and design are obvious. Otherwise, there is little the aquaculturist can do, unless it would be to invest in snow removal equipment or many tons of lime.

Acid rain, perhaps more than any other form of pollution, points out the need for aquaculturists to be environmentalists as well. (Though consideration of sewage, silt, pesticides, PCB's, heavy metals, mine wastes, nuclear pollution, etc., should

Environmentalism

eventually lead one to the same conclusion.) It is perfectly true that is is an honorable endeavor and a contribution to society to grow aquatic foods, for sale or for your own consumption. But my opinion is that, threatened as all aquaculture is by various forms of environmental pollution, the aquaculturist who does not become involved in at least one environmental cause is simply not doing a responsible job.

CHAPTER VIII-5

Legal Requirements

Having placed this chapter in the section entitled "Special Problems in Aquaculture" may reflect something about the author's psychology. Be that as it may, my opinion is that, at this point in the evolution of North American aquaculture, the law does more to hinder than to help the aquaculturist. This situation will improve as the importance of aquaculture increases, but even where laws are favorable or reasonable, it will remain quite a hassle for the aquatic farmer to find out just what laws and regulations are in effect, and how to comply with them.

The current state of confusion is in part due to the newness of the field. Government agencies have small experience in dealing with aquaculture, and it is not always clear whether federal, state or provincial, or local governments should exercise authority. Further, aquaculture does not fit neatly into established categories of human activity. The regulation of aquaculture is to some extent the legitimate concern of agencies charged with administering agriculture, land and water use, conservation and environmental pollution, business, and public health. These various authorities are still trying to sort out their powers and responsibilities at all levels.

Usually, non-commercial aquaculturists have an easier time complying with legal requirements than commercial growers. Among commercial growers, large operators are usually more able to cope with regulations than small-scale producers. This probably reflects a general deplorable trend in our legal and economic life rather than anything intrinsic to aquaculture.

This chapter will have more to say about how to comply with legal requirements than with what I think they should be, but I will indulge in some editorializing. If you finish this chapter feeling that aquaculture is not receiving the legislative consideration that it should, you may be interested in joining one or more of the organizations devoted to influencing aquaculture-related legislation. At the national level these include the Catfish Farmers of America, The Trout Farmers of America, and The Canadian Fish Farmers' Association. (See Appendix IV-4 for addresses.) There are also numerous regional organizations. While some of these groups are dominated by large producers, in general they work on behalf of all aquaculturists.

The rest of this chapter is little more than an annotated list of the types of regulations which could apply to a given aquaculture enterprise. To attempt more would require another book—one which would soon be out of date. The key piece of advice is this: Ask. I hope the chapter will give you some idea of whom to ask, or at least where to begin asking. It will also help you to cultivate a relationship with your local conservation officers, whether or not you already have questions for them. In addition to official information sources, do not neglect to inquire of nearby aquaculturists, if any. They may understand the legal pitfalls which await you better than the agencies which create them.

Types of laws affecting aquaculture

Regulations which affect aquaculturists include:
- Fishing regulations not created with aquaculture in mind, but which may apply to aquatic farmers.
- Licenses and permits.
- Restrictions on species which may be imported, raised, or sold.
- Regulations designed to protect public water supplies.
- Regulations intended to protect the health of fish stocks.
- Public health regulations.
- Zoning ordinances.
- Restrictions on pond construction, principally for safety's sake.

Fishing regulations

Farm pond owners wanting to maximize food fish production are sometimes frustrated by being required to comply with state-established seasons, minimum sizes, bag limits, etc., even though these may be at odds with good management of ponds as a food resource. In some places there is a potential for an absurd situation. For example, the law in Louisiana regards all fish, regardless of where found or how maintained, as the property of the state. Though this law has not been interpreted strictly enough to prevent Louisiana from becoming the third-ranking state in total commercial aquaculture production, the legal threat remains.

Potentially as problematical as fishing regulations are the laws regarding public access. There are places where, if you accept free fish for stocking, or other services from the state, you will be required to permit public access for fishing. Be sure you understand the rules governing public access before you make any commitment to fish culture.

Licenses and permits

Some sort of licensing is almost always required. License fees for noncommercial growers are often nominal, and there are rarely requirements beyond payment of the fee and filling out the necessary forms. Commercial aquaculturists, however, may find licensing more complicated and expensive. Certain types of licenses, for example scientific collecting permits, are a mere formality for research institutions, but difficult or impossible for private individuals to obtain. It is often the cumulative effect of licensing requirements which is discouraging. In some states, certified compliance with up to thirty regulations is required! Among the types of licenses and permits which may be required are:
- Sport or commercial fishing licenses.
- Permits to import or export certain species, or aquatic animals in general.
- A fish propagator's license.
- A scientific collecting permit, if you plan to take young fish from public waters for stocking.
- Licenses to build or maintain a private fish pond.

- Licenses to operate a fee fishing establishment. (See Chapter IX–1.)
- A bait dealer's license.
- Certification that fish stocks are free from disease.
- Licenses to sell food fish, especially if fish are processed on the premises.
- Permits related to an environmental impact statement, where required.

Some of these licenses and permits, while not imposing onerous fees or requirements to obtain them, require fairly detailed annual reporting. Note where this is the case and maintain adequate records. (You will usually find these to be useful in other ways.)

Importing fish

Restrictions on the importation of animals and plants have been established by virtually all federal, state and provincial governments. The chief reason for these laws is to prevent the introduction of exotic species which might damage natural or agricultural ecosystems. Some states and provinces forbid the importation of virtually anything, while others single out particular species or groups; a few are very lenient. Most often prohibited are grass carp (*Ctenopharyngodon idellus*), the tilapias (*Tilapia* spp.), walking catfish (*Clarias* spp.), common carp (*Cyprinus carpio*) and goldfish (*Carassius auratus*). Often the name "carp" is treated as a "bogey man," with importation of all "carps" (an imprecise term including dozens, perhaps hundreds, of species) prohibited, while other fish, with equal potential for damage, are admitted. Some states, while not forbidding the import of particular animals, require a special permit. Obtaining such a permit may be a formality, or it may be nearly impossible. All I can say here is that it never hurts to ask.

In some cases the laws governing fish importation are patently ridiculous. For example, one New Alchemy associate has been unable to bring blue tilapia (*Tilapia aurea*) into Maine, where there is not the slightest chance tilapia could survive the winter outdoors. However, the general principle behind importation laws is a good one, and I can sympathize with state officials who hope to avoid individual review of dozens of cases by imposing a blanket ban on exotic fishes.

The matter of "exotic" organisms points up the artificiality of our political boundaries. I have heard aquaculturists who should know better expound on their "native" fish communities when none of the fish were native to their watershed—as if the issue was citizenship. Until that enlightened day when we have government by watersheds, we shall have to deal with such confusion between the political and natural worlds. Meanwhile, we should be as conscientious about bringing organisms into a new watershed as we would importing them from a foreign country, even though there may be no laws governing such actions.

It is fairly common in the ornamental fish business to simply ignore the laws on importation. I urge food aquaculturists not to emulate this practice. While an exotic is often the best species for a given situation, always begin with a hard look at local species. If you do want to stock an exotic, you may be able to obtain an individual exemption from import regulations, especially if the fish are to be used for research purposes. Before seeking such cooperation from the local authorities be sure you understand what the likelihood is that the species in question could survive in your region. And do be honest about the likelihood of escape into natural waters. At one extreme, if fish are cultured in cages or in ponds created by diverting surface waters, a few fish are almost certain to escape eventually. At the other, a closed system, situated on high ground remote from natural bodies of water, is virtually escape-proof.

Existing restrictions on the species which may be cultivated and sold are, to my mind, a good deal less sensible than the importation laws. The premise of these restrictions on fish culture is that certain predatory fishes (trout, salmon, pike, bass, pike-perch, sunfish, etc.) are both more popular with sport fishermen and more subject to depletion from overfishing than species occupying lower trophic levels. Therefore, commercial fishing for these species is generally prohibited. Prohibiting the cultivation of these fishes for food, or their sale for that purpose, is either an outcome of the wording of fishing regulations or an attempt to simplify enforcement by doing away with the necessity to discriminate between cultured and captured fish of "game" species.

The uselessness of most of these regulations is demonstrated by the trout situation. Commercial cultivation of trout is legal everywhere, simply because it is a recognized business of considerable importance. As far as I know, clandestine commercial fishing for trout to be marketed as cultured fish is a rare occurrence, and the states and provinces feel they are capable of adequately regulating both trout culture and trout fishing. If this is true, there is then no necessity to prohibit the commercial cultivation of other game or pan fish.

Sometimes all that is required to set things right is one determined individual. For example, a few years ago, Ohio fish farmer Bill Gressard persuaded his state to legalize sale of cultured bluegills (*Lepomis macrochirus*) as food fish simply by demonstrating that he could produce and sell bluegill fillets in quantities sufficient to justify a commercial enterprise.

Restrictions on use of public water supplies may be designed to prevent interference with other uses, depletion of water, or pollution. The only types of aquaculture which normally conflict with other uses of public waters are raft culture (not done in fresh water) and cage culture, which may interfere with fishing, recreation or navigation. Although the legal questions raised by placement of cages in public waters have been dealt with for years in other countries, notably Japan, the United States and Canada are still in a bit of a muddle over the concept. It seems that for the foreseeable future decisions on cage culture in public waters will have to be made case by case, and that cage culture should never be attempted in such waters without the express permission of the appropriate authorities.

Wherever aquaculturists draw on relatively small sources of water or where there are large concentrations of aquatic farms the depletion of water supplies is a matter of concern. In arid regions, regulations may be enforced by the state or by smaller entities, such as local irrigation districts. Permits to use water may hinge on whether a project will "borrow" water and return it for use downstream, or whether ponding will lead to significant evaporation loss. Elsewhere, restrictions on the quantity of water an aquaculturist may use exist most often where trout farmers tap natural streams which support recreational trout fisheries. Here, the normal practice is to specify a maximum percentage of the total summer flow that can be diverted.

The quality of effluent from an aquaculture system is important to everyone downstream. In theory, chemical pollution by aquaculture is prevented through the restriction of chemical use by aquaculturists. As a result, aquaculture effluent is usually subject to being monitored only for temperature, suspended solids and dissolved oxygen content. Just what may be considered "pollution" with regard to these depends on the condition of water before it enters the aquaculture system, and the ecology and use of the downstream waters. It also depends on standards which may have been arbitrarily set, with minimal knowledge of local situations. Aquaculture use may increase the temperature of water (by exposing it to solar radiation for a greater time), but it is also possible for effluent to be colder than inflowing water, as for example when a warm water stream is dammed and the resulting pond drains from the bottom. Warming of water may be considered pollution, but I have never heard of the opposite effect being called pollution, though it conceivably could damage a warm water fishery.

Aquaculture does not usually generate suspended solids in the form of silt, though pond construction may cause temporary silt discharges. In fact, impoundment is more likely to reduce downstream silt loads. Pollution by suspended solids is generally in the form of fish wastes and uneaten feed. This is important largely as a sign of high organic content leading to oxygen depletion. A heavy organic load may fertilize downstream waters so that "rough" fish, such as suckers, chubs, catfish, etc., move in and compete with or displace trout or other game fish. Dissolved oxygen in aquaculture effluent is proportional to the rates of stocking and feeding. (I have given some suggestions on how to deal with oxygen depletion in Chapter VIII–4.)

For some aquaculturists, one of the more difficult obstacles is finding out just what are the applicable standards in thermal or organic pollution. Point sources of water pollution from most industries in the United States come under the Federal Water Pollution Control Act of 1972, administered by the Environmental Protection Agency (EPA), or state agencies designated by the EPA. Industries must apply for a National Pollutant Discharge Elimination System (NPDES) permit, which may specify the types of pollution control to be carried out. So far the EPA has not set specific effluent standards for aquaculture (except for facilities where on-site processing of cultured aquatic animals is a source of pollution), but has left the matter of awarding NPDES permits to regional EPA administrators and/or state authorities. The result is that the aquaculturist may not know to whom to apply or what rules to obey, and may wind up as the bystander in a dispute between the EPA and state authorities. There is a growing tendency to try to cut through this fog by forming coordinating agencies at the state level. For example, in New Jersey, prospective aquaculturists can inquire of the Division of Environmental Coordination in the Department of Environmental Protection to receive information on what sorts of reviews, environmental impact statements and permits may be required.

Effluent standards

In all, the situation is confused. Some sectors of the aquaculture industry have tried to use this fact, along with the lack of information on treatment technologies for aquaculture pollutants, to call for preferential treatment from the EPA. This is no more defensible than similar stances which might be taken by other industries. Clearly, the aquaculturist's responsibility for the environment is the same as anyone else's.

A real injustice to the aquaculturist arises when separate permits are required for each point of discharge. While it is perfectly reasonable that aquaculture effluent at every discharge point be required to conform to standards, this ruling subjects some aquatic farmers to the considerable legal expense and red tape of permit application several times over. It would be more fair to require a single permit application for each farm, incorporating data collected at each point of discharge and stipulating that every discharge must meet the same standards.

Most non-commercial aquaculturists and the smallest commercial growers will find that their operations are insignificant, or even beneficial, in terms of downstream water quality. As the size or intensiveness of the operation increases, so does the potential for pollution. This fact was recognized in 1976, when the EPA redefined the term "aquatic animal facility" to exclude certain small operators. Nevertheless, you will be wise to check with state authorities or the regional EPA office before proceeding with even a small aquaculture operation.

The above remarks on protection of public waters from pollution due to aquaculture refer to the United States. My impression is that in Canada fresh water aquaculture is not generally recognized as an industry in this sense. It is also worth noting that, while cage culture in public waters can result in some pollution of water beneath and around cages, neither country has enacted legislation restricting cage culture on this basis.

Disease prevention

Some of the restrictions on importation of fish are imposed, not to prevent introduction of exotics, but to check the spread of fish disease. Thoughtless introduction of fish leads to proliferation of diseases and parasites, or the breakdown of genetic resistance to pathogens. A recent case in point is the appearance in the United States of the cyprinid fish tapeworm, believed to have been brought in with grass carp from Asia.

The spread of disease through transfer of stock has been especially serious in the trout industry. This may be partly due to the relative lack of hardiness of trout as compared to many warm water fishes. But it is also true that trout stock have been far more extensively exchanged than other fish. As a result, we can expect greater restriction of salmonid importation in the future. Though this may seem harsh to some trout farmers, it really is not when you consider the alternatives. Many of the diseases of salmonid fish, especially virus diseases, are contagious and incurable, and may result in near-total mortality. Some of these diseases are still unknown in North America, or are not widely distributed here. If such a disease is found, checking its spread may require the total destruction of a culturist's stock and repeated chemical treatment of the water suppply. If this happens, the culturist will be put out of business for one to several years, if not permanently.

Drugs and chemicals

One method of dealing with fish diseases is through drugs or other chemicals, but legal restraints make it almost impossible for commercial aquaculturists in the United States to employ chemical therapy. They are also hindered in applying chemicals for other purposes. At present, the only chemical substances which may legally be used where fish are grown for sale as food are Terramycin, Sulfamerazine, sodium chloride and vinegar. Canadian growers and non-commercial culturists are for the most part free to use whatever chemicals are not altogether illegal, though they could be held liable for any damages resulting to public waters.

The restrictions on chemicals in aquaculture are imposed to protect public health. The present situation in the United States shows how well-intentioned lawmaking administered through ponderous bureaucracies can retard both good and bad development. The stickiest part of the regulations concerning use of chemicals in aquaculture is the Delaney Clause of the Federal Food, Drug and Cosmetic Act (FDCA). This states that carcinogenic substances shall not be added to food products. As presently defined, any chemical is presumed to be carcinogenic until it passes a battery of tests and receives clearance from the Food and Drug Administration (FDA) and the EPA. The tests are expensive and the costs must be borne by industry, so only a chemical with large commercial potential is likely to be cleared. As small as the aquaculture industry is, it is easy to see why companies are not likely to try to obtain clearance for chemicals manufactured specifically for aquatic use. It is also easy to see why patentable synthetic chemicals are more likely to be submitted for testing than common, inexpensive substances.

In my opinion, the assumption that synthetic chemicals are carcinogenic until proven otherwise is justified in terms of protecting the public. We already know enough about some synthetics to justify great caution, even if it does retard development in new fields like aquaculture. But when a commercial aquaculturist cannot legally treat stock with a substance commonly administered directly to human beings the situation becomes absurd.

Another absurdity to which some commercial aquaculturists are subjected in the name of public health comes in the form of regulations concerning food processing. As long as you deal exclusively in whole live fish you need not worry about food processing regulations. But scale, skin, gut or fillet one fish for a customer and you may come up against a whole new set of regulations. In some places, "processing" of fish requires that you invest in all stainless steel equipment, sterilization facilities, and an inspector on the premises at all times when fish are being processed. Where these regulations are not in effect, some fish which cannot be profitably marketed whole can be sold at a profit as fillets or cleaned fish. But where such "public health" laws apply, the effect has often been to exclude small-scale commercial aquaculture altogether. Clearly a legal distinction needs to be made between, for example, canning plants, which present a serious threat to public health if they are allowed to operate unsupervised, and an aquaculturist who cleans fresh fish for customers.

Processing fish

Zoning may present one of the first legal obstacles to your getting started in aquaculture. Most existing zoning ordinances were drawn up without a thought of aquaculture. The would-be fish farmer may find that, while aquaculture is very much in the spirit of a particular zoning law, it goes against the letter. It is sometimes possible to have a zoning ordinance declared invalid if it is unreasonable, arbitrary, discriminatory or confiscatory. Or you may be able to obtain a variance if it can be shown that aquaculture will not alter the essential character of a locality. Whatever your current situation, I urge all aquaculturists to be aware of zoning and to be active in determining zoning policy. Aquaculture, as a new and little understood activity, is likely to be adversely and unfairly affected by zoning if its proponents do not take it upon themselves to speak out and educate the public.

Zoning

Even where zoning and water use restrictions present no problem, you may still be required to seek local permission to build a pond. The justification for this is primarily in terms of public safety. The U.S. Soil Conservation Service or local building contractors should be able to advise you on this.

Pond construction and public safety

Farm ponds and the like may be overpoweringly attractive to some fishermen, and you may have to resort to posting, whether you like it or not. In many cases posting will not give adequate protection, and a sturdy, forbidding fence may be necessary. These precautions are not only to prevent theft or damage to the facilities. There is also the possibility that someone will be drowned or injured, in which case you may be held liable. Posting and fencing make it clear that the public is not invited, and reduce the risk of financial liability. Nevertheless, you should see to it that your homeowner's insurance covers water-related accidents.

Insurance

A final, legal aspect of aquaculture concerns research. Small-scale aquaculture is a new phenomenon in North America and—as I am embarrassingly aware in writing this book—there are still great gaps of information on the subject. Some of this information will grow out of the experience of practical aquaculturists, but it

Research legislation

should also be the legitimate concern of scientific research, including research funded by the government. To date, however, both private and public support for aquaculture research has been woefully inadequate. Both of the North American federal governments have failed to resolve the question of just who is responsible for aquaculture research, and the same is true of all but a very few of the states and provinces.

The organizations mentioned earlier in this chapter have tried to bring the cause of aquaculture to the attention of the federal governments, but the legislation so far passed has not resulted in much research money being made available. Further, such government money as does go into aquaculture research is almost entirely geared to be of use to aquabusiness, or to production of sport fish by government hatcheries. This is why we have good nutritional data for channel catfish (*Ictalurus punctatus*) and a few species of trout and salmon, which are important to aquabusiness and/or as hatchery-reared sport fish. But we have none at all for sunfishes, bullheads, carps, and others which could eventually be of greater importance as cultured food fish. This is also why there is a great proliferation of scientific papers on diets which use readily industrializable fish feed ingredients, like fish meal and grains, and next to none dealing with feeds like live insects and earthworms, which, while they may be less suited for industrialization, are readily usable by the small-scale grower.

The irony is that it is the small-scale grower who is most in need of government help if appropriate research is to be financed. We can scarcely expect aquabusiness to support development which may lessen its share of the economic pie, and small-scale growers are in no position to subsidize much research themselves. I urge all readers to keep themselves informed on governmental policy with regard to aquaculture research, and to use whatever political power they have to promote the funding of research in aquaculture—particularly in small-scale aquaculture.

PART IX

Aquaculture for Non-Food Purposes

Introduction

Certainly the production of food is the most important task for aquaculture today, and it is the focus of this book as well. But it is not the only economic role open to the aquatic farmer. In the following short chapters I will do no more than outline some of the economic options in non-food aquaculture, and open the door to written and other information sources.

You can look at some of the processes discussed here (fee fishing ponds, culture of bait animals, and sale of eggs and young to other culturists) as part of a chain leading eventually to food on the customer's table. But, even where this is not so, you need not view cultivation of these aquatic cash crops strictly as an alternative to cultivation of food organisms. You will note that many of the species mentioned in this section are also useful as food animals. Cultivating them may be particularly appropriate to the homesteader or small farmer, who may retain a small portion of the harvest for home table use, while also realizing a profit from creatures perhaps not economically feasible to grow for sale as food.

CHAPTER IX-1

Fee Fishing Ponds

To an avid sport fisherman like myself, the thought of paying for the privilege of hauling naive, hand-fed fish out of a specially stocked pond is not appealing. In some of the smaller trout fish-out ponds in particular the process approaches the proverbial "shooting fish in a barrel." Certainly fee fishing is not to be confused with the sport, the craft and the nature experience that is angling.

Yet fee fishing has a place in the scheme of things, and not just as a money-maker for the proprietor. After all, no one objects to "pick your own" fruit farms, so why should we object to "catch your own" fish farms?

From the aquaculturist's point of view, fee fishing is often one of the most profitable ways to market a fish crop, or to utilize over-size or otherwise left-over fish. It also eliminates the problem of unsalable harvests, and may allow the farmer to make additional money on services and supplies. Fee fishing can alleviate the processing problem, though many pay pond operators find it good business to dress fish for a slight additional charge.

Advantages of fee fishing

Fee fishing may also represent a real service to the customer who views it for what it is, and not as a substitute for sport fishing in the natural environment. A well-managed fee fishing pond can be at one time a surer supply of fish than natural waters, and a cheaper one than the usual commercial sources. Many families who patronize such establishments realize real savings on food bills.

Fee fishing ponds are sometimes stocked with purchased or wild-caught fish, but the fish culturist is in a favored position to ensure the continuous presence of a desirable catch. Among the commonly cultured fish popular for this purpose are channel, blue and white catfish (*Ictalurus punctatus, I. furcatus* and *I. catus*), bullheads, rainbow trout (*Salmo gairdneri*), largemouth bass (*Micropterus salmoides*), bluegills (*Lepomis macrochirus*), hybrid sunfishes and crappies (*Pomoxis* spp.).

Popular fee fishing species

As in many businesses, the difference between commercial success and failure often lies in complementary goods, services and facilities (dressing fish, bait, soft drinks, picnic tables, parking, etc.). These and other aspects of the business are well treated, though with the usual bias toward large-scale operations, in Grizzell's paper (1972). (See Appendix IV-1.)

CHAPTER IX-2

Culturing Bait Animals

Bait "minnows"   In most parts of North America raising "minnows"[1] commercially as bait for hook and line fishing predates commercial food fish culture. In Arkansas, the original center of American fresh water fish culture, and still the leader among the states and provinces in total production, bait minnow culture accounts for more acreage (21,217 acres or 8,699 ha) and brings in more dollars (12 million annually) than catfish farming. In the United States as a whole, the annual crop of bait fish is valued at forty-three million dollars (wholesale).

The two predominant species are the fathead minnow (*Pimephales promelas*) and the golden shiner (*Notemigonus crysoleucas*), but a myriad of other species are locally important. At least two commonly grown bait species, the golden shiner and the goldfish (*Carassius auratus*) (sometimes known in the bait trade as "Baltimore minnow") reach considerable sizes, feed on plankton, and can be produced in respectable quantities (over 500 pounds per acre or 552 kg/ha using current techniques). If it were not for the low quality of their flesh, they might have potential for culture as food fish. More attractive to me is the possibility of achieving high production at low cost in a short growing season of a small species and coming up with a "fresh water sardine." This has yet to be attempted.

A secondary use of some "bait species," notably the fathead minnow, is as forage for cultured catfish and bass. (See Chapter III–5.)

Other bait animals   Among the other aquatic animals used as fish bait are frogs, larval salamanders, leeches, insect larvae, and crustaceans. Of these animals only crayfish have so far been cultured on a significant scale for this purpose. Since the culture systems used for bait crayfish can also be applied to production of food crayfish for home use and perhaps for sale as well, they are discussed in Chapter II–10.

Mosquitofish   The mosquitofish (*Gambusia affinis*) is often found in considerable quantities on

---

[1] Strictly speaking, the word "minnow" should be used only for various small fishes of the family Cyprinidae, including the chubs, daces and shiners. In North American popular usage it is indiscriminately applied to almost any very small fish, including young of species which attain large size.

bait minnow farms. As the name implies, it feeds on mosquito larvae and is sometimes used to control mosquitoes. The mosquitofish is too small (maximum length 2½ in, or 6.25 cm) to be of much use as a bait minnow, and bait farmers usually treat it as a pest. But as chemical control of mosquitoes wanes, the mosquitofish may become a common control agent, and could assume commercial importance. In recent experiments up to 180 pounds per acre (198 kg/ha) of mosquitofish have been produced in a season in polyculture with catfish, buffalofish and Chinese carps, and up to 460 pounds per acre (506 kg/ha) have been produced in monoculture.

CHAPTER IX-3

Culturing Aquatic Laboratory Animals

A variety of aquatic creatures are sold, live or preserved, as laboratory animals for experimental or educational purposes. Students most frequently encounter frogs, crayfish, "mud dogs" (*Necturus*, a large salamander) and yellow perch (*Perca flavescens*). Live frogs and goldfish (*Carassius auratus*) are often purchased for experimental use.

Economic success has so far been achieved with goldfish, crayfish and frogs. Goldfish are also cultured for use as bait fish, but are more often used as ornamentals. (See Chapter IX–4.) Crayfish are grown both for food and fish bait. (See Chapter II–10 for culture methods.)

Frogs are the only aquatic animals supporting a culture industry based primarily on sale to research institutions and schools. In fact, commercially raising frogs as food or for bait has not yet been proved feasible. (See Chapter II–9 for a brief discussion of frog biology and culture.)

Commercial culture of frogs and/or crayfish (crayfish are a favored food of frogs when small) may be attractive to some small farmers, as it offers the possibility not only of appreciable income from a small area, but of skimming a few gourmet meals off the top.

CHAPTER IX-4

Culturing Ornamental Fishes

The ornamental fish industry is a substantial one. The diffuse nature of the business makes it difficult to come by accurate figures, but it is estimated that the wholesale value of ornamental fishes imported into the United States is about eighty million dollars annually. Canada imports between seven and eight million dollars' worth. The retail value of imported and domestically produced ornamental fishes, plus supplies for their maintenance, sold annually in the United States is estimated at seven hundred million dollars. The retail value of ornamental fishes actually raised in Florida, the principal producing state, amounts to nearly two hundred million dollars. Florida production alone makes ornamental fish culture the single most economically significant form of aquaculture in North America.

Ornamental fish industry's scope

As environmental degradation and overfishing take their toll in Latin America, Africa and Southeast Asia, and as restrictions on the import of exotic fishes become more stringent, the commercial cultivation of ornamental fishes in North America will become more important. A fortuitous by-product of this situation may be that more attention will be paid by aquarists to our beautiful native North American fishes, but for the time being at least, the only commercial possibilities in ornamental fishes are with exotics.

Aquaculture's role

Of the over 1,000 species of ornamental fishes which have been commercialized to date, by far the leader in cash value is the goldfish (*Carassius auratus*). Originally native to China, it is used for fishing bait and as a laboratory animal as well. The goldfish industry and culture methods are well established and have been described in print many times. (See Appendix IV-1.)

Goldfish industry

The tropical fish breeding industry is much less organized. Some individuals have made substantial sums of money operating primarily or entirely with indoor aquaria. Such successes may entail many years of experience and high degrees of skill. Understandably, the possessors of these accomplishments are often secretive about their methods.

Tropical fish industry

Less exacting methods of ornamental tropical fish cultivation are practiced in outdoor facilities, especially near Tampa, Florida, where there is an abundance of isothermal spring water perfectly suited for many of the popular aquarium fishes. Some of the Florida growers get by simply by placing selected fish in outdoor pools and letting nature take its course. Certain geothermal waters offer similar, but unexploited opportunities.

Options for growers

Three courses are open to the prospective commercial grower of ornamental fishes:

(1) Producing "bread-and-butter" fish such as goldfish or easily bred tropicals (notably the live-bearers, family Poeciliidae): Since these fish bring rather low prices, a fair amount of water, high rates of production, and considerable efficiency are necessary.

(2) Producing hard-to-breed species: Supplies of many of the more popular tropicals are still partially or totally dependent on imports. A consistently successful breeder of one of these species can be in a commercially advantageous position. Breeding pairs of some prized species have sold for five hundred dollars or more, but volume sales at moderate prices is a more likely route to profit. Anyone trying this approach will need time for experimentation, plus experience and skill in the aquarium hobby.

(3) Selective breeding of highly prized varieties: Certain ornamental fishes have shown a high degree of genetic plasticity. That is, they can be bred, through selection, to produce many forms rarely or never seen in wild stock. Notable in this respect are the goldfish, the guppy (*Poecilia reticulatus*), the swordtails and platies (*Poecilia hellerii, Poecilia maculatus*, etc.) and the Siamese fighting fish or betta (*Betta splendens*). An example of the commercial potential in this field is the achievement of a friend who has developed unique and beautiful strains of guppy, and is now able to make a living on the basis of only three steady customers.

Shipping

All commercial growers of ornamental fishes must develop a high degree of competence in shipping live fishes. (See Chapter VI–3.)

Ecological dangers

Exotic fish breeders must take every precaution to avoid escape of fish into the natural environment. This problem is especially severe in the southern U.S., where natural conditions may approximate the native haunts of tropical fishes. Florida serves as an example of the scope of the problem: Because of the tropical climate, the established ornamental fish culture industry, and thoughtlessness on the part of many, at least twenty-five species of uninvited exotic fishes are now more or less established in the fresh waters of the state. Some of these fishes may prove benign or even beneficial in the long run, but others already seem to be detrimental to the native fauna.

As an illustration that the problem is not confined to semi-tropical areas, consider the goldfish, which may now be found wild in almost all of the states of the U.S. and in at least three Canadian provinces (Ontario, Alberta, British Columbia). Goldfish, in their place, have their virtues, but in North American natural ecosystems they generally lose their ornamental qualities and revert to a short-finned, drab olive-brown form, and compete with more desirable food and game species. They have also been accused of roiling waters and eating spawn of other fishes.

CHAPTER IX-5

Hatchery Operation:
Selling Eggs and Young Fish

Aquaculture relates to the fish hatchery business exactly as agriculture does to the seed business. In some instances it is easier for culturists to breed their own fish, but it will often prove necessary or more economical to purchase stock. Such was the situation of the early Chinese fish culturists who, until the perfection in the 1960's of techniques for spawning Chinese carps by pituitary injection, were forced to rely on purchasing eggs or young fish from specialists who captured them in the large rivers where natural spawning occurs. To this day in China, North America, and elsewhere it often proves convenient for certain individuals favored by skill, interest, location and/or water supply to specialize in production of "seed" for distribution to other aquaculturists.

Role of hatcheries

Two types of "seed" material are generally sold: "eyed" eggs and "fingerlings." The newly fertilized egg is too delicate to tolerate much handling, but when the "eyed" stage is reached (when the eye of the embryonic fish is clearly visible as a black dot), the egg develops a tough membrane and is capable of withstanding considerable shock. With air transport, it has become possible to ship eyed fish eggs almost anywhere in the world.

Types of "seed" material

The next two life stages, the larval fish, or "alevin," and the early "fry," are again delicate and not suited for shipping. The next opportunity for sale comes with the "advanced fry" and "fingerling" stages. From the breeder's point of view, the sooner he can get rid of these fish, the better, so as to reduce the cost of labor and the possibility of loss. For the customer, there is a trade-off. Smaller fish are cheaper, while large fingerlings will reach harvestable size sooner, with lower mortality. But breeders can demand a higher price for growing larger fish and absorbing the cost of mortality.

For certain polyculture schemes, or where an optimum rather than a maximum size food fish is desired for marketing at a preset time, it may be better to begin with smaller stock. Some Chinese culturists know how to indefinitely maintain fish without growth, so that growers may purchase stock of a certain size at any time of year. These fish, though sometimes called "stunted," are not truly stunted, since they will resume growth when placed under suitable conditions. Maintaining "stunted" fry and fingerlings requires a rather precise understanding of the maintenance (as opposed to growth) requirements of fish which is uncommon in North America.

Choice of species

The greatest likelihood of commercial success in the hatchery business is with trout or channel catfish (*Ictalurus punctatus*). Breeding methods for these fish have been written about extensively. (See Appendix IV–1.) Ornamental fishes also offer interesting prospects for the breeder.

The hatchery industry is also important in the supply of hybrids and selected strains of fish, since most food fish growers have neither the time nor the space to devote to hybridization and selection. (See Chapter V–2 for a discussion of hybridization and selective breeding.) To date, selection and maintenance of selected strains is most important in trout culture, but a trade is beginning to develop in sunfish hybrids. (See Chapters II–6 and II–1.)

In an earlier chapter I distinguished between animals which can and cannot be easily bred by the farmer. You might think that, excepting for hybrids, there would be little market for young of those species which reproduce naturally in culture ponds. This is not necessarily so with respect to "farm pond" fishes. Some years ago, the U.S. Federal Government and many of the states provided free fish for stocking farm ponds. However, these programs have been greatly curtailed, and many farm pond owners find it convenient to buy new or replacement stocks of fish from private producers.

Markets

Potential markets for hatchery operators include not only food fish growers, but fee fishing operators and recreational fisheries. Sometimes, government fish and game agencies contract with private hatcheries to buy trout for stocking in public fishing waters. (But note that the trend in trout fishery management is away from stocking.) Angling clubs, children's fishing "derbies," and pond owners whose main interest is recreational represent additional markets for trout, catfish, bass, sunfish and other fishes.

If fish culture in North America develops as I would like, we will eventually see small regional hatcheries providing "seed" for clusters of fish farmers. It will then be even more true than now that commercial hatchery operation is most likely to succeed where numbers of farmers are already growing the species produced by the hatchery. Some hatcheries do succeed on the basis of long-distance shipment, but this is more risky and costly.

While there is stiff competition in the trout and catfish hatchery business, stock of other fishes is often difficult to obtain. For the adventurous entrepreneur, there is the exciting possibility of performing a real service by pioneering with other species. There is a developing market for hybrid sunfishes in the Midwest and South, which several breeders are already beginning to meet. In the future, suppliers of top quality monosex tilapia or selected carp stock may be in an advantageous commercial position. Now is the time to begin such enterprises if you are willing to take the risk that markets will not develop.

Opportunities in the hatchery business may be great, but so are the risks. Even more than food fish growers, hatchery operators must be sure of adequate water and feed supplies, sanitation, and ready markets. Required are first-rate facilities and expertise in shipping, as well as a good understanding of fish biology and a high level of skill. It is sometimes possible for the "average" fish culturist to make a go of food fish growing. In the commercial hatchery business only the superior operators succeed in the long run.

PART X

Physical and Chemical Aspects of the Water

CHAPTER X-1

Physical Aspects of the Water: Temperature, Flow Rate, and Light

Temperature

In the broad sense, an aquatic organism's physical environment includes the body of water in which it lives, the plants and animals with which it shares the environment, the food supply and perhaps certain pollutants—all of which were dealt with in preceding chapters. Here I will discuss three physical factors which can be measured in any aquatic environment: temperature, flow rate, and light. Of these the most important to the aquaculturist is temperature.

Temperature and fish habitats

Fishery biologists are accustomed to the division of aquatic animals into two or three classes, depending on their temperature requirements. The categories of "warm water" and "cold water" fish are widely recognized. Among the major fresh water aquaculture species, only trout fall into the cold water category. Other cold water animals I discuss in this book are salmon, whitefish, mooneyes, some of the suckers (but not the buffalofishes), and some crayfishes and clams. Some would refer to all the other species mentioned here as warm water fish, but others recognize a third, intermediate, group referred to as "cool water" fishes. Just which species might be included depends on whom you talk to, but certainly the smallmouth bass (*Micropterus dolomieui*), yellow perch (*Perca flavescens*) and walleye (*Stizostedion vitreum*) would qualify. These categories are merely a convenience; there is much overlap and warm, cool and cold water species may sometimes be found in the same place at the same time. The categories reflect only the temperatures at which these species do best, and the upper limits of their temperature tolerances.

Also, these categories apply to fishes of the temperate zones. For the sake of discussion, we need a fourth category for species which may die of cold at temperatures above the freezing point: "tropical fishes." Such species are found in nature in North America only in isolated thermal springs and a few portions of the extreme southern United States. However, aquaculturists can grow tropical species elsewhere during the summer months or in heated environments. Table X–1–1 lists what is known of the thermal requirements of some important or potentially important aquaculture animals.

Temperature and growth rate

A few comments are needed to explain Table X–1–1: "Lethal" temperatures are determined experimentally by acclimating animals at a given temperature, then raising or lowering it. The temperature when fifty percent of the animals have died is the "LD 50" or "lethal" temperature. It varies greatly with the acclimation temperature. For most aquaculture purposes, the lower lethal temperature of those fish native to environments with regular freezing is at or near the freezing point, since in nature freezing usually happens gradually enough that fish can acclimate nearly to the end. Occasionally, fish will even survive being frozen solid. So I have assumed 32°F (0°C) as the lower lethal temperature of all northern fishes, even though this has not been experimentally determined in most cases.

The "breeding" temperatures listed are the extremes at which spawning has been observed in nature and culture. Usually, breeding at a temperature somewhere near the midpoint of the listed range will result in the best hatching rates, survival and larval growth.

Though there is a fair body of experimental data on "preferred" temperatures, this is not much use to the culturist. Fish sometimes select temperatures which do not result in maximum growth, and may even be drawn to water which is lethally warm. I have listed the "temperature range for maximal growth." This is not intended as precise scientific data, but rather as a guide, being a compilation of experimental data on growth rate and feed conversion, plus information collected from practical aquaculturists.

The fish in the table are grouped roughly according to the temperature categories I have mentioned. Some are listed as intermediate between the categories. In trying to make sense of these groupings, be conscious of genetic variation. An extreme example is the common carp (*Cyprinus carpio*). There are strains of common carp which are adapted to climates as diverse as Siberia and Java.

Finally, here is one example to prepare you for the full complexity of considering temperature factors. In nature, the pumpkinseed sunfish (*Lepomis gibbosus*) is more commonly found in cool water than the bluegill (*Lepomis macrochirus*). So you might then consider it as a substitute for the bluegill in cool farm ponds. Yet the table shows that the "temperature range for maximal growth" of pumpkinseeds is higher than it is for bluegills. This may be because pumpkinseeds spawn at lower temperatures than bluegills. So, at high temperatures, adult bluegills may "waste" energy in reproductive behavior, while pumpkinseeds continue growing.

For each species in Table X–1–1, the range of temperatures producing maximal growth is much nearer the upper lethal limit than the lower. Figure X–1–1 further illustrates this relationship.

Taking advantage of heightened growth at high temperatures is not as simple as it might seem. There is a certain brinkmanship involved, due to the risk that the upper lethal temperature will be approached and that at temperatures just a few degrees short of that level, the fishes' resistance to stress and disease is substantially

Table X-1-1. Temperature requirements, in degrees Fahrenheit (degrees Celsius) of some cultured or potentially culturable fishes.

Category	Species	Lower lethal temperature	Upper lethal temperature	Breeding temperature	Temperature range for maximal growth
Cold water	Lake whitefish (*Coregonus clupeaformis*)			32.9–50.0 (0.5–10.0)	
	Arctic char (*Salvelinus alpinus*)			46.4 (8)	53.6–60.8 (12–16)
	Brook trout (*Salvelinus fontinalis*)		73.4–77.5 (23.0–25.3)	53.1 (11.7)	44.6–64.4 (7–18)
	Golden trout (*Salmo aguabonita*)			44.6–50.0 (7–10)	
	Rainbow trout (*Salmo gairdneri*)		75.2–82.4 (24–28)	41.9–62.6 (5.5–17)	55.4–69.8 (13–21)
	Brown trout (*Salmo trutta*)			44–50.0 (6.7–10)	65–75 18.3–23.9)
	Longnose sucker (*Catostomus catostomus*)			80.6 (27)	
Cold water to cool water	White sucker (*Catostomus commersoni*)	32–42.8 (0–6)	78.8–88.2 (26–31.2)		
Cool water	Razorback sucker (*Xyrauchen texanus*)			53.6–71.6 (12–22)	
	Yellow perch (*Perca flavescens*)	32–38.9 (0–3.7)	70.3–85.5 (21.3–29.7)	44–54 (6.7–12.2)	66.2–69.8 (19–21)
	Walleye (*Stizostedion vitreum*)			68 (20)	
Cool water to warm water	Pike (*Esox lucius*)			44.6–51.8 (7–11)	
	Chain pickerel (*Esox niger*)		over 98 (36.7)	42.8–60.8 (6–16)	68 (20)
	Smallmouth bass (*Micropterus dolomieui*)		84.2–95 (29–35)	55.8–65 (13.2–18.4)	78.8–84.2 (26–29)
	Rock bass (*Ambloplites rupestris*)		87.1–95.0 (30.6–35)	60.1–78.8 (15.6–26.0)	
	American eel (*Anguilla rostrata*)		95 or more (35)		
	Redear sunfish (*Lepomis microlophus*)			68–75.2 (20–24)	
	Pumpkinseed (*Lepomis gibbosus*)		82.4–86.4 (28.0–30.2)	55.4–84.2 (13–29)	77–86 (25–30)
	Bluegill (*Lepomis macrochirus*)	32–34 (0–1.1)	82.4–102.2 (28–39)	62.6–89.6 (17–32)	68.9–86 (20.5–29)

Table X–1–1. *Continued.*

Category	Species	Lower lethal temperature	Upper lethal temperature	Breeding temperature	Temperature range for maximal growth
	Black crappie (*Pomoxis nigromaculatus*)		93.2 or more (34)	39.9–68 (4.4–20)	
Warm water	White crappie (*Pomoxis annularis*)			57.2–73.4 (14–23.5)	
	Green sunfish (*Lepomis cyanellus*)		68.0–82.4 (20–28)	68–75.2 (20–24)	82.4 (28)
	Largemouth bass (*Micropterus salmoides*)	32–53.2 (0–11.8)	84–102 (28.9–38.9)	52.7–84.2 (11.5–29)	73–86 (22.8–30)
	Bigmouth buffalo (*Ictiobus cyprinellus*)			57.9–69.8 (14.4–21)	
	Brown bullhead (*Ictalurus nebulosus*)		83.5–99.5 (28.6–37.5)	70 or more (21.1)	
	Yellow bullhead (*Ictalurus natalis*)		91.4 or more (33)		82.4–84.2 (28–29)
	Black bullhead (*Ictalurus melas*)		95 (35)	69.8 (21)	
	Channel catfish (*Ictalurus punctatus*)	32–42.8 (0–6)	86.5–95.0 (30.3–35)	59.9–85 (15.5–29.5)	69.8–80.6 (21–27)
	Blue catfish (*Ictalurus furcatus*)			71.6 or more (22)	
	Fathead minnow (*Pimephales promelas*)	32–51.8 (0–11)	82.4–91.8 (28–33.2)	60–86 (15.6–30)	
	Mosquitofish (*Gambusia affinis*)	34.7–58.1 (1.5–14.5)	95.7–99.1 (35.4–37.3)		
	Golden shiner (*Notemigonus crysoleucas*)	32–58.1 (0–11.2)	86.9–95 (30.5–35)	64.4–86 or more (18–30)	
Warm water to tropical	Common carp (*Cyprinus carpio*)	32–33.3 (0–0.7)	87.8–96.3 (31–35.7)	62.6–82.4 (17–28)	80–85 (26.7–29.4)
	Goldfish (*Carassius auratus*)	32–48.2 (0–9)	84.2–105.8 (29–41)	59–85.6 (15–29.8)	
Tropical	Java tilapia (*Tilapia mossambica*)	48.2–53.6 (9–12)		70 or more (21.1)	71.6–86 (22–30)
	Blue tilapia (*Tilapia aurea*)	48–55 (8.9–12.8)	107.6 (42)	73.9 or more (23.3)	85 (29.4)
	Fresh water shrimp (*Macrobrachium rosenbergi*)	40–53.6 (10–12)			
	Walking catfish (*Clarias batrachus*)	55–65 (12.8–18.4)		85–92 (29.5–33.4)	85 or more (29.4)

reduced. Also, what applies to the metabolism of fish applies to other metabolisms as well, so that the rate of transpiration of plants and decay of organic materials is increased, and with them, the danger of oxygen depletion. In addition, while the increased metabolic activity of the cultured fish and that of other aquatic animals requires more oxygen, water's capacity to retain dissolved oxygen is less at higher temperatures. I will explore this relationship in the next chapter but here I wish to show that temperature alone is of next to no value as a guide to stocking rates or species combinations.

Thermometers   Among the first tools you buy should be one or more accurate thermometers. Any thermometer for use in normal air temperatures is suitable for use in water, but there are a few types which may be particularly useful to aquaculturists. Most are available with either Fahrenheit or Celsius scale; which to get depends mostly on whether you expect to get most of your information from popular or scientific sources. Thermometers with both scales are most convenient.

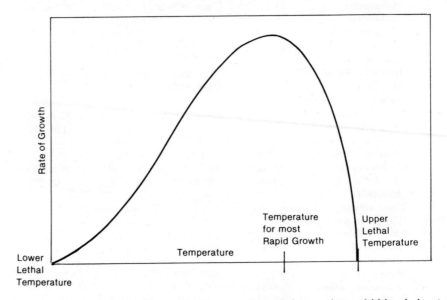

FIGURE X–1–1. *Typical growth rate curve for a fish (or other cold-blooded animal) at temperatures ranging from minimum to maximum tolerance (rate of feeding held constant).*

Types of thermometers   Aquarium thermometers, variously designed to float, clip on to tank walls or rest upright on the bottom, may be handy for use in small enclosures. Be warned, though, that many aquarium thermometers are of low quality. You may do better to adapt some other type of thermometer to float or attach. Soil thermometers, which can be read while looking down on them are particularly convenient where a floating instrument is needed.

Many aquaculture systems are shallow enough so that knowing only the surface temperature is sufficient for management purposes. You can read subsurface temperatures with any thermometer by sinking it with a line attached, hauling it up rapidly and reading it immediately, but you can achieve greater accuracy and convenience with a fisherman's thermometer. A variety are available from fishing tackle suppliers; some have built-in depth gauges. You can obtain simultaneous readings of temperature at the surface and at any depth with an indoor-outdoor thermometer.

If you are concerned about daily temperature variation, maximum-minimum thermometers are available through scientific supply houses. You can obtain greater detail with a recording thermometer, but these expensive instruments are necessary only for some types of aquaculture research.

You will find a use for thermometers even before the aquaculture system is constructed. If you can wait a year—I don't blame you if you can't—it would be useful to have a year's temperature data on your water source. Of course the temperature of the water in ponds, raceways, tanks, etc., will not always be the same as that of the source. The following are the chief natural sources of variation:

Sources of temperature variation

• Air temperature: Water changes temperature more slowly than air, but is affected by the air above it. This is particularly important in broad, shallow ponds where, for example, water which enters at a temperature suitable for cold water fish may warm up and become unsafe for them. Air temperature is the main influence on seasonal variation in water temperature, which in turn is a major limitation on aquaculture in northern climates.

• Solar radiation: Sunlight not only heats water but causes evaporation, reducing depth and volume, and accelerating the rate of heating. Shade moderates these effects.

• Depth: Shallow bodies of water tend to warm and cool more rapidly, and usually have the same temperature from top to bottom. Deeper water may stratify, with colder, heavier water on the bottom. In some cases this makes it possible to culture animals with different temperature requirements in the same pond. For fishermen and fishery managers an important aspect of stratification is the occasional formation of three distinct layers known respectively, from top to bottom, as the epilimnion, thermocline and hypolimnion. Usually, only cage culturists need be concerned with this form of stratification, which is discussed in Chapter IV-3.

• Color of the water and bottom: As with solid objects, dark water, or colorless water over a dark bottom will absorb heat faster than their opposites. Turbidity, which colors while reducing transparency, may concentrate solar heating in the surface layer. (Turbidity is discussed later in this chapter.)

• Circulation: Gravity-induced flow is the most important form of circulation, but wind circulation may cause cooling and convection currents which serve to equalize pond temperatures. Where a fairly rapid flow can be maintained through a system, the effects of air temperature, solar radiation, shade, depth and color are reduced so that a greater volume of water may be kept at a temperature closer to that of the source.

The logical time to begin putting together this information—the temperature of the water at the source and the ways in which the environment will affect it—is when you are just beginning to design the system (though improvements can be made in pre-existing systems as well).

Water source temperature

Most aquaculturists have to choose the species to match the water temperature, but if you are lucky enough to have your choice of several sources of water you may be able to do the opposite. The best source is an isothermal spring, that is a spring which provides water at a constant temperature year-round. Large isothermal springs are responsible for two of the major concentrations of fish farms in the United States: the trout farms of Idaho's Magic Valley, and the ornamental tropical fish producers of west-central Florida. A few privileged persons in the West have access to the ultimate: hot and cold springs which can be mixed to provide any desired temperature.

If your water source, isothermal or not, is ground water with a suitable temperature, you will want to shade it until it enters the growing area. If ground water is too cold for your purpose, you may want to expose it to the sun before use, perhaps in a broad, shallow, dark-bottomed pool.

All ponds receive some water from precipitation, and this will often have a significant effect on temperature. Snowmelt may have a particularly important effect on temperature. You can minimize the influence of snowmelt and run-off, respectively, by snow removal and by building embankments to divert run-off.

Choosing sites and species in relation to temperature

Some seasonal temperature variation will certainly occur in all but very small systems fed by isothermal sources. If you are shopping for land to begin aquaculture you may want to take this into account by choosing a site where the temperature pattern is best for a particular species. Essentially, this is what has happened with the catfish industry: though it began in central Arkansas, the focus of production has shifted to the Mississippi Delta and Louisiana, due to the longer growing season in these more southerly regions.

Most small-scale aquaculturists will not have the luxury of scouring the country for suitable sites, and more can be gained by selecting species to fit a given set of temperatures. This becomes more important as you move northward and the growing season becomes shorter. For example, the standard advice given to farm pond owners is to stock largemouth bass and bluegills. But this advice is based on research done in the central and southern United States. In some northern waters, a combination of for example, smallmouth bass and rock bass, might enjoy a slightly longer growing season and be more productive. Similarly, in very cold water, brook trout might out-produce the rainbow trout grown by nearly all commercial trout culturists.

Don't overlook the possibility of strictly seasonal aquaculture. For example, in many places where tilapia would not survive the winter outdoors, they can be grown to edible size during the summer. In experiments in northern Georgia, rainbow trout, which would not survive summer temperatures, were successfully grown in cages during the winter. The same cages were used during the six warmest months to grow channel catfish, which would survive but not grow appreciably during winter.

Depth and temperature control

The deeper a body of water in relation to its other dimensions, the more stable will be its temperature regime. For this reason, added depth is usually advantageous. In some cases, though, a shallower pond will provide a longer growing season by heating up sooner in the spring. Some deep ponds stratify enough that trout can pass the summer in the deep parts while warm water fish are growing in the shallows.

If a pond is deep enough to show a temperature gradient between top and bottom, take this into account in designing the outlet. A device like the L-shaped turn-down drainpipe illustrated in Figure VII–1–13 normally draws cold water off the bottom, but can also be used to draw warm water off the top simply by removing the outer sleeve.

Solar heat

You can prevent overheating of water by planning for natural shade in siting ponds, or creating artificial levees, hedgerows, etc. However, shading will also cut down on solar energy for photosynthesis, which may be disadvantageous.

In conventional aquaculture deliberate use of solar radiation to heat water is largely limited to siting ponds in full sun and perhaps building them shallower than they might otherwise be. Also, pond bottoms can be covered with dark materials to absorb heat.

More sophisticated use of solar heat has so far been confined to indoor and closed systems. The New Alchemy Institute has pioneered the use of small pools in solar-heated greenhouses or outdoor solar courtyards. (See Chapter IV–4.)

A 350 square foot (32.5 m²) solar greenhouse and aquaculture facility designed and built by the Amity Foundation, Eugene, Oregon. (Photo by William Head, Amity Foundation.)

Getting ready for night at the Amity Foundation greenhouse. The foil-covered styrofoam insulation-reflector panels reflect up to 25% additional light into the greenhouse on clear days and insulate it at night. (Photo by William Head, Amity Foundation.)

Inside the Amity Foundation greenhouse. Fish are being harvested by crowding them into one end of the tank with a rigid net. (Photo by William Head, Amity Foundation.)

It should also be possible to design passive solar heaters for small outdoor ponds, not in order to drastically alter the thermal regime, but to increase growth by adding a few degrees to the temperature at any given moment and extending the growing season.

You can incorporate separate solar heaters as one step in closed systems. One such device used at New Alchemy is shown in Figure X–1–2. A simpler method is to pass water through a coiled black hose in the sun.

Water circulation and temperature

You can regulate the rapidity of circulation of water, and hence its rate of heating or cooling, by designing your facilities with greater or lesser slope or wind exposure. Aerators, particularly those releasing bubbles at the bottom, though used primarily for oxygenation (see Chapter X-2), also serve to circulate and cool water.

Freezing

A possible temperature-related problem which fortunately is very rare is an entire pond's freezing solid. Except in very shallow ponds, this is prevented naturally by an interesting physical property of water: While most natural substances become progressively more dense as they cool, water reaches its maximum density at about 39.2°F (4°C), and at freezing temperatures the very coldest water, in the form of ice, floats on the surface. If water behaved like other substances, during a severe winter ice would form first at the bottom and eventually the entire water column might solidify.

While nature prevents the disaster of a pond's freezing solid, you will sometimes have to prevent freezing over, which can cause catastrophic losses by preventing oxygen absorption from the air and, if the ice becomes opaque or covered with

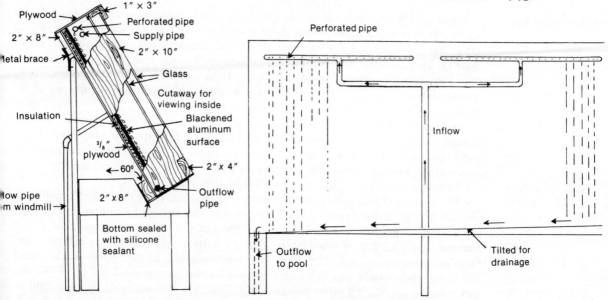

FIGURE X–1–2. *Solar water heater for use in closed aquaculture system.*

snow, will interfere with photosynthesis as well. You can laboriously achieve a crude and temporary cure with an axe, and floating logs on the pond surface may retard freezing over. But the best prevention is to circulate the water. Circulation is particularly effective in controlling freezing if bottom water is exchanged with surface water. One way to accomplish this is through aeration. Another is with a wind-powered turbine called a Pond Master®. (See Appendix IV–4.)

The effect of most of the heating and cooling methods mentioned here, if you measure simply in terms of degrees at any given moment, might seem insignificant. But, if you measure in degree-days, and take into account the change in degree-days over a whole season, you will see that the daily addition of a fraction of a gram to the growth of each cultured animal, the extension of the growing season by a week or two, or the prevention of some temperature related problem may be significant indeed.

Importance of small temperature changes

A final method of temperature control which may occur to some is the use of electrical or fuel burning heaters. Small closed systems can be maintained indoors and heated by constant contact with heated air, but heating water directly can be justified only for the overwintering of small numbers of fish in aquaria. Heating on a larger scale will always prove to be prohibitively expensive and wasteful of energy.

Artificial heating

The current of a stream has an important effect on water temperature. The speed of moving water can be expressed as a simple velocity, as in measuring the speed of an object travelling over land or through the air. However, most of the effects of moving water are functions of its volume as well as its velocity, and so aquatic biologists and aquaculturists usually speak, not of velocity, but of flow rate. The customary units for measuring flow rate are cubic feet per second (cfs) or liters per second.

Flow rate

Trout are the only fish commonly grown in visibly flowing water. A rule of thumb for rainbow trout farmers is that there should be an available water supply of 0.18 cfs (5 liters per second) for each ton of fish maintained, assuming "average" water quality. This quantity will actually be needed only as the crop approaches harvestable size, but this often coincides with the time when the quantity and quality of available water are lowest. The rate may be adjusted downward or upward in accordance with dissolved oxygen concentration.

Growers of other fish (and some trout culturists) more often use "static" waters, though streams, raceways and closed recirculating systems are also used. Even pond culturists, more often than not, have to contend with some flow rate at inlets and outlets. Among the factors to consider in connection with flow rate are:

• Temperature effects (discussed earlier in this chapter).

• The species of fish to be grown: Though circulating water may often benefit fish growth, it also has a negative effect because of the energy fish must consume in swimming or holding a position against the current. You can judge how much current a species can tolerate both from its natural habitat and its morphology. Slender, streamlined fish like trout are well adapted to deal with fairly rapid currents, while broad, flat fish like the sunfishes or less streamlined ones like the catfishes are less so. All other things being equal, any fish will gain more weight in slack water than in flowing water. The effect of current on fish flesh quality is uncertain but exercise (as in resisting a current) is likely to be conducive to high quality.

• Water quality: The main advantage of flowing systems is that pollutants can be eliminated and a high concentration of dissolved oxygen maintained more readily

Daily cleaning of screens, as in this trout rearing facility in Washington, is a must if water quality is to be maintained in raceways. (Photo courtesy Meredith Olson, WCCA, Waverly, Minnesota.)

than in standing water. (See Chapter X–2.)

• Feeding methods: Feed not consumed immediately will likely be lost in flowing water. Slow eaters, such as common carp, are therefore inefficient in flowing water. Plankton-based food chains cannot be maintained in flowing water.

• Quantitative water supply: A flowing water system presupposes a sufficient supply of water during the driest season. Where water conservation is an issue, flow may have to be restricted.

• Erosion: Because flowing water systems must be built to resist erosion they are more expensive to construct. Do not begin construction of raceways or inlet and outlet structures which are expected to withstand high or constant flows without consulting an engineer.

Flowing water allows stocking at greater densities while maintaining adequate water quality. So, your choice of a high, low, or static flow rate, though partly determined by biological and topographical factors, will also depend on the quantity of fish you wish to produce. In weighing the advantages of flowing water against the costs and engineering problems involved, my recommendation for those who can choose from a range of flow rates (particularly for non-commercial growers of warm water fishes) is to opt for minimal flow. If you must increase production and can choose either to increase the flow rate or the volume of your system, the latter tactic is usually safer and cheaper—assuming space is available.

Light is essential in the aquatic environment both to sustain photosynthesis and to permit fish and other animals to see. Light in water is limited by external factors (the diurnal cycle, shade, weather conditions) but also by water itself which, at its clearest, is many times less transparent than air. Both the amount of time during which water is lighted and the intensity of light are important. *Light*

The hours of light over a given body of water are determined entirely by the natural diurnal cycle, unless the water can be covered or artificially lighted. As on land, some wild aquatic animals are active primarily at night, others during the day. While some visual feeders cannot adapt to nocturnal feeding, nocturnal species usually adapt to daylight feeding. But the time of feeding may affect the efficiency of feed conversion. In one experiment, goldfish in an artificially lighted environment grew most rapidly when fed immediately after the onset of light; after this, the rate of growth declined steadily through a twenty-four hour period. Traditionally, early morning is used as a feeding time by professional aquaculturists, but whether all fish operate on the same biological clock as the experimental goldfish is still uncertain. *Diurnal cycle*

Just as day length is important for plant growth, it also affects fish, particularly their sexual maturity. For example, salmonid fishes are either spring or fall spawners, but by artificially altering day lengths over a period of months, fall spawners can be induced to breed in the spring and vice versa.

Shade and cloudy weather affect aquaculture mainly through their influence on the growth of plants, including phytoplankton. Shade may also help to induce feeding in shy or normally nocturnal animals.

How much light is transmitted in water is determined by the color and turbidity of the water. Water need not be "crystal clear" to be transparent. For instance, bog water may have the deep red-brown color of strong tea and still be transparent enough that objects may be clearly perceived at a considerable distance. Color does *Light transmission: color and turbidity* reduce light transmission, but there is little to be done about the color of water short of massive chemical treatment.

More serious interference with light is caused by turbidity. Particles in water scatter light, so that even low level turbidity affects visibility. At higher levels it seriously interferes with light transmission, so that photosynthesis is inhibited.

Fine instruments for measuring turbidity are available, but the standard tool is the Secchi disc (described in Chapter III-2). How much turbidity is tolerable depends partly on the form of aquaculture being practiced, and also on the source of the turbidity. Almost all turbidity is due either to plankton—usually phytoplankton—or to silt particles. Plankton turbidity is either green or in the case of zooplankton blooms, composed of millions of usually red-brown animals which can be seen moving. Silt turbidity in water may appear in any of the colors of soil.

Whether turbidity from plankton is a problem will depend largely on the kind of aquaculture system. In the rare case of reduced visibility or light transmission due to zooplankton, fish will usually crop the "bloom" in a few days. The only danger is that in the interim the sudden increase in aerobic organisms may cause oxygen depletion. Turbidity due to phytoplankton is always an indication of fertility. Assessing and dealing with this problem was discussed in Chapters III-2 and VIII-1.

Silt turbidity

Ideally, silt turbidity should be zero, though there will always be measurable turbidity over certain soils, or if many benthic fish are present. In those parts of the United States having clay soils, significant turbidity is almost unavoidable. I have found it interesting that the fish I have taken from these waters were always of good texture and flavor. Just how much turbidity is too much is debatable, but you should be concerned if silt turbidity exceeds 20,000 parts per million, or if Secchi disc readings of under one foot (30 cm) are obtained.

Turbidity remedies

Preventive measures for siltation were discussed in Chapter VIII-4. If, despite these, you should find yourself with a silt turbidity problem, there are various measures you can take. Turbidity remedies are based on acidifying the water, so that the positive electrical charge on the silt particles becomes negative, causing them to precipitate out. The substance most commonly used for this purpose is gypsum. Alum may be less expensive, but is said to have the side effect of reducing the availability of phosphorus. Recommended dosages for both substances vary widely. Gypsum is prescribed at 100 to 1,000 pounds per acre (110 to 1,100 kg/ha) while the recommended rates for alum are from 15 to 250 pounds per (16.5 to 275 kg/ha). It is not essential to know the precise rate for a given body of water, since either one may be added incrementally until precipitation of silt occurs.

A third turbidity cure employs old hay at seven to ten bales per acre (17 to 25 bales/ha), broken up and scattered in shallow water. (Hay should not be used during cloudy weather, or at other times when there is a high chance of oxygen depletion.)

You can lessen turbidity caused by suspended organic materials by liming. (See Chapter III-2.) For this reason lime is often applied just before, during, or just after manuring.

CHAPTER X-2

Water Chemistry

Water (H_2O) is the "universal solvent"; it reacts with everything with which it comes in contact. The diversity of chemical substances found dissolved in water, and their reactions with each other and with other aspects of the aquatic environment, can be bewildering. Books much larger than this have been written on fairly narrow aspects of water chemistry, and revision of previous thinking and expansion of the field go on constantly. Here I will try only to select some of the chemical substances and reactions with practical importance for aquaculturists and suggest means of measuring and managing them. I will present theory only when it is necessary for an understanding of the chemical processes discussed. References which go deeper into water chemistry are listed in Appendix IV-1.

First, I should mention certain important aspects of water chemistry not dealt with in this chapter. The fertility of a body of water is based on the concentration of organic materials and nutrient minerals in the water and its soil substrate. Since this is the first link in the aquatic food chain, fertility and its chemical constituents are covered in the section on feeding. (See Chapter III-2.) Also, as the concentration of calcium salts is critical to the action of fertilizers, calcium and liming were discussed in Chapter III-2, though their effect on D.O., pH, and alkalinity is covered here.

Any substance not normally present in water could intentionally or accidentally be added to it, and the subject of water pollution by foreign chemicals has become one of increasing concern. Some of the chemical pollutants which have already caused problems are discussed in Chapter VIII-4.

Because I have chosen to confine myself to fresh water aquaculture, salinity is scarcely mentioned in this book. Salinity is defined as "the total amount of solid materials, in grams, contained in one kilogram of sea water when all the carbonate has been converted to oxide, the bromine and iodine replaced by chlorine, and all organic matter completely oxidized." (F. W. Wheaton, Aquacultural Engineering, 1977.) Here, we can think of salinity as the concentration of dissolved metallic salts (excluding calcium salts). The most familiar of these salts, and the one found in the highest proportion in sea water, is sodium chloride (NaCl). Salinity is usually measured in parts per thousand (ppt or ‰). Pure sea water generally varies from 33 to

37‰ salinity. Salt lakes sometimes exceed 40‰, but most inland waters have a salinity of 3‰ or less. There is of course a complete series of intergrades, especially where fresh water meets the sea. Here, I will discuss only waters with salinity under 3‰.

The other aspects of water chemistry which are important enough to us to be discussed here are dissolved oxygen concentration (D.O.), nitrogen compounds, pH, alkalinity, hydrogen sulfide, and iron. Carbon dioxide (CO_2) should also be on the list, but since it is almost impossible to discuss dissolved oxygen or pH without extensive reference to CO_2 it is not treated separately.

Table X–2–4 (p. 472) provides a quick guide to these chemical characteristics, their effects on aquatic animals and how to manage them successfully.

Oxygen-function

By far the most important chemical constituent of natural waters is dissolved oxygen, and only a few specialized microscopic organisms can survive in water without it. Many other chemical conditions causing problems in aquatic systems do so indirectly, by consuming oxygen or interfering with its use by animals. Where there is pollution, as often as not the cause of mortality, disease, or loss of growth is scarcity of dissolved oxygen, not the pollutant itself.

While in the chemical formula for water (H_2O), the "O" stands for oxygen, this oxygen is completely distinct from dissolved oxygen. The "O" in H_2O is chemically bound to the two hydrogen ions together with which it makes up a molecule of H_2O. Dissolved oxygen is pure gaseous oxygen (O_2) in the same form as is found in air. In perfectly pure water not exposed to air it is absent. D.O. is usually measured in mg/l or ppm. (The two units are identical.)

Though everyone knows that if a certain minimum amount of dissolved oxygen is not present, aquatic animals cannot survive, the aquaculturist's task is not so simple as merely maintaining a certain minimum D.O. plus a margin of safety. This might be satisfactory if you only wanted the survival of your aquatic stock. But the aquaculturist is usually interested in maximizing growth, and at barely tolerable D.O. levels, aquatic animals may survive yet fail to grow at all. From there on up, increases in D.O. will usually result in increased growth, until the saturation point is reached. It is possible to have too much oxygen, but such cases are rare, and you can go on the assumption that "more is better than less"—with exceptions to be noted later.

Sources of O_2

Ultimately the only source of O_2 in the biosphere is photosynthesis, the chemical process whereby green plants convert CO_2, water, and solar energy into their own food (sugars) with O_2 as a by-product. Green plants exist in most aquatic environments, but they do not produce all the dissolved oxygen. A significant amount of the D.O. in a body of water, often the majority, comes from land plants, via the air. So the D.O. content of water can be enhanced either by encouraging green aquatic plants or by bringing the water in contact with air. (Pure oxygen has its place in shipping fish, but is otherwise uneconomic for aquaculture use. See Chapter VI-3.)

Preventing O_2 depletion

As important as the oxygenation of water is preventing O_2 depletion. The great majority of all animals and many non-green plants, especially bacteria, are aerobes. That is, they respire by "burning" or oxidizing O_2 to CO_2 to provide energy for their life processes. Green plants also respire, by oxidizing the sugar produced during photosynthesis to provide energy. So green plants, especially at night, can cause temporary D.O. depletion. While in most natural situations the D.O. exceeds the demand, in densely stocked aquaculture systems this may not be so. This tells the aquaculturist three things: (1) Avoid overstocking. (2) Remember that "weed" animals will compete with crop species for the available O_2. (3) Carefully evaluate the role of green plants.

Oxidation is not only a characteristic of living things, since all organic matter is broken down by oxidation. So fertilizers, dead organisms, uneaten feed, organic pollutants, etc., are also in competition with aquatic animals for O_2. There are also inorganic oxidation processes, such as the rusting of iron.

Before discussing how to manage aquatic systems to maximize D.O. we must consider the importance of temperature in oxygen dynamics. I have mentioned that the metabolic rate, and hence feeding and growth rates of aquatic animals, increases with temperature up to a point near the maximum the animal tolerates. This is true, all other things being equal, but the difficulty is that the solubility in water of all gases, including O_2, is reduced as temperature increases. This places a considerable limitation on the aquaculturist, particularly in densely stocked or highly fertile systems. The relation of D.O. to temperature is shown in Table X-2-1.

Temperature and D.O.

Table X–2–1. Maximum concentration of dissolved oxygen in water of 0 salinity at 760 mm Hg pressure.

Temperature in °F (°C)	Dissolved oxygen in ppm	Temperature in °F (°C)	Dissolved oxygen in ppm
32.0 (0)	14.6	68.0 (20)	9.2
33.8 (1)	14.2	69.8 (21)	9.0
35.6 (2)	13.8	71.6 (22)	8.8
37.4 (3)	13.5	73.4 (23)	8.7
39.2 (4)	13.1	75.2 (24)	8.5
41.0 (5)	12.8	77.0 (25)	8.4
42.8 (6)	12.5	78.8 (26)	8.2
44.6 (7)	12.2	80.6 (27)	8.1
46.4 (8)	11.9	82.4 (28)	7.9
48.2 (9)	11.6	84.2 (29)	7.8
50.0 (10)	11.3	86.0 (30)	7.6
51.8 (11)	11.1	87.8 (31)	7.5
53.6 (12)	10.8	89.6 (32)	7.4
55.4 (13)	10.6	91.4 (33)	7.3
57.2 (14)	10.4	93.2 (34)	7.2
59.0 (15)	10.2	95.0 (35)	7.1
60.8 (16)	10.0	96.8 (36)	7.0
62.6 (17)	9.7	98.6 (37)	6.9
64.4 (18)	9.5	100.4 (38)	6.8
66.2 (19)	9.4	102.2 (39)	6.7
		104.0 (40)	6.6

In order to make intelligent decisions about oxygen management, you must know your animals' requirements and have the means to measure the D.O. Table X-2-2 lists the D.O. requirements of some cultured fish. To understand the table, you must understand that the amount of oxygen required by a fish varies with a number of factors, perhaps the most important being the rate of activity. The more active the fish, the greater the need for oxygen. And, as temperature rises, the need for oxygen increases as the water's ability to supply it decreases. This is a severe limiting factor in intensive culture systems. Other factors determining a fish's need

Animals' D.O. requirements

for oxygen include the age and health of the fish, genetic differences between individual fish within a species, the presence or absence of other stress factors, and the duration of a given D.O. level.

In Table X-2-2, the "limiting" level of activity is defined as that level below which survival cannot be assumed. A "critical" level is where activity and growth are restricted, but survival can be counted on if other aspects of the environment are favorable. The "full" activity level is one where normal or better growth can be attained.

Table X–2–2. Limiting, critical and full activity levels of D.O. for certain species of fish. (See text for explanation of activity terms.)

Species	Temperature in °F	(°C)	D.O. in ppm	Activity Level
Largemouth bass	32.0–39.2	(0–4)	2.3–4.8	limiting
(*Micropterus salmoides*)			0.38–2.3	limiting
Smallmouth bass	60.1	(15.6)	0.63–0.98	limiting
(*Micropterus dolomieui*)				
Bluegill	32.0–39.2	(0–4)	0.2–0.6	limiting
(*Lepomis macrochirus*)			0.56–0.8	limiting
Pumpkinseed	32.0–39.2	(0–4)	0.3–1.4	limiting
(*Lepomis gibbosus*)			0.3–0.9	limiting
Green sunfish	32.0–39.2	(0–4)	1.5–3.6	limiting
(*Lepomis cyanellus*)				
Black crappie	32.0–39.2	(0–4)	1.4–1.5	limiting
(*Pomoxis nigromaculatus*)			1.4	
Rock bass	32.0–39.2	(0–4)	2.3–3.2	limiting
(*Ambloplites rupestris*)				
Brown bullhead	32.0–39.2	(0–4)	0.2–0.3	limiting
(*Ictalurus nebulosus*)	86.0	(30)	0.3	limiting
Black bullhead	32.0–39.2	(0–4)	0.3–1.1	limiting
(*Ictalurus melas*)			0.2–0.3	limiting
Channel catfish			1.0	limiting
(*Ictalurus punctatus*)			3–5	full
Common carp	32.0–39.2	(0–4)	0.8–1.0	limiting
(*Cyprinus carpio*)	86.0	(30)	1.1	limiting
Israeli carp	60.8	(16)	0.59–2.5	limiting
(*Cyprinus carpio*, var. specularis*)				
Rainbow trout	60.8	(16)	2.4–3.7	limiting
(*Salmo gairdneri*)	65.3	(18.5)	1.05–2.06	limiting
	55.4–68.0	(13–20)	2.5	limiting
	68.0	(20)	5–5.5	limiting
	68.0	(20)	9	full
Rainbow trout (eggs)	46.4–55.4	(8–13)	7	critical
Brown trout	43.5–50.0	(6.4–10)	1.13–1.6	limiting
(*Salmo trutta*)	62.6–64.4	(17–18)	1.62–2.48	limiting
Brook trout	41.0–50.0	(5–10)	6–7	critical
(*Salvelinus fontinalis*)	68.0	(20)	9.0	critical
	50.0	(10)	2.0	limiting
	59.0–60.1	(15–15.6)	1.35–2.35	limiting
	66.2–71.6	(19–23)	2.4–2.5	limiting
			1.1	limiting

Table X–2–2. *Continued.*

Species	Temperature in °F	(°C)	D.O. in ppm	Activity Level
Yellow perch	32.0–39.2	(0–4)	0.3–4.8	limiting
(*Perca flavescens*)	59.9	(15.5)	0.37–0.88	limiting
	60.8	(16)	1.1–1.3	limiting
	68.0–78.8	(20–26)	2.25	limiting
			0.3–1.5	limiting
	68.0	(20)	7.0	critical
			4–5	full
Goldeye	50.0	(10)	9.0	critical
(*Hiodon alosoides*)	59.0	(15)	11.0	critical
American eel	62.6	(17)	1.0	limiting
(*Anguilla rostrata*)				
Pike	32.0–39.2	(0–4)	0.3–3.2	limiting
(*Esox lucius*)			0.3–2.3	limiting
Goldfish	50.0	(10)	0.5	limiting
(*Carassius auratus*)	68.0	(20)	0.6	limiting
	86.0	(30)	0.7	limiting
	68.0	(20)	2.5	critical
Golden shiner	32.0–39.2	(0–4)	0.2–0.3	limiting
(*Notemigonus crysoleucas*)			0.2	limiting
Blue tilapia	77.0	(25)	1.0	critical
(*Tilapia aurea*)				

In Table X-2-2 many of the limiting D.O. values were measured at 32.0-39.2°F (0-4°C), in order to determine the absolute minimum D.O. at which a species could survive. More useful to the aquaculturist would be a greater range of limiting and critical D.O. values determined at summer temperatures. Still, you can infer that whenever D.O. values fall below 2 ppm for long an emergency may be at hand. As a rule of thumb, one can say that whenever D.O. falls below 7 ppm in warm water culture or 9 ppm in salmonid culture, remedial action should be taken. Regardless of temperature or species, one hundred percent saturation (See Table X-2-1.) should be the goal.

Before discussing methods and equipment used to measure D.O., it will be useful to consider how much precision is important in measuring the chemical parameters discussed in this chapter. You can make fairly precise measurements of physical parameters with simple, inexpensive instruments. However, for many fish farmers a sizable investment in technology for chemical analysis will repay itself in problems avoided. Though aquaculturists whose equipment does not include up to date water chemistry technology need not operate totally in the dark, they will have to be more conservative in stocking, feeding, and fertilizing, as well as paying more heed to preventive measures. There are inexpensive test kits for most chemicals, as well as visual and behavioral indicators of the chemical health of a system. But, most of these are useful only in determining whether or not a problem exists, while the chief value of more sophisticated test equipment is in detecting dangerous trends early on, so they can be corrected before a serious problem arises.

How valuable sophisticated test equipment will be varies with the kind of operation. Managers of traditional farm ponds and other low-intensity, warm water systems can prob bly get by without any investment in chemical testing. Trout are

Measuring D.O.

Degrees of precision

more chemically sensitive, and even low intensity trout culturists should consider monitoring D.O. unless they are certain of constant high water quality. Cage culturists generally have little need of chemical monitoring. However, where "turnover" is a possibility (see Chapter IV-3) equipment for measuring D.O. is a must.

As culture intensity increases, so does the need for chemical sophistication. Most commercial catfish farmers routinely monitor D.O., and some are beginning to analyze nitrogen compounds as well. I recommend both wherever fish are densely stocked and/or heavily fed. Algal systems, such as manure-fed ponds or New Alchemy's solar-algae ponds are subject to rapid fluctuations in D.O., various aspects of nitrogen chemistry and pH, so all three should be monitored routinely. Solar-algae ponds, because of their very small size, are particularly unstable and require frequent surveillance, though with experience, you can learn to foresee problem periods. High intensity closed recirculating systems are at least as sensitive chemically. Some important differences between these bacterially filtered systems and equally intensive algal purified systems will be discussed individually for the various chemical variables.

Signs of oxygen deficiency

The earliest sign of sublethal oxygen deficiency is uneaten food. Loss of appetite does not always mean there is oxygen stress, but if there is not a more obvious explanation, D.O. should be checked. A more serious sign is fish congregating at the surface or near inflows; sometimes they can be seen gasping for air. (In some fish, notably tilapia and goldfish, this looks exactly like a type of filter feeding behavior.) Fish which die from oxygen deprivation sometimes, but not always, die with the mouth popped open and the gills flared. Other aquatic animals may abandon a deoxygenated pond. For example, if you find crayfish or snails crawling out of the water in numbers, you may have an oxygen problem.

Except in small, very well mixed systems, oxygen depletion does not occur simultaneously and equally at all points in the pond. Depletion usually progresses from the bottom up, so that an indicator of lethal conditions in the bottom water may be useful in preventing fish kills. You can conduct a simple test for severe depletion of oxygen in the bottom layer of a pond with nothing more than an oak strip or pole at least as long as the water is deep. Clean and sand the surface of the wood and stick it vertically into the bottom mud at the deepest point. Wait at least forty-five minutes and remove. If the D.O. is 0.2 ppm or less, iron (present in all natural waters) will react with the tannic acid in the oak and stain it. The part of the pole which is stained indicates the depth at which there is severe oxygen depletion.

It is better to anticipate and prevent D.O. depletion than to measure it. Anything which causes great or sudden increase in B.O.D.—increase in numbers or weight of fish or their feeding rate, fertilization with natural materials, or any of the forms of organic pollution mentioned in Chapter VIII–4, reduction of water volume, loss of circulation, or increase in temperature—is grounds for concern.

Algal systems are prone to sudden catastrophic increases in B.O.D. just when D.O. is at its highest. Whenever algal turbidity leads to Secchi disc readings in the 3.1–4.7 in (8–12 cm) range, the system should immediately be diluted or drained, even though D.O. is at the saturation point. Otherwise, there is virtual certainty of a "crash" with a massive die-off of phytoplankton. A brown or yellow turbidity, or a detergent-like odor are symptoms that an algal system is already experiencing low D.O. and emergency measures should be taken.

Oxygen supersaturation

Oxygen depletion is a fairly common occurrence. Its opposite, supersaturation, occurs when O_2 is produced or added more rapidly than it can be used and in quantities greater than the water can hold in solution. In nature supersaturation is rare, but we have observed it in solar-algae ponds, and it has been found in rapidly flowing cold waters with high green plant populations and in heavily aerated tanks. You can sometimes detect supersaturation by the presence of myriad tiny bubbles of oxygen. Supersaturation is dangerous in that the blood of aquatic animals may also become supersaturated. If this progresses to the point where oxygen bubbles are formed in the blood, it is called "gas bubble disease" and may be lethal, particularly to very young fish.

Emergency treatment for oxygen supersaturation consists of stopping aeration, vigorous stirring so that oxygen escapes to the atmosphere, and partial replacement of water.

While D.O. is lowest at the bottom, it is normally highest at the surface (and around inflows and aerators). It is best to measure D.O. at a variety of points, but to allow comparison with other reported figures the usual practice is to measure D.O. at six inches (15 cm) below the surface. As a form of trouble-shooting, the best time to measure D.O. is at its daily low point. Some photosynthetic activity occurs in most aquatic systems, even those with clear water, so readings are usually lowest just at dawn, before photosynthesis begins. For the same reason, D.O. problems often arise during cloudy weather, especially if the weather is hot.

You can choose from three types of equipment for measuring D.O.: colorimetric test kits, titration apparatus, and battery powered meters. Colorimetric kits, including the well known Hach kits, are inexpensive, but not very precise. (Probably their accuracy is related to one's ability to discriminate between colors. I must admit that while the colors I tend to see in the test tube resemble nothing on the charts provided for comparison, others have less trouble.) With a little experience, you can predict or "sense" trends in D.O. just as well as you can measure them colorimetrically.

D.O. measuring equipment

Titration gives greater accuracy at comparable cost. (Equipment and chemicals for two hundred tests can be had for under $50.) However, it is very much more time consuming, since it requires careful individual collection of water samples, followed by time spent in the laboratory. If you will make few and infrequent D.O. tests you are probably well advised to save money by titrating. Others will want to consider purchasing a meter.

Good quality oxygen meters can be had for between $500 and $1,000. (I've heard that if you are an electronics buff you can get by cheaper, but this is something I know nothing about.) For more money you can get great accuracy, but fish farmers do not need measures finer than to the nearest 0.1 ppm, and often the nearest ppm will suffice. The great advantage of oxygen meters is that they are portable and automatic, so that you can obtain any number of measurements at pondside without waiting. Most meters also include a thermometer, and some can be set to compensate for variables such as temperature, salinity and altitude. Keep a titration or colorimetric kit on hand as a back-up in case the meter breaks down.

Management for high D.O.

I have discussed precautions and preventive measures against oxygen depletion in various parts of this book, especially Chapters III–2, III–7, IV–4, VII–1, VIII–4 and X–1. Here I will discuss ways of providing more oxygen than you can achieve simply by good design and avoiding mistakes. I will also discuss measures you can take when D.O. has already dropped below the desired level. The types of measures you can take to increase the D.O. of an existing system, or at least to retard oxygen depletion, are:

• Reduce the demand for oxygen.
• Alter physical or chemical factors related to oxygen supply and use.
• Add oxygen to the water.

Reducing a system's oxygen demand

You can reduce oxygen demand (nearly always B.O.D.) by reducing the animal population, removing specific organic materials, or drawing off the most polluted water and replacing it with uncontaminated water. The last tactic is very much less effective than removing the organic solids, since it does not deal with the root of the problem. In most aquaculture systems the animal population consists mainly of desired species, so you can use the first tactic only where thinning the crop is feasible. If the problem is excess vegetation, physical removal or other control measures may be possible, but most organic pollution is caused by fine particulate matter which you can remove only together with a substantial volume of water.

In very small systems, sediments and suspended particulates can be removed by some sort of filter. Filter media can be fine mesh, spun fiberglass, and sand. The bacterial filters in closed recirculating systems, while primarily designed to treat toxic chemicals, also serve as mechanical filters. But while filters in recirculating systems remove sediments from the growing area, they do not remove them from the entire system. All filters require periodic cleaning, which is more or less of a nuisance, but bacterial filters which double as mechanical filters must be cleaned very frequently to be effective. In very fertile systems, and especially in algal systems where a filter may remove desired algae, it is better to periodically use a siphon or some sort of pump to "vacuum clean" sediments. I have mentioned the importance of being able to draw deoxygenated water off the bottom of larger bodies of water in Chapter VII-1.

Physical factors

Among the physical factors of the aquatic environment related to D.O., depth and surface area can only be slightly modified, and in obvious ways. Temperature is perhaps more adjustable in very small systems and less so in larger ones, while volume is somewhat more manipulable. Often your best emergency measure is to add clean water. In this way, you not only reduce the quantity of organic material relative to the total volume, but you immediately increase the D.O. This does nothing about the source of the problem, whatever it may be, but often your first priority is to rescue fish which are in trouble and/or to gain time to study the problem.

Chemical factors

One chemical approach to deoxygenation is to add an oxidizing agent so that excess organic matter is "burned up" rapidly. You can do this by applying hydrated lime at rates of 30 to 100 pounds per acre (33 to 110 kg/ha). Apply hydrated lime at a low rate if the pond is not well buffered—that is, if pH fluctuates greatly. This technique will not solve the root problem of excess organic matter. It may even make it worse, since nitrogenous compounds are converted to nitrates, which, while less toxic than ammonia or nitrites, are harder to remove from water and may eventually build up to dangerous levels.

You can temporarily increase D.O. by adding superphosphate fertilizer at 40 to 100 pounds per acre (44–110 kg/ha) to trigger a phytoplankton bloom. But if the bloom later dies off, the original problem may return. Under no circumstances use synthetic nitrogenous or natural fertilizers when D.O. is low.

Another treatment for oxygen deficiency, using potassium permaganate ($KMnO_4$) is controversial, not because it is considered harmful, but because results are inconsistent. Some very experienced culturists consider potassium permanganate at

1 ppm to be almost a panacea for temporary D.O. problems, but others have been disappointed.

You can add oxygen to water either as a remedial measure after D.O. is depleted, as a precaution when you are expecting depletion (for example, if hot, cloudy weather is forecast), or routinely, to increase holding capacity or improve water quality. Inexperienced aquaculturists often hope to enhance D.O. by stocking aquatic plants. While in nature aquatic plants are a source of oxygen, we do not yet know enough to use them as a tool in D.O. management, and they may actually lead to reduced D.O. Though phytoplankton make an important contribution to the chemistry of some very intensive aquaculture systems, it appears that they are more important in processing ammonia than as a source of oxygen. (See the following discussion of nitrogenous compounds.) So we are left with aeration.

Aeration of water

Some aeration takes place by diffusion wherever water that is not saturated with O_2 is exposed to air, but oxygen diffuses much more slowly within water than when it enters from the atmosphere into the surface layer. It has been estimated that to raise D.O. at a depth of 33 feet (10 m) from 0 to 0.4 ppm by diffusion would require six hundred years. Oxygenation of natural waters then depends, apart from photosynthesis by aquatic plants, on processes which bring sub-surface water in contact with air. Artificial aeration works to increase the air-water interface by creating currents, turbulence, or air bubbles.

In an emergency you can use any crude method which results in agitation and aeration of water. An outboard motor works very well. Even more crude techniques, such as pouring water back and forth between the pond and a bucket, or stirring and beating the surface with a paddle, have kept fish alive for hours while more permanent remedies were being sought.

Emergency aeration

Before you stir a body of water as a remedy for low D.O., be sure you know just where the problem is. If there is depletion on the bottom, too deep or vigorous stirring may make matters worse by mixing deoxygenated bottom water and organic sediments with the surface water.

Makeshift emergency measures aside, aeration implies the use of some sort of mechanical device to most North American aquaculturists. Before seriously considering mechanical aeration, ask yourself whether you have done all that is possible to maximize D.O. by system design, moderation of B.O.D. through careful stocking and feeding, pollution prevention and manipulation of other chemical and physical factors as described in this and the preceding chapter. If the answer is yes, and D.O. is still below optimum, you may want to consider constructing or purchasing a mechanical aeration system.

Mechanical aeration

There is a great deal of technical writing on aeration of water, requiring some knowledge of chemistry and engineering to understand. Much of it comes from research on sewage treatment systems, which differ from aquaculture systems in a number of ways. Perhaps the most significant is that sewage treatment uses aeration to maintain a minimum D.O. of 1 ppm, whereas aquaculture requires much higher concentrations. Eventually we will have a comparable body of research in aquaculture aeration, but the work has barely begun. The following paragraphs are not intended to provide answers about aeration so much as to help you wade through advertising materials, and perhaps some of the technical literature, and ask the right questions.

Aerators can be divided into four types, based on the way in which the air-water interface is increased. They are gravity, surface, turbine and diffuser aerators.

Gravity aerators

Gravity aerators exist in nature as waterfalls. Water falling through air is broken up into drops, greatly increasing the surface area. Further aeration occurs with the turbulence of falling water striking the horizontal water surface, and from air bubbles trapped in the fall. (Gravity aeration is particularly useful if your water supply is at a higher elevation than your culture facility, since here the energy used is free.)

The effectiveness of gravity aeration depends on the volume of water dropped, the distance it falls and how much it is broken up in falling. Breaking up falling water is easily done, and all kinds of structures have been created to do so. The simplest is no more than a flat surface or "splash board" on which water falls before striking the water surface beneath. You can improve on this considerably by perforating the splash board, so that the water is broken up into drops and/or by splashing the water down a series of steps. Since no mechanical devices are involved, these improvements cost relatively little, and you would be wise to use a fairly complex splash device.

The degree of oxygenation is also directly proportional to the height of the fall, but oxygenation increases at a slower rate as height increases. This fact, along with the possible increase in energy costs proportional to height, suggests that you find an optimum, rather than a maximum height.

Gravity aeration almost never produces supersaturation with oxygen when water falls less than thirty-three feet (10 m). For this reason, gravity aeration is preferable in some systems.

Surface aerators

Surface aerators use some sort of rotor to break up or agitate surface water, increasing the surface area of water exposed to air. Commercial aerators sold for aquaculture use are usually of this type. Most manufacturers of surface aerators provide fairly accurate data on their performance, so you can readily compare the various models. Two factors which may go unnoticed, however, are the way in which the rotor effects aeration and the motor's location. Some rotors pick up water and spray it over the surface, while others merely create turbulence and waves near the rotor. Some of the more vigorous machines of the first type resemble a fountain, casting water for several yards on all sides. This may not be desirable, depending on how the body of water is used. The location of the motor becomes important where winter aeration is desired. For winter use, a submerged motor is best, since motors located above water may freeze.

Turbine aerators

Turbine aerators also involve a rotor, but the rotor is submerged more deeply and operates by creating a vertical current, constantly bringing water to the surface. Turbine aerators may deal more effectively with oxygen depletion at the bottom than either gravity or small surface aerators. Turbine aeration is rare in aquaculture, though the Pond Master®, used to prevent freezing over (described in Chapter X-1) is a sort of turbine. The main problems with powerful turbine aerators in aquaculture are that currents may be strong enough to interfere with fishes' normal swimming, and that they may be struck and injured by the rotor.

Diffuser aerators

Gravity, surface, and turbine aeration all function primarily by moving water to bring it into contact with air. In diffuser aeration, it is mainly compressed air that is moved, though as a by-product water is always moved vertically, enhancing the aeration effect. (Be sure to use only a compressor designed to deliver oil-free air.) The simplest sort of diffuser aeration involves no more than bubbling air through a pipe into the water. This is extremely inefficient since the bubbles produced are few and large. Anything which will convert the same amount of compressed air into

Fabricating a small Savonius rotor windmill to be used in aerating a fish pond. (Photo courtesy Meridith Olson, W.C.C.A., Waverly, Minnesota.)

"Pond Master" Savonius rotor windmill (*manufactured by Wadler Mfg., Co., Galena, Kansas*) mounted on the roof of the "Six-Pack" passive solar greenhouse at the New Alchemy Institute, Hatchville, Massachusetts. The Pond Master is designed to keep outdoor ponds from freezing over, but here it is adapted for aeration. (Photo by Robert Sardinsky.)

Left: The shaft of the Pond Master passes through the greenhouse roof and into a sunken pond. In addition to the submerged rotor that comes with the Pond Master, an aerating device, consisting of a wok cover perforated on the rim and sides, has been added at the water line. (Photo by Robert Sardinsky.)

Right: When wind turns the rotor, water is drawn into the bowl and sprayed out again, providing aeration. (Photo by Robert Sardinsky.)

more and smaller bubbles will increase oxygenation. (The disadvantages of small bubbles are that they require greater pressure to produce, and that small holes are more prone to clog than larger ones.) Most readers are familiar with "air stones" used in aquariums to break air introduced into water into thousands of tiny bubbles. Similar but larger stones are available for aquaculture use. Though you can fabricate a variety of crude devices to substitute for air stones, I had not seen anything comparable in terms of number or diameter of bubbles produced until Elmer Hedlund showed me a diffuser used by a Wisconsin minnow farmer. This device can easily be made with no more than a length of surgical hose and a needle. (See Figure X–2–1.) The number of tiny holes is limited only by the length of the coiled tube and the patience of those doing the perforating.

More sophisticated diffusers include aspirators and ejectors which mix air and water under varying amounts of pressure. Some of the more powerful of these create considerable turbulence, which further increases aeration, but may also have disadvantages.

The depth at which you introduce compressed air is also important in determining the effectiveness of aeration. Often, the most serious D.O. depletion occurs at the bottom, and air introduced higher up may have no effect on it. In highly fertile waters, air introduced at the bottom may very quickly be used up in the mineralization of organic matter. Although oxygen used in this way is not available to fish, increase in the rate of mineralization may lead to improved growth. If you aerate the bottom water you also create a current so that it is exchanged with surface water. Remember, too, that the further a bubble travels, the more time it has to diffuse oxygen into the water. The main disadvantage of bottom diffuser aeration is that the amount of energy needed to pump compressed air into water increases with the depth to which it must be delivered.

You can further increase the time spent by a bubble between its release into water and arrival at the surface by providing structures to detain bubbles. Some

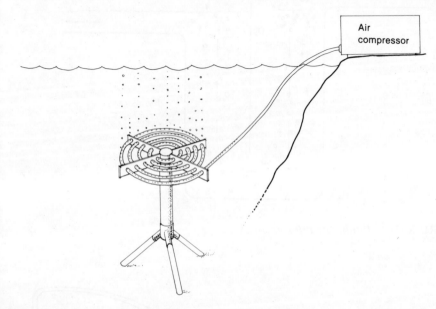

FIGURE X–2–1. *Air diffuser fabricated from surgical hose.*

diffuser aerators release bubbles into an inverted cone for this purpose, and you can place such cones over air stones.

A specialized and very effective aerator based on this principle is the U-tube diffuser, which combines diffuser aeration with running water. (Figure X–2–2) Water passes into one leg of the U at a volume which is calibrated to exceed the rise velocity of air bubbles in still water. A bubbler of some sort is operated at or near the surface of that leg. The result is that bubbles are carried to the bottom of the U and temporarily trapped there before exiting via the other leg. In this way they lose more than the usual amount of oxygen to the water. U-tube diffuser aerators are often capable of producing supersaturation.

The four types of aerator are shown schematically in Figures X–2–3 through X–2–6. Though I have mentioned some of the advantages and disadvantages of each type, it still may be difficult to choose the proper type of mechanical aeration (or perhaps none at all) for your aquaculture system without engineering analysis and/or testing in the system. First, ask the following questions in choosing both a type and a specific system of aeration.

• What is the need for additional oxygen? With what you already know about your aquaculture system, it should be possible to roughly calculate the increase in D.O. you will achieve with a given aeration system. If you need a 2 ppm margin of safety for a small, low intensity warm water fish pond, then a simple, inexpensive system should do. At the opposite extreme, maintaining D.O. saturation in a large,

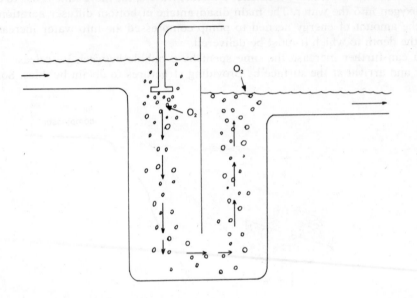

FIGURE X–2–2. *U-tube air diffuser.*

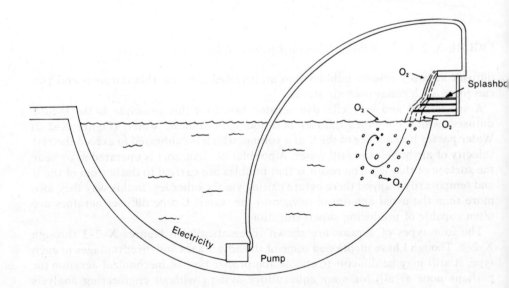

FIGURE X–2–3. *Gravity aeration system with perforated step-wise splash board.*

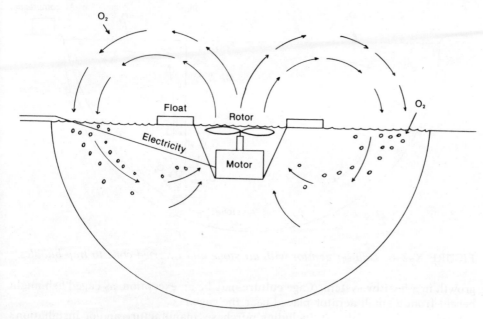

FIGURE X–2–4. *Spray type surface aerator.*

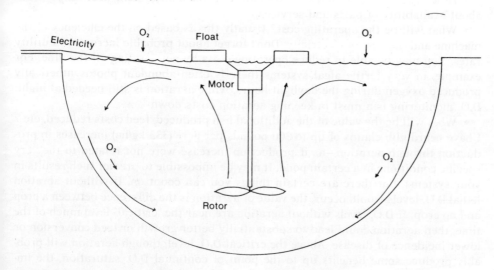

FIGURE X–2–5. *Turbine aerator.*

heavily stocked and fed commercial trout culture pond could require the most refined equipment and could still justify the great expense.

• What special characteristics of the aquaculture system need to be taken into account?

• How effective is the aerator? The bottom line is the increase in ppm of D.O. which can be expected in a body of water as a whole. The increase in D.O. right around an aerator may be helpful in emergencies, but will have little or no effect on

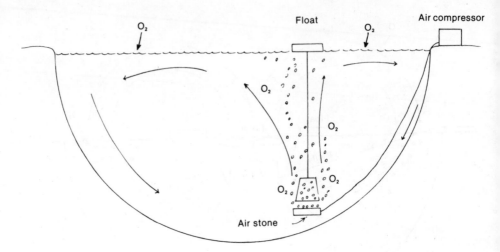

FIGURE X-2-6. *Diffuser aerator with air stone and inverted cone to trap bubbles.*

growth in a healthy system. (Cage culture may be an exception, as caged fish might benefit from a small aerator placed near the cage.)

• What is the initial cost, including purchase, manufacture and/or installation?

• How reliable is the aeration system? Is a back-up system necessary? What about availability of parts and service?

• What will be the operating cost? Usually this is based on the efficiency of the machine and the cost of electricity. Don't forget about probable increases in utility rates. You may be able to reduce operating costs by not aerating all the time. For example, in very fertile algal systems there is often abundant photosynthetically produced oxygen during the daylight hours, while aeration is still needed at night. D.O. monitoring is a must in keeping aeration costs down.

• What will be the value of the additional fish produced, feed costs reduced, etc.? I have heard glib claims of up to 300 pounds per acre (333 kg/ha) increases in production through aeration—as if production increase were not related to the very specific conditions in a certain pond. It may be impossible to predict such results in your systems, but there are certain things you can count on. If without aeration lethal D.O. levels would occur, the value of aeration is the difference between a crop and no crop. If D.O. levels without aeration are near the "critical" level much of the time, then aeration could lead to substantially better growth or feed conversion or lower incidence of disease. Above the critical D.O. level, though aeration will probably produce some benefits up to the point of continual D.O. saturation, the improvement may not justify the increased cost.

• What will be the environmental cost? Nearly all the machines used for aquaculture aeration require electrical energy. How the environment is affected will depend on the source of your power, how you interpret related technical environmental information, and your own conscience—but this cost should be considered. In emergencies, you can create gravity aeration by using liquid fueled pumps, but they are not suitable for continuous operation nor are they environmentally preferable to electrical machines.

Wind aeration   Wind aeration is an alternative which has scarcely been explored. Its major drawback is that wind is not available on a constant basis. Wind is better for raising D.O. levels which are already adequate than to provide oxygen for intensive systems where mass mortality would occur when aeration stops. In such situations, though, a wind aerator with electrical back-up would result in substantial savings.

Nitrogen

Air contains 78 percent nitrogen gas (N_2), and only about 21 percent oxygen. Similarly, there is more dissolved nitrogen in water than dissolved oxygen, but it is much less important. Gaseous nitrogen becomes important for fish culturists only in certain flowing water situations. Where water drops from a considerable height into deep water, bubbles may be carried to some depth. As they rise there is enough time for the surrounding water to become supersaturated with nitrogen. For nearby fish, the result may be their blood becoming supersaturated with nitrogen, leading to the same gas bubble disease that occurs with oxygen supersaturation. The only way to deal with this is to modify the design of the inflow.

The nitrogen cycle

Nitrogen is tremendously important in the aquatic environment both as a component of living tissue and of pollutants. Although some gaseous nitrogen does enter water through the activity of nitrogen-fixing aquatic plants, it is quickly converted to organic forms, where it is recycled many times before returning to the atmosphere or the terrestrial environment. For a better understanding of the management of nitrogen compounds I will first describe the workings of the natural nitrogen cycle.

Natural nitrogen cycle

The main source of all the nitrogen compounds normally found dissolved in water is wastes produced by aquatic organisms, and their bodies after they die. Some nitrogen, in various forms, also enters from the land. In a healthy natural system, this input is moderate, and is balanced by nitrogen which leaves naturally as animals leave or are removed from the system.

The bodies of dead organisms and other organic wastes are decomposed chemically and by bacteria. The great majority of the nitrogen in this organic matter is converted to ammonia (NH_3), which exists in equilibrium with its ionized form, ammonium (NH_4^+). Ammonia and ammonium are reduced to nitrite (NO_2^-), chiefly by the activity of nitrifying bacteria such as *Nitrosomonas* sp.; algae may also be involved in this process. Nitrite is then oxidized to nitrate (NO_3^-) by the bacterium *Nitrobacter* sp. Certain diatoms, algae and bacteria may reverse this process, and some nitrite may in turn be converted to nitrogen gas and lost to the atmosphere, but in nature the overall direction of the cycle is from nitrite to nitrate. Nitrate acts as a fertilizer, being the form of inorganic nitrogen most useful to higher plants and many bacteria as well. From there the nitrogen re-enters the food chain, eventually to be recycled or perhaps to enter into a terrestrial cycle, as for example when a fish is caught for food.

Nitrogen toxicity

All three of the intermediate forms of nitrogen can become toxic to aquatic animals. Though toxicity levels of nitrogen compounds have not been studied as thoroughly as those for temperature and D.O., certain generalizations may be made. Ammonia is potentially the most dangerous nitrogenous pollutant. If you find NH_3 concentrations of 0.1 ppm or more, you should probably take remedial action. The 24-hour lethal level of NH_3 is 2.8 ppm for channel catfish, and is likely to be similar for most fish—though blue tilapia can withstand as much as 5 ppm. Blue tilapia are also the only fish known to be able to develop a resistance to NH_3 with constant exposure. Ammonium is 75 to 100 times less toxic than ammonia. I will shortly discuss the $NH_3 - NH_4^+$ equilibrium. Any system where nitrite is detectable by

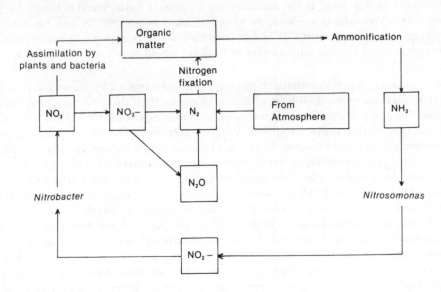

FIGURE X–2–7. *The nitrogen cycle in the aquatic environment.*

normal methods can be considered polluted, and a concentration of 0.5 ppm is definitely critical. For channel catfish lethal nitrite levels are in the 20–40 ppm range. The critical level for nitrate is probably over 20 ppm.

Measuring for nitrogen   In uncontaminated natural systems, ammonia, nitrite, and nitrate concentrations rarely approach the critical level, and never lethal levels. But aquaculture begins by increasing the population density and adds still more nitrogenous matter with feed and fertilizer. The result is often that the natural nitrogen cycle's capacity is exceeded, causing pollution.

There seems to be an absolute upper limit to the tolerable ratio of nitrogen (including aquatic organisms, their wastes, feeds and fertilizers) to water. This limit is around 1 pound (0.45 kg) to 250 gal (947l) of water. Though it is difficult to accurately determine the total weight of nitrogen in a system, it may be helpful to know that fish are about 3.8 percent nitrogen and that proteins contain 7 to 9 percent nitrogen.

The non-chemical signs of nitrogen pollution are less certain than the clues to D.O. depletion, and I will discuss specific indicators for ammonia, nitrite, and nitrate separately. There is a single colorimetric kit which tests for all three, and you can buy kits and enough chemicals for 200 tests for under $500.

As with D.O., how important it is to have sophisticated test equipment will depend on the intensity of culture. Low-intensity operations can get along without it, since there is little likelihood of nitrogenous pollution. Though there is a growing recognition in the catfish and trout industries that it is important to monitor nitrogen chemistry, this practice is still not common. Ammonia should be checked routinely in all very small or intensive systems. Determining nitrite levels is advisable in such systems and absolutely necessary in closed recirculating systems with bacterial filters. Nitrate is not likely to reach harmful levels in systems where sediment is removed regularly. (A possible source of nitrate pollution is overdosage of synthetic nitrogenous fertilizer. See Chapter III–1.)

You cannot detect dangerous concentrations of ammonia without chemical testing. (If there was enough to produce a recognizable odor the fish would long since be dead.) The physiological symptoms of ammonia toxicity are also not distinctive. The pH of the water can give one clue as to whether ammonia might have contributed to either mortality, reduced growth, or lowered disease resistance. As already mentioned, ammonia in solution exists in equilibrium with its ionized form, ammonium, and ammonia is the very much more toxic of the two. It happens that the proportion of the two substances in the equilibrium is determined by pH. Table X–2–3 shows that at an acid or neutral pH, ammonia makes up a very small proportion of the total, while at a pH of 9.5 it may predominate, making ammonia toxicity much more likely. One of the commonest mistakes made by beginners in intensive aquaculture is to use chemical test equipment to measure the total amount of ammonia and ammonium present and then interpret the results without allowing for pH-related differences in ionization.

Table X–2–3. Percentage of total ammonia in the un-ionized form (NH_3) as a function of water temperature and pH.

pH	Temperature in °F (°C)							
	32 (0)	41 (5)	50 (10)	59 (15)	68 (20)	77 (25)	86 (30)	95 (35)
6.0	0.008	0.01	0.02	0.03	0.04	0.06	0.08	0.1
6.2	0.01	0.02	0.03	0.04	0.06	0.09	0.1	0.2
6.4	0.02	0.03	0.04	0.07	0.1	0.1	0.2	0.3
6.6	0.03	0.05	0.07	0.1	0.2	0.3	0.3	0.4
6.8	0.05	0.08	0.1	0.2	0.3	0.4	0.5	0.7
7.0	0.08	0.1	0.2	0.3	0.4	0.6	0.8	1.1
7.2	0.1	0.2	0.3	0.4	0.6	0.9	1.3	1.8
7.4	0.2	0.3	0.5	0.7	1.0	1.4	2.0	2.8
7.6	0.3	0.5	0.7	1.1	1.6	2.2	3.1	4.3
7.8	0.5	0.8	1.2	1.7	2.5	3.5	4.8	6.7
8.0	0.8	1.2	1.8	2.7	3.8	5.4	7.5	10.2
8.2	1.3	1.9	2.9	4.1	5.9	8.3	11.3	15.2
8.4	2.0	3.0	4.5	6.4	9.1	12.5	16.9	22.1
8.6	3.2	4.7	6.9	9.8	13.7	18.4	24.3	31.0
8.8	5.0	7.3	10.5	14.7	20.1	26.4	33.7	41.6
9.0	7.7	11.1	15.7	21.4	28.5	36.2	44.7	53.0
9.2	11.6	16.5	22.8	30.2	38.7	47.4	56.1	64.2
9.4	17.2	23.9	31.8	40.7	50.0	58.8	67.0	73.9
9.6	24.8	33.2	42.5	52.1	61.3	69.3	76.3	81.8
9.8	34.3	44.1	54.0	63.2	71.5	78.2	83.6	87.7
10.0	48.3	55.5	65.0	73.2	79.9	85.0	89.0	91.9

There are three ways to approach the problem of ammonia toxicity. The most important is to avoid overloading the system with organic materials. A glance at Figure X–2–7 suggests the main sources of ammonia in a fish culture system: uneaten food, fish excreta, and dead organisms. Though with proper feeding there will not be uneaten food, accidents and mistakes do occur. The solids in fish excreta can be removed by filtration, draw-off of bottom water or "vacuuming," but when a protein-rich food is used, enough ammonia to cause a problem may be excreted directly from the gills. Lastly, though some systems have few dead organisms, algal systems constantly accumulate dead algal cells which must be removed if ammonia is to be controlled.

In nature, the second anti-ammonia approach is accomplished by *Nitrosomonas* sp. In closed recirculating systems, these bacteria and *Nitrobacter* sp. are concentrated in a bacterial filter where ammonia is converted to nitrite. (See Chapter IV–4.)

Algae will also convert ammonia to nitrite, but few aquaculture systems have enough algae to do the job. Even in dense algal populations there are problems. The algae remove ammonia, lowering pH as they do so. But the high photosynthetic rate also removes CO_2, raising pH and intensifying ammonia problems. If the ammonia concentration exceeds what the algae can deal with, the situation cannot be improved without sediment removal, together with filtration or large-scale replacement of water. In algal systems this problem is especially severe in the early part of a growing season when animal populations are small and little CO_2 is produced.

Bacterial filters and algae both fail to remove nitrogen from the system, but leave it in the form of nitrate. Though higher plants can be used to remove nitrate, they are usually not practical in aquaculture systems. On all counts the single most effective method of controlling ammonia, nitrite, and nitrate is to remove organic solids—mainly sediments. It would be wise to clean very intensive systems as often as twice a week. Though less frequent cleaning removes just as much material by weight, this will eliminate less nitrogen and more carbon, since waste materials have more time to break down.

Where a bacterial filter is used, sediment should be removed before the water reaches the filter. Even so, it will be necessary to periodically shut down and backflush the filter.

Later in this chapter I will discuss how to control ammonia concentration by manipulating pH.

Nitrite management   If you can detect any nitrite colorimetrically it should be considered a problem. You can also detect nitrite toxicity by examining affected fish. Nitrite reacts with hemoglobin to produce methemoglobin, so that the blood turns a chocolate brown color. (Healthy fish blood is a brilliant red.) Fish suffering from nitrite toxicity tend to rest quietly on the bottom, but may begin to swim erratically just before they die. The ultimate cause of death from nitrite toxicity is lack of oxygen in the blood, since methemoglobin cannot absorb dissolved oxygen. But though fish killed by lack of oxygen in the water sometimes die with mouth and gills flared open, victims of nitrite toxicity often die with mouth open and opercles (gill covers) closed.

You might assume that since ammonia precedes nitrite in the nitrogen cycle, where ammonia levels are low nitrite need not be monitored. In fact, it is possible to have low ammonia and low nitrate levels with high nitrite. This is most likely in closed recirculating systems, particularly new ones. What happens is that *Nitrosomonas* sp. develops much more rapidly than *Nitrobacter* sp., which is inhibited by ammonia. So, until the population of *Nitrosomonas* sp. becomes adequate to deal with ammonia, nitrite builds up. The problem may repeat should *Nitrobacter* sp. die off, as may happen in the presence of certain chemicals, including antibiotics.

In algal systems the rule that ammonia toxicity will appear before nitrite toxicity is more reliable. Usually, nitrite will not present a problem in green water if dead algal cells are removed regularly. Though ammonia may affect growth in intensive pond culture, nitrite toxicity is unknown in large open systems.

Nitrate management   Nitrate toxicity problems are rare. They may develop, though, if organic matter is allowed to build up indefinitely, since nitrate is the end product when nitrogenous organic compounds decompose. Toxic levels of nitrate are 50 to 100 times those of nitrite.

You have probably noticed that while ammonia, nitrite and nitrate are to a certain extent separate problems, nitrogen chemistry is best managed as a whole. (with exceptions as noted). Though you can bring high concentrations of ammonia and nitrite down to acceptable levels, and though the physiological effects of nitrogen toxicity are reversible, the best "medicine" is, as always, preventive. The total amount of nitrogen a system can deal with can be increased by adding enough calcium chloride ($CaCl_2$) to bring salinity up to 3‰, but regardless of salinity you must still control the amount of organic nitrogen entering the system.

Preventing nitrogen problems

Your best opportunity to control organic nitrogen inputs is in fertilization and feeding. Do not add nitrogenous fertilizers, especially natural manures, in excess of what the food chain can assimilate, or ammonia pollution will result.

Overfeeding is obviously to be avoided, but there is another, more subtle aspect of water chemistry related to feeding. It is well known that all animals need dietary protein, which contains nitrogen, and that some animals need more protein than others. But it has been widely assumed that even a fish with a low protein requirement would grow faster on a high protein diet. The decision whether, for example, to feed tilapia on low protein rabbit feed or high protein trout feed has been considered mainly as an economic one, but there is an interesting chemical aspect to the choice.

Clearly, if low protein content is compensated by a higher feeding rate then the total amount of organic matter added to a system will be higher than if the same amount of fish is produced on a high protein feed. Even though in both cases no food is left uneaten, the former strategy will produce a greater volume of fish metabolites. However, studies carried out at New Alchemy suggest that a given species of fish can only assimilate a certain percentage of the nitrogen in its diet, regardless of the amount fed. If this is true, then using a high protein diet could actually lead to a higher input of ammonia and nitrite, which after a time could suppress growth.

David Engstrom of New Alchemy is working to develop a method for choosing feeding strategies based on the efficiency of nitrogen assimilation. Given the percentage of nitrogen in a feed and the efficiency with which it is assimilated by a certain species of fish it should be possible to predict, for any feeding regime, how long it will take to reach a given concentration of ammonia. This information would be most useful in projecting growth and conversion rates. Though Engstrom's method is not yet perfected, he has learned enough to be able to suggest that, if nitrogenous pollutants are a recurring problem (even though management seems to be good), a change of feed is worth trying.

pH

Even chemically pure water does not exist solely in the form of molecules of H_2O, but a portion is always dissociated into ions of hydrogen (H^+) and hydroxide (OH^-). The relative proportion of these ions is important in the chemical action of water. It is a convention to measure this characteristic by the quantity of hydrogen ions. Since this number is always very small, it is expressed, for convenience sake, as its logarithm without the negative sign. Values of this exponent, commonly referred to as pH, vary along a scale of 0 to 14, and a pH of 7 is considered "neutral." Lower values are acidic, and higher values are basic. (People often speak of pH values above 7 as "alkaline," but this is confusing, as another important chemical parameter of water is termed alkalinity.) I have seen references to natural waters with pH values ranging from 1.7 to 12.0, but extreme cases will not interest aquaculturists, since the forms of life adapted to such waters are few and usually tiny. Critical pH levels cited for fish include 9.2 for trout and 10.8 for common carp. My impression is that naked fishes, such as catfishes and eels, are more sensitive to extremes of pH than scaled fish, but this has not been confirmed experimentally. The great majority of natural bodies of fresh water have a pH between 6 and 9, and aquaculturists should try to maintain a pH between 7.0 and 8.5.

Determining pH

In aquaculture, simple litmus paper testing of pH is not adequate, but many fish growers can get by with a simple colorimetric test system designed for aquarium use which costs about five dollars. pH meters are available for about one hundred dollars, and though more precise, they are not necessarily more convenient than colorimetric kits.

Be sure to determine the pH of your water, at least in the beginning. Note that the pH of the water source and the growing area may not be the same. From then on, how often to test will depend on the intensity of culture and the amount of manipulation being done to the water chemistry. In very low intensity systems, if growth is satisfactory, it may be enough to check pH every few months. Intensive pond systems should be monitored every few weeks. If you fertilize or lime, monitor pH not only before and after, but constantly during treatment. (See Chapter III–2.) Small algal systems are particularly subject to pH fluctuation and may need to be checked daily or even more often. Whenever you do check pH it is a good idea to test at different times during a single day.

Buffering pH

Though it is important to stay close to an optimal pH, it is even more important to prevent sudden fluctuations. As with temperature, the shock of going abruptly from one pH to another can be harmful to aquatic animals even where neither of the extremes reached is in itself harmful. In some aquaculture systems there is much more pH fluctuation than in natural systems. To understand this and how to prevent problems, it is first necessary to understand the concept of buffering.

Several groups of chemicals can be involved in buffering pH, but to aquaculturists carbonates are the only important ones. Two chemical substances are essential to carbonate buffering. One is carbon dioxide (CO_2), which enters water from the atmosphere (mainly via rain), through respiration of living organisms, and by decomposition of non-living organic matter. The other is calcium (Ca), present in aquatic environments chiefly as calcium carbonate ($CaCO_3$).

Carbon dioxide goes into solution in water as carbonic acid (H_2CO_3), which reacts with calcium carbonate to form calcium bicarbonate ($CaHCO_3$). This in turn dissociates to form ions of hydrogen and carbonate (CO_3^{--}). This chemical buffer system tends to resist changes in pH so long as sufficient calcium carbonate and carbon dioxide are present. If something is added which tends to make the water more acid (i.e. an increase in hydrogen ions), calcium carbonate will dissolve, according to the following reversible equation: $CaCO_3 + H^+ \leftrightharpoons Ca^+ + HCO_3^-$ tying the hydrogen up in bicarbonate ions (HCO_3^-). On the other hand, if something is added which tends to make the water more basic (i.e., an increase in hydroxide ions), essentially the reverse happens: $Ca^+ + HCO_3^- + OH^- \leftrightharpoons CaCO_3 + H_2O$ The bicarbonate is converted to carbonate and water molecules are formed.

This is fine as long as there is a desirable pH in the system to begin with and conditions do not develop to overcome the buffering system. Only a very few bodies of water are too basic for aquaculture but a significant minority are too acid.

Fortunately, it is easier to correct excess acidity than its opposite. The traditional method, liming, also results in setting up a new buffer system which resists further change. (In my experience, wild fish taken from tea-colored bog waters, which may be assumed to be quite acid, are superior in texture and flavor. Some culturists might opt for these qualities as a trade-off for lower production.)

The problem of overcoming the buffer system is a more serious one. Contamination by highly acidic or basic substances is rare in natural waters and even rarer in controlled aquaculture systems, but the processes of respiration and photosynthesis tend, respectively, to drive pH down and up. Since intensive aquaculture begins by increasing the amount of organic material in a system many times over what would naturally be present, there is an obvious potential for trouble.

To explain the problem I will use an extreme case: a solar-algae pond with a dense population of fish, heavy feeding, a dense phytoplankton bloom and a limited supply of carbonate and bicarbonate. In such a system there is extremely intense photosynthetic activity during the daylight hours. The amount of hydrogen ions produced soon overcomes the buffer system and pH rises. At dusk, photosynthesis ceases, and the intense respiration rate of the algae, plus that of the fish and decomposition of non-living organic matter uses up hydrogen ions and drives the pH back down. Figure X–2–8 shows the behavior of such a system.

pH management

Undesirable, though less drastic changes in pH can occur in other types of aquaculture systems. Most often they are downward changes, associated with accumulation of carbon dioxide and organic materials. This is another reason for bacterial filters in closed recirculating systems, since the calcium carbonate substrate not only encourages the establishment of desirable bacteria, but also buffers pH. Catastrophic declines in pH can also occur after calcium-poor waters are diluted by heavy rain or snowmelt.

Usually, you can solve pH problems by preventive management, primarily by

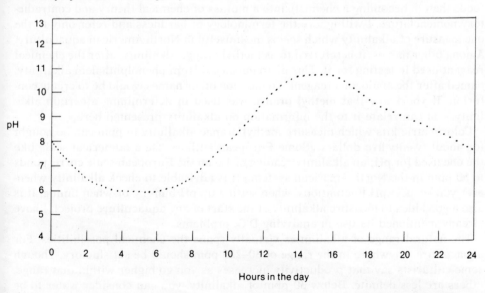

FIGURE X–2–8. *Pattern of diurnal pH fluctuation in a solar-algae pond with a dense population of fish, heavy feeding, a dense phytoplankton bloom and a limited supply of carbonate and bicarbonate.*

avoiding organic overloading, and by adding calcium carbonate either through liming or in the process of filtration. In non-algal systems, control of algae and other green plants will keep photosynthetic production of hydrogen ions within acceptable bounds. Algal systems, particularly small ones such as solar-algae ponds, and the tiny minority of excessively basic natural waters, present much more difficult problems. In solar-algae ponds, pH fluctuation may turn out to be the greatest limitation on their usefulness.

Theoretically you could adequately buffer solar-algae ponds with calcium carbonate. But in fact, the amount of buffer material needed would require an impractical amount of space. You could construct a suitable bacterial filter, but that would turn the system into an algal recirculating system, negating one of the chief advantages of solar-algae ponds—the low mechanical energy input. Similarly, you could perhaps improve waters with naturally high pH by adding weak acids, but the amounts you would need to overcome the high pH buffer system on anything approaching a permanent basis would be economically prohibitive.

Alkalinity   When you add some source of calcium to an aquaculture system to improve or stabilize the pH, you are, in fact, adjusting the alkalinity, which in turn affects the pH. Alkalinity is then a measure of the pH buffering capacity of a body of water, and the higher the alkalinity, the more stable the pH.

You can also use alkalinity to indicate the source of D.O. problems. If D.O. is low or falling, and you do not know the source, check alkalinity. If it is rising, it is probably due to an increase in free CO_2, which in turn suggests that decomposing organic material is the underlying problem.

What has been written on alkalinity and hardness of water is extremely confusing. Rather than drag the reader (who I assume is more interested in producing foods than in becoming a chemist), into a morass of chemical theory and contradictory nomenclature, I will ignore the terminology of hardness and refer only to the one measure of alkalinity which seems most useful in North American aquaculture. Among other names, it is referred to as methyl orange alkalinity, after the chemical reagent used in testing for it. This distinguishes it from phenolphthalein alkalinity, named after the analogous reagent. To mention other names would be to create confusion. If you know that methyl orange was used in determining a certain alkalinity, you can relate it to the information on alkalinity presented here.

Measuring alkalinity   Colorimetric kits which measure methyl orange alkalinity in ppm can be bought for about twenty-five dollars. (Some European authors use a numerical scale like the one used for pH; an alkalinity reading of 1.0 on the European scale corresponds to 50 ppm in the North American system.) It is advisable to check alkalinity whenever you suspect pH fluctuations, when setting up pH buffers, or when liming. It is also a good idea to measure alkalinity at the start of any aquaculture project. I have already mentioned its use in analyzing D.O. problems.

The natural range of alkalinities virtually spans the chemical possibilities. For aquaculture, anywhere in the range of 30–225 ppm should be satisfactory. Though some culturists say that productivity increases as you go higher within that range, others are less definite. Below 30 ppm of alkalinity, you can consider water to be poorly buffered, while above 225 ppm precipitation of carbonates may coat submerged objects with a white "fur" which interferes with biological and mechanical processes. (Adjusting alkalinity through liming is discussed in Chapter III–2.)

Hydrogen sulfide   Hydrogen sulfide (H_2S) is a very toxic substance produced by the decomposition of organic materials in an anaerobic state on the bottom. To detect hydrogen sulfide you need only a properly functioning nose. Just remove a small sample of the bottom sediment, stir, and check for the characteristic rotten egg odor. Even small amounts of hydrogen sulfide can be discovered in this way, and you can consider any detectable amount an excess.

Sometimes hydrogen sulfide is present in well water and should be removed before the water is used for aquaculture. The best way to do this is by gravity aeration (described earlier in this chapter). If hydrogen sulfide is present in the culture water, aeration will drive it off, but that may not solve the underlying problem. Since hydrogen sulfide is produced only on the bottom under anaerobic conditions, its presence implies a D.O. deficiency in the bottom sediments. Removal, aeration, or mechanical or biological stirring of the sediments is the short-term answer and may be a permanent solution in some cases. As discussed earlier in this chapter some recommend potassium permanganate as a treatment for D.O. problems. If the root problem proves to be some sort of organic overloading, the long-term solution becomes obvious.

Iron

Iron (Fe) is present in all natural waters, but in some places it is abundant enough to cause problems for aquaculturists. Iron problems are almost always linked with the use of ground water. Ground water is often anaerobic; under such conditions iron is in the ferrous form (Fe^{++}). When ferrous iron comes into contact with oxygen, it is oxidized to the ferric form (Fe^{+++}), which may appear, combined with other inorganic substances, as a flocculent red-brown precipitate. You may have seen this substance where springs emerge from the ground.

Where iron is sufficiently abundant this precipitate can coat the bottom, interfering with natural processes and even clogging the gills of young fishes. It may also speed up rusting of submerged metals. In extreme cases, the oxidation of ferrous iron as it enters a pond is so intense that it causes oxygen depletion.

The primary solution to the problem is aeration at the inflow point or before the water enters the culture area. Where iron precipitation is heavy you can build a separate settling pool with a graveled bottom. Severe iron problems may require rapid filtration through sand. (See Figure VII–1–7, p. 339.)

General water chemistry management

If the material presented in this chapter seems overwhelming, remember it is reference material to consult in setting up a system, and again if you experience problems or anticipate changes. If you faithfully hold to the following rules, you should be able to manage water chemistry with minimal reference to the remainder of this chapter.

• Get baseline data on temperature, D.O., pH, and alkalinity before you start, and check your water supply for presence of hydrogen sulfide or iron.

• Monitor chemistry (and appropriate physical variables) according to the recommendations made in this chapter. Keep notes, and do not wait until there is a problem to do chemical testing. Data collected when things are going well will enable you to determine what has changed if a problem occurs.

• Always know what is going into your water. (Do not overlook outside sources if you do not control your water supply.) Do not add any substance without considering possible effects on all important chemical parameters, and do chemistry tests after any change in the system.

• Avoid organic overloading. Be cautious with feed and fertilizer, and avoid overloading from the start by choosing the least intensive system that will satisfy your needs.

• Monitor your stock's growth. It is the biomass, not the number of individuals, which determines B.O.D. and volume of wastes.

• Do everything you can in terms of diet, disease control, etc., to keep your fish healthy. A healthy, well-nourished animal is more resistant to any stress than a poorly cared-for one.

• Give priority to D.O. management. Most common chemical problems affect aquatic animals indirectly, by reducing D.O. or interfering with the animals' ability to utilize it. While a high initial D.O. may not solve another chemical problem, it will increase the ability of the animals and the system to withstand it and buy time while you search for remedies.

• Before you start, find out who is the closest person who can help you if you encounter a problem you cannot diagnose or treat—perhaps an experienced fish farmer, a state biologist, or someone in a university. Keep this person's phone number handy at all times.

Summary

I offer Table X-2-4, not as a substitute for the material in the text, but as a quick guide to the ways in which chemical factors interact, how they can be manipulated and their effects on aquatic animals.

Table X-2-4. Summary of important chemical characteristics of the aquatic environment, their determining factors, their effects on aquatic animals, and management criteria.

Chemical Characteristic	Increases with:	Decreases with:	Effects on Aquatic Animals
D.O.	Aeration, circulation, plant population (daytime), phosphate fertilization, $KMnO_4$, liming.	Temperature, organic matter, animal population, plant population (night), iron content.	Health improves up to saturation point. Supersaturation may result in mortality.
N_2	Aeration	Temperature.	Unimportant, unless supersaturated, in which case may cause mortality.
NH_3	pH, aerobic decomposition of organic matter, animal population.	NH_4^+, bacterial filtration, algal population.	Toxic, should be maintained at lowest possible level.
NH_4^+	Aerobic decomposition of organic matter, animal population.	NH_3, pH, bacterial filtration, algal population.	Harmless, except at very high levels.
NO_2^-	Nitrification of organic matter, animal population, $NH_3 \rightleftharpoons NH_4^+$	Bacterial filtration, phytoplankton population.	Toxic by interference with oxygen metabolism, should be maintained at lowest possible level.
NO_3^-	Nitrification of organic matter, nitrogenous fertilization, animal population, bacterial filtration, N_2	Plant population.	Normally harmless or may be beneficial by increasing fertility, toxic at very high levels.

Table X–2–4. *Continued*

Chemical Characteristic	Increases with:	Decreases with:	Effects on Aquatic Animals
pH	Liming, plant population, salinity.	Organic matter, animal population.	Optimum varies with species, neutral or slightly basic values preferred. Stability important, maintained by high alkalinity.
Methyl orange alkalinity	Liming, aerobic decomposition of organic matter.	Bacterial filtration.	High values preferred because buffers pH.
H_2S	Anaerobic decomposition of organic matter.	D.O., circulation, liming, $KMnO_4$	Toxic, should be eliminated.
Fe^{++}	–	D.O., mechanical filtration.	Depletes D.O. (Turns to Fe^{+++} under aerobic conditions.)
Fe^{+++}	D.O.	Mechanical filtration.	Undesirable.
Salinity (0–3‰ only)	Evaporation.	Dilution with fresh water.	Varies with species, but generally desirable within range specified, increases tolerance for nitrogen compounds.

APPENDICES

APPENDIX I

Fish Cookery

APPENDICES

Individual consumption of fish in North America is low by the standards of much of the world. Though to some degree this is due to price and availability, it is primarily an expression of preference. One reason North Americans prefer poultry and red meat to fish is simply that most of us have little access to *fresh* fish. Along the coasts you may find reasonably fresh seafoods for sale, but the majority of the population lives inland. In most areas fresh water commercial fisheries and aquaculture are either small or non-existent, and few anglers are reliable providers. Many of the fish bought by North American consumers have spent several days on ice aboard ship and more time on ice enroute and in the store. Though they may not be spoiled, in no way can they be called "fresh" fish.

Since aquatic foods are much more delicate than meat, people who can afford to be choosy about what they eat are likely to prefer a high quality steak to a second rate fish. My experience is that many of the people who "don't like fish" can be converted if offered well prepared, truly fresh fish. Homestead scale or decentralized commercial aquaculture can offer truly fresh aquatic foods for everyone.

A second reason North Americans eat few fish is that, by and large, we are terrible fish cooks. Apart from some (by no means all) of the fancier seafood restaurants and a few isolated coastal areas, the best fish cookery in North America is found in the southern United States. Southern fish cookery is extremely simple, consisting mainly of expert frying. Latin Americans have the same skill, and also a way with aquatic creatures in soups. Old World cuisines, particularly those of France and Asia, treat fish in a much more elaborate manner, with results that can be a revelation. So, if you are dissatisfied with the fish you eat, don't be afraid to seek improvements from beyond your own back yard.

This is not the place for a list of North American and exotic fish recipes, though I hope the annotated list of cookbooks which closes this appendix will be helpful. My intention here is to offer some general suggestions on fish cookery with a few comments on the animals likely to be grown by the small scale aquaculturist. My purpose is to increase your enjoyment of aquaculture crops. But I also hope that by upgrading the quality of North American fish cookery, we will to some degree widen the acceptance and popularity of aquaculture on this continent.

As with any produce, the provision of a first class fish dinner begins even before the harvest. Among the factors which may affect the table quality of aquaculture crops are health, diet, temperature, flow rate, turbidity, pH, pollution, manner of harvesting, handling during and immediately after harvest, and the way in which fish are processed for the table or for preservation. Disease prevention is discussed in Chapter VIII–2, but the subject of health really encompasses everything related to good aquaculture practice. Diet is the subject matter of Part III. Temperature, flow rate and turbidity are taken up in Chapter X–1. Chemical aspects of water quality, including pH and some pollutants, are dealt with in Chapter X–2; Chapter VIII–4 has more to say about pollution. Chapters VI–1 and 2 deal with harvesting, handling and processing of fish. If you are concerned with crop quality, study the appropriate portions of the chapters just mentioned, as well as appropriate chapters of Part II.

Fish cookery is a subject that could fill volumes, but there are two rules of paramount importance. The first is that the fish must be fresh, or very well preserved. Freshness matters with any food, but it is perhaps more important with fish than with any other. If there will be appreciable time between harvest and preparation, whoever is in the kitchen should treat the fish with great care.

If the fish is not live, check to make sure it is fresh. Lack of bad odor or other obvious evidence of spoilage does not prove freshness. Where gills are left on they should be bright red. Another clue is the eye; if it has a collapsed or cloudy appearance quite unlike the clear, sparkling, protruding eye of a live fish, it is not fresh. Scales should be shiny and adhere to the flesh, which should spring back up when pressed with a fingertip. If the fish is not fresh, I suggest you take up the matter with the family cat.

For continued freshness, clean the fish carefully right away, removing all soft organs, including the gills (See Chapter VI–2.) If the fish has already been cleaned, check it under a good light and make any necessary finishing touches. If the fish will not be cooked immediately, refrigerate it. Sealing it in a plastic bag is neat

and acceptable, but it is better to use a more porous wrapper which will still retain moisture Some recipes call for a marinade before cooking, which can be quite good, but don't try to store fish in water. If you plan to keep the raw fish for more than twenty-four hours, freeze it to keep it as fresh as possible.

Sometimes fish cannot be refrigerated or frozen, in which case you will need some other form of short term storage. Salt is very effective when used properly. Score the fish deeply as shown in Fig. App. I–1. Then apply copious amounts of salt with your fingertips, taking care to work the salt into the cuts. Don't neglect to salt the body cavity, too. Then wrap the fish in green leaves and place it in the shade in a covered container. Some of this salt can be washed or wiped off before cooking. As for the rest, you will be surprised how much salt can be absorbed without adversely affecting the flavor. I have used this method many times in hot, humid, tropical climates, with never a disappointing result.

Some aquatic foods are eaten raw, but the great majority are cooked. With the exception of shellfish and a few of the more delicate finfish, fish should always be well cooked to obtain good texture. This is the second rule of fish cookery. Strive for perfection in getting the fish "done," but if you must err, err slightly on the side of overcooking with most fresh water fish.

The texture of properly cooked fish is usually described as "flaky." That is, when probed with a fork, the muscle fibers tend to separate, rather than cling together. Well cooked flesh also becomes opaque, as contrasted to the slight translucence of the raw flesh. Ideally, there will also be a certain degree of moistness, which is lost in overcooked fish. Some experienced cooks say that the internal temperature of the flesh should never exceed 145°F (62.7°C). If you cannot achieve a flaky texture, and the fish turns mushy instead, it was probably not fresh when you started or you may have cooked it too fast.

Many North Americans who are quite discriminating about meats and other foods tend to see "fish" as a single food, as if there were

FIG. App. I–1. *Scoring a fish to apply salt or spices or to speed up cooking and insure even cooking.*

no interesting differences in flavor. While fish flavors are more subtle than meat flavors, the false notion of a monolithic "fish" comes from the lack of attention paid to fish in the North American diet. Much of the interest in fish dishes is due to sauces, spices, cooking tricks, etc., but even when fish are simply pan fried, with no frills, there is a world of difference in taste between and among species. To the practiced palate, the difference between, say, trout, perch and suckers from the same pond is as great as the difference between beefsteak, ham and chicken.

Complaints of objectionable flavors in fish are quite common, and such flavors are seldom imaginary. But, while tastes differ, none of the animals mentioned in this book has a characteristically objectionable flavor. Most often one hears undesirable flavors described as "lakey," "musty," and "fishy." The first two almost certainly reflect the environment or diet of the fish. Sometimes you can deal with them by skinning or filleting the fish, as they may be concentrated in the skin or next to the body cavity. If such tactics fail, the short term remedy is holding fish live in a clean environment, giving them selected feeds for a while before slaughter. The long term solution is to change the diet and/or environment. A strong, "fishy" flavor almost always means lack of freshness. Still another kind of "off" flavor, in the form of various "chemical" tastes, implies some sort of pollution problem and potential health hazard. (See Chapters VIII–1 and 4.)

Fish can be prepared for the table by frying, baking, broiling, steaming or poaching, and cold cooking. They can also be smoked, pickled, and made into soups or salads. Of these, frying and baking are the commonest, and the choice between them is often a question of size. Small fish, or steaks or other small portions of large fish are best fried, while large fish are best baked.

Fish can be sauteed, deep fried, or stir fried, but pan frying is the best method for the beginner dealing with small fish. The first secret of successful pan frying is to preheat the oil so that the fish sizzles when it hits the pan. When frying small panfish a good indication of "doneness" is when the fins become crisp like a potato chip. (Don't neglect to eat them, they're good!) Another clue is when the muscle tissue pulls away from the backbone. Have some paper or other absorbent material available to absorb excess oil when the fish are done.

An important question in frying is whether to use a bland or flavored oil. Perhaps the best oil for fish which require little cooking is butter, but it will not do for the longer cooking required by most fin fish. Fish with a bland flavor will benefit from a flavored oil such as olive oil, or the coconut oil used in West Indian cooking.

The particulars of baking fish are well covered in cookbooks. In my opinion, a baked fish should always be stuffed or served with a sauce. (Bear in mind that baked fish will continue to cook in the pan for a few minutes after it is removed from the oven.)

Broiling is particularly suited for oily fish and for fillets. For a fish dinner combining speedy preparation with high quality, nothing beats fillets broiled with butter.

In my opinion, most steamed or poached fish is an insipid product, but it has its advocates. The best use of boiled fish is as a first step in making soup. You can make a good fish soup by adapting any chowder recipe. You can use whole, cleaned fish, but fish heads, or the carcasses left over after filleting also make good soups. Simply simmer the heads and carcasses until the meat starts to fall off the bones. Cool, pick off the meat, discard the bones and other inedible materials, recombine with the broth, and go ahead and make soup. (Brains, eyes, etc. will not affect the flavor of the broth while entrails or gills would.)

"Cold cooking" is what really happens with such "raw fish" dishes as the ceviche of Latin America and sashimi of Japan. The acidity of the citrus brings about a change in the muscle tissue similar to that induced by heat. The "secret" of cold cooking, apart from recipes, is to have the fish in very small or thin pieces. My impression is that shellfish and marine fishes are better suited to cold cooking than fresh water fin fish, but this is not based on great experience.

Smoking fish is valuable both for preservative action and gourmet results. It is best suited for oily fishes, and is also a good way of making otherwise ordinary fish extremely palatable.

Pickling is an underused method of preparing and preserving fish, practiced chiefly in Scandinavia. Like smoking it is best suited to oily fishes. It may be the only way to render extremely oily and bony fishes palatable. The best collection of fish pickling recipes I have found is in Vern Hacker's *A Fine Kettle of Fish.*

Fish salads are a valuable means of dealing with the leftovers from large fish meals. These can be safely refrigerated for several days.

People often complain about the bones in fish. While it is true that fish do contain types of bones not found in other animals, it is also true that North Americans are culturally deprived about the mechanics of eating fish. The trick is to learn where the bones are and how to get around them.

If a fish is well cooked, the meat will often fall off the bones; many times you can easily remove the great majority of the bones by splitting the cooked fish and lifting out the backbone, with all the ribs attached. With some fish, this will still leave a considerable number of Y-shaped and potentially dangerous intermuscular bones. If you know where to look for these you can usually remove them all by hand before eating. (If you do get one of these bones stuck in your throat, abandon all dignity and swallow the biggest wad of bread you can get down whole. This usually dislodges it.)

The bones along the base of the dorsal fin and around the head are less annoying and dangerous, but cause some people to waste substantial amounts of meat. It would behoove us to learn to suck and slurp morsels of flesh from these bones as some of our Third World neighbors do.

While the diner often has to deal with bones, much can be done ahead to simplify matters Filleting is the best known method of eliminating bones but it is wasteful unless the carcasses are used. The most consistent protection the cook can offer is in thorough cooking. Well done fish falls clear of the bone much more readily than undercooked fish. In some forms of cooking, such as steaming, bones become softer and therefore less dangerous. At the opposite extreme, hard frying may make the bones brittle, so that they can be crunched up with the flesh. On camping trips, when larger fish were not available, I have used this method to render 3 inch (7.6 cm) fish not merely edible, but delightful. You can put very bony fish through a meat grinder and make patties, sausages, etc.

Before going on to discuss the individual food fishes, I want to take up the matter of size. It is usually easier to get small fish than large ones, and a greater weight of young than of old fish can usually be harvested from a body of water. On the other hand, it is far more work to clean five one pound fish than one five pound fish.

Quality may also be a consideration in choosing the size of fish that goes on the table. While age and size do not affect the quality of fish flesh as profoundly as in warm blooded animals, for at least some fish there is a size range at which taste and texture are best. (I will mention these later.) A final factor which is becoming increasingly important is energy use. Since small fish cook faster than large ones, a given amount of small fish can be cooked with less fuel than would be needed to cook the same weight of large fish.

The following paragraphs deal with cooking methods for specific aquaculture crops, and I must warn the reader that this material, as befits questions of taste, is liberally laced with opinion.

Usually, most of the fish harvested from a well-managed farm pond will be bluegills (*Lepomis macrochirus*). There is no better way to prepare bluegills than to dredge or batter and pan fry them until they are brown and crispy. The tastiest of all the sunfishes is the redbreast sunfish (*Lepomis auritus*), usually neglected by aquaculturists on account of its small size. The other *Lepomis* spp. I have eaten have been good, but not up to bluegills. Sunfishes can be eaten at almost any size, but you get more fish for your efforts, and better flavor, it seems to me, by waiting until they are "filled out." (See Fig. App. I–2.)

What applies to *Lepomis* spp. applies to most of the other small Centrarchids. But crappies (*Pomoxis* spp.), though at least equal in flavor to bluegills, are softer in texture. Summer-caught crappies may be disagreeably soft for some tastes. Unlike other Centrarchids, crappies never really fill out, but their slab-sided shape makes them easier to cook thoroughly. For faster and more even frying, split small Centrarchids by making an incision starting at the anal fin and proceeding toward and along the back, using the backbone as a guide. Lay the split fish open in the pan to cook.

Some people consider largemouth bass (*Micropterus salmoides*) inferior table fare, and I must admit I have eaten some that were pretty bad. The difference appears to be mainly due to the environment, while some ponds consistently produce delicious largemouths, others tend to produce a "lakey" taste. Why this is so I can't say.

To eliminate "lakey" flavors, try skinning and/ or filleting. Sometimes it is better to discard the flank piece (from the pectoral fin back to the vent, including all the muscle tissue surrounding the gut). To me the secret to really enjoying largemouths up to two pounds (0.9 kg) or the fillets from larger fish is to fry them in deep oil. Smallmouth bass (*Micropterus dolomieui*) are firmer textured and more consistently good flavored, so they can usually be baked or fried with none of the precautions described for largemouths.

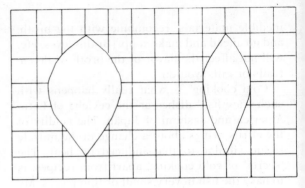

FIG. App. I–2. *Cross-sections showing shape of a "filled out" sunfish, ready for harvest (left), and an underdeveloped or underfed one (right).*

Many pond owners and fishermen scorn the chain pickerel (*Esox niger*). I consider it a better flavor gamble than largemouth bass, and the texture is superb. Unless my pickerel are small, I prefer them baked. In my opinion, the chain pickerel is the tastiest of the pikes (family Esocidae), all of which have relatively many intermuscular bones. Pike flesh is more agreeable cold than most fish, and I recommend Esocids for fish salads.

There are no finer fresh water fish than the bullheads (*Ictalurus natalis*, *Ictalurus nebulosus* and *Ictalurus melas*). Their distinctive sweet flavor is so good that I sometimes fry them without batter or flour, though this increases the danger of burning. Skinning is traditional but may not always be necessary. Properly cooked bullheads present the least bone problems of any fish. Though the texture may deteriorate in the summer, I have never eaten a bullhead which was disagreeably soft.

The best size bullheads to pan fry are in the quarter to half pound (0.1–0.2 kg) range. Larger fish are just as tasty, but harder to cook evenly. The wedge shape of the bullhead's body tends to result in the front part being underdone while the tail burns. Matters may be improved by scoring the flesh at the head end.

The channel catfish (*Ictalurus punctatus*) and the other large *Ictalurus* spp. have excellent texture. Their flavor, while good, is less distinctive than the bullheads'. Unlike most other fish, cultured channel cats fed on concentrates are said to be better and/or more consistent in flavor than wild fish.

As channel catfish are more streamlined than bullheads, fish of up to one pound (0.45 kg) may be pan fried. They can be prepared in almost any way, and the catfish culture industry has capitalized on this by sponsoring catfish cooking contests which have resulted in an upsurge of creativity in catfish cookery.

I have never eaten buffalofish (*Ictiobus* spp.), but I have heard that they are not as good as most of the other suckers (family Catostomidae). I have heard that they are a good fish to steam, and the National Marine Fisheries Service has attempted to use them as the base for a commercial fish sausage. Hacker offers the only recipe I have seen for home preparation of fish sausage. Buffalofish are traditionally eaten in the lower Mississippi Valley, and that would be the place to research buffalofish cookery.

Cold and cool water suckers are among the most flavorful of fresh water fishes. Their white, flaky flesh recalls the sweetness of bullheads, or some of the drier ocean fish. They are also perhaps the boniest of fish, and their texture is very likely to deteriorate with warm temperature, so I don't bother with suckers in the summer. But in the spring, I like nothing better than finely ground suckers mixed with parsley, onion, etc., made into patties and fried. Like their relatives, the pikes, suckers are agreeable to eat cold.

The much maligned common carp (*Cyprinus carpio*) is perhaps not a great food fish, but it can be a good one. Any carp from clean water can be made palatable, but cultivated varieties taste better than wild. To enjoy carp at their best, hold them in clear water for at least a week after harvest and feed them grains.

Some cooks prefer to cut out the strip of dark meat which runs the length of the carp's body (the "mud vein") before cooking. I almost always marinate carp. They are among the best fish for smoking, are good baked, and often used to make gefilte fish (a dish brought to North America by European Jewish settlers). Hacker offers a wide variety of carp recipes.

As many anglers know, nothing beats the taste of wild trout, particularly brook trout (*Salvelinus fontinalis*) or sea-run trout of any species. If you wish to approach that flavor, the secret is not in the cooking but in the feeding: the more natural food, particularly insects or crustaceans, your trout get, the more "wild" they will taste.

Trout are among the few fish which usually do not need to be either scaled or skinned for the table. They are also the exception to the rule of erring on the side of overcooking. These oily but delicate fish should be lightly fried, baked or broiled, but never overdone. Many more or less elaborate trout recipes exist, but if you have wild quality trout, unless you have more than I've ever had access to, I can't imagine bothering to do more than frying or baking them, preferably in butter, with no more frills than a sprig of parsley and a dash of salt, pepper and/or lemon juice.

White-fleshed cultured trout may inspire more creativity. I will not attempt to add to the literature on trout cookery except to say that they are superb smoked, and to mention that one of the best trout I ever ate was dredged in paprika and fried by a Quechua Indian woman in the mountains of Peru.

The Pacific salmons (*Oncorhynchus* spp.) may be treated like trout, with the recognition that they are more oily. Broiling seems especially well suited to salmon.

The yellow perch (*Perca flavescens*) can be prepared much like the sunfishes, but its flesh is firmer and somewhat blander. The very best perch tend to be the largest individuals with well filled out bodies and deep profiles.

Grass carp (*Ctenopharyngodon idellus*) are among the best fresh water fish with respect to flavor and texture, but are nearly as bony as suckers. For this reason larger individuals are better for cooking. Since grass carp are native to China, they should be suitable for many Asian fish dishes.

Tilapia are usually of excellent quality, similar to sunfishes or perch, only better. However, tilapias' quality seems to vary more than native North American fishes. The few bad-to-mediocre tilapia I have eaten (but also some of the good ones) have come from highly fertile waters with a thick layer of soft bottom sediments. Sediment removal or aeration is a wise precaution in tilapia culture.

Tilapia can be baked, smoked or cooked in almost any manner, but I generally fry them. Due to their slab-sided shape tilapia weighing up to two pounds (0.9 kg) can conveniently be pan fried. Score the flesh of medium and large size individuals before frying. For a better appearance you may prefer to remove the thin black lining of the gut wall.

Whitefishes may be considered as similar to the trouts, though with less pronounced flavor. They are excellent smoked, and are the preferred fish for gefilte fish.

In North America almost no attention has been paid to cyprinids as food fish—apart from the common carp and the grass carp. Two widespread species with some advantages for culture, the goldfish (*Carassius auratus*) and the golden shiner (*Notemigonus crysoleucas*) are only good for emergency fare. The former has watery flesh, while the latter is utterly devoid of texture or flavor. If you plan to experiment with North American cyprinids, and since most of Chinese fish culture is based on Cyprinids, you might find inspiration in that direction, or possibly look for European recipes for tench (*Tinca tinca*).

While many are repelled by eels, I consider them to be among the most delicious and versatile of food fish, with perhaps the lowest dressing loss of any fish. Eels are extremely oily, and excellent for smoking, for which the skin may be left intact. (Normally they are skinned.) My favorite use of eel is in stir fries. As they take a while to cook, they should be put in the frying pan or wok before most of the other ingredients. When the tiny pieces of eel turn white and opaque, and curl up like small shrimp, they are done. A good preparatory trick learned from a Portuguese neighbor is to simmer the skinned, whole eel in a water-vinegar mixture, then cook it in sauce.

Among the fresh water Serranids, I have considerable experience cooking white perch (*Morone americana*), some with white bass (*Morone chrysops*) and none whatever with yellow bass (*Morone mississippiensis*). On the basis of this incomplete experience, I suggest that these good textured, agreeable, but bland fish are well suited to cooking with strong spices, sauces or flavored oils. As with largemouth bass the flank meat of white perch is sometimes unpalatable.

Many people consider the walleye (*Stizostedion vitreum*) best among fresh water fish. In my opinion it is unbeatable for texture, but a bit bland as compared to bullheads, trout, suckers, etc. My favorite way to prepare walleye is as broiled or fried fillets with butter and paprika.

Fresh water drum (*Aplodinotus grunniens*) are coarse textured and of mediocre flavor. They are probably best used in soups.

The eggs, or roe, of most fish are edible, although gar roe (family Lepisosteidae) is poisonous. However, most roe is nearly impossible to cook because the fluids in the eggs are under such high pressure that the cooking heat makes them explode. Among the roes you can cook fairly safely are sturgeon, herring, shad, trout, salmon and suckers.

To prepare roe for the table remove the ripe ovaries as intact as possible and gently remove any veins. Place in a frying pan with a little butter and saute very slowly until the eggs just turn opaque. Never overcook roe. Though you will find more elaborate recipes for dealing with roe, as far as I am concerned the only way to eat it is lightly sauteed with a sprig of parsley and perhaps a touch of lemon. Roe is an acquired taste, and the quantity you can get will never by itself justify an aquaculture venture, but I urge you all to give it a try.

You can hardly go wrong with frogs, crayfish or fresh water shrimp. Frog meat is best sauteed in butter, or cooked in wine. It is said to be very low in protein in comparison to other animals. The first step in cooking crayfish is nearly always to boil them, after which the delicate white meat can be savored plain or used in all kinds of recipes. It takes some practice to learn to "peel" the tails. You can prepare fresh water shrimp as you would ocean shrimp, but they are better. I like to simply boil them in liberally spiced water.

In my experience, fresh water clams have been either completely bland or horrid. You can render horrid clams bland by keeping them in a healthy phytoplankton bloom for a while. The one exception in my experience, though I have heard there are others, is the introduced Asiatic clam *Corbicula*, which often has a pleasant, buttery taste. Fresh water clams are steamed to open them, then prepared as you would prepare marine clams.

Fresh water turtle is truly a gourmet food, but its preparation for the table should be approached as an epic adventure. One such tale is referenced below. (*The Trash Fish Cookbook Rides Again*.)

Every fish cook is well advised to keep the classic *Joy of Cooking* at hand. (Recommended cookbooks are listed below.) Unlike many general cookbooks, it treats fish knowledgeably, recognizing the diversity of edible fishes. The basic fish cookbook is James Beard's *Fish Cookery*. The major flaw of these and many others is the use of imprecise names, which are those of the fish market, and not of the fish farmer, angler or biologist. The knowledge that you have a mess of bass or perch to cook is next to useless when confronting most cookbooks.

The best information I have found on preserving fish has come from *Putting Foods By*.

You are likely to learn of the best cookbooks on ethnic cooking from experts in the various cuisines, but one we use extensively in our fishy kitchen is Charmain Solomon's *Complete Asian Cookbook*. To me, perhaps the most amazing fish cookbook is Alan Davidson's *Fish and Fish Dishes of Laos*, which is probably not available in North America. You can find several good Portuguese ways of cooking fish in *The Martha's Vineyard Cookbook*. Other helpful ethnic cookbooks we have consulted have been French, Mexican, West Indian, Brazilian and Scandinavian in origin. A good Cajun cookbook would be handy too, particularly for crayfish.

Two methods of preparation seldom covered well in cookbooks are smoking and pickling. One of the best descriptions of smoking I have seen, starting with smokehouse construction, is in *McClane's Standard Fishing Encyclopedia*. You can find similar material at much less cost in Herb Ludgate's *How to Smoke Fish*. Ludgate also covers pickling, canning and drying. The best pickling recipes I have seen, and the only recipe I know of for home preparation of fish sausage, are in Vern Hacker's *A Fine Kettle of Fish*.

Any number of cookbooks and pamphlets have been devoted to individual types of fish. I am particularly fond of *The Eel Cookbook*, put out by The Institute for Anguilliform Research at the University of Bridgeport. Prize winning trout and catfish recipes can be obtained from the United States Trout Farmers Association and the American Catfish Marketing Association, respectively. (Their addresses are listed below.)

Not to be outdone in versatility, the New Alchemy Institute has published the two part *Trash Fish Cookbook*, covering bullheads, eels, pickerel, black bass, fresh water clams and turtles. Our own prize winning tilapia recipes were revealed in The New Alchemy Newsletter for Fall-Winter, 1978, which issue can be ordered. Hacker offers a greater diversity of recipes for "rough fish" of the Midwest.

A list of publications cited in this appendix follows:

Beard, J. 1954. *James Beard's Fish Cookery*. Paperback Library, New York, 412 pp.

Davidson, A. 1975. *Fish and Fish Dishes of Laos*. Imprimerie Nationale Vientane, Vientane, Laos. 203 pp.

Hacker, V. 1982. *A Fine Kettle of Fish*. State of Wisconsin Department of Natural Resources, Madison, Wisconsin. 64 pp.

Herzberg, R., Vaughan, B., and Greene, S. 1973. *Putting Food By*. Stephen Greene Press, Brattleboro, Vermont. 360 pp.

King, L.T. and Wexler, J.S.. 1971. *The Martha's Vineyard Cookbook*. Harper and Row, New York. 305 pp.

Ludgate, H.T. 1945. *How to Smoke Fish*. Netcraft Co., Toledo, Ohio. 48 pp.

McClane, A.J. 1965. *McClane's Standard Fishing Encyclopedia and International Angling Guide.* Holt, Rinehart and Winston, New York. 1057 pp.

McLarney, B., and Butler, B. 1976. *the Trash Fish Cookbook.* The Journal of the New Alchemists, No. 3: 17-21.

McLarney, B., and Ervin, S. 1977. *The Trash Fish Cookbook Rides Again.* The Journal of the New Alchemists, No. 4: 40-42.

Rombauer, I.S., and Becker, M.R. 1973. *Joy of Cooking.* Signet, New York, 849 pp.

Solomon, C. 1976. *The Complete Asian Cookbook.* McGraw-Hill, New York. 511 pp.

The Institute for Anguilliform Research. 1976. *The Eel Cookbook.* The Institute for Anguilliform Research, Bridgeport, Connecticut. 48 pp.

Address to write for pamphlet of crayfish recipes:

Cities Service Company
Fairbanks, Louisianna

Addresses to write for prize winning trout and catfish recipes are:

United States Trout Farmers Association
Communications Division
111 East Wacker Drive
Chicago, Illinois 60601

American Catfish Marketing Association
P.O Box 1609
Jackson, Mississippi

APPENDIX II

Characteristics of Important Cultured Animals Summarized

The following pages will present, in capsule form, the key characteristics of animals most likely to be of value. Most of those included have been proven useful as food or forage animals in small scale aquaculture in North America. A few are included which do not strictly meet these standards. For example, the brown trout, though seldom grown for food, is of great value to both small and large scale hatchery operators. I included the chain pickerel because I am convinced that it has a place in farm pond stocking. The American eel and striped mullet claim space because of their successful cultivation in other parts of the world.

This Appendix describes first food fishes, then forage fishes, and finally other food animals. Space does not permit inclusion of forage invertebrates or bait and ornamental fishes. (See Chapter III–5 and Part IX). It is intended only to familiarize you at a glance with the most important species in North American aquaculture, and to help you select species for further study, either in this book or in other appropriate sources.

In using this Appendix it will help to be aware of the following:

The *Description* does not ensure proper identification. For this a taxonomic key and/or expert assistance is needed. (Color is a particularly untrustworthy guide to species identification.) If you obtain your stock from reputable dealers you can usually avoid identification problems.

Family and Taxonomic Names are intended to be an aid, not a stumbling block (Some people make scientific nomenclature seem more forbidding than it needs to be.) The Latin names are included here, not only for the benefit of the scientifically oriented reader, but to make the book more useful for everyone. Since the taxonomic name of a species is used worldwide, it serves as the "court of last resort" when common names fail. Latin names will also be useful where you find it necessary to dig into the scientific literature.

The name following the first, or genus, and second, or species name is that of the scientist who first described the species. When placed in parentheses, it indicates that the species was originally described in a different genus. For example: the black basses comprise the genus *Micropterus*. *Micropterus salmoides* (Lacépède) indicates the species commonly called largemouth bass, which Lacépède originally named *Labrus salmoides*. The smallmouth bass, *Micropterus dolomieui* Lacépède, was so named by Lacépède, who was the first to describe it. In most cases you can identify an animal and find appropriate references with the genus and species name alone.

The *Common Name* given for each fish species at the top of the page is the name officially adopted by The American Fisheries Society. Where this does not apply (exotic fishes not established in North

America, frogs, invertebrates) I have used what I consider to be the most widely used and understood name. I have also made two exceptions with fish, which are explained in the text.

Common names usually lack precision. For example, the term "perch," when properly used, designates fish of the family Percidae. In North America it is also applied to various members of the Centrarchidae, Serranidae and Cichlidae, plus a number of marine families. Only two of the dozens of fishes in the Percidae are commonly referred to as perch. "Sheepshead" may refer to the fresh water drum, *Aplodinotus grunniens* (family Sciaenidae), a marine wrasse (Labridae), or porgy (Sparidae), or a killifish (Cyprinodontidae), depending on what part of the continent you are in.

The common name listings are by no means complete, and include only names which are needed to ensure communication. For example, sport fishermen like to refer to smallmouth bass as "bronzebacks," more for the sake of color than anything else. But you are not likely to encounter anyone referring to bronzebacks who would not also know them as smallmouth bass.

Range and Habitat descriptions are drawn from a variety of sources. As alteration of our environment continues to take place, and many species of fish have been introduced widely outside their native range, do not be surprised to find a fish outside the given range, or conversely, not to find it where it "should" be. (Temperature and D.O. requirements are summarized in Chapters X–1 and X–2, respectively.) Also, in nature fish are limited to environments which can supply *all* their needs, while culturists can maintain and grow fish in waters which would not sustain natural populations. For example, trout normally need flowing water to spawn. But we can stock and grow trout in standing water, provided we can provide a suitable temperature, food, etc.

Size and Growth data are drawn from reports on natural populations (unless otherwise indicated). With fertilization or feeding, the culturist should aim to grow fish faster than nature.

In considering *Breeding Requirements*, keep in mind the categories described in Chapter V–1. Breeding temperatures are listed in Chapter X–1.

Information on *Food and Feeding* is drawn from other's writings and my own experience. Many animals in captivity will consume a variety of foods they rarely if ever encounter in nature.

My comments on *Table Qualities* are, of course, subjective.

Food Fishes

Largemouth Bass
CHAPTER II-1

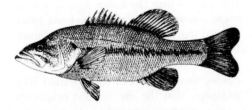

Family: Centrarchidae
Taxonomic Name: *Micropterus salmoides* (Lacépède)
Common Names: "Black bass" comprises all the *Micropterus* spp. In North America, "bass" used without modifers, usually means the same. Also "green trout" (Louisiana), "achigan à grande bouche" (French Canada).

Description and Identifying Characteristics: (See also smallmouth bass.) Dark or bright green to olive above, yellow-green to silvery on sides, milk white below. Many populations exhibit a longitudinal black band; this is more distinct on young fish, which are also more silvery. No tendency toward vertical barring. Problems often arise in distinguishing between this species and the smallmouth bass. The surest way is to draw a vertical line past the posterior tip of the maxillary. If this line misses the eye, the fish is a largemouth; if it passes through the eye, it is not.

The trait of mouth size can't be used to identify juvenile fish. In juvenile largemouth bass, the posterior third of the tail is darkened. In juvenile smallmouth bass, the middle portion of the tail is darkest and the basal third is yellow-orange. You can usually see these characteristics quite easily looking down on the fish in the water.

Range: Native to the Great Lakes (except Lake Superior), Ontario from the Ottawa River system south, the Mississippi River system throughout its length, and west to central Nebraska, to Texas, northeastern Mexico, and the entire southeastern

United States north to Virginia. It has been successfully introduced almost everywhere in the United States and southern Canada where suitable habitat exists, and in many other countries.

Habitat and Environmental Tolerances: Shallow, weedy, soft-bottomed areas of lakes, ponds and slow-moving streams. Also found in deep waters of reservoir lakes, particularly in the southern United States. Ideally adapted to the typical farm pond environment. Territorial and almost always found near some sort of cover. Prefers clear water. Much more associated with shallow, weedy environments than the smallmouth bass.

Size and Growth: Grows much more rapidly in the southern part of its range. The subspecies *M. salmoides floridanus*, found in peninsular Florida, appears to grow more rapidly on a genetic, as well as an environmental basis. Young of the year (fish less than one year old) in Ohio reach lengths of 2–5 inches (51–127 mm); comparable figures for Louisiana are 7–9 inches (178–229 mm). Ten inches (25.4 cm) and 3/4 lb. (0.34 kg) are considered "keeping" size. Hook and line catches generally run up to 19 inches (48.3 cm) or 5 pounds (2.3 kg). Largest known individual (angling record): 22¼ pounds (10.1 kg). Males mature at 2 to 4 years, females at 3 to 5.

Breeding Requirements: Reproduces naturally in shallow standing water, usually over soft bottoms. Spawns in the spring in northern areas. Parents guard the eggs and young.

Food and Feeding: Exclusively carnivorous, and capable of ingesting surprisingly large animals. Young under 2 inches (5.1 cm.) eat mostly insects and crustaceans. Larger fish eat almost anything of suitable size and are often cannibalistic. Feeds at all levels, from surface to bottom, often at night. May be trained to take non-living, moist food.

Table Qualities: Poor to good. Value sometimes reduced by infestation with "black spot" and "yellow grub" parasites (harmless to humans).

Comments on Culture: Traditional farm pond fish. Rarely intensively cultured, though now being studied for this purpose.

Harvesting: Easy to take on hook and line with natural bait or artificial lures. Other harvesting methods give unpredictable results.

Smallmouth Bass
CHAPTER II-2

Family: Centrarchidae

Taxonomic Name: Micropterus dolomieui Lacépède

Common Names: "Achigan à petite bouche" (French

Canada). See also largemouth bass.

Description and Identifying Characteristics: (See also largemouth bass). Light yellow-green to dark olive-green dorsally, with bronzy reflections. Sides lighter, white ventrally. Usually three dark streaks on side of head; eye reddish. Can be distinguished from largemouth bass on the basis of these characteristics. Also, often has a tendency to be vertically barred and appears more brownish than the largemouth. The surest indicator, though is mouth size. (See largemouth bass). Remember that, if any of the less widely distributed *Micropterus* spp. are present, mouth size will establish that a fish is not a largemouth, but not that it is a smallmouth. In this case, you will need an appropriate taxonomic key. The trait of mouth size cannot be used to identify young fish. (See largemouth bass.)

Range: Native to the Great Lakes, St. Lawrence system and the upper Ohio, Tennessee and Mississippi valleys. Now generally found from Nova Scotia south to northern Georgia, west to Oklahoma, and north to southern Manitoba. Introduced widely and successfully to the west of that range.

Habitat and Environmental Tolerances: Rock, gravel, and sand-bottomed lakes, ponds and streams. Found in shallow water in spring and fall, deeper in summer and winter. Usually found around cover, though less frequently around weed beds than the largemouth bass. Much more often found in swiftly flowing water than the largemouth. Prefers cooler water than the largemouth.

Size and Growth: Young of year reach 1.5–4.5 inches (3.8–11.4 cm); may reach 7.0 inches (17.8 cm) with force feeding. Ten inches (25.4 cm) and 3/4 lb. (0.34 kg) are considered "keeping" size. Maximum size appears to be around 15 pounds (6.8 kg), and the angling record is 11 pounds., 15 ounces (5.4 kg). Males mature in 2 to 5 years, females in 3 to 6.

Breeding Requirements: Sometimes reproduces naturally in ponds. Spawn in late spring and early summer, over sandy, gravel or rock bottom near some sort of cover. Parents guard the eggs and young.

Food and Feeding: Exclusively carnivorous. Young under 2 inches (5.1 cm) eat first zooplankton, then mostly insects. Older fish eat fish and other animals, and appear to be particularly fond of crayfish.

Table Qualities: Much superior to largemouth bass, but more subject to unattractive parasites.

Comments on Culture: Occasionally stocked in cool water farm ponds.

Harvesting: More difficult, by all methods, than large-mouth bass.

Bluegill
CHAPTER II–1

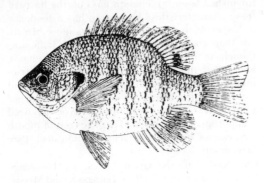

Family: Centrarchidae

Taxonomic Name: Lepomis microchirus Rafinesque

Common Names: "Sunfish," or "sunny" is more or less indiscriminately applied to all the *Lepomis* spp. of which the bluegill is one of the most widely distributed and the most commonly caught. To a lesser degree, the same comment applies to "bream" or "brim," as well as "bluegill." Also called "roach" or, in French Canada, "crapet arlequin."

Description and Identifying Characteristics: Green to brown dorsally and paler, with a blue sheen on the sides, often with several more or less distinct dark vertical bars. Breast yellow copper to orange in breeding males. Opercular flap very dark blue to black. A dark blotch on the posterior portion of the dorsal fin. The combination of a *pointed* pectoral fin and the lack of any red or orange coloration on the opercular flap serve to distinguish the bluegill from any other sunfish likely to reach edible size. If seen in the water from above, the end of the tail has a pale blue cast.

Young bluegills may be separated from pumpkinseeds, at least, by their distinct and regular vertical bars.

Range: Native from the St. Lawrence River (downstream to about Montreal) and south, west of the Appalachians, to Georgia, west to Texas and northeastern Mexico, and north to southwestern Ontario; also in coastal drainages from Virginia through Florida. Successfully introduced practically everywhere in North America east and west of this range, as well as in other continents.

Habitat and Environmental Tolerances: Shallow, weedy waters of lakes, ponds and slow moving streams in spring, summer and fall. To depths of 20 feet in winter and midsummer. Small individuals sometimes found in pools of very small streams. Naturally adapted to the farm pond environment. Prefers clear water. Usually, but not always, found near cover.

Size and Growth: 1.0–4.0 inches (2.5–10.2 cm) in first year. Adults generally 3.5–10.0 inches long (8.9–25.4 cm). Minimum harvestable size is generally taken to be about 6 inches (15.2 cm) and 3 ounces (85 g). In stunted populations very few fish ever attain this size. Largest individual known (angling record) 4 ¾ pounds (2.2 kg), but 1 pound (0.4 kg) is considered very exceptional. Farm pond fish exhibit the same range of sizes and growth rates as "wild" fish. Quarter pound (0.1 kg) wild fish may be 3–5 years old, but intensively fed cultured fish may reach that size in 1–1½ years. Males mature in 1–3 years, females in 2–4.

Breeding Requirements: Reproduces naturally in shallow waters of ponds and lakes, mostly in spring and early summer, but may spawn any time the water is not too cold. Hybridizes with other *Lepomis* spp. Individual males defend nests, but nests are placed in compact groups. Parents guard the eggs and young. Very high reproductive potential often leads to overpopulation and stunting.

Food and Feeding: Young feed on zooplankton. Adults eat primarily insects. Aquatic plants are often found in bluegill stomachs, but it is debatable whether they are eaten deliberately. Can feed at all levels, but generally does not feed off the bottom. An avid surface feeder, and also feeds off the undersides of water lilies and the like. Quickly learns to take floating or sinking artificial feeds, though some individuals never learn.

Table Qualities: Good to excellent flavor, slightly soft texture.

Comments on Culture: The most important food species in conventional farm pond culture, though the tendency to stunt is a limitation. Beginning to be used in many kinds of intensive culture. Probably the best warm water insectivore for North American ponds. Crossed with green sunfish to produce fast-growing hybrids.

Harvesting: Easy to take on hook and line with natural bait, flies or small lures. A good children's fish. Somewhat difficult to net or trap. Must be heavily harvested.

Green Sunfish
CHAPTER II–1

Family: Centrarchidae

Taxonomic Name: Lepomis cyanellus Rafinesque

Common Names: Green perch; "crapet vert" (French Canada).

Description and Identifying Characteristics: Dorsally olive-green; lighter, often yellowish on sides, white to yellow ventrally. Opercular flap black with lighter border and no red marks. The entire flap is stiff, rather than having a flexible tab on the end, as in most sunfishes. The mouth is larger and the body thicker, but less deep than in other *Lepomis* spp.

Range: Various authorities list slightly different native ranges. According to the more generous, the green sunfish is native from the Lake Ontario drainage of New York south, west of the Appalachians, to the Gulf Coast of Alabama, west to northwestern Mexico and central New Mexico, north to the Hudson Bay drainage of Ontario and east through the Great Lakes. It has been introduced to the Florida panhandle, California and Utah.

Habitat and Environmental Tolerance: Almost any shallow, still or slow moving water, including swamps, overflow ponds and temporary ponds. Frequent sunken brush heaps. More tolerant of turbidity than other *Lepomis* spp.

Size and Growth: 1.0–3.2 inches (2.5–8.1 cm) at 1 year. Adults usually 2.5–7.5 inches (6.4–19.0 cm), 0.3–10.0 oz. (0.09–0.3 kg). Rarely reaches "harvestable' size for sunfishes. (See bluegill.) Often becomes stunted. Maximum size probably around 12 inches (30.5 cm). Matures in 1–3 years.

Breeding Requirements: Spawns in late spring and summer. Parents guard the eggs and young. Hybridizes naturally with other *Lepomis* spp.

Food and Feeding: Insects, crustaceans, molluscs and small fish. The relatively large mouth of the green sunfish indicates that it ingests larger food organisms than the other *Lepomis* spp. Readily learns to take all kinds of artificial feeds. Very aggressive feeder.

Table Qualities: The poorest of the sunfishes.

Comments on Culture: Usually regarded as a pest, but used to produce fast growing hybrids with bluegills and redear sunfish.

Harvesting: Very easy to take on hook and line. Moderately difficult to net.

Redear Sunfish
CHAPTER II-1

Family: Centrarchidae

Taxonomic Name: Lepomis microlophus (Günther)

Common Names: "Shellcracker," "stumpknocker." Also see bluegill.

Description and Identifying Characteristics: Olive-green speckled with darker green dorsally. Sides lighter with brassy reflections: usually 5–10 dark vertical bands. (These are more distinct in young and females.) Ventrally yellow to brassy. Opercular flap black, pale bordered, with an orange to blood red spot posteriorly. Breeding males colored more intensely. Young are more silvery and the opercular spot is gray or yellow.

Range: Lower Ohio and Tennessee valleys; everywhere south of the Virginia-North Carolina border and east of central Oklahoma except southern Florida. Widely and often successfully introduced outside this range, particularly in the Midwest.

Habitat and Environmental Tolerances: Ponds, lakes, and slow moving streams in weeded and open water. Congregates around brush, stumps and logs.

Size and Growth: 1.0–4.0 inches (2.5–10.2 cm) at 1 year; adults usually 4.5–9.0 inches (11.4–22.9 cm) and 1–11 ounces (0.028–0.3 kg). Maximum size probably about 12 inches (30.5 cm) or less. May mature in their first year of life.

Breeding Requirements: Similar to bluegill, but with much lower reproductive potential, hence much less prone to stunting.

Food and Feeding: Like the bluegill, but rarely surface feeds. Seems particularly fond of molluscs. Will learn to take sinking artificial feed.

Table Qualities: Said to be similar to bluegill, but superior.

Comments on Culture: Sometimes substituted for bluegills in farm ponds. Less suited to intensive culture. Crossed with green sunfish to produce fast growing hybrids.

Harvesting: More difficult to take on hook and line than the bluegill; equally difficult to net or trap.

Pumpkinseed
CHAPTER II-1

Family: Centrarchidae

Taxonomic Name: Lepomis gibbosus (Linnaeus)

Common Names: "Common sunfish," "yellow sunfish," "sun perch," "kiver" or "kibby," "crapet-soleil" in French Canada. Also see bluegill.

Description and Identifying Characteristics: One of the most beautiful of North American fishes. Dorsally golden brown to olive green. Sides lighter with spots and reflections of yellow, orange, red, blue, green and/or olive. Many individuals have a more or less chain-like series of darker bars on the sides, but these are neither as distinct or as regular as in the bluegill. Ventrally yellow, bronze or various shades of orange. Opercular flap black with white, yellow, orange or blue border, a small spot of red (sometimes pink, orange or yellow) posteriorly. All these colors are more intense in males, particularly at breeding time.

Young are less brilliant, but may show the red spot on the opercular flap. They appear more greenish or yellowish and less distinctly banded than young bluegills.

Range: Native east of the Appalachians from New Brunswick to South Carolina, and west through the Great Lakes basin at least to Minnesota and Iowa. Successfully introduced in Wyoming, Montana, the Pacific Coast states and southern British Columbia.

Habitat and Environmental Tolerances: Shallow waters of ponds and lakes and quiet waters of streams. More restricted to weedy areas than other sunfishes; likes sunshine and clear water. Appears to do better in cool waters than the bluegill or redear sunfish.

Size and Growth: 1.2–3.5 inches (3.0–8.9 cm) at 1 year. Adults usually range from 2.5 inches (6.4 cm) in stunted populations to 8.0 inches (20.3 cm) and weigh from 0.3 oz–9.0 oz (0.09–0.26 kg). Stunting is not as prevalent as with bluegills, probably due to the pumpkinseed's greater tendency toward cannibalism, but is still a problem. Maximum size appears to be around 1½ pounds (0.68 kg.)

Breeding Requirements: Spawns like bluegills in spring and summer at depths of 6–12 inches (15.2–30.5 cm). Generally spawn earlier than bluegills. Do not spawn in groups. Sexual maturity is usually reached by age 1 or 2. Stunting occurs (see above). Hybridizes naturally with other *Lepomis* spp. and with warmouth (*Chaenobryttus gulosus).*

Food and Feeding: Insects and other small invertebrates, some small fish. Fond of snails. Feeds at all levels, but does not surface feed as much as the bluegill.

Table Qualities: Similar to bluegill.

Comments on Culture: Rarely stocked in farm ponds.

Harvesting: Very easily taken on hook and line. Good children's fish. Moderately difficult to harvest with nets or traps.

White Crappie
CHAPTER II-1

Family: Centrarchidae

Taxonomic Name: Pomoxis annularis Rafinesque

Common Names: Most fishermen do not distinguish between this and the following species, lumping both as "crappie" or "croppie." Also known as "papermouth" or "tin mouth," "strawberry bass," and "calico bass." French name: Marigane blanche.

Description and Identifying Characteristics: Dorsally green to brown, with bluish overcast, sides lighter, more iridescent green, shading to silver then white ventrally. Five to ten vertical black bands (sometimes indistinct) on sides. Dorsal, anal, and caudal fins with black vermiculations or spots. Extremely flattened laterally (flatter than the *Lepomis* or *Micropterus* spp.) Difficult to distinguish from the black crappie. The white crappie tends to be more slender, and to have a less concave forehead. The black markings on the sides are more regularly arranged, forming bars in the white crappie. White crappie usually has 5 or 6 spines in the dorsal fin; the black crappie 7 or 8. The surest indicator is to measure the length of the base of the dorsal fin and project that length forward along the back. If the projection reaches or nearly reaches the level of the eye it is a black crappie; if it only reaches a point above the opercle it is a white crappie.

Range: Native to New York state west of Rochester, south, west of the Appalachians to the Gulf Coast of Alabama and the Rio Grande River, north through central Texas, Oklahoma, Kansas and Nebraska to southeastern South Dakota, and east through the southern halves of Lake Michigan and Huron and all of Lake Erie. Introduced and well established in Atlantic coast drainages from northern Florida to North Carolina, and in scattered locations elsewhere, definitely including the Pacific Coast states and Connecticut.

Habitat and Environmental Tolerances: More characteristic of ponds and lakes over 5 acres (2.0 ha) than smaller ponds. Also found in slow-moving streams. Frequents brush piles. Very tolerant of siltation and turbid water. Often most abundant over silted bottoms, but this may be more an expression of its inability to compete with sunfishes and black bass than of an environmental preference. In shallow water in spring, deeper at other times.

Size and Growth: Young of year in Ohio reach 1.0–3.8 inches (2.5–9.7 cm) in October; 1.2–5.0 inches (3.0–12.7 cm) in first year. Adults 5.0–14.0 inches (12.7–35.6 cm), 1 ounce–2 pounds (0.03–0.9 kg). Reaches harvestable size (7 inches or 17.8 cm) in about 3 years. Subject to stunting in marginal habitat or when overpopulated. Maximum size around 6 pounds (2.7 kg). Matures in 2–4 years.

Breeding Requirements: Spawns in late spring and early summer. May nest as isolated pairs or in colonies, in the open or under banks. Hybridizes with black crappie.

Food and Feeding: Young first eat zooplankton, then larger invertebrates. Adults eat mostly small fish, and some invertebrates.

Table Qualities: Excellent flavor, soft texture. Flat shape simplifies cooking.

Comments on Culture: Occasionally used as a substitute for largemouth bass in farm ponds, particularly where turbidity is a problem. Mainly used in low intensity polyculture in the South Central states, also in fee fishing ponds. May become stunted.

Harvesting: Easily caught on hook and line using natural or artificial baits, particularly minnows. Good children's fish. Easier to net than other Centrarchids; sometimes fyke nets are very effective.

Black Crappie
CHAPTER II–1

Family: Centrarchidae

Taxonomic Name: Pomoxis nigromaculatus (LeSueur)

Common Names: "Speckled perch" (Florida), "marigane noire" (French Canada). Also see white crappie.

Description and Identifying Characteristics: Very similar to white crappie, but generally with a greater amount of black pigment on sides and fins, and not regularly patterned. Fish from turbid waters are paler and less clearly marked than fish from clear, weedy waters. Looked down on in the water, they often have a distinctly "lacy" appearance.

Range: From the limit of tidewater in the St. Lawrence River south, west of the Appalachians, to the Gulf Coast of Alabama, in Atlantic coastal drainages north to Virginia, west to central Texas, north to southern Manitoba, and east through the Great Lakes (excluding the Lake Superior drainage). Introduced, often successfully, almost everywhere in the United States and in the lower Fraser River valley of British Columbia.

Habitat and Environmental Tolerances: Similar to the white crappie, but prefers cooler, clear, weedy waters; not as tolerant of siltation and turbidity. Tendency to form schools.

Size and Growth: Young of year reach 1.0–3.0 inches (2.5–7.6 cm) by October; 1.1–4.5 inches (2.8–11.4 cm) in first year. Adults usually 5.0–12.0 inches (12.7–30.5 cm), 1 ounce–1 ¾ pounds (0.03–0.79 kg). Reaches harvestable size in 2–3 years. Subject to stunting. Maximum size over 5 pounds (2.3 kg). Matures in 2–4 years.

Breeding Requirements: Reproduces naturally in ponds. Spawns in late spring and early summer in water 10 inches–2 feet (25.4–61.0 cm) deep, near vegetation or under a bank. Hybridizes with white crappie. Nest in colonies. Parents guard the eggs and young.

Food and Feeding: Young filter feed on zooplankton until they are as large as 6 inches (15.2 cm) long, after which they eat invertebrates. But small fish (seldom over 2.5 inches, or 6.4 cm) predominate.

Table Qualities: See white crappie.

Comments on Culture: See white crappie.

Harvesting: See white crappie.

Rock Bass
CHAPTER II–1

Family: Centrarchidae

Taxonomic Name: Ambloplites rupestris (Rafinesque)

Common Names: "Redeye" or "redeye bass" (not to be confused with *Micropterus coosae*, officially known as redeye bass, found only in Alabama), "goggle eye," or "crapet de roche" (French Canada).

Description and Identifying Characteristics: Brown to olive with bronzy reflections, shading to silver to white below. Breeding males much darker. Dark saddle markings or marblings may be seen on small fish or on adults taken at night, but are frequently absent or indistinct in the daytime. Scales below the lateral line each have a single black spot, usually with black edges on dorsal, anal, and caudal fins. Orange to red eye. Easy to distinguish from other common centrarchids.

Range: Native from St. Lawrence River and Lake Champlain drainage south, west of the Appalachians, to the Gulf Coast of Alabama and Florida, west to southeastern Louisiana, north through Arkansas, northeastern Oklahoma and eastern Kansas, Nebraska and the Dakotas to southeastern Saskatchewan, central Manitoba, and east generally in Ontario and Quebec along a line from Lake of the Woods to Montreal. Has been introduced or has migrated through canals to areas slightly further north in Canada, much of the Atlantic coastal plain in the United States, and in Colorado, Wyoming, and probably other western states.

Habitat and Environmental Tolerances: Found in a variety of habitats, including lakes, ponds and streams of low to moderate gradient over most types of bottom. Often associated with smallmouth bass. Most abundant over rocky bottoms. Tends to aggregate.

Size and Growth: Young of year reach 0.8–2.0 inches (2–5 cm) by October in Ohio, 1.1–3.5 inches (2.8–8.9 cm) in a year. Adults usually 4.3–10.5

inches long (10.9–26.7 cm); 1–14 ounces (0.03–0.4 kg). May require up to 6 years to reach harvestable size (7 inches or 17.8 cm). Smaller individuals generally found in small streams: Subject to stunting in marginal habitats. Rarely exceeds ½ pound (0.22 kg), but maximum size is around 4 pounds (1.8 kg).

Breeding Requirements: Reproduces naturally in ponds. Spawns in late spring and early summer over almost any type of bottom. Often nests in colonies. Parents guard the eggs and young.

Food and Feeding: Aquatic insects, other invertebrates (particularly crayfish) and small fish. Often feeds nocturnally.

Table Qualities: Fair to excellent flavor, soft texture.

Comments on Culture: May be useful for monoculture or polyculture in cool water farm ponds, but this idea has not been well tested.

Harvesting: Very easy to catch on natural or artificial baits. Good children's fish. Easier to net than most centrarchids.

Chain Pickerel

CHAPTER II-1

Family: Esocidae

Taxonomic Name: Esox niger LeSueur

Common Names: The term "pickerel" used without modifier, generally refers to this species, though it may also apply to two subspecies of *Esox americanus,* the "little pickerels." Large specimens sometimes mistakenly called "pike." Also "eastern pickerel," "mud pickerel" (also applied to *E. americanus*), "reticulated pickerel," "brochet maillé" (French Canada). In areas where pickerel are not appreciated, "jack," "jackfish" and even "snake" are applied as derogatory terms.

Description and Identifying Characteristics: Adults green to dark brown dorsally, yellow-green, yellow or gold on sides with a dark chain-like pattern, white below. A black vertical line below the eye which is often obscure on large individuals. May be distinguished from larger Esocids by the fully scaled cheeks and opercles and the mark below the eye, and from little pickerels by the chain markings.

Young under about 8 inches (20.4 cm) are vertically barred in green to black on yellow-green,

with yellow or gold sides and may lack the chain markings. Viewed from above they display a pronounced golden mid-dorsal stripe. This stripe is chestnut brown and not nearly so vivid as in the little pickerels.

Range: Native to Atlantic coast drainages from New Brunswick to central Florida, Gulf Coast drainages to extreme east Texas, up the Mississippi River to southern Missouri, and in the Tennessee drainage to Alabama. Successfully introduced in Nova Scotia, the Lake Erie drainage of New York, the Ohio River drainage of Ohio and Pennsylvania, in Colorado, Washington, and probably elsewhere.

Habitat and Environmental Tolerances: Ponds of all sizes, lakes, and quiet waters of streams. Usually found in shallow water, but may go into deeper water in winter. Associated with weed beds, brush and logs. Solitary and territorial. Active at all temperatures; very tolerant of salinity (up to 15 ‰, and pH (as acid as 3.8).

Size and Growth: Young of year in nature reach 4–5 inches (10.2–12.7 cm) in September; hatchery reared fish may reach 8 inches (20.3 cm). Minimum harvestable size usually considered to be 12–14 inches (30.5–35.6 cm), which is normally reached in about 4 years. Most pickerel taken by fishermen run 15–18 inches (38.1–45.7 cm) and 1–2 pounds (0.45–0.9 kg). Maximum size over 30 inches (76.2 cm) and 10 pounds (4.5 kg). Angling record ·9 pounds, 6 ounces (4.3 kg). Maturity reached in 1 to 4 years. Stunted populations are occasionally seen.

Breeding Requirements: Reproduces naturally in ponds. Spawning takes place in shallow water (temporarily flooded areas, if available) in early spring, winter or occasionally fall. No nest is built.

Food and Feeding: Exclusively carnivorous. Zooplankton in first week, then changes to invertebrates and small fish. Adults eat mostly fish, but also take frogs, crayfish, etc. Often cannibalistic. One of the few North American food fishes which it has so far proved impossible to induce to eat artificial feeds.

Table Qualities: Generally underrated. Superior texture, moderately bony, flavor varies from poor to excellent.

Comments on Culture: Seldom used, but recommended for farm ponds, particularly in very acid or weedy water.

Harvesting: Easy to take on hook and line, artificial lures are usually faster than live bait.

Yellow Bullhead

CHAPTER II-1

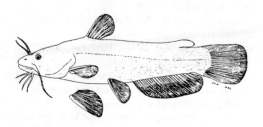

Family: Ictaluridae

Taxonomic Name: Ictalurus natalis (LeSueur) (formerly *Ameiurus natalis*)

Common Names: Non-fishermen often lump the bullheads and the other three ictalurids listed here together as "catfish." Fishermen distinguish "bullheads" from other catfish, but seldom distinguish between them. Also known as "chucklehead" (South), "horned pout" (New England) or "barbotte" (French Canada). Dressed and skinned bullheads are known commercially as "red cats."

Description and Identifying Characteristics: Dorsally yellow-olive to black, lighter on sides, rarely mottled, ventrally yellow to white. Fins paler than body. Upper barbels brown, chin barbels pale; no black pigment on barbels.

Young are dark brown or black dorsally, white ventrally with *white chin barbels.*

Difficult to distinguish from the black and brown bullheads. The pale, unpigmented chin barbels distinguish it from the brown bullhead, but not always from the black bullhead. The surest indicator is the pectoral fin spine. In this species, the posterior margin is serrated, as you can determine by running a finger along it. In the brown bullhead it is provided with more stout projections, and in the black bullhead it is smooth.

Range: Native from the Hudson River drainage south through Florida, west to south Texas and northeastern Mexico, north to central North Dakota, and east through the Great Lakes (except Lake Superior), including southern Ontario. Successfully introduced in the Connecticut River valley, California, and elsewhere.

Habitat and Environmental Tolerances: Lakes, ponds, and quiet waters of streams, usually in shallow, clear, weedy water. There is disagreement as to whether the yellow or brown bullhead is more tolerant of turbidity, but it is clear that removing weeds and other cover in natural waters often has a more detrimental effect on this species than on the brown or black bullheads.

Size and Growth: Young of the year reach 2.0–3.5 inches (5.1–8.9 cm) in October in Ohio; 2.5–5.0 inches (6.4–12.7 cm) in first year. Adults usually 5.5–15.0 inches (14.0–38.1 cm); 2 ounces–2 pounds (0.06–0.9 kg). Reaches harvestable size (6 inches or 15.2 cm) in 2 years. Maximum size probably around 20 inches (50.8 cm) and 4 pounds (1.8 kg). Probably reaches maturity in 2–3 years.

Breeding Requirements: Reproduces naturally in ponds. Spawns in shallow water in the spring, possibly earlier than other bullhead species. Parents guard the eggs and young.

Food and Feeding: Capable of feeding at all depths and on the surface, but usually a bottom feeder. In nature feeds on invertebrates (including molluscs), fresh dead animal matter, and, occasionally, fishes. Cultured or captive fish readily learn to surface feed and to accept artificial feeds, pieces of fish or meat, etc. Often feeds at night. Primarily a chemosensory feeder.

Table Qualities: Excellent.

Comments on Culture: Sometimes stocked in farm ponds, though there is danger of stunting. Excellent for cage culture, and probably some other intensive systems.

Harvesting: Easily taken on hook and line using natural bait. Good children's fish. Fairly easy to seine. In small ponds best harvest method, at least in spring, is trapping.

Brown Bullhead

CHAPTER II-2

Family: Ictaluridae

Taxonomic Name: Ictalurus nebulosus (LeSueur) (formerly *Ameiurus nebulosus*)

Common Names: See yellow bullhead. The subspecies *I. nebulosus marmoratus*, found in the southern portion of the range is known as "speckled bullhead," or "marbled bullhead."

Description and Identifying Characteristics: Dorsally yellow, olive, gray brown, or almost black; sides often mottled with brown. Ventrally yellow to white. Barbels dark brown or black, usually noticeably darker than body.

Young darker, white ventrally. Fish under about two inches (5.1 cm) often appear black, or nearly so.

Difficult to distinguish from other bullheads. I have noticed that when captured, brown bullheads often produce a grunting or croaking noise by stridulation while yellow bullheads do not. (Also see yellow bullhead.)

Range: Native from Maritime Provinces of Canada

south through Florida, west through Alabama, northern Mississippi and Arkansas to eastern Oklahoma, and north to central Minnesota (not including central Missouri), northwest to eastern North Dakota, southern Manitoba and southeastern Saskatchewan, and east through the Great Lakes (except Lake Superior) to about Montreal. Successfully introduced in the Connecticut River drainage, Idaho, California, Vancouver Island, the lower Fraser River valley, British Columbia and elsewhere, including Europe.

Habitat and Environmental Tolerances: Lakes, ponds and quiet waters of streams, usually in shallow water, but is found down to forty feet (12 m). Generally found in deeper water than the yellow bullhead. Survives large concentrations of organic and industrial pollution; can withstand dissolved oxygen concentration as low as 0.2 ppm in winter. May burrow into mud to avoid adverse conditions.

Size and Growth: Young of the year in Ohio reach 2.0–4.8 inches (5.1–12.2 cm) by October. To 2.7–6.0 inches (6.9–15.2 cm) at 1 year. Adults usually 6.0–16.0 inches (15.2–40.6 cm); 2 ounces–2.5 pounds (0.06–1.1 kg). Reaches harvestable size (6 inches or 15.2 cm) in 2 years or less. Maximum size may be over 20 inches (50.8 cm) and 8 pounds (3.6 kg). Stunted populations, with few fish over 5 inches (12.7 cm) are quite common. Cultured speckled bullheads in Alabama reached 0.2–0.5 pound (0.09–0.2 kg) in 1 year or less. Usually attains maturity by age 3 or earlier.

Breeding Requirements: Reproduces naturally in ponds. Spawns in late spring and summer. In the South, spawning may continue through September and individuals may spawn more than once a year. Nest near cover; parents guard eggs and young.

Food and Feeding: See yellow bullhead. Sometimes eats filamentous algae. In close confinement is capable of catching substantial numbers of small fishes. Cultured fish have been fed on pure grain products, as well as more complex feeds. However, simple grain feeds may lead to vitamin deficiency diseases.

Table Qualities: Excellent.

Comments on Culture: See yellow bullhead. Has been intensively cultured in ponds, but seems unsuited for cage culture.

Harvesting: See yellow bullhead.

Black Bullhead
CHAPTER II-2

Family: Ictaluridae

Taxonomic Name: Ictalurus melas (Rafinesque) (formerly *Ameiurus melas*)

Common Names: See yellow bullhead. Also "yellow belly bullhead."

Description and Identifying Characteristics: Healthy individuals are usually more stout than the other bullheads. Dorsally yellowish, greenish, brownish, olive, brown, or almost black. Sides yellow-brown to yellow or white, never mottled. Ventrally bright yellow to white; the color of the sides does not usually merge into that of the belly but is sharply demarcated. Barbels gray, black, or spotted. Fins with darker edges and membranes. Breeding males often very dark above.

Young are black and white ventrally.

Difficult to distinguish from other bullheads; see yellow bullhead.

Range: Native from Lake Ontario drainage of New York south, west of the Appalachians, to the Mobile Bay drainage of Alabama, west through most of Texas, north to eastern Montana and southern Saskatchewan, southeast through southwestern Manitoba, Minnesota, Michigan and extreme southwestern Ontario. Successfully introduced and has spread to all suitable habitat in the Columbia River drainage; also in Arizona, California, Connecticut and perhaps elsewhere.

Habitat and Environmental Tolerances: Found in all but very cold or fast flowing waters. More often found in streams, swamps, overflow ponds, deep or cold waters, and heavily polluted waters than the brown or yellow bullheads. Seldom found with them, but often replaces them if the environment deteriorates. Very tolerant of siltation and turbidity.

Size and Growth: Smaller than yellow or brown bullheads. Young of year reach 1.2–4.0 inches (3.0–10.2 cm) in Ohio by October; 1.5–5.0 inches (3.8–12.7 cm) at 1 year. Adults usually 4.5–12.0 inches (11.4–30.5 cm), 1 ounce–15 ounces (0.03–0.4 kg). May reach harvestable size (6 inches or 15.2 cm) in 3 years. Fish over 2 pounds (0.9 kg) are rare, but the angling record is 8 pounds (3.6 kg). Stunted populations, with few fish over 5 inches (12.7 cm) are quite common. With intensive feeding will reach 1 pound (0.45 kg) in 1 year.

Breeding Requirements: Reproduces naturally in ponds. Spawns in spring and/or summer.

Food and Feeding: See yellow bullhead.

Table Qualities: Excellent.

Comments on Culture: See yellow bullhead.

Harvesting: See yellow bullhead.

Channel Catfish

CHAPTER II–3

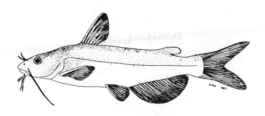

Family: Ictaluridae

Taxonomic Name: Ictalurus punctatus (Rafinesque)

Common Names: The name "catfish" or "cat," without modifers applies equally to this fish and the other large *Ictalurus* spp. The natural morphological and color variation in this species caused it previously to be described under a variety of names. Among the most confusing is "blue catfish," the official name of *Ictalurus furcatus*. In parts of its range, the channel catfish has more of a tendency to be bluish than the blue catfish, thus the name. Adding to the confusion is the name "blue channel catfish." Albino channel catfish, sometimes stocked as a novelty in fee fishing ponds, are called "albino catfish" or "golden catfish," but also "white catfish" (not to be confused with *Ictalurus catus*, officially known as white catfish). Young channel catfish are quite distinctively marked and were once thought to be a different species. They are still sometimes referred to as "willow catfish," "silver catfish," "spotted catfish," "ladycats," or "squealers." Also known as "lake catfish," "Great Lakes catfish" or in French Canada as "barbue de rivière."

Description and Identifying Characteristics: Tail deeply forked; definitely the most "streamlined" of the North American catfishes. Young fish are pale blue to olive dorsally, silvery with small black spots on the sides, and white or silvery ventrally. As the fish mature they lose the spots and often become darker dorsally and less silvery or even yellowish ventrally. Breeding males are noticeably bluish, darker dorsally, and darkest on the head. There is also an albino race, pinkish-white with pink or red eyes.

Can be distinguished from the bullheads by the forked tail, from the blue catfish by the rounded outline of the anal fin and the position of the eyes in the upper half of the head (the eyes of the blue catfish are located more nearly in the middle of the head), and from the white catfish by the more sharply forked tail and the longer anal fin.

Range: Native from the St. Lawrence River in Quebec south, west of the Appalachians, to Georgia and on into central Florida, west to Texas and northeastern Mexico, northwest to central Montana, and east through the Dakotas and the Red River system of Manitoba to the Great Lakes (except Lake Superior) and southern Ontario. Present in the Potomac River system, probably as a result of immigration via canals. Introduced in the South Atlantic States, Connecticut, California, Utah, Idaho and probably elsewhere, including several Latin American countries.

Habitat and Environmental Tolerances: Found wild in lakes and slow moving waters of medium to large rivers. Smaller fish more commonly found in small or swift streams. Adapts well to ponds, raceways, etc. Prefers sand, gravel, or rubble bottoms to those that are silted or vegetated. Characteristically a bottom dweller. Cultured fish tend to be diurnal. Can survive dissolved oxygen content as low as 1 ppm. Occasionally found in brackish water with salinities up to 20‰.

Size and Growth: Young of year in Ohio, 2.0–4.0 inches (5.1–10.2 cm) in October; 3.5–7.5 inches (8.9–19.0 cm) at 1 year. Adults usually 11.0–30.0 inches (27.9–76.2 cm); ¾–15 pounds (0.3–6.8 kg). Growth rate varies greatly in nature, and is usually more rapid in warmer water. A length of 12 inches (30.5 cm) may be reached in 2–8 years. Maximum weight 60 pounds (27 kg) or more; angling record 58 pounds (26.3 kg).

Cultured fish reach marketable size of ¾ pound–1 pound (0.3–0.45 kg) in 1½ years, or 1½–2 pounds (0.7–0.9 kg) in 2 years. Reaches maturity in 2–3 years.

Breeding Requirements: Will not breed in most ponds, especially clear ones, unless special structures are provided. Most culturists purchase stock annually. In nature spawns in late spring or summer in sheltered locations. Parents guard eggs and young.

Food and Feeding: Young eat first zooplankton, then aquatic insects. Adults eat almost any live or fresh animal matter, also green algae, higher aquatic plants, and tree seeds. Like all catfishes, primarily a nocturnal, chemosensory, bottom feeder, but appears more adaptable to diurnal, visual, and midwater feeding than the others.

Both wild and adult fishes, at all ages, easily adapt to many kinds of prepared feeds or simple grain feeds. A variety of commercial feeds are manufactured specifically for the channel catfish.

Table Qualities: Texture is good, flavor poor to excellent. Cultured fish are considered to be more consistently good in flavor than wild fish.

Comments on Culture: The most important commercially cultured fish in the United States. Main culture method is intensive monoculture in ponds, but also used in high and low intensity polycultures, closed systems, cage culture, farm ponds, and fee fishing ponds. The industry is concentrated in Arkansas, Louisiana and Mississippi, with some commercial culture in the South Atlantic States, the Great Plains states, Texas, California and Idaho.

Harvesting: Usually by drawdown and seining, but results are not always satisfactory. Trapping methods are being developed. Hook and line, usually with natural bait, is satisfactory in non-commercial culture.

Blue Catfish
CHAPTER II-3

Family: Ictaluridae

Taxonomic name: *Ictalurus furcatus* (LeSueur)

Common Names: See channel catfish. In some localities, it is more whitish than the white catfish (*Ictalurus catus*), and is called "white catfish." Also "forktail catfish" or "Mississippi catfish."

Description and Identifying Characteristics: Dorsally pale bluish, lighter on sides, shading to silvery or white ventrally. Breeding adults are darker; male is blue-back on dorsal surface and head. A more stream-lined fish than the white catfish, with a longer, straight-edged anal fin. (See channel catfish.)

Range: Native from Gulf Coast drainages of Mexico, north through south and east Texas, and east to extreme west Florida. To the north found in and near the main channels of rivers, including the Alabama, the Red to southwestern Arkansas, the White to northeastern Oklahoma, the Tennessee, the Ohio, the Mississippi to Minneapolis, and the Missouri to central South Dakota. May have been introduced outside this range in regions where channel catfish are cultured.

Habitat and Environmental Preferences: In nature even more a fish of large rivers than the channel catfish. Found in faster water than is the channel catfish, over bottoms of rock, rubble, gravel, or sand, and in deeper water in winter. Adapts well to pond life.

Size and Growth: may reach weights over 100 pounds (45 kg). Sizes of cultured fish same as channel catfish. (See channel catfish.)

Breeding Requirements: See channel catfish.

Food and Feeding: See channel catfish. Adapts more readily to surface feeding than the channel catfish.

Table Qualities: See channel catfish.

Comments on Culture: See channel catfish.

Harvesting: See channel catfish.

White Catfish
CHAPTER II-3

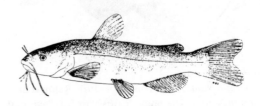

Family: Ictaluridae

Taxonomic Name: *Ictalurus catus* (Linnaeus)

Common Names: See channel catfish and blue catfish.

Description and Identifying Characteristics: Form is intermediate between bullheads and channel and blue catfishes. Tail forked, but body "chunkier" than channel or blue catfish. Milky-gray to dark blue dorsally; sides lighter, sometimes mottled. White ventrally. Breeding fish have blue-black heads and are generally darker. (See channel catfish and blue catfish.)

Range: Native to Atlantic coastal drainages from the Delaware River to central Florida, and west to western Florida and southern Alabama. Introduced to Lake Erie, Massachusetts and to various localities where channel catfish are cultured, including California.

Habitat and Environmental Preferences: Intermediate in most respects to the preferences of the bullheads on the one hand and the channel and blue catfishes on the other. Adapts well to pond life. Sometimes found in brackish water.

Size and Growth: Adults usually 10.0–18.0 inches (25.4–45.7 cm); 8 ounces–3 pounds (0.2–1.4 kg). Maximum size reported to be 24 inches (61 cm). Sizes of cultured fish same as channel catfish. (See channel catfish.)

Breeding Requirements: See channel catfish. Less often available for sale than channel catfish.

Food and Feeding: See channel catfish.

Table Qualities: See channel catfish. Much higher dressing loss than in channel or blue catfish.

Comments on Culture: See channel catfish. Much less commercially important than channel or blue catfish.

Harvesting: See channel catfish.

Bigmouth Buffalo
CHAPTER II-4

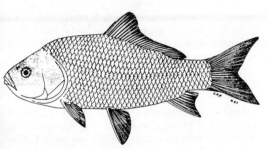

Family: Catostomidae

Taxonomic Name: *Ictiobus cyprinellus* (Valenciennes)

Common Names: "Buffalofish," "buffalo sucker," and "buffalo" all are used indiscriminately to refer to this species, the smallmouth buffalo (*Ictiobus bubalus*) and the black buffalo (*Ictiobus niger*) This species is also called "redmouth buffalo," "common buffalo," "gourd-head buffalo," and "buffalo à grande bouche" (French Canada).

Description and Identifying Characteristics: Brown, bronze, or olive dorsally, lighter on the sides, white below. Fins brown or grey. Breeding adults show more blue or green tints. Could be confused with such other heavy-bodied, large-scaled fish as the common carp, the goldfish or some of the other members of the Catostomidae. Differs from all of them except the goldfish in having a definitely terminal mouth. But the goldfish (as well as the common carp) has a single dorsal spine.

Range: Native almost throughout the low gradient portions of the Mississippi-Missouri system, except in the western reaches of Missouri tributaries. Also in the Red River drainage of Manitoba and Saskatchewan. Present in Lake Erie, but it is not clear whether this is the result of introductions. Also introduced in Arizona, California and Israel.

Habitat and Environmental Tolerances: Inhabits mostly shallow and quiet waters of large and medium-sized rivers, though also found in lakes. In spawning season also found in ditches and flooded areas. May be kept in ponds. Highly tolerant of turbidity and siltation. Tends to school.

Size and Growth: Young of the year reach 1.7–4.0 inches (4.3–10.2 cm) in Ohio by October, 5.0–7.0 inches (12.7–17.8 cm) at 1 year. Adults usually 15.0–30.0 inches (38.1–76.2 cm) 2–30 lb. (0.9–13.6 kg). Maximum size 80 pounds (36.3 kg) or more. Grows much more rapidly in the southern part of its range. Maturity may be attained at ages of anywhere from 1–12 years.

Cultured fish reach the preferred market size of about 4.4 pounds (2 kg) in 2 years, or 2.2 pounds (1 kg) in 18 months. The hybrid ♀ *I. niger* × ♂ *I. cyprinellus* is said to grow about 33% more rapidly.

Breeding Requirements: Culturists spawn their own fish through a simple process involving regulation of population density and manipulation of the water level. Temperatures of 64.4–69.8°F (18–21°C) are considered optimal.

Natural spawning takes place during a very short period in spring, and is triggered by rising water. Hybridizes naturally with other *Ictiobus* spp.

Food and Feeding: A filter feeder, feeding both in mid-water and on the bottom, mainly on small invertebrates. Cultured fish have been successfully fed on grain-based feeds.

Table Qualities: Palatable, but not generally considered a superior quality food fish. Bony.

Comments on Culture: One of the few large, native North American filter feeders, and as such potentially very valuable as a cultured fish. Well suited for culture in rotation with rice. Culture methods were developed in Arkansas in the 1950's, but channel catfish became much more important economically. Mostly used in low-intensity polyculture, but beginning to see use as a supplemental fish in conventional catfish culture systems.

Harvesting: Rarely taken on hook and line. Easily harvested with seines.

Common Carp
CHAPTER II-5

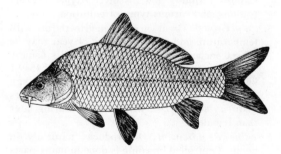

Family: Cyprinidae

Taxonomic Name: *Cyprinus carpio* Linnaeus

Common Name: Also known as "European carp," "German carp," and "carpe" (French Canada), as well as by a great variety of humorous and/or derogatory names.

The variety usually cultured in North America is the "mirror" or "Israeli" carp (*C. carpio* var. *specularis*). Since the carp is a truly domesticated fish, many other varieties have been named, including "leather carp" and "line carp." The species of Chinese carps most commonly referred to in fish culture literature are the grass carp, silver carp, big head carp, mud carp, and black carp or snail carp. These should not be confused with *C. carpio*.

Description and Identifying Characteristics: Shape and color vary tremendously with habitat, and also because carp have been selectively bred for virtually every imaginable character. The species should be distinguishable on the basis of the humped back, large scales, long dorsal fin with a single spine, subterminal mouth, and two pairs of barbels. Wild fish are generally slimmer than cultured stock, and vary in color from olive to reddish-bronze.

Mirror carp have only scattered scales of exceptionally large size. Line carp have a single horizontal row of scales. Leather carp have no scales at all.

Range: Native to Eurasia, introduced virtually throughout Europe, Asia and North America, and in other continents as well. In the United States found wherever there is suitable habitat except in Maine, Alaska, and peninsular Florida. In Canada, they are found in the St. Lawrence River, the Great Lakes and tributaries, the Lake Winnipeg area, and many other waters near the U.S. border.

Habitat and Environmental Tolerances: Found in virtually all types of water which are not excessively cold or swift-flowing; usually in the shallows. Adapts well to pond life. Favors warm, fertile water, and soft bottoms. Relatively tolerant of salinity, turbidity, low dissolved oxygen concentrations, and various types of pollution.

Size and Growth: Great variation in growth rate with temperature and fertility. Young of year may be 0.7–8.0 inches (1.8–20.3 cm) long. Adults usually 12–30 inches (30.5–76.2 cm); 1–20 pounds (0.45–9.1 kg). The angling record is 55 pounds (24.9 kg); maximum size certainly exceeds that. Reaches maturity at ages 2–5, with males maturing later than females.

Breeding Requirements: Reproduces naturally in ponds. Controlled breeding is done in many parts of the world, using a variety of methods, some of them apparently suitable for North American small scale culture. In tropical areas may be spawned several times annually. Selectively bred for a great variety of characteristics.

In nature in North America spawns in spring and early summer over vegetation, often in flooded areas. Hybridizes naturally with goldfish, producing fertile offspring.

Food and Feeding: One of the most omnivorous fishes. Feeds chiefly by sucking up detritus, spit-

ting it out, and selecting the food items, but will also feed in midwater or at the surface. Natural foods include aquatic plants of all kinds, seeds, insects, annelids, molluscs, fish eggs, and fresh dead fish. Mirror carp are said to be more herbivorous than other strains. Cultured carp will eat almost any kind of invertebrate or grain-based feeds.

Table Qualities: Though often considered inedible in North America, carp are among the world's most valued and important food fish. Quality of wild carp varies greatly; cultured fish are generally better and may be excellent. Carp are particularly well suited for smoking, or for preparation of gefilte fish.

Comments on Culture: Though originally introduced to North America as a food fish for culture, the carp is rarely cultured here. However, it may be the most adaptable and universally recommendable fish for culture in North America where marketability is not a major consideration. Responds very favorably to fertilization. Not recommended for farm ponds where hook and line fishing for bass, bluegill, and the like is highly valued.

In the United States carp are occasionally cultured as bait "minnows" or stocked in fee fishing ponds. In Europe and Asia carp are the objects of a great variety of high and low intensity monoculture and polyculture methods including growth in sewage treatment facilities, cage culture, closed systems, etc. Also important in fish culture as the universal "donor" for induced spawning by means of pituitary injection.

Harvesting: Easily harvested by seining or drainage. Moderately difficult to take on hook and line.

Rainbow Trout
CHAPTER II-6

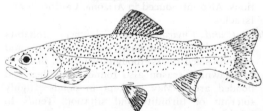

Family: Salmonidae

Taxonomic Name: *Salmo gairdneri* Richardson (formerly *Salmo irideus*)

Common Names: Anadromous populations are often referred to as "steelheads." In the past, a number of varieties of rainbow trout named after their native waters were recognized, the most renowned being the "McCloud River trout." Many of these have since been lost through mongrelization. An exception which may become important to culturists is the "Kamloops trout" (see below). The rainbow trout is a truly domesticated fish, so that a number of new varieties have been developed and named, perhaps the most important being the "super trout" or "Donaldson trout." An albino variety sometimes called "golden trout" should not be confused with *Salmo aguabonita*, officially known as golden trout. French name "truite arc-en-ciel."

Description and Identifying Characteristics: Varies greatly with habitat; in general lake or sea-run fish are more silvery and less colorful. Dorsally, various shades of blue and green to brown, silvery on the sides, and white ventrally. A pale pink to brilliant red band usually extends from the gill cover to the tail. Back, sides and dorsal, adipose, caudal, and anal fins with small black spots. No colored spots.

Young similar, but with 5–10 oval black blotches ("parr marks") on the sides.

Range: Native to the eastern Pacific ocean and its tributary drainages from the Kuskokwim River, Alaska to northwestern Mexico. Also probably native to the Peace and Athabasca Rivers of the Arctic drainage of Alberta. Introduced widely on all continents. In the United States it is found almost everywhere with suitable habitat, particularly in New England, New York, the Appalachians, the upper Great Lakes Basin, the upper Mississippi valley, the Ozarks, and the Rocky Mountains. In Canada, successfully introduced in Newfoundland's Avalon Peninsula, Prince Edward Island, Nova Scotia, the vicinity of Montreal, the Great Lakes basin, the Lake of the Woods drainage, southern Manitoba, southeastern Saskatchewan, the Rocky Mountains north to northern British Columbia, parts of the Yukon Territory, and probably elsewhere.

The range is being further extended as a result of culture in isolated areas with suitably cold water, and by seasonal culture in waters where winterkill or summer temperatures would prevent trout populations reproducing naturally.

Size and Growth: Natural growth rates are too variable to mention here. Minimum harvestable size is generally considered to be 6–7 inches (15.2–17.8 cm), and adults usually range between that size and 20 inches (50.8 cm). Largest known specimen was 52 pounds (23.6 kg).

Intensively fed cultured fish reach marketable size of 8–14 inches (20.3–35.6 cm) and ½–1 pound (0.22–0.45 kg) in 7–14 months. Non-intensive growers achieve similar growth at much lower population densities. Specialists produce 1–6½ inch (2.5–16.5 cm) fingerlings for sale to other cul-

turists in 1–8 months. The Kamloops strain and certain artificially selected strains may grow much more rapidly.

Breeding Requirements: Ordinarily does not spawn naturally in ponds. In nature spawns in fast-flowing water over gravel, usually in the spring, but sometimes in the fall.

Most culturists do not breed their own fish, but purchase eyed eggs or fingerlings from specialists, who do not rely on anything like natural reproduction. Eggs are artificially fertilized and hatched in flowing water.

Habitat and Environmental Tolerances: Needs flowing water for natural reproduction, but can survive and grow in suitable standing waters. Requires cold water saturated with dissolved oxygen.

Food and Feeding: In nature, exclusively carnivorous. Eats mainly insects in most environments, but also other invertebrates, fish and fish eggs. Most cultured fish are fed on commercial feeds prepared specifically for this species. Such feeds may contain some vegetable matter, but must contain high percentages of animal matter. Some culturists also use meat or fish offal.

Table Qualities: Good to excellent. Wild fish, or cultured fish with a high percentage of insects or crustaceans in their diets, are preferred to fish fed exclusively on concentrates. The latter's flesh is white, rather than red, and is less flavorful.

Comments on Culture: The world's major cold water cultured food fish. The most important commercial aquaculture crop in Canada, the second most important in the United States, and similarly important in other temperate zone countries. Most commercially cultured trout are monocultured in flowing spring water, but cages, closed systems, and various forms of seasonal culture are also used. In Europe some trout are polycultured. They are often stocked in cold water farm ponds and can be used in fee fishing.

Some culturists realize profits by selling eyed eggs or fingerlings, rather than edible size trout. This requires considerable skill and a particularly good water supply.

Harvesting: Readily harvested by seining, trapping, gill netting, or draining. For partial harvests umbrella nets or hook and line may be used.

Brown Trout

CHAPTER II-6

Family: Salmonidae

Taxonomic Name: *Salmo trutta* Linnaeus

Common Names: Many varieties of brown trout have been introduced to North America, and such names as "German brown trout," "English brown trout," "Loch Leven trout" and "von Behr trout" persist, but have become almost meaningless through mixing of stocks. French name "truite brune." Sea-run fish are sometimes called "salters."

Description and Identifying Characteristics: In streams, brownish or greenish on the back, shading to white, yellow, or silvery ventrally. Some fish have a golden cast. There are both black and red or orange spots. The latter, found only on the body, are usually surrounded by a halo of light blue. Sea-run fish, lake fish, and many hatchery fish are more white or silvery in color, and the colored spots, and particularly the blue halo, may be obscured.

Young with fewer spots and 9–14 dark blotches ("parr marks") along the sides.

Range: Native to Europe and western Asia. Successfully introduced on all continents, but particularly North America. Its range in the United States is similar to that of the rainbow trout except that it is absent from Alaska. In Canada, found in southern Ontario, southern Alberta, and scattered locations in all the other provinces except Prince Edward Island, the Yukon, and Northwest Territories.

Habitat and Environmental Tolerance: Similar to rainbow trout.

Size and Growth: Similar to rainbow trout. Largest known fish 39 pounds 4 ounces (17.8 kg) (angling record).

Breeding Requirements: Similar to rainbow trout, but spawns in the fall. May spawn in standing water. Professional hatcherymen use a variety of tricks to induce brown trout to spawn out of season.

Food and Feeding: Similar to rainbow trout. Large individuals are said to be more piscivorous. Unlike the rainbow trout and brook trout, brown trout often feed nocturnally.

Table Qualities: See rainbow trout, but not considered to be quite as high quality as rainbow or brook trout.

Comments on Culture: See rainbow trout. Seldom grown as a food fish, but often grown in hatcheries for stocking.

Harvesting: See rainbow trout. More difficult to capture, especially with hook and line.

Brook Trout

CHAPTER II-6

Family: Salmonidae
Taxonomic Name: *Salvelinus fontinalis* (Mitchill)

Common Names: "Squaretail," "speckled trout," "brookie," "speckled char," "mud trout," "mountain trout." French names "omble de fontaine," "truite mouchetée." The name "native trout" is applied to the brook trout in eastern North America to distinguish it from the introduced brown and rainbow trouts, but this term may also be used to distinguish naturally spawned trout of whatever species from hatchery plants. Searun fish are called "sea trout," "salters," or "truite de mer" (French Canada). In the Great Lakes, lake-run fish are sometimes called "coasters." As with the rainbow and brown trouts, a number of local varieties have been given names, for example "Aurora trout," but these have become largely meaningless due to mixing of stocks.

Description: Dorsally olive green to dark green or brown, heavily vermiculated with a lighter shade. On the sides the vermiculations break up and become spots. There are also red spots with blue halos on the sides, but no black spots. A suggestion of vertical bars (remnants of parr marks) is often present. White ventrally. Anal, pelvic, and pectoral fins orange or red, with a distinct white leading edge followed by a black line. Breeding fish of both sexes are more brightly colored: in males, the belly becomes red-orange with a black area on either side.

Sea-run fish, those inhabiting lakes, and some hatchery fish are more or less silvery and lack the light spots and blue halos on the sides. Young are more silvery, less colorful, with 8 to 12 dark blotches on the sides ("parr marks"), and a few yellow or blue spots.

Range: Native to the western Atlantic Ocean, and its tributaries from Ungava Bay, Quebec, south to Cape Cod (Massachusetts), in the Appalachians south to northern Georgia, west through the upper Great Lakes to the upper Mississippi River drainage in Minnesota, and north to Hudson Bay. Introduced on all the continents. In North America successfully introduced in the lower peninsula of Michigan, Wisconsin, Manitoba, Alaska, the Rocky Mountains, and probably elsewhere.

Habitat and Environmental Tolerances: Usually found in small streams, but may be found in many other types of environment where temperatures are low and D.O. high. Of the trouts I have described this is the least tolerant of pollution, heat, turbidity or interspecific competition, but the most tolerant of acid pH.

Size and Growth: In nature usually smaller than the rainbow or brown trouts. Largest known specimen 14.5 pounds (6.6 kg).

Size of cultured fish is about the same as rainbow trout's.

Breeding Requirements: In nature, spawns in late summer and fall. See rainbow trout.

Food and Feeding: See rainbow trout. Less prone to feed at the surface than brown or rainbow trout.

Table Qualities: The best of the three trouts described here. See rainbow trout.

Comments on Culture: Not as important as a food fish as rainbow trout, but is monocultured in some places where the water is too cold for optimum growth of rainbow trout, or to satisfy a specialty market. Also raised in hatcheries for stocking.

Harvesting: The easiest of the trouts to take on hook and line. See rainbow trout.

Yellow Perch
CHAPTER II-7

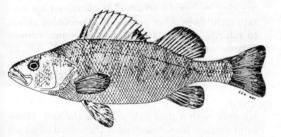

Family: Percidae

Taxonomic Name: *Perca flavescens* (Mitchill). (May be synonymous with the European perch, *Perca fluviatilis* Linnaeus, in which case that name would supersede *P. flavescens*.)

Common Names: In North America "perch," used without modifiers generally refers to this fish. Also called "American perch," "lake perch," "green perch," "red perch," "coontail perch," "ring perch" (Florida), and "perchaude" (French Canada). Large yellow perch in the Great Lakes and the Finger Lakes of New York are sometimes called "jack perch."

Description: Dorsal surface and head bright green to greenish-brown. Vertical bars of this color extend onto the yellow-green or yellow sides. White ventrally, with brilliant yellow-orange to crimson pelvic fins. Dorsal and caudal fins yellow to green. Young are paler and transparent or silvery. May sometimes be distinguished in the water by the jauntily erect carriage of the dorsal fin.

Range: Native in Atlantic coastal drainages from Nova Scotia to South Carolina. Beyond that, there is disagreement on where it is native and where introduced. West of the Appalachians, it is found from Southwestern New York southwest to northeastern Kansas, northwest to Great Slave Lake,

and southeast to James Bay and New Brunswick. Successfully introduced in Georgia, Florida, Montana, Idaho, Washington, British Columbia, Oregon, California, Utah, New Mexico, Texas, and probably elsewhere.

Habitat and Environmental Tolerances: The main characteristics of suitable perch habitat seem to be found in ponds, lakes and streams, over all bottom types, in clean and polluted waters, in salinities up to 15‰ and at all depths up to 150 feet (46 m), though usually at depths of less than 30 feet (9 m). Strong tendency to school.

Size and Growth: In Ohio young of the year reach 1.8–4.0 inches (4.6–10.2 cm) and 2.0–4.5 inches be clear water and considerable aquatic vegetation. Where these are present, yellow perch are (5.1–11.4 cm) in 1 year. Adults usually 4.5–12.0 inches (11.4–30.5 cm). Commercial size (about 8 inches or 20.3 cm) usually reached in 3–6 years. Stunting often occurs, with hardly any fish ever reaching 6 inches (15.2 cm). Largest known individual 4 pounds 3½ oz. (1.9 kg) (angling record).

Breeding Requirements: Reproduces naturally in ponds. Spawns in shallow water of ponds, lakes, or rivers in spring or early summer.

Food and Feeding: Exclusively carnivorous. Natural foods mainly include zooplankton, insect larvae, other aquatic invertebrates, and fish, but also terrestrial insects and fish eggs.

Table Qualities: Texture is usually superb. Flavor varies from poor to excellent.

Comments on Culture: Not recommended for farm ponds, but may eventually find some use in cool water farm ponds. Currently being studied for intensive closed system culture.

Harvesting: Fairly easy to harvest by seining. Gill nets used in commercial fishing. Easy to take on hook and line with natural bait. Good children's fish.

Grass Carp
CHAPTER II-8

Family: Cyprinidae

Taxonomic Name: *Ctenopharyngodon idellus* Valenciennes

Common Names: "White amur," "herbivorous carp."

Description and Identifying Characteristics: A power-fully built fish, with scales almost as large as those of the common carp. Gray to silvery in color, with no markings. Can be distinguished from common carp or most large suckers by lack of the downward projecting fleshy lips. More slender than common carp or bigmouth buffalo.

Range: Native to major river basins of northern China and eastern Soviet Union. Introduced throughout Asia and Europe. First introduced in Mexico in 1960 and the United States in 1963. The state of Arkansas has an active stocking program and permits private individuals to obtain grass carp. Whether this has resulted in reproducing populations is uncertain, but grass carp have been caught at many points north and south of Arkansas on the Mississippi River. Private importation of grass carp is banned in Canada and 43 of the states of the United States.

Habitat and Environmental Tolerances: Characteristically a fish of large, turbid rivers, but capable of surviving, if not reproducing, in ponds and lakes, and in a great variety of temperatures and salinities.

Size and Growth: Among the fastest growing fish in the world. In temperate zones may reach 1 pound (0.45 kg) in 1 year and continue to grow 2–3 pounds (0.9–1.4 kg) per year. In tropical regions can reach 2.–2.5 pounds (0.9–1.4 kg) in 6 months and and gain 1 pound (0.45 kg) per month for the next 6 months. Maximum weight over 100 pounds (45 kg).

Breeding Requirements: In nature, breeds annually in large, turbid rivers. Spawning is triggered by rising water.

Does not reproduce in ponds, and culturists throughout the world must obtain stock from specialists. Can be spawned successfully by means of pituitary injection.

Food and Feeding: Omnivorous. Young eat mainly zooplankton, after which they consume many kinds of invertebrates and occasionally fish, but also great amounts of soft aquatic plants.

Cultured grass carp are fed aquatic or terrestrial plants. They also accept most types of commercial feed.

Rate of feeding increases with temperature.

Table Qualities: Among the finest of fresh water fish in terms of taste and texture, but bony.

Comments on Culture: An integral part of traditional Chinese pond polyculture. In North America used primarily in weed control, and secondarily as a food fish.

Harvesting: Difficult to seine, owing to its great agility and vigor. Also difficult to catch on hook and line. Gill netting seems best.

Java Tilapia
CHAPTER II-8

Family: Cichlidae

Taxonomic Name: *Tilapia mossambica* (Peters).

Common Names: Here again I depart from the American Fisheries Society list; their official name for *T. mossambica* is "Mozambique mouthbrooder." (It is also sometimes called "Mozambique tilapia.") Aquaculturists use "tilapia" to designate all the fishes included in the genus *Tilapia*. So it seems appropriate to use names which distinguish among the *Tilapia* spp., while keeping the generally accepted word "tilapia." "Mouthbreeder" or "mouthbrooder" is also used for certain cichlids in other genera. It is probably safest to refer to all tilapias by their taxonomic names. Tilapia are marketed in the United States as "Saint Peter's Fish," or "African Perch" (not to be confused with the Nile perch [*Lates niloticus*], a member of the Serranidae.)

Description and Identifying Characteristics: Greenish-gray with a dull gold edge to the tail. Three faint vertical bars. Sometimes shows red edges on the fins, but not with the intensity of *Tilapia aurea*. (See blue tilapia.)

Range: Native to East Africa; widely introduced on all the continents. May be established in Florida and/or California. Several states of the United States prohibit its importation.

Habitat and Environmental Tolerances: Characteristically a fish of shallow, fertile lakes and ponds. Does well in small ponds. Very tolerant of chemical stresses, including D.O. as low as 1 ppm. Does not do well in acid water.

Size and Growth: May reach ½ pound (0.2 kg) in 6 months. Maximum size 5 pounds (2.3 kg) or more. Subject to stunting.

Breeding Requirements: Reproduces naturally wherever there is adequately warm water, access to the bottom, and a dark corner. Hybridizes with other mouthbreeding tilapias.

Food and Feeding: Omnivorous; more carnivorous when young. Filter feeds on phyto- and zooplankton, also eats filamentous algae, soft green plants, and invertebrates. In captivity quickly learns to accept many kinds of food, including terrestrial plants, insects, grains, and commercial fish feeds.

Table Qualities: Poor to excellent, depending chiefly on environment.

Comments on Culture: One of the world's most important cultured food fish, though less important in North America. In tropical and semitropical areas polycultured in fertilized ponds. In North America grown mostly in small closed systems. Also used to control filamentous algae and mosquito larvae, as an aquarium fish, and as one parent in most of the monosex tilapia hybrids.

Harvesting: Cast nets are best; umbrella nets will do for small harvests. Difficult to seine, gill net, or take with hook and line. Will burrow into the mud when ponds are drained.

Blue Tilapia
CHAPTER II-8

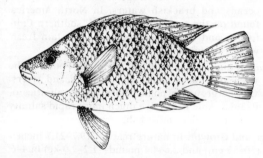

Family: Cichlidae

Taxonomic Name: Tilapia aurea (Steindachner).

Common Names: See Java tilapia.

Description and Identifying Characteristics: Bluish-gray and iridescent. Blue bar across cheek at high temperatures only. Displays black vertical bars when chilled. Healthy blue tilapia have vivid crimson borders on the dorsal and caudal fins.

Range: Native to the Mideast, the Nile valley and east and west central Africa. Widely introduced outside that range. In United States established in west central Florida.

Habitat and Environmental Tolerances: See Java tilapia.

Size and Growth: See Java tilapia.

Breeding Requirements: See Java tilapia.

Food and Feeding: See Java tilapia. Fond of filamentous algae, and appears to utilize phytoplankton and vascular plants better than Java tilapia. Occasionally eats small fish.

Table Qualities: See Java tilapia.

Comments on Culture: In tropical and semitropical areas polycultured in fertilized ponds. In North America grown mostly in small closed systems. Experimentally stocked in "phosphate pits" in Florida. A parent species for at least one monosex tilapia hybrid.

Harvesting: See Java tilapia. Hook and line fishing with live bait selects for fish under 6 inches (15.2 cm). In Florida, harvested with cast nets and snag hooks.

Walking Catfish
CHAPTER II-8

Family: Clariidae

Taxonomic Name: *Clarias batrachus* (Linnaeus)

Common Names: "Clarias" is used in the aquarium trade to indicate all *Clarias* spp.

Description and Identifying Characteristics: Grey-brown dorsally and white ventrally. There is an albino strain which is pinkish-white with red eyes. It is quite unlike any North American catfish in form, and has an especially long dorsal fin.

Range: Native to southeast Asia. Established in southern Florida.

Habitat and Environmental Tolerances: In its native lands found in ponds, swamps, rivers and ditches. Its auxiliary air-breathing organ allows it to survive for long periods in waters without dissolved oxygen, or even out of water. Will migrate across land.

Size and Growth: Maximum size is said to be just over 16 inches (40.6 cm). Marketable size in Thailand is about 10½ inches (26.7 cm); this is reached in 4 months by cultured fish.

Breeding Requirements: Can be spawned in ponds using methods similar to those used in channel catfish culture.

Food and Feeding: Omnivorous and voracious. Though accepts grains, cheese, etc., it grows best on animal food.

Table Qualities: Good to excellent, but "tough" texture.

Comments on Culture: A major and extremely productive pond monoculture crop in Thailand, but not cultivated in Florida.

Harvesting: Most methods, including hook and line, should work.

American Eel
CHAPTER II-13

Family: Anguillidae

Taxonomic Name: *Anguilla rostrata* (LeSueur)

Common Names: "Eel" used without modifiers generally refers to this species. Young eels are called "elvers" or "grass eels." Also "common eel," "freshwater eel," "anguille" (French Canada). The color of eels differs at various ages, and these are sometimes distinguished by names like "silver eel," "bronze eel," "yellow eel," and "black eel."

Description and Identifying Characteristics: Dorsally yellow, green, brown, or almost black. Ventrally white, silvery, or yellow. Young are transparent as glass in salt water, but turn greyish green on entering brackish water. The only fish likely to be confused with *A. rostrata* are the conger eels (family Congridae), which are restricted to deep marine environments.

Range: Native to the western Atlantic Ocean and its drainages from southwestern Greenland to the Carribean. Reaches Minnesota in the Mississippi drainage, South Dakota in the Missouri, and Colorado in the Rio Grande. Became established in the Great Lakes above Niagara Falls after the Welland Canal was opened. Now being introduced in Taiwan and Japan.

Habitat and Environmental Tolerances: Extremely tolerant of variations in temperature, salinity, turbidity, and dissolved oxygen concentration.

Size and Growth: Elvers in North America are 2.5–3.5 inches (6.3–8.9 cm) long, and are 1 year old or more by the time they reach North American waters. Adults are usually less than 40 inches (101.6 cm) long and weigh 2 ounces–5 pounds (0.06–2.3 kg). Maximum size at least 52 inches (132 cm) and 16 pounds (7.3 kg).

Breeding Requirements: Spawns at unknown locations in mid-ocean. For the foreseeable future, eel culture will depend on capture of wild stock.

Food and Feeding: Exclusively carnivorous and voracious. Eats virtually any live or fresh animal matter. Cultured eels are fed on meat offal, fish and fish offal, invertebrates, and prepared feeds which incorporate grains as well as animal protein. Often feed nocturnally.

Table Qualities: One of the finest of food fish, but unpopular in North America.

Comments on Culture: Infrequently cultured in North America. An important cultured food fish in Taiwan and Japan and, to a lesser extent, in Europe. Grown in monocultures and polycultures in ponds, running waters and closed systems, with intensive feeding.

Harvesting: Usually trapped. Small harvests may be made with hook and line, using live bait, or with a spear at night.

Striped Mullet

CHAPTER II-13

Family: Mugilidae

Taxonomic Name: *Mugil cephalus* Linnaeus

Common Names: "Black mullet" (Florida) or "grey mullet." Not to be confused with the several species of suckers (family Catostomidae) commonly called "mullet" in North America.

Description and Identifying Characteristics: You can distinguish mullets from other fishes by the upturned mouth, the pectoral fin placed very high on the body, and the two small dorsal fins. The striped mullet is silvery or bluish, with several distinct dark lateral stripes.

Range: World-wide in tropical and semitropical oceans and brackish waters. In North America found off the south Atlantic, Gulf, southern California, and Hawaii coasts, and in certain fresh waters draining into them.

Habitat and Environmental Tolerances: Found in the open seas, bays, estuaries, marshes and rivers. Can be grown in ponds. Withstands salinities of 0–38‰. Able to survive much more rapid salinity transitions than most fish.

Size and Growth: In nature reaches 19.7–21.7 inches (50–55 cm) and 2.6–4.4 pounds (1.2–2.0 kg) in 4–6 years. When stocked as fry, mullet cultured in Israel reach harvestable size in 120–150 days.

Breeding Requirements: Spawns at sea. Mullet culturists still rely mainly on naturally produced stock—which may sometimes be induced to enter ponds constructed so as to receive tidal flow. Methods of spawning with pituitary injection have been developed.

Food and Feeding: Almost exclusively herbivorous; feed on phytoplankton, benthic algae and, in fresh water, decaying plants. Cultured fish accept grain feeds, but fertilization, rather than feeding, is the usual practice.

Table Qualities: Excellent, but not well liked in North America.

Comments on Culture: Important cultured food fish in southeast Asia and the Mediterranean. Seldom cultured in the United States, though more or less accidental stocking of tidal ponds has produced good yields. Mostly grown in polyculture at all levels of intensity.

Harvesting: Never taken on hook and line, except by snagging. Can be seined, but tendency to leap lessens efficiency. Cast nets may be best.

Bait and Forage Fish

Goldfish

PART IX

Family: Cyprinidae

Taxonomic Name: *Carassius auratus* (Linnaeus)

Common Names: "Golden carp," "poisson doré" (French Canada). Sometimes known as "Baltimore minnow" in the bait trade. Ornamental varieties bear many fanciful names, such as "black moor," "shubunkin," "telescope," "lion head," "veiltail," "doubletail," "comet," etc.

Description and Identifying Characteristics: There are ornamental varieties with almost all imaginable combinations of body and fin shape, and shades of orange, red, pink, white, black, gray, and silver. They are not likely to be confused with any other fish. Most wild fish have reverted to their natural coloration: olive-green to bronze dorsally, lighter on the sides, and white ventrally. They may be distinguished from superficially similar stout-bodied, large-scaled fishes by the presence of a single spine in the dorsal fin, the terminal mouth, and the lack of barbels.

Range: Native to China. Introduced in most parts of the United States, often unsuccessfully, but there are scattered wild populations, notably in Lake Erie. In Canada most likely to be encountered in southern Ontario, and near the U.S. border. Available from pet dealers nearly everywhere. Introduction into natural environments prohibited in many places.

Habitat and Environmental Tolerances: Seem to do best in small, weedy ponds. Otherwise similar to carp, but appears less tolerant of current, cool water, high turbidity, siltation, and organic or industrial pollution.

Breeding Requirements: Reproduces naturally in ponds. In nature spawns on aquatic plants in spring. Easily bred by culturists using well-established methods. Hybridizes naturally with carp, producing fertile offspring.

Food and Feeding: Omnivorous, consuming invertebrates, phyto- and zooplankton, and aquatic plants. Readily learns to surface feed and to take prepared feeds.

Comments on Culture: Attains sufficient size and grows well enough to be grown as a food fish but, perhaps due to the inferior quality of the flesh, it is not grown for food in North America, though it

may play a minor role in pond polyculture in Europe and Asia. Easily cultivated in ponds for ornamental, bait, and experimental purposes.

Harvesting: Easily harvested by seining or draining.

Golden Shiner

PART IX

Family: Cyprinidae

Taxonomic Name: *Notemigonus crysoleucas* (Mitchill)

Common Names: Seems to have a collection of names properly applied to other fishes, including "roach," "bream," "butterfish," "sunfish," "dace," "gudgeon," "shad," "chub," and "goldfish." The situation is further confused by the general imprecision in the use of "minnow." Also known as "bitterhead," "windfish," and "chatte de l'est" (French Canada).

Description and Identifying Characteristics: Stouter bodied and more laterally flattened than most cyprinids. Dorsally golden-olive to golden brown. Sides brassy or silvery. Individual scales distinctly dark edged. Fins yellow to bright red. Young less than 3 inches (7.6 cm) long usually have a dusky lateral band, and are more silvery.

If in doubt, look for a narrow naked strip on the belly where the scales do not cross. This will distinguish the golden shiner from the great majority of other small cyprinids.

Range: Native from Nova Scotia south through Florida, west to central and southern Texas, northwest to eastern Wyoming and Montana, northeast to northern Manitoba, and southeast through the James Bay and Great Lakes drainages to the St. Lawrence River. Sporadically introduced west of this range in the United States and in Prince Edward Island.

Habitat and Environmental Tolerances: Clear, weedy, quiet, and fertile waters. Ranges over open water, as well as near cover and shoreline. Schools.

Breeding Requirements: Spawns in the spring over aquatic weeds, or in nests of Centrarchids.

Food and Feeding: A midwater and surface feeder, consuming mainly zooplankton, also aquatic and terrestrial insects, filamentous algae, and molluscs. In culture accepts grain-based foods.

Comments on Culture: Large enough and productive enough to be at least marginally useful as a food fish, but not grown for that purpose. Probably the most important cultured bait fish, with methods well established. Also used as a forage fish in farm ponds and intensive bass culture.

Harvesting: Easily harvested with seines or umbrella nets.

Fathead Minnow
PART IX

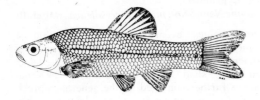

Family: Cyprinidae

Taxonomic name: *Pimephales promelas* Rafinesque

Common Names: "Blackhead minnow," "tuffy," "tête-de-boule" (French Canada).

Description: Small (to 3.5 inches or 8.9 cm), fairly short-bodied. Dorsally olive-yellow, olive-green or brown, with copper, purple, or silvery sheen. Sides lighter, yellowish. Ventrally white, yellowish, or silvery. Usually with a broad, distinct, dark lateral band. Some individuals are said to exhibit a "saddle" pattern in which two light vertical bands extend from dorsal to ventral surfaces, one behind the head, and the other beneath the dorsal fin. This pattern is said to occur most often on breeding males, which are otherwise very dark (including the fins), so that the lateral band is often obscured. Breeding males also develop light tubercles on the head, and a spongy bluish or gray pad in front of the dorsal fin.

Most easily confused with the bluntnose minnow (*Pimephales notatus*), or the bullhead minnow (*Pimephales vigilax*). Can be distinguished from the former by the terminal mouth (subterminal in bluntnose minnow). Can be distinguished with certainty from the bullhead minnow only by opening the fish. If the peritoneum is brown or black it is a fathead (or bluntnose) minnow; if silvery, it is a bullhead minnow.

Range: Native from New Brunswick through the entire Great Lakes basin and northern New England. In the South, west of the Appalachians to northern Alabama and Mississippi. Southwest through Texas to Chihuahua, Mexico, north to the Great Slave Lake drainage, and southeast through the James Bay drainage to the St. Lawrence River. Probably introduced outside this range.

Habitat and Environmental Tolerances: Ponds, small lakes, ditches, and small streams where competition from other fishes (especially the bluntnose minnow) is not great. Very tolerant of extremes in pH and turbidity; can withstand salinities at least to 10%.

Breeding Requirements: Spawns under logs and the like in spring and summer. Easily bred in ponds.

Food and Feeding: Zooplankton, insect larvae, algae, detritus and bottom mud.

Comments on Culture: A commonly cultured bait fish, also used as a forage fish in intensive and farm pond culture of catfish and bass.

Harvesting: Easily seined.

Mosquitofish
PART IX

Family: Poeciliidae

Taxonomic Name: *Gambusia affinis* (Baird and Girard)

Common Names: "Gambusia" is sometimes used as a common name for this species, but this is confusing as there are several other species in the genus *Gambusia*.

Description and Identifying Characteristics: Small: females to 2.3 inches (5.8 cm), males to 1¼ inches (3.2 cm); olive dorsally with a dusky mid-dorsal stripe. Dusky edge to scales, giving a cross-hatched appearance. Lighter on the sides, and white or silvery below. Many small black spots. Male has anal fin modified into an intromittent organ.

Range: Native to Atlantic coast drainages (also in brackish water) from New Jersey through Florida, Gulf Coast drainages to northeastern Mexico, and in the lower Mississippi valley to southern Illinois. Successfully introduced in Hawaii and in the vicinities of Toledo, Ohio and Ann Arbor, Michigan and perhaps elsewhere. Sometimes available in the aquarium trade.

Habitat and Environmental Tolerances: Low-gradient and standing waters, including ponds, small pools, ditches, marshes, and creeks. Does best where there is clear water and aquatic vegetation.

Breeding Requirements: Can reproduce virtually anywhere it can survive.

Food and Feeding: Feeds at or near the surface, principally on insect larvae.

Comments on Culture: Formerly compulsory in fish ponds in parts of the southern United States to control *Anopheles* mosquitoes, and still sometimes stocked for that purpose. Limited use as an aquarium fish and a forage fish in farm ponds.

Harvesting: Dip nets, seines, and minnow traps.

Other Food Animals

Bullfrog
CHAPTER II-9

Family: Ranidae

Taxonomic Name: Rana catesbiana Shaw

Common Names: "Bullfrog" seems to be universal.

Description and Identifying Characteristics: More or less greenish especially on the head, often with brown spots. Dusky crossbars on the back legs. Often dark mottlings on the ventral surface. Color and markings are very variable, but the bullfrog differs from other frogs in that the eardrums are as large or larger than the eyes.

Tadpoles larger than other tadpoles, olive green with well defined black spots above, white or yellowish below.

Range: Native from Gulf of St. Lawrence west through the entire Great Lakes drainage to central Minnesota and Iowa, up the Missouri river into Kansas, and south to eastern Texas. Found everywhere south and east of these points, except in southern Florida. Widely introduced in the western United States, especially California; also in extreme southwestern British Columbia, Hawaii, northern Mexico and Cuba.

Habitat and Environmental Tolerances: Bullfrogs may be found in almost any permanent body of water, but they prefer cover, such as overhanging banks or weeds. Require soft mud to overwinter in the North. Only occasionally found more than a few feet from water.

Size and Growth: Tadpoles up to 5½ inches (14.0 cm) long; adults to 8 inches (20.3 cm) long (body length only, not including legs). The bullfrog spends from 5 months to 2 years or more in the tadpole stage. One to several years more is required to reach marketable size of ¼ to ½ pound (0.11–0.22 kg). The bullfrog is an exception to the general rule for cold blooded animals, in that northern bullfrogs are larger and grow more rapidly than southern ones.

Breeding Requirements: Lays eggs at night in shallow water, usually near plants, whenever the weather is warm.

Food and Feeding: Tadpoles eat mostly algae, but will take any soft plant food, as well as dead animals. Adults will eat anything that moves, but most of their natural diet is comprised of insects and, in some places, crayfish. They will accept non-living food of many kinds, but only if it is made to move.

Table Qualities: Outstanding flavor and texture, but lower in protein than most animal flesh.

Comments on Culture: Easy to maintain and reproduce, but never cultured successfully as a commercial food crop. Sometimes cultured commercially as laboratory animals.

Harvesting: Spear, hook and lines and hands are used, preferably at night with a light.

Red Crayfish
CHAPTER II-10

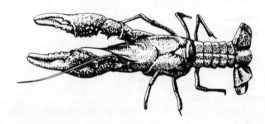

Family: Astacidae

Taxonomic Name: Procambarus clarki (Girard)

Common Names: "Red swamp crayfish" or "Louisiana red." "Crawfish" or "crawdad" are often substituted for crayfish. "Ecrevisse" in French.

Range: Mississippi and Ohio valleys to Missouri and Tennessee, Gulf drainages to Florida. Introduced in southern California, Nevada and Hawaii.

Habitat and Environmental Tolerances: Found in shallow, weedy swamps and ponds. Hard water preferred. Very sensitive to chemicals.

Size and Growth: Minimum size for harvest 0.35–0.50 ounces (10–14 g); preferred size 0.1 pound (45 g). Such large crayfish are 8–14 months old; maximum age is 2 years.

Breeding Requirements: Mate in late spring, young hatch a few weeks thereafter. Need soft banks, above water level, to burrow in.

Food and Feeding: Omnivorous. Cultured crayfish are fed mainly on aquatic plants.

Table Qualities: Superb.

Comments on Culture: Commercially important in French Louisiana, but underdeveloped elsewhere. Grown in swamps, artificial ponds, and in rotation with rice.

Harvesting: Caught in special traps.

White Crayfish
CHAPTER II-10

Family: Astacidae

Taxonomic Name: *Procambarus blandingi* (sometimes referred to as *Procambarus acutus*)

Common Names: "White river crayfish."

Range: Mississippi valley from Louisiana to southern Wisconsin, through the Great Lakes to the Atlantic Coast, and south to Florida.

Habitat and Environmental Tolerance: In nature more often found in rivers than the red crayfish, but the same culture environment is suitable.

Size and Growth: See red crayfish.

Breeding Requirements: See red crayfish.

Food and Feeding: See red crayfish.

Table Qualities: Superb.

Comments on Culture: See red crayfish.

Harvesting: See red crayfish.

Fresh Water Shrimp
CHAPTER II-11

Family: Atyidae

Taxonomic Name: *Macrobrachium rosenbergi* (de Man)

Common Names: "Fresh water prawn," or "Giant fresh water prawn."

Description and Identifying Characteristics: *Macro-brachium* spp. can be distinguished from marine shrimps by their exceedingly long first legs, but I have not discovered how to distinguish *M. rosenbergi* from the many other species.

Range: Coastal rivers and estuaries throughout the tropical Pacific, including Hawaii. Widely introduced, including in Florida, Puerto Rico, and South America.

Habitat and Environmental Tolerances: Needs warm water saturated with dissolved oxygen; usually found in flowing water in nature. Adults are truly fresh water animals, though they can withstand quite high salinity, but young more than one week old need water of 8 to 22%₀ salinity.

Size and Growth: Minimum time required to reach commercial size (0.1 lb. or 0.04 kg) is 9 months. There is great individual variation, and even under ideal conditions some individuals will take much longer. Maximum length 9.4 inches (24 cm) exclusive of front legs and claws.

Breeding Requirements: Spawn in warm, fresh or slightly saline water at any time of year.

Food and Feeding: Larvae eat both phytoplankton and zooplankton in nature. Cultured larvae are fed on brine shrimp nauplii, minced fish, shrimp, clams, or worms and egg custard. Adults are carnivorous; in culture the principal feed source is fertilization of the pond, but they also receive chicken mash.

Table Qualities: Superb, combining the best traits of marine shrimp and lobster.

Comments on Culture: Commercial and subsistence culture is underway in Hawaii. The main problems are in larval rearing, which is done by the state.

Harvesting: Mass harvests by seining; small harvests may be made by hook and line, or at night with dip nets or spears.

APPENDIX III

Glossary

Acetone-dried: dried using acetone (CH_3COCH_3) as an agent to absorb water.

Acid: Having a pH below 7.0.

Acre-foot: That volume of water which will cover one acre (0.45 ha) to a depth of one foot (0.3 m); 325,851 gallons or 1,231,716 liters.

Adipose fin: Small tab of fatty tissue found behind the dorsal fin on some fishes, including salmonids and catfishes.

Adsorption: adhesion of molecules of a gas, liquid or dissolved substance to a surface (as disstinguished from absorption, in which the absorbed molecules are taken into the absorbing substance).

Aerator: Any device which adds air or oxygen to water.

Aerobe: requiring oxygen to sustain life.

Aerobic: An environment which contains free oxygen, or an organism which requires such an environment.

Aflatoxin: A toxic metabolite of a mutant blue-green mold (*Aspergillus flavus*), sometimes resulting from poor storage of fish feeds containing oilseed meals.

Agribusiness: the complex of farming and associated businesses (the fertilizer industry, processing and packaging of farm products, etc.); usually used to connote large scale corporate farming as opposed to the traditional family farm.

Air lift pump: Water pump which operates by injecting air into a submerged tube, causing water to rise in the tube.

Air stone: A type of diffuser made of porous carborundum.

Alevin: Larval fish which has not yet developed the morphological characteristics of the adult.

Algae: Green plants composed of only one type of cell, lacking true roots, stems and leaves, ranging from microscopic species to giant seaweeds. and including the phytoplankton.

Algaecide (or algicide): Chemical used to kill unwanted algae.

Alkalinity: The capacity of water to neutralize acids, measured as the amount of carbonates and bicarbonates.

Amides: Nitrogen compounds formed by the reaction of ammonia with organic acids.

Anadromous fish: Fish which lives in salt water but enters fresh water to spawn.

Anaerobic: Opposite of aerobic.

Anal fin: The single fin on the ventral surface of a fish, ahead of the tail.

Angling: Fishing with hook and line.

Annelids: Any of a phylum of worms (Annelida), including the earthworms and leeches, with a body made up of joined segments or rings.

Anoxia: Total lack of dissolved oxygen.

Aquabusiness: Analogous to agribusiness, which see.

Aquaculture: The rearing of aquatic organisms under controlled conditions.

Aquarist: Person who maintains aquaria, particularly an aquarium hobbyist.

Aquatic: (1) Living in or frequenting water. (2) The fresh water environment.

Aquiculture: Variant spelling of aquaculture.

Balanced population: Population in which the numbers of individuals of the various species are in the proper proportion to give the desired results in terms of growth and production.

Barrage pond: Impoundment.

Basic: Having a pH above 7.0.

Benthic: Living primarily on the bottom of a body of water.

Benthos: Plants and animals inhabiting the bottom of a body of water.

Biochemical oxygen demand: An index of the rate of consumption by natural processes of dissolved oxygen in a body of water.

Biodegradable: Capable of being broken down by natural biological processes.

Biofilter: Device in which toxic wastes are transformed to useful or inocuous forms by living organisms, usually bacteria.

Biological oxygen demand: See biochemical oxygen demand.

Biomass: (1) The total weight of living organisms in a given environment or portion of an environment. (2) The total weight of living or non-living organic material. (The first definition is more common in aquaculture publications.)

Bivalve: Mollusk, such as a clam, with a two-valved shell.

Black body: In physics, an ideal surface that can absorb completely all the radiation striking it. Loosely used to mean any dark object used to capture solar radiation.

Bloom: Dense and highly visible growth of microorganisms; usually applied to phytoplankton. (Eg. Algal Blooms.)

Blue-green algae: Algae of the class Cyanophyta; usually, but not always blue-green in color.

B.O.D.: See biochemical oxygen demand.

Brackish water: Water having a salinity between that of fresh water and sea water.

Brailles: The poles used to pull a seine.

Bryozoan: Member of the class Bryozoa, the sponges.

Buffer: A solution of weak acids and their salts which tends to stabilize the pH of water.

Cage culture: Cultivation of fish in floating mesh cages.

Cannibalistic: Eating its own species.

Capture fisheries: Fishing which depends on the hunting of wild animals, as opposed to culture fisheries or aquaculture.

Carcinogenic: causing cancer.

Carotenid: Any of a number of yellow, orange, red, or purple pigments occurring in plant and animal fats.

Cast net: Type of net which is thrown over the water surface and closed by means of a drawstring.

Catadromous fish: Fish which lives in fresh water but goes into the ocean to spawn.

Cation: A positively charged ion; as opposed to negatively charged ions, or anions. For example, when ammonium hydroxide (NH_4OH) dissociates in water, it breaks down into cations of NH_4^+ and anions of OH^-.

Caudal fin: The tail of a fish.

Caudal peduncle: The muscular, flexible portion of a fish's body immediately forward of the caudal fin.

C conversion: Conversion ratio based on the complete dietary intake of an animal, including both natural feeds and those supplied by the culturist.

Centrarchid: Fish of the family Centrarchidae, including the black basses, sunfishes, crappies, etc.

Centrifuge: Machine using centrifugal force to separate particles of varying density.

cfs: Abbreviation for cubic feet per second, a measure of flow rate.

Chelate: Chemical compound containing a

metal ion attached to a hetero-cyclic ring (ring formed of two or more elements) in two or more places.

Chemical feeder: Fish which relies primarily on the chemical senses (taste and smell) to locate food.

Chemical fertilizer: See synthetic fertilizer.

Chemosensory feeder: Using primarily the chemical senses (smell and taste) to locate food.

Chironomid: Any of a large number of small, two-winged, gnat-like insects belonging to the family Tendipedidae or Chironomidae.

Chlorinated hydrocarbon: Any of many synthetic organic chemicals derived from petroleum and containing one or more chlorine atoms, including many of the synthetic pesticides.

Chorionic gonadotrophin: Substance derived from urine of female humans or other animals; used as a source of pituitary hormones to stimulate maturation of eggs or sperm in induced breeding of fish.

Cichlid: Fish of the family Cichlidae, including the tilapias and a number of popular aquarium fishes.

Cladocerans: Any of an order (Cladocera) of crustaceans. including the daphnia, with a folded upper shell covering the body.

Closed system: System in which water is neither added nor removed during operation.

Cold water fish: Fish which do not do well at water temperatures over 70°F (21.1°C).

Coliform count: Count of the number of *E. coli* bacteria in a sample of water; used as an index of sewage pollution.

Colloidal mud: Semi-liquid mud, composed of particles from 0.005 to 0.2 micron in diameter.

Colonial bryozoans: Those members of the phylum Ectoprocta or Bryozoa (minute aquatic and marine animals including the sponges) found in groups produced by budding from a single parent and usually so fused together that demarcations between individuals are indistinguishable.

Colorimeter: Instrument for measuring the intensity of color; used in a variety of chemical tests of water.

Columnar denitrification: The removal or chemical reduction of nitrogen or nitrogen compounds in water by passing the water vertically through a column containing filtering agents.

Complete feed: A feed which supplies one hundred percent of the dietary requirements of an animal, used when there is little or no access to natural food.

Compost: Fertilizer produced by the more or less controlled decomposition of natural organic materials.

Conversion ratio: Ratio of the weight of feed supplied to the amount of weight gained by the animal being fed.

Cool water fish: Ill-defined term referring to fish intermediate in their thermal requirements to the cold and warm water fishes.

Copepods: Any of a subclass (Copepoda) of small, often parasitic aquatic or marine crustaceans.

Core: A layer of impermeable material placed in the center of a dam or levee to prevent leakage.

Core trench: Excavation made in order to install a core.

Crustacean: Member of the large Arthropod class Crustacea, including crayfish, crabs, shrimp, and many kinds of zooplankton.

Current: The speed at which water passes a point, independent of the volume of water. See flow rate.

Cyprinid: Fish of the family Cyprinidae, including the carps and minnows.

Dam board: Boards used to regulate water level in a sluice.

Dam pond: Pond formed by damming a watercourse.

Degree day: Measure of the difference between an arbitrarily selected base temperature and the temperature of a given place averaged over a day; used as a measure of the energy required in heating or cooling.

Demand feeder: Device which dispenses small quantities of feed when activated by animals.

Denitrification: The chemical reduction of nitrate to nitrite, and then to elemental nitrogen.

Deoxygenation: Loss of dissolved oxygen from water.

Detritivore: Animal feeding primarily on detritus.

Detritus: Fine solid materials suspended in the water column.

Diatom: Any of numerous algae, belonging to the phylum Chrysophyta, whose cell walls consist of two valves and which contain silica; diatoms are among the most important components of many aquatic food webs.

Diffuser: Any of a variety of devices for breaking a stream of air or oxygen up into fine bubbles.

Dinoflagellate: any organism of the order Dinoflagellata; they are mobile, like animals, but photosynthesize like plants and are often considered to be intermediate between the two kingdoms.

Dispersing agent: Any of a number of salts which disperses clay particles in water, used in sealing ponds.

Diversion pond: Pond constructed by diverting a watercourse.

D.O.: Dissolved oxygen concentration. (Usually indicated in milligrams per liter or parts per million.)

Dorsal: The back surface of an animal.

Dorsal fin: One, two, or three fins located on the back of a fish, but not including the adipose fin.

Dragline: A special rope pulled by a tractor, used in excavating.

Draw down: Intentionally lowering the water level.

Dry feed: Feed with most of the water removed from the ingredients to improve storage qualities.

Dry weight: The weight of a substance minus the weight of water in its natural makeup.

Dugout: Pond constructed by excavating earth.

Ecology: The science which deals with the relationships among living organisms, as well as their relation to non-living components of the environment.

Ecosystem: A given physical environment, such as a pond, considered together with all the plants and animals which inhabit it.

Electrofishing: Fishing by means of passing an electrical field through the water so that fish are stunned.

Embankment: Any raised earthen structure, including dams and levees, for the purpose of holding back or containing water.

Emergency spillway: A channel cut in the top of a dam or levee so that in times of extremely high water, water passes over at only one point.

Emergent plant: Plant which grows rooted in water, but with part of the plant normally extending above water.

Epilimnion: The upper layer of a thermally stratified body of water. See thermal stratification.

Epizootic: Outbreak of disease affecting many individuals of a species at one time; epidemic.

Essential element: Any one of the twenty chemical elements considered necessary for the maintenance of life.

Estuary: The brackish water environment formed where a fresh water stream meets the sea.

Eutrophication: The natural process of "aging" whereby a body of water becomes more fertile and, usually, shallower as a result of erosion and the death of plants and animals within it.

Excavation pond: See dugout.

Exotic species: Species not native to a given environment or region.

Extensive culture: The opposite of intensive culture; characterized by relaively low population and little management or control by the culturist.

Farm pond: Artificially constructed pond, usually between ¼ acre and 5 (0.1–2 ha) acres in size, located on a farm and used for a variety of purposes, usually including recreational or subsistence fishing; characteristically stocked with largemouth bass (*Micropterus salmoides*) and bluegills (*Lepomis macrochirus*), though other species may be used.

Fecundity: The number of eggs produced per female fish, or per unit body weight of female fish.

Feeding ring: A band of fine mesh fastened around the inside of a fish cage near the waterline to prevent floating feed from being carried out of the cage.

Feed quotient: See conversion ratio.

Fee fishing: Practice of charging admission, or charging for fish caught in a privately owned and stocked pond opened to the public for purposes of angling.

Filamentous algae: Any of a number of species of algae in which individual cells are connected in long, hair-like filaments.

Filter feeder: Fish which obtains its food by swimming with open mouth and straining out food particles on its gill rakers.

Finfish: Any member of the class Pisces; includes the vertebrate animals commonly referred to as "fish," as opposed to the invertebrate shellfish.

Fingerling: Young fish larger than a fry but not adult; not rigidly defined but usually denotes a fish between 0.8 and 10 inches (2–25 cm) long.

Finishing: Practice of giving fish special feeds in the period just before slaughter, so as to produce desirable flavor or texture.

Fishery biology: Science dealing with the exploitation of aquatic and marine animals by humans.

Fishery products: Salable products of the commercial fishing industry, including whole and processed fish and shellfish.

Float line: The top line of a seine, gill net or trammel net, supported by floats.

Floating plant: Plant supported on the surface of the water, usually but not always by its leaves, with the roots not in contact with the bottom.

Flocculent: Composed of fine soft particles which adhere one to another.

Flow rate: The volume of water moving past a point in a given time, usually expressed in cubic feet per second (cfs) or liters/second. Also see current.

Food chain: A sequence of organisms, each of which provides food for the next, beginning with primary producers and extending to carnivores.

Food web: The complex feeding relationships, made up of interlocked food chains, which exist in any natural system.

Forage fish: Fish species stocked or cultured as food for other fish.

Freeboard: Distance between the water level and the top of an embankment.

Fresh water: Variously defined, but most commonly considered as water having a salinity of less than 0.5 parts per thousand. (This book considers water with salinities up to 3 parts per thousand.)

Fry: Very young, but post-larval, fish; intermediate between alevins and fingerlings.

Fyke net: Net made up of a series of hoops of diminishing size, each with a funnel shaped entrance.

Game fish: Fish commonly taken on hook and line for sport, especially species which can be taken on artificial lures.

Gefilte fish: Chopped fish, mixed with onion, egg and/or seasonings and boiled.

Genetic plasticity: Capacity of a species to evolve different characteristics; the more plastic a species, the more specimens from different environments are likely to differ, and the greater the possibilities of selective breeding.

Genital papilla: A small flap of flesh located just forward of the vent on some fishes; used in sexing.

Gill arches: The cartilaginous arches which support the gills.

Gill covers: The bony outer covers of the gills.

Gill filaments: See gill lamellae.

Gill lamellae: The thin-walled, blood-filled, visibly red filaments on a fish's gills which take up oxygen from the water.

Gill net: A net which is hung vertically in the water so that fish attempting to pass through become caught by their gill covers.

Gill rakers: The protusions on the opposite side of the gill arches from the gill lamellae; used as "sieve" by filter feeding fishes.

Gill slits: The openings at the back of the gills through which water is expelled.

Gleization: The process of producing gley.

Gley: A so-called biological plastic produced by bacterial action when organic matter is composted anaerobically on the bottom of a pond; used to seal porous soils.

Glochidium: Larval stage of fresh water Unionid clams; parasitic on the gills of fishes.

GNP: Gross national product; the total value of a nation's annual output of goods and services.

Grader: Device used to sort fish by size.

Green manure: Fresh plant material used as a fertilizer.

Ground water: Water contained underground in an aquifer.

Growing out: Practice of feeding fish until they reach a desired size for harvest.

Growing pond: Pond used primarily to grow animals to harvestable size.

Growth coefficient: Conversion ratio.

Gular area: "Throat" region of a fish, on the ventral surface, just behind the gills.

Habitat: That portion of the environment where an organism normally lives.

Half life: Time required for the disintegration of half the atoms in a radioactive substance; used as a measure of the relative degradability of such substances.

Hardness: The concentration of divalent cations, especially calcium and magnesium ions, in water.

Harvest basin: A basin deeper than the rest of a pond, located near the outlet, so that when the pond is drained fish are concentrated to facilitate harvesting.

Harvestable fish: Fish of size desired for harvest.

Hatchery: An aquaculture facility where the main activity is breeding of animals and rearing of the early life stages.

Heavy metal: Any of a number of chemical elements, including copper, lead, mercury, etc. with toxic properties, commonly found as pollutants in water, and capable of being concentrated in animal flesh.

Herbicide: Chemical used to kill unwanted plants.

Herbivore: Animal which eats plants.

High technology: A relative term; in general a technology can be said to be "high" or "hard" (as opposed to "low" or "soft") in comparison to another if it uses more energy from nonrenewable sources, requires more specialized personnel or facilities to operate or is less amenable to management on a small scale.

Higher plant: See macrophyte.

Hoop net: See Fyke net.

Hydrological cycle: The cycle by which water passes from the atmosphere onto the earth's surface, into the ground and back again.

Hydrology: The science of water on and within the earth and in the atmosphere; study of the hydrological cycle.

Hydroponics: The cultivation of plants, including normally terrestrial forms, rooted in an aqueous nutrient solution rather than in soil.

Hypervitaminosis: Disorder resulting from an excess of one or more vitamins.

Hypohysation: See pituitary injection.

Hypolimnion: The bottom layer of a thermally stratified body of water. See thermal stratification.

Ice jigger: Device for setting a gill net under ice.

Impermeable (impervious) soil: Soil through which water will not pass, suitable for making ponds.

Impoundment: Body of water made by damming a watercourse.

Induced breeding: Breeding brought about through manipulation of the environment or treatment of the animal by the culturist.

Inorganic fertilizer: Usually used to mean synthetic fertilizer. See synthetic fertilizer and natural fertilizer.

Insectivore: Animal which eats insects.

Integrated system: Artificial ecosystem in which a number of processes and cultivated organisms are made to work to mutual advantage. This exists, for example, when pigs are penned over a fish pond, fertilizing the aquatic food web with their feces, and being fed on excess fish, while water and sludge from the pond is used to grow vegetables as human food and/or to feed the fish and/or pigs.

Intensive culture: The opposite of extensive culture; characterized by relatively high population density, and a high degree of management and control by the culturist.

Intermuscular bones: Fine, Y-shaped bones found in the muscle tissue of some fishes; they "float" unconnected to the skeleton.

Intraperitoneal injection: Injection into the lining of the body cavity, known as the peritoneum.

Intromittent organ: Male copulatory organ, analogous to the penis in mammals.

Invertebrate: Any animal without a backbone; any animal other than a fish, amphibian, reptile, bird or mammal.

Ionization: Process whereby a chemical compound breaks down into positively and negatively charged ions.

Isothermal: Having a constant temperature.

Jar method: Hatchery technique by which eggs are incubated in jars into which water is passed from below.

Juvenile: General term used to indicate a young animal or plant which has not matured sexually. May differ in other respects from the adult, but does not differ grossly in form. (See larva.)

Key: See core.

Lamellae: Thin platelike layers forming the gills of bivalve mollusks.

Larvae: Early, free-living stage of an animal differing grossly in form from the adult (e.g. a caterpillar is a larval butterfly, a tadpole is a larval frog).

LD 50: The point in an experiment at which fifty percent of the experimental organisms have died; used as measure of lethality.

Lead line: The bottom line of a seine, gill net or trammel net, held down by lead weights.

Levee: An embankment constructed above ground level to serve as the wall of a pond.

Levee pond: A pond totally or partially surrounded by levees.

Limnology: The science of inland waters, including their biology, chemistry, geology, etc.

Liner: A thin sheet of plastic or rubber, placed on the bottom of a pond or pool, to prevent leakage.

Lipid: Any of a group of chemical substances including the fats and others of similar properties.

Live car: A mesh cage used to temporarily hold live fish; also called live box.

Live hauler: Someone who purchases fish to be hauled to market alive.

Loading rate: The rate at which organic matter is introduced to an aquatic system.

Macrophyte: A "higher" plant, with specialized cells; includes all the green plants other than the algae.

Malachite green: A dye composed chiefly of basic copper carbonate, ($CuCO_3.Cu[OH]_2$), used in treatment of diseases in aquatic organisms.

Mariculture: Aquaculture in salt water.

Marine: Of or pertaining to the salt or brackish water environments, and *not* to fresh water.

Maturity: Capability of sexual reproduction; unlike higher vertebrates, fish are often said to "mature" annually or more often, though they become "adult" only once.

Measuring board: Tool for holding a fish for accurate length measurement.

Mechanical filtration: Any filtering process which functions by separating out physical particles, as distinguished from chemical or biological filtration.

Menhaden oil: Oil processed from whole menhaden (*Brevoortia tyrannus*), a herring-like ocean fish.

Metabolism: The aggregate of processes in a living organism by which nutrition, respiration, and growth are achieved.

Metabolite: Any substance produced by an organism as part of its life processes.

Metamorphosis: More or less abrupt physical transformation of an animal as, for example, when a tadpole metamorphoses to become a frog.

Microbial: Having to do with a microbe, or microscopic organism.

Micronutrient: See trace element.

Microorganism: Organism too small to be seen with the naked eye.

Milt: The secretion produced by the testes of male fish, including the sperm.

Mollusk (also spelled **mollusc**): Any member of the phylum Mollusca, comprising clams, snails, and all the shellfish other than the crustaceans.

Monk: A device for draining and regulating the water level of a pond, constructed in the pond rather than in the pond bank.

Monoculture: Cultivation of only one species at a given time and place.

Monosex hybrid: Offspring of a cross which produces one hundred percent or nearly one hundred percent of one sex.

Montmorillonite: Any of a group of related clay minerals with the general formula $Al_2Si_{14}O_{10}(OH)_2$, which swell greatly when moistened.

Morphology: The study of the physical form and structure of organisms.

MS-222 (also known as tricaine or medicaine): ethyl-m-aminobenzoate methane sulfonate ($C_{10}H_{15}NO_5S$), used as an anesthetic for fish.

Mucket: Any of a number of North American fresh water clams (e.g. the fat mucket, *Lampsilis radiata*) belonging to the family Unioniidae.

Mud line: The bottom line of a seine specially designed to be used on very muddy bottoms; constructed of heavy rope without the usual lead weights.

Mutagenic: Capable of causing mutations.

Natural fertilizer: Fertilizer in which the active ingredients are not chemically altered from forms found in nature.

Neutral: Having a pH of exactly 7.0.

Niche: The precise and unique role played by a species in an ecosystem; includes habitat, feeding, and other aspects of behavior.

Nitrification: The oxidation of nitrogen from ammonia through nitrite to nitrate.

Nitrogen fixation: The process by which certain bacteria are able to take nitrogen from the air and "fix" it in soil or water.

Nuptial tubercles: Horny protrusions which appear on the heads of the males of certain species of fish at breeding time.

Nutrient: Any substance which plays a role in the food chain.

Nutritive ratio: See conversion ratio.

Omnivore: Animal which eats both animal and vegetable matter.

Open system: Aquaculture system which participates in the natural hydrological cycle; i.e. inflow and outflow are only partially under the control of the culturist.

Opercle (= **Operculum**): the bony covering of the gills of fishes.

Opercular lobe: fleshy flap extending from the opercle of some fishes.

Organic agriculture: School of agriculture which does not use synthetic fertilizers or pesticides.

Organic fertilizer: See natural fertilizer.

Overflow: Device to permit water to escape from a body of water once a certain maximum level is reached.

Oxidizing agent: Substance which uses up dissolved oxygen or speeds the process of oxidation.

Oxygenation: The addition of oxygen to water by aeration or other means.

Panfish: Any of a number of fishes, including the sunfishes, bullheads, and perches, commonly harvested at a size which will fit into an ordinary frying pan.

Parasite: An organism which lives on or in another organism, and obtains its food at the expense of that organism.

Parr marks: Dark blotches characteristically found on the sides of juvenile salmonid fishes.

Particulates: Small particles of matter suspended or floating in water.

Pathogen: Organism causing disease.

Pathology: Study of diseases.

PCB: Abbreviation for polychlorinated biphenyl, a class of synthetic compounds which are almost ubiquitous as toxic pollutants.

Pectoral fins: The paired fins found on either side of a fish's body just behind the gills.

Pelvic fins: The paired fins found on the ventral surface of a fish's body, between the pectoral fins and the single anal fin.

Permeable (pervious) soil: Soil through which water can pass; must be treated or covered if used in pond construction.

Pesticide: Chemical used to kill unwanted organisms.

pH: The negative logarithm of the hydrogen ion concentration expressed in gram equivalents; used as an index of the acidity or basicness of water.

Pharyngeal teeth: Teeth located in the pharynx, or throat of a fish.

Pheromone: Chemical substance released by an animal for purposes of communication with others of its species.

Photolytic: Chemically changed by light.

Photosynthesis: The process by which green plants produce food for their own growth—plus oxygen—from carbon dioxide and water, in the presence of light.

Phytoplankton: That portion of the plankton made up of algae.

Piscivore: An animal that eats fish.

Pituitary injection: The practice of injecting fish with substances derived from the pituitary gland for the purpose of inducing reproduction.

Plankter: An individual planktonic animal or plant.

Plankton: Aquatic organisms suspended in the water column with little or no power of locomotion.

Point discharge: Discharge of a substance into an aquatic system at a single definable point; as opposed, for example, to silt, which commonly originates as soil eroded from a wide area.

Polyculture: Cultivation of more than one species in a single place.

Pothole: (1) A small, shallow pond used for waterfowl breeding. (2) Small, natural lakes occurring in the northern Great Plains, and used for trout farming.

ppm: Abbreviation for parts per million, a common measure of concentration of chemicals in liquids; equivalent to milligrams per liter in water.

Precipitation: Process occurring when a chemical is removed from solution in water and settles to the bottom in solid form.

Predaceous, or **predatory:** Capturing and eating live animals.

Primary production: That portion of total production of biomass in a system attributable to photosynthesis by green plants.

Production: The gain in weight accomplished by cultured organisms over a given period of time.

Productivity: Capacity of a system to support production.

Protein sparing: Use of lipids, carbohydrates, etc. to supply energy in the diet so that protein, which is generally more costly, is used only for growth.

Pupation: Enclosure in a cell or cocoon, during which an animal ceases feeding and undergoes metamorphosis. For example, a caterpillar pupates to become a butterfly.

Raceway: Aquaculture chamber through which water flows, usually rapidly, or at least fast enough so that the flow can be seen.

Recirculating system: Aquaculture system in which at least some of the water is recycled one or more times.

Respiration: Process by which chemical energy in food is transformed into other kinds of energy by plants and animals, as oxygen is consumed and carbon dioxide produced.

Ripe: Containing fully developed eggs; ready to spawn.

Riprap: Rocks placed on the bank of a body of water to prevent erosion.

Rivaldi valve: Flexible tube connected to a rigid tube extending through a dam; used for drainage and to control water level.

Roe: The eggs of a fish, especially when considered as food.

Roiling (of water): Dirtying, as by stirring up silt.

Rosary ponds: Ponds set up in series, so that water passes from one into the next, and so on.

Run-off: That portion of precipitation which does not soak into the ground but runs off, eventually entering bodies of surface water.

S conversion: Conversion ratio based only on food provided by the culturist, omitting naturally available food.

Salinity: Measure of the total amount of dissolved salts in water.

Salmonid: Any fish of the family Salmonidae, including the trouts, chars, salmons and graylings; since the whitefishes and ciscoes were added to the Salmonidae they may be called Coregonids or Salmonids.

Saturation: The maximum amount of a substance which can ordinarily be dissolved in water at a given temperature and pressure.

Scap net: Square-cornered, shallow dip net used for rapid handling of small fish.

School: Strictly speaking, a school of fish is a permanent grouping; the members of the school exhibit the same behavior and respond to stimuli in the same way. Survival of a member outside the school is improbable. In common usage any sizable group of fish is referred to as a school.

Scientific name: The Latin name of an organism, consisting of the genus, species and sometimes subspecies, in that order, for example *Ictalurus nebulosus marmoratus.*

Secchi disc: A black and white disc used as a tool in measuring water transparency.

Sediments: Solid matter which has settled out of suspension, found on the bottom of a body of water.

Seine: A long, narrow net with floats on the top edge, and usually with weights on the bottom, used by hauling it through the water.

Semiclosed system: "Closed" system in which some amount of water is periodically lost or removed, and then replaced.

Sexing: Process of determining the sex of an animal.

Shellfish: Invertebrates used as food, including mollusks and crustaceans, excluding fiinfish.

Silt: Soil particles less than 1/16 millimeter in diameter.

Siltation: Deposition of silt on the bottom of a body of water.

Skewed sex ratio: with a predominance of one sex of the other, i.e. one would ordinarily expect a 50–50 sex ratio; where this expectation is not met the ratio is said to be skewed.

Slope: Ratio of the difference in elevation of two points to the horizontal distance between them.

Sluice or sluice gate: A device built into a dam or channel for draining or regulating the flow of water.

Snag hooking or snatch hooking: Fishing by casting out a hook and jerking it so that fish become impaled.

Solar-algae pond: Transparent or translucent fiberglass cylinder designed as an above-ground tank to maximize phytoplankton production.

Solvent: Substance, usually a liquid, which can dissolve another substance.

Spawn: To produce or deposit eggs, sperm or young. Usually, but not always applied to aquatic or marine animals.

Spawning mop: Synthetic "weeds" on which cyprinid fishes deposit their eggs.

Spillway: Structure over which water passes at the outlet of a pond.

Splash board: Flat surface off which water is splashed to oxygenate it on entering a pond.

Spoils: Earth removed in construction of a pond.

Standard length: Measure of fish length most often used by biologists; consists of the distance between the tip of the snout and the end of the caudal peduncle.

Standing crop: The total biomass of a particular organism, or all organisms, present in a body of water at a given time.

Station feeding: Practice of feeding at a single spot, so that fish are concentrated.

Stratum: Layer distinguishable from other layers by its composition.

Stridulation: Production of sound by rubbing together of hard body parts.

Stress: Any condition inimical to the health or growth of an organism.

Stripping: Process of artificially removing the eggs and milt from mature fish.

Stunted: Condition of being undersized and not growing appreciably; this condition is usually permanent in warm-blooded animals but is reversible in cold-blooded animals.

Sublethal: A damaging or dangerous condition (of temperature, concentration of a chemical, etc.) but not reaching a level which, in and of itself, would result in death.

Submerged plant: Plant which exists entirely underwater.

Subsistence aquaculture: Aquaculture for purposes of food but not for sale or barter.

Substrate: The bottom layer, that on or in which benthic organisms are found.

Subterminal mouth: Opening on the ventral surface rather than at the very end of the snout (terminal mouth). See Figure III–1–1.

Summerkill: Loosely used as a counterpart to winterkill; describes any mass mortality occurring during warm weather, but usually attributable to low D.O. reflecting heavy B.O.D., warm water, and/or low water levels.

Supersaturation: Condition in which a substance is present in amounts above that which can normally be dissolved in water at a given temperature and pressure.

Supplemental feeding: Feeding not to provide a complete diet, but to enhance growth by qualitively and/or quantitatively improving on the natural diet.

Surface water: All water found on the earth above ground, including streams, lakes and the oceans.

Suspended solids (also, **in suspension**): Particles of solid matter present in the water column.

Swale: A shallow trench constructed around the perimeter of a body of water to collect solids carried by run-off.

Swim-through feeder: Feeder designed so that small fish can swim in and out, but larger fish must feed outside.

Synergy: The combined action of two substances or processes to produce a result that could not be obtained by means of either alone.

Synthetic fertilizer: Fertilizer produced by a controlled chemical reaction resulting in substances not normally found in nature.

Tailwaters: The water immediately below the outlet of an aquaculture system or natural or artificial body of water.

Taxonomic key: logical device used to identify organisms according to their scientific classification. Keys ordinarily list characteristics of organisms in couplets. At each couplet, the user decides whether or not the organism in question exhibits certain characteristics and is thus referred to another numbered couplet. Eventually, by elimination, one arrives at a single species (or other taxonomic grouping).

Taxonomy: The science of classifying organisms.

Teratogenic: Capable of causing deformity.

Thermal shock: Shock resulting from sudden change of temperature.

Thermal stratification: Condition found in some bodies of water, where there are three distinct layers, based on temperature. The upper warm layer, usually well supplied with oxygen, is called the epilimnion. The bottom colder layer, often low in oxygen, is called the hypolimnion. They are separated by a band which changes in temperature from its top to bottom, called the thermocline.

Thermocline: The middle layer of a thermally stratified body of water. See thermal stratification.

Tile: Pipe installed in a field to collect run-off.

Titration: Process of determining the concentration of a substance in solution in terms of the smallest amount of it required to bring about a given effect in relation to another known substance.

Total length: Common measure of fish length used by fishermen; consists of the distance from the tip of the snout to the farthest protrusion of the caudal fin.

Toxaphene: A commercial chlorinated hydrocarbon insecticide ($C_{10}H_{10}Cl_8$), made by chlorinating camphene.

Trace element: Chemical element used by organisms in minute quantities, but believed necessary to their health.

Trammel net: Net similar to a gill net, but composed of two or more layers of netting, between which fish become entangled.

Transpiration: Loss of water to the atmosphere from the leaves of green plants.

Triploid: Having three haploid sets of chromosomes in each nucleus.

Trophic level: The position an organism occupies in the food chain.

Tropical fish: Fish which cannot tolerate temperatures near freezing.

Turbidity: Degree to which the penetration of light into water is limited by the presence of suspended or dissolved matter.

Turnover: Process during which thermally stratified bodies of water become totally mixed at certain times of year as a result of convection currents.

Umbrella net: Rectangular net suspended from its four corners, placed underwater and hauled up rapidly when fish swim over it.

Vascular plant: Plant containing xylem and phloem, the specialized tissues which distribute food and water within the plant.

Ventral: The lower surface of the body of any animal.

Ventral fins: See pelvic fins.

Venturi drain: Type of standpipe installed in the center of a tank or pool.

Vermiculation: Worm-like markings.

Visual feeder: Animal which relies on vision to locate its food.

Vitamin premix: Package of synthetic vitamins added to prepared animal feeds.

Warm water fish: Fish which can tolerate freezing water as well as water above 70°F (21.1°C).

Watershed: The land area contributing to the water supply of a body of water.

Wet weight: The weight of a substance in its natural state, including the weight of water.

Winterkill: Die-off of fish in winter associated with snow cover or opaque ice cutting off both photosynthesis and surface uptake of oxygen.

Zooplankton: That portion of the plankton made up of animals.

APPENDIX IV

Access to Aquaculture Resources

Appendix IV-1: Bibliography

CHAPTER I-1

General texts on aquaculture:

BARDACH, J.E.; RYTHER, J.H.; and McLARNEY, W.O. 1972. *Aquaculture: The Farming and Husbandry of Freshwater and Marine Organisms.* New York: Wiley Interscience. 868 pp. (World survey of aquaculture.)

BROWN, E.E. 1977. *World Fish Farming: Cultivation and Economics.* Westport, Connecticut: Avi Publishing Co. 397 pp.

CHAKROFF, M. 1976. *Freshwater Fish Pond Culture and Management.* Manual Series No. 36 E. VITA. Mt. Rainier, Maryland: VITA Publications. 191 pp. (Small scale and low technology methods.)

FAO. 1966. *FAO World Symposium on Warmwater Pond Fish Culture.* FAO Fisheries Reports 44 (4). (A collection of 123 technical papers by aquaculturists from all over the world.)

HICKLING, C.F. 1971. *Fish Culture.* London: Faber and Faber. 317 pp.

HUET, M. 1970. *Textbook of Fish Culture: Breeding and Cultivation of Fish.* Surrey and London, England: Fishing News (Books). 436 pp. (Best treatment of European methods.)

LOGSDON, G. 1978. *Getting Food from Water.* Emmaus, Pennsylvania: Rodale Press. 371 pp. (Simple and non-technical.)

McCRIMMON, H.R.; STEWART, J.E.; and BRETT, J.R. 1974. *Aquaculture in Canada: The Practice and the Promise.* Ottawa: Department of the Environment, Fisheries and Marine Service. 85 pp.

NATIONAL ACADEMY OF SCIENCES. 1978. *Aquaculture in the United States: Constraints and Opportunities.* Washington: National Academy of Sciences. 123 pp. (Concerned only with commercial aquaculture.)

STICKNEY, R.R. 1979. *Principles of Warmwater Aquaculture.* New York: Wiley Interscience. 375 pp. (Concerned only with commercial and experimental aquaculture.)

Texts dealing with Asian aquaculture:

BARDACH et al. 1972. See under "General texts on aquaculture," above.

DREWS, R. 1951. *The Cultivation of Food Fish in China and Japan: A Study Disclosing Contrasting National Patterns for Rearing Fish Consistent with the Differing Cultural Histories of China and Japan.* Ph.D. thesis, University of Michigan. Ann Arbor, Michigan and London: University Microfilms International. 263 pp.

HICKLING 1971. See under "General texts on aquaculture," above.

HORA, S.L., and PILLAY, T.V.R. 1962. *Handbook on Fish Culture in the Indo-Pacific Region.* FAO Fisheries Biology Technical Paper 14. Rome: FAO. 204 pp.

LING, S.W. 1977. *Aquaculture in Southeast Asia: A Historical Overview.* Seattle and London: University of Washington Press. 198 pp.

TAPIADOR, D.D.; HENDERSON, H.F.; DELMENDO, M.N.; and TSUTSUI, H. 1977. *Freshwater Fisheries and Aquaculture in China.* Rome: Food and Agricultural Organization of the United Nations. 84 pp.

Publications of historical interest:

DREWS 1951. See under "Texts dealing with Asian aquaculture," above.

FAN LEE. 5th century, B.C. *The Chinese Fish Culture Classic.* Translated by T.S.Y. Koo. Contribution No. 489. Solomons, Maryland: Chesapeake Biological Laboratory University of Maryland.

HUET 1970. See under "General texts on aquaculture," above. (Lists a number of references of historical interest, particularly in European aquaculture.)

LING 1977. See under "Texts dealing with Asian aquaculture," above.

RAVENET-WATTELL, C. 1914. *La Pisciculture Industrielle.* Paris: G. Doine et Cie. 408 pp.

SCHAPERCLAUS, W. 1933. *Lehrbuch der Teichwirtschaft.* Berlin: Paul Parey. 289 pp.

TOWNSEND, C.H. 1914. The Private Fish Pond—a Neglected Resource. *Transactions of the American Fisheries Society* 43 : 87–92.

Publications on mariculture:

BARDACH, J.E. 1968. *Harvest of the Sea.* New York: Harper and Row. 301 pp.

BARDACH, RYTHER, and McLARNEY 1972. See under "General texts on aquaculture," above.

GLUDE, J.B. 1978. Waterfront Aquaculture. *Commercial Fish Farmer* 4(6): 35–37. (Small scale mariculture.)

HICKLING, C.F. 1970. Estuarine Fish Farming. *Advances in Marine Biology* 8 : 119–213.

IVERSON, E.S. 1968. *Farming the Edge of the Sea.* London: Fishing News (Books). 436 pp.

SCHWIND, P. 1977. *Practical Shellfish Farming.* Camden, Maine: International Marine Publishing Co. 91 pp.

Publications on the need for aquaculture and comparing aquaculture to other means of protein production:

BARDACH, J.E. 1968. Aquaculture. *Science* 161 : 1098–1106.

McLARNEY, W.O. 1979. Aquaculture. *Biology Digest,* October, 1979 : 11–30.

RYTHER, J.H. 1975. Mariculture: How Much Protein and for Whom? *Oceanus* 18(2) : 10–22.

SHANG, Y.C. 1974. Comparison of the Economic Potential of Aquaculture, Land Animal Husbandry and Ocean Fisheries: The Case of Taiwan. *Aquaculture* 2(4) : 187–95.

SHPET, G.I. 1972. Comparative Efficiency of Fish Culture and Other Agricultural Activities per Unit of Area Used. *Hydrobiological Journal* 8(3) : 46–51.

WEATHERLEY, A.H.; and COGGER, B.M.G. 1977. Fish Culture: Problems and Prospects. *Science* 197 (4302) : 427–30.

WHEATON, F.W. 1977. *Aquacultural Engineering.* New York: Wiley Interscience. 708 pp.

WINDSOR, M., and COOPER, M. 1977. Farmed Fish, Cows and Pigs. *New Scientist,* 22 September 1977 : 740–742.

Texts elucidating the economic, social and ecological views which underlie this book:

MERRILL, R., ed. 1975. *Radical Agriculture.* New York: Harper and Row. 459 pp.

SCHUMACHER, E.F. 1973. *Small is Beautiful: Economics as if People Mattered.* New York: Harper and Row. 290 pp.

VAN DRESSER, P. 1972. *A Landscape for Humans: A Case Study of the Potentials for Ecologically Guided Development in an Uplands Region.* Albuquerque, New Mexico: Bio-technic Press. 125 pp.

Non-technical periodicals dealing with aquaculture:

Aquaculture Magazine (not to be confused with the scientific journal *Aquaculture;* formerly *Commercial Fish Farmer* and before that *American Fish Farmer* and *World Aquaculture News*), P.O. Box 2329, Asheville, North Carolina 28802.

Commercial Fish Farmer (See *Aquaculture Magazine.*)

Farm Pond Harvest, Professional Sportsmans Publishing Co., Box AA, Momence, Illinois 60954.

New Alchemy Quarterly, New Alchemy Institute, 237 Hatchville Road, East Falmouth, Massachusetts 02536.

Organic Gardening and Farming, Rodale Press, Emmaus, Pennsylvania 18049.

Rodale's Network, Rodale Press, Emmaus, Pennsylvania 18049.

The Journal of the New Alchemists, New Alchemy Institute, 237 Hatchville Road, East Falmouth, Massachusetts 02536.

Scientific journals dealing with aquaculture:

Aquaculture, Elsevier Scientific Publishing Company, Amsterdam, The Netherlands. (Not to be confused with Aquaculture Magazine, see above.)

Bamidgeh: Bulletin of Fish Culture in Israel, Nir-David, Israel 19150. (Much of the material is directly useful in North America.)

Canadian Fish Culturist, Department of Fisheries of Canada, Ottawa.

Journal of the Fisheries Research Board of Canada, Government of Canada, Fisheries and Oceans,

Scientific Information and Publications Branch, Ottawa, Ontario K1H 0E6.

Proceedings of the Southeastern Association of Game and Fish Commissioners. Generally published annually after organization's meeting.

Progressive Fish Culturist, U.S. Government Printing Office, Washington, D.C. 20402.

Sport Fishery Abstracts, U.S. Fish and Wildlife Service, Editorial Office, Aylesworth Hall, Colorado State University, Fort Collins, Colorado 80521.

The Journal of the New Alchemists, New Alchemy Institute, 237 Hatchville Road, East Falmouth, Massachusetts 02536.

Transactions of the American Fisheries Society, American Fisheries Society, 5410 Grosvenor Lane, Bethesda, Maryland 20014.

Computerized information retrieval service:

The National Aquaculture Information Service has compiled most of the recent technical literature in aquaculture, along with some key earlier and non-technical works. This information is available commercially on microfiche. To inquire phone 800–227–1960 (toll free), or write Lockheed Missile and Space Company, Code 5020/201, 3251 Hanover St., Palo Alto, California 94304.

Directory of libraries specializing in aquaculture:

Aspens System Corporation. 1982. Directory of Aquaculture Information Resources. Beltsville, Maryland: U.S. Department of Agriculture, National Agriculture Library. 53 pp.

CHAPTER I–2

DENDY, J.S. 1963. Farm Ponds. In Frey 1963. See below. (Excellent summation of the state of the art, and review of the literature on farm ponds.)

EVERHART, W.H.; EIPPER, A.E.; and YOUNGS, W.D. 1975. *Principles of Fishery Science.* Ithaca, New York : Cornell University Press. 288 pp.

FREY, D.G., ed. 1963. *Limnology in North America.* Madison, Wisconsin: The University of Wisconsin Press. 734 pp. (Regional and historical accounts of limnology in North America.)

HUET 1970. See Chapter I–1.

LAGLER, K.F. 1956. *Freshwater Fishery Biology.* Dubuque, Iowa: Wm. C. Brown Co. 421 pp.

PORTER, K.G. 1977. The Plant-Animal Interface in Freshwater Ecosystems. *American Scientist* 65 : 159–170. (Excellent article on an aspect of aquatic biology too little understood by aquaculturists.)

RUTTNER, F. 1953. *Fundamentals of Limnology.* To-

ronto: University of Toronto Press. 295 pp.

SWINGLE, H.S. 1970. History of Warmwater Pond Culture in the United States. In *A Century of Fisheries in North America*, American Fisheries Society Special Publication no. 7 : 95–105.

WETZEL, R.G. 1975. *Limnology.* Philadelphia: Saunders. 743 pp.

CHAPTER II–1

Farm pond stocking and management:

DAVISON, V.E. 1947. Farm Fishponds for Food and Good Land Use. *USDA Farmers Bulletin 1983.* 29 pp.

DAVISON, V.E. 1955. Managing Farm Fishponds for Bass and Bluegills. *USDA Farmers Bulletin 2094.* 18 pp.

DYCHE, L.L. 1914. *Ponds, Pond Fish and Pond Fish Culture.* Kansas State Department of Game and Fish. 208 pp.

EDMINSTER, F.C. 1947. *Fish Ponds for the Farm.* New York: Charles Scribner's Sons. 114 pp.

GABELHOUSE, D.W., JR. 1978. Redear Sunfish for Small Impoundments? *In* Novinger, G.D., and Dillard, J.G., editors. *New Approaches to the Management of Small Impoundments.* Dearborn, Michigan: North Central Division, American Fisheries Society. pp. 109–23.

KRUMHOLZ, L. A. 1952. Management of Indiana Ponds for Fishing. *Journal of Wildlife Management* 16(3) : 254–57.

LEWIS, W.M., and HEIDINGER, R.C. 1973. Fish Stocking Combinations for Farm Ponds. *Southern Illinois Fisheries Bulletin* 4 : 1–17.

REGIER, H.A. 1963. Ecology and Management of Largemouth Bass and Bluegills in Farm Ponds in New York. *New York Fish and Game Journal* 10(1) : 1–89.

SWINGLE, H.S. 1949. Experiments with Combinations of Largemouth Black Bass, Bluegills and Minnows in Ponds. *Transactions of the American Fisheries Society* 76(1946) : 46–62.

SWINGLE, H.S. 1951. Experiments with Various Rates of Stocking Bluegills, *Lepomis macrochirus* (Rafinesque) and Largemouth Bass *Micropterus salmoides* (Lacepede) in Ponds. *Transactions of the American Fisheries Society* 80(1950) : 218–30.

SWINGLE 1970. See Chapter I–2, in Appendix.

Up-to-date information on farm ponds may be found in the farm pond handbooks put out by most of the states and the province of Ontario.

Calculation and management of farm pond populations:

EVERHART, EIPPER, and YOUNG 1975. See Chapter I–2, in Appendix.

MacARTHUR, R.H., and CONNELL, J.H. 1966. *The Biology of Populations*. New York: Wiley Interscience. 200 pp.

SWINGLE 1970. See Chapter I–2, in Appendix.

Monoculture and other aspects of black bass management:

ROBBINS, W.H., and MacCRIMMON, H. 1977. *The Blackbass in America and Overseas*. Sault Ste. Marie, Ontario, Canada: Biomanagement and Research Enterprises. 196 pp.

Hybrid sunfish:

HEIDINGER, R.C., and LEWIS, W.M. 1972. Potentials of the Redear Sunfish Hybrid in Pond Management. *Progressive Fish Culturist* 34(2) : 107–9.

Lewis, W.M., and HEIDINGER, R.C. 1971. Aquaculture Potential of Hybrid Sunfish. *The American Fish Farmer* 2(5) : 14–16.

LEWIS, W.M., and HEIDINGER, R.C. 1978. Use of Hybrid Sunfishes in the Management of Small Impoundments. *In* Novinger, G.D., and Dillard, J.G., editors. *New Approaches to the Management of Small Impoundments*. Dearborn, Michigan: North Central Division, American Fisheries Society. pp. 104–8.

CHAPTER II–2

BENNETT, G.W. 1943. Management of Small Artificial Lakes. *Illinois Natural History Survey Bulletin* 22(3) : 357–76.

BENNETT, G.W. 1952. Pond Management in Illinois. *Journal of Wildlife Management* 16(3) : 249–53.

McLARNEY, B., and PARKIN, J. 1980. Cage Culture. *The Journal of The New Alchemists* 6 : 83–89.

NOVINGER, G.D., and DILLARD, J.G., editors. 1978. *New Approaches to the Management of Small Impoundments*. Dearborn, Michigan: North Central Division, American Fisheries Society. 132 pp.

SWINGLE, H.S. 1954. Experiments on Commercial Fish Production in Ponds. *Proceedings of the Annual Conference of Southeastern Game and Fish Commissioners* 1954 : 69–74.

SWINGLE, H.S. 1957. Commercial Production of Red Cats (Speckled Bullheads) in Ponds in the Southeast. *Proceedings of the Annual Conference of Southeastern Game and Fish Commissioners* 10 (1956) : 162–69.

CHAPTER II–3

BARDACH, RYTHER, and McLARNEY 1972. See Chapter I–1, in Appendix.

BROWN, E. E. 1977. *World Fish Farming: Cultivation and Economics*. Westport, Conneticut: Avi Publishing Co. 397 pp.

GRIZZELL, R., JR.; DILLON, O., JR.; and SULLIVAN, E. 1975. Catfish Farming. USDA Farmers Bulletin 2260. 21 pp.

LEE, J.S. 1973. *Commercial Catfish Farming*. Danville, Illinois: The Interstate Printers and Publishers. 263 pp.

LOVELL, R.T., and STICKNEY, R.R. 1977. Nutrition and Feeding of Channel Catfish. *Southern Cooperative Series Bulletin No. 218*. Auburn, Alabama: Auburn University. 66 pp.

MACK, J. 1971. *Catfish Farming Handbook*. San Angelo, Texas: Educator Books. 195 pp.

MEYER, K.; SNEED, K.; and ESCHMEYER, P. 1973. *Second Report to the Fish Farmers. (The Status of Warmwater Fish Farming and Progress in Fish Farming Research)*. Resource Publication 113, U.S. Bureau of Sport Fisheries and Wildlife. 123pp.

MITCHELL, T.E., and USRY, M.J. 1967. *Catfish Farming —a Profit Opportunity for Mississippians*. Jackson, Mississippi: Mississippi Research and Development Center. 83 pp.

PRETTO, R., and SMITHERMAN, R. 1974. Polyculture Systems Utilizing Channel Catfish, *Ictalurus punctatus*, as the Principal Species. Auburn University Agricultural Experiment Station, *Fisheries Research Annual Report* 11 : 104–18.

REGIER, H.A. 1963. Ecology and Management of Channel Catfish in Farm Ponds in New York. New York Fish and Game Journal 10(2) : 170–85.

TIEMEIER, O.W., and DEYOE, C.W. 1967. *Production of Channel Catfish*. Kansas State University of Agriculture and Applied Science, Agriculture Experiment Station Bulletin 508. 24 pp.

WOLTMAN, R. 1974. *The Layman's Book of Catfish Farming*. Newberry Springs, California: Band D Fish Farms. 80 pp.

CHAPTER II–4

BARDACH, RYTHER, and McLARNEY 1972. See under Chapter I–1, in Appendix.

BRADY, L., and HULSEY, A.H. 1959. Propagation of Buffalo Fishes. *Report of the South East Association of Game and Fish Commissioners, 13th Annual Conference* : 80–89.

MARTIN, M. 1973. Practical Farmer's Guide to Buffalo Cross-Breeding. *American Fish Farmer* 4(2) : 6–8.

MEYER, SNEED, and ESCHMEYER 1973. See under Chapter II–3, in Appendix.

SWINGLE, H.S. 1957. Revised Procedure for Commercial Production of Bigmouth Buffalofish in Ponds in the Southeast. *Proceedings of the 10th Annual Conference of the South East Association of Game and Fish Commissioners:* 162–65.

CHAPTER II–5

ALIKUNHI, K.H. 1966. Synopsis of Biological Data on Common Carp *Cyprinus carpio* (Linnaeus) 1758, Asia and the Far East. *FAO Fisheries Synopsis* 31.1.

BARDACH, RYTHER, and McLARNEY 1972. See Chapter I–1, in Appendix.

BROWN, E.E. 1969. The Fresh Water Cultured Fish Industry of Japan. University of Georgia College of Agriculture Experiment Stations, Research Report 41. 57 pp.

BROWN 1977. See Chapter II–3, in Appendix.

BUCK, D.H.; THOITS, C.F., III; and ROSE, C.R. 1979. Variation in Carp Production in Replicate Ponds. *Transactions of the American Fisheries Society* 99 (1) : 74–79.

HUET 1970. See Chapter I–1, in Appendix.

McLARNEY, W. 1972. Why Not Carp? *Organic Gardening and Farming*, February, 1972 : 76–81.

NAMBIAR, K.P.P. 1970. Carp Culture in Japan: A General Study of the Existing Practices. Indo-Pacific Fisheries Council, Occasional Papers 1970/1. 41 pp.

SARIG, S. 1966. Synopsis of Biological Data on Common Carp *Cyprinus carpio* (Linnaeus) 1758, Near East and Europe. *FAO Fisheries Synopsis* 31.2.

SCHAPERCLAUS, W. 1965. Lehrbuch der Karpfenertrage in Teichen durch Stickstoffdungung. (N. Dungungsversche in Karppa, 1965). *Deutsche Fischerei Zeitung* 13(1) : 6–14. (In German.)

STEFFENS, W. 1969. *Der Karpfen.* Wittenberg, Lutherstadt, East Germany: A. Ziemsen Verlag. 156 pp. (In German.)

CHAPTER II–6

Books, booklets and papers

BARDACH, RYTHER, and McLARNEY 1972. See under Chapter I–1, in Appendix. (Synopsis of commercial, non-commercial and hatchery methods in various countries.)

BREGNBALLE, F. 1963. Trout Culture in Denmark. *Progressive Fish Culturist* 25(3) : 115–20. (Commercial culture in Denmark.)

BROWN 1969. See under Chapter II–3, in Appendix. (Commercial culture in Japan.)

BROWN 1977. See under Chapter II–3, in Appendix. (Commercial culture in various countries.)

BROWN, E.E.; HILL, T.K.; and CHESNESS, J.L. 1974. Rainbow Trout and Channel Catfish—A Double-cropping System. *Georgia Agricultural Experiment Station Research Report 196.* (Commercial cage culture and double cropping in the United States.)

COLLINS, R.A. 1972. Cage Culture of Trout in Warm-water Lakes. *American Fish Farmer* 3(7) : 4–7.

DAVIS, H.S. 1953. *Culture and Diseases of Game Fishes.* Berkeley: University of California Press. 332 pp. (Hatchery methods.)

GREENBURG, D.B. 1960. *Trout Farming.* Philadelphia: Chilton Company 197 pp. (Commercial culture.)

HUET 1970. See under Chapter I–1. (Commercial culture and hatchery methods in Europe.)

KLOONTZ, G.W., and KING, J.G. 1975. *Aquaculture in Idaho and Nationwide.* Boise, Idaho: Idaho Department of Water Resources. 86 pp. (Commercial culture in the United States.)

LEITRITZ, E., and LEWIS, R.C. 1976. *Trout and Salmon Culture.* California Department of Fish and Game, Fishery Bulletin 164. 197 pp. (Hatchery methods.)

MARRIAGE, L.D.; BORELL, A.E.; and SCHEFFER, P.M. 1976. *Trout Ponds for Recreation.* USDA Farmers' Bulletin 2249. 13 pp. (Non-commercial culture.)

MYERS, G.L., and PETERKA, J.J. 1976. Survival and Growth of Rainbow Trout (*Salmo gairdneri*) in Four Prairie Lakes, North Dakota. *Journal of the Fisheries Research Board of Canada* 33(5) : 1192–95. (Pothole culture.)

OLSON, M. 1979. The Second Wave: The Application of New Alchemy Aquaculture Techniques to a Remote, Small-scale Trout Farm. *The Journal of the New Alchemists* 5 : 110–15. (Small scale culture.)

ROBERTS, R.J., and SHEPHERD, C.J. 1974. *Handbook of Trout and Salmon Diseases.* London: Fishing News (Books). 168 pp. (I haven't seen it, but said to be a good treatment of culture methods as well as diseases.)

Salmonid. (Periodical published by U.S. Trout Farmers' Association, Lake Ozark, Missouri.)

SCHEFFER, P.M., and MARRIAGE, L.D. 1969. *Trout Farming: Could Trout Farming be Profitable for You?* USDA Leaflet 552. 8 pp. (Commercial culture in the United States.)

SEDGWICK, S.D. 1976. *Trout Farming Handbook.* London: Seeley Service and Co. Distributed in the U.S. by Scholium International, Flushing, New

York, 163 pp. (Commercial culture and hatchery methods.)

SWANSON, M. 1975. *Aquaculture in Saskatchewan.* Saskatchewan Department of Tourism and Renewable Resources. 19 pp. (Pothole culture.)

SWANSON, M.E. 1979. Pothole Aquaculture in Saskatchewan. *Commercial Fish Farmer* 5(2) : 23–25.

U.S. Trout News. (A periodical for commercial trout growers, published by U.S. Trout Farmers' Association. See Appendix I–5.)

WESTERS, H., and PRATT, K. 1977. Rational Design of Hatcheries for Intensive Salmonid Culture, Based on Metabolic Characteristics. *Progressive Fish Culturist* 39(4) : 157–65.

Bibliography:

MEADE, J.W. 1971. *A Reference List for Trout Culturists.* Pennsylvania Fish Commission, Division of Fish Hatcheries, Bellefonte, Pennsylvania: Benner Spring Research Station. 5 pp.

CHAPTER II–7

CALBERT, H.E. 1976. *Fundamentals of Fish Farming.* University of Wisconsin: Sea Grant Publication (WIS–56–76–126).

CALBERT, H.E., and HUK, H.T. 1976. Raising Yellow Perch, *Perca flavescens*, under Controlled Conditions for the Upper Midwest Market. *Proceedings of the Seventh Annual Workshop* San Diego, California: World Mariculture Society 10 pp.

CHAPTER II–8

Tilapia:

There is no shortage of literature dealing with tilapia culture, but most of it deals with tilapia in the context of a particular culture method. Check the reference lists for chapters III–2, IV–2, IV–3, and IV–4. Scientific publications containing much material on tilapia culture include the following:

Bamidgeh: Bulletin of Fish Culture in Israel. See under Chapter I–1.

Symposium on Culture of Exotic Species, Fish Culture Section, American Fisheries Society, Atlanta, Georgia, January 4, 1978.

The Journal of the New Alchemists. (Closed systems.) See under Chapter I–1, in Appendix.

World Symposium on Warm Water Pond Fish Culture. See under Chapter I–1, in Appendix.

Books covering tilapia culture deal mostly with methods for tropical countries. They include:

BARD, J.; DE KIMPE, P.; LAZARD, J.; LEMASSON, J.; and

LESSENT, P. 1976. *Handbook of Tropical Fish Culture.* Nogent-sur-Maren, France: Centre Technique Forestier Tropical. 165 pp.

BARDACH, RYTHER, and McLARNEY 1972. See under Chapter I–1, in Appendix.

CHAKROFF 1976. See under Chapter I–1, in Appendix.

HICKLING 1971. See under Chapter I–1, in Appendix.

MAAR, A.M.; MORTIMER, A.E.; and VAN DER LINGEN, I. 1966. *Fish Culture in Central East Africa.* Rome: FAO. 158 pp.

MILES, D. 1977. Theoretical and Practical Aspects of the Production of the All Male Tilapia Hybrids. *Bamidgeh* 29(3) : 94–101.

References dealing with control of reproduction in tilapia:

CROSS, D.W. 1979. Methods to Control Over-breeding of Farmed Tilapia. *Fish Farming International* 3(1) : 27–29.

FRAM, M. 1977. Production of *Tilapia nilotica* (Linnaeus) and of the all-male hybrid, *T. nilotica* (female) x *T. hornorum* Trewavas (male) in earthen pond monoculture systems, with observations on their effects on water quality. M.S. thesis, University of Puerto Rico. 19 pp.

PAGAN-FONT, F.A. 1975. Cage Culture as a Mechanical Method for Controlling Reproduction of *Tilapia aurea* (Steindachner). *Aquaculture* 6(3) : 243–47.

The issue of introduction of exotics is so important that I urge every prospecitve tilapia culturist to become informed on the subject, and therefore recommend the following paper:

HUBBS, C. 1968. An Opinion on the Effects of Cichlid Releases in North America. *Transactions of the American Fisheries Society* 97 : 197–98.

The best single paper on tilapia culture:

HICKLING, C.F. 1963. The Cultivation of *Tilapia.* *Scientific American* 208 : 143–52.

Grass carp:

The situation with grass carp is similar to that with tilapia. See under Chapters III–1, III–2, and VIII–1.

BARDACH, RYTHER, and McLARNEY 1972. See under Chapter I–1.

Commercial Fish Farmer. 1976. Two Names—Same Fish: Good or Bad? Controversy Heats up over White Amur-Grass Carp. *Commercial Fish Farmer* 2(5) : 16–21. (Article presents viewpoints from various commercial fish farmers, fishery agencies and scientists.)

HICKLING 1962. See under Chapter I–1, in Appendix.

HICKLING, C.F. 1967. On the Biology of a Herbivorous Fish, the White Amur or Grass Carp *Ctenopharyngodon idella* Val. *Proceedings of the Royal Society of Edinburgh, Section B (Biology)*, vol. LXX, Part 1 (No. 4.)

HORA and PILLAY 1962. See under Chapter I–1.

KILGEN, R.H., and SMITHERMAN, R.O. 1971. Food Habits of the White Amur Stocked in Ponds Alone and in Combination with Other Species. *Progressive Fish Culturist* 33 : 123–27.

SUTTON, D.L. 1977. Grass Carp (*Ctenopharyngodon idella* Val.) in North America. *Aquatic Botany* 3 (1977): 157–64. (Includes extensive bibliography)

VAN ZON, J.C.J. 1977 Grass carp (*Ctenopharyngodon idella* Val.) in Europe. *Aquatic Botany* 3(1977) : 143–55. (Includes extensive bibliography.)

VINOGRADOV, V.K., and ZOLOTOVA, Z.K. 1974. The Influence of the Grass Carp on Aquatic Ecosystems. *Hydrobiological Journal* 10(2) : 72–78.

Walking catfish:

BARDACH, RYTHER and McLARNEY 1972. See under Chapter I–1.

SIDTHIMUNKA, A.; SANGLERT, J.; and PAWAPOOTANEN, O. 1966. The culture of catfish (*Clarias* spp.) in Thailand. *FAO World Symposium on Warmwater Pond Fish Culture*. FAO Fisheries Reports 44(4).

CHAPTER II–9

BARDACH, RYTHER, and McLARNEY 1972. See under Chapter I–1, in Appendix.

CULLEY, D.D., and CRAVOIS, C.T. 1970. Frog culture. *American Fish Farmer* 1(10) : 5–10.

NACE, G.W. 1968. The Amphibian Facility of the University of Michigan. *BioScience* 18(8) : 767–75.

PRIDDY, J.M., and CULLEY, D.D. 1972. Frog Culture Industry, Past, Present, Future? *American Fish Farmer* 3(9) : 4–7.

U.S. BUREAU OF SPORT FISHERIES AND WILDLIFE. 1965. *Frog Raising*. Fishery Leaflet 436.

Good general texts on frogs are:

DICKERSON, M.C. 1906. *The Frog Book: North American Toads and Frogs with a Study of the Habits and Life Histories of those of the Northeastern States*. New York: Doubleday, Page & Co. 253 pp.

OLIVER, J.A. 1955. *Natural History of North American Amphibians and Reptiles*. Princeton, New Jersey: Van Nostrand. 359 pp.

CHAPTER II–10

Louisiana methods:

AVAULT, J.W., JR. 1973. *Crawfish Farming in the United States*. Center for Wetland Resources, Louisiana State University, Baton Rouge, Louisiana.

AVAULT, J.W., JR., ed. 1975. Freshwater Crayfish. *Papers from the Second International Symposium on Freshwater Crayfish*. Louisiana State University, Baton Rouge, Louisiana. 676 pp.

BARDACH, RYTHER, and McLARNEY 1972. See under Chapter I–1, in Appendix.

BROWN 1977. See under Chapter I–1, in Appendix.

HAM, B.G. 1971. Crawfish Culture Techniques. *American Fish Farmer* 2(5) : 5–6, 21, 24.

HUNER, J.V., and BARR, J.E. 1980. *Red Swamp Crawfish: Biology and Exploitation*. Sea Grant Publication No. LSU–T–80–001. Center for Wetland Resources, Louisiana State University, Baton Rouge, Louisiana 70803. 148 pp.

LaCAZE, C. 1976. *Crawfish farming*. Louisiana Wild Life and Fisheries Commission, Fisheries Bulletin No. 7. 27 pp.

LOUISIANA STATE UNIVERSITY. *List of LSU Crawfish Publications*. Louisiana State University, Baton Rouge, Louisiana 70803.

Descriptions of crayfish capture fisheries outside of Louisiana:

MILLER, G.C., and VAN HYNING, J.M. 1970. The Commercial Fishery for Freshwater Crawfish, *Pacifastacus leniusculus* (Astacidae) in Oregon, 1893–1956. *Oregon Fisheries Commission Research Report* 2 : 77–89.

THREINEN, C.W. 1958. A Summary of Observations on the Commercial Harvest of Crayfish in Northwestern Wisconsin, with Notes on the Life History of *Orconnectes virilis*. *Wisconsin Conservation Department, Fish Management Division, Miscellaneous Report* 2. 14 pp.

Culture methods for other crayfish, including bait species:

CROCKER, D.W., and BARR, D.W. 1968. *Handbook of the Crayfishes of Ontario*. Toronto: University of Toronto Press. 158 pp.

FORNEY, J.L. 1968. *Raising Bait Fish and Crayfish in New York Ponds*. Cornell Extension Bulletin 986. 31 pp.

HUNER, J.V. 1976. Raising Crawfish for Food and Fish Bait: A New Polyculture Crop with Fish. *The Fisheries Bulletin* 1(2) : 7–9.

HUNER, J.V. 1978. Crawfish Culture in Small Ponds.

Farm Pond Harvest, Winter, 1978 : 8–10.

HUNER, J.V., and AVAULT, J.W. No date. *Producing Crawfish for Fishbait.* Sea Grant Publication LSU-T1-76-001, Center for Wetland Resources, Louisiana State University, Baton Rouge, Louisiana 70803. 23 pp.

LANGLOIS, T.H. 1935. Notes on the Habits of the Crayfish, *Cambarus rusticus* Girard, in Fish Ponds in Ohio. *Transactions of the American Fisheries Society* 65 : 189–92.

MASON, J.C. 1974. Aquaculture Potential of the Freshwater Crayfish (*Pacifastacus*). I. Studies during 1970. Fisheries Research Board of Canada Technical Report 440. 43 pp.

RICKETT, J.D. 1974. Trophic Relationships Involving Crayfish of the Genus *Orconectes* in Experimental Ponds. *Progressive Fish Culturist* 36 : 207–11.

TACK, P.I. 1941. The Life History and Ecology of the Crayfish *Cambarus immunis* Hagen. *American Midland Naturalist* 25 : 420–46.

Books and papers on crayfish in North America:

CREASER, E.P. 1931. The Michigan Decapod Crustaceans. *Papers of the Michigan Academy of Sciences, Arts and Letters* 19 : 581–85.

CREASER, E.P. 1932. the Decapod Crustaceans of Wisconsin. *Wisconsin Academy of Sciences, Arts and Letters* 27 : 321–38.

CREASER, E.P., and ORTENBURGER, A.I. 1933. The Decapod Crustaceans of Oklahoma. *Publications of the University of Oklahoma Biological Survey* 5 : 14–80.

CROCKER, D.W. 1957. *The Crayfishes of New York State.* New York State Museum and Science Service, Bulletin 355. 97 pp.

CROCKER and BARR 1968. See under "Culture methods for other crayfish" above.

EDMONDSON, W.T., ed. 1959. *Ward and Whipple's Fresh Water Biology.* New York: Wiley Interscience. 1248 pp.

HOBBS, H.H., JR. 1972. Crayfishes (Astacidae) of North and Middle America. *Biota of Freshwater Ecosystems, Identification Manual* 9 : 1–173.

HOBBS, H.H., JR. 1974. A Check List of the North and Middle American Crayfishes (Decapoda: Astacidae and Cambaridae). *Smithsonian Contributions to Zoology* 166 : 1–161.

LYLE, C. 1938. The Crawfishes of Mississippi, with Special Reference to the Biology and Control of Destructive Species. *Iowa State College Journal of Science* 13 : 75–77.

MEREDITH, W.G., and SCHWARTZ, F.J. 1960. *Maryland Crayfishes.* Maryland Department of Resource Education, Educational Series No. 46. 32 pp.

PENN, G.H. 1942. An Illustrated Key to the Crawfishes of Louisiana with a Summary of their Distribution within the State. *Tulane Studies in Zoology* 7 : 3–20.

PENNAK, R.W. 1978. *Fresh Water Invertebrates of the United States.* New York: Wiley Interscience. 803 pp.

RHOADES, R. 1944. The Crayfishes of Kentucky, with Notes on Variation, Distribution and Descriptions of New Species and Subspecies. *American Midland Naturalist* 31 : 111–49.

TURNER, C.L. 1926. The Crayfish of Ohio. *Ohio Biological Survey* 13 : 145–95.

WILLIAMS, A.B. 1954. Species and Distribution of the Crayfishes of the Ozark Plateau and Ouachita Provinces. *University of Kansas Science Bulletin* 36 : 803–918.

CHAPTER II–11

BARDACH, RYTHER, and MCLARNEY 1972. See under Chapter I–1, in Appendix.

Commercial Fish Farmer 3(2), 1977. (Special issue devoted to freshwater shrimp.)

LING, S.W. 1977. See under Chapter I–1, in Appendix.

SHANG, Y.C., and FUJIMURA, T. 1977. The Production Economics of Freshwater Prawn (*Macrobrachium rosenbergi*) farming in Hawaii. *Aquaculture* 11 (1977): 99–110.

CHAPTER II–12

BAKER, F.C. 1928. *The Fresh Water Mollusca of Wisconsin. Part II. Pelecypoda.* Madison, Wisconsin: Wisconsin Academy of Science, Arts and Letters. 495 pp.

BURCH, J.B. 1975. *Fresh Water Sphaeriacean Clams (Mollusca: Pelecypoda) of North America.* Hamburg, Michigan: Malacological Publishing. 204 pp.

CHAMBERLIN, R.V., and JONES, D.T. 1929. A Descriptive Catalog of the Mollusca of Utah. *Bulletin of the University of Utah* 19 : 1–203.

CLARKE, A.H., and BERG, C.O. 1959. The Fresh Water Mussels of Central New York. *Memoirs of the Cornell University Agricultural Experiment Station* 307 : 1–79.

CLENCH, W.J., and TURNER, R.D. 1956. Freshwater Mollusks of Alabama, Georgia and Florida from the Escambia to the Suwanee River. *Bulletin of the Florida State Museum* 1(3): 99–239.

CONNOR, M.S. 1980. Biological Filters, Water Quality and Fresh Water Clams. *The Journal of The New Alchemists* 6 : 90–92.

GOODRICH, C., and VAN DER SCHALIE, H. 1939. Aquatic Mollusks of the Upper Peninsula of Michigan. *Museum of Zoology of the University of Michigan Miscellaneous Publication* 43 : 1–45.

GOODRICH, C., and VAN DER SCHALIE, H. 1944. A Revision of the Mollusca of Indiana. *American Midland Naturalist* 32(2) : 257–326.

HEARD, W.H. 1962. Distribution of Sphaeriidae (Pelecypoda) in Michigan, U.S.A. *Malacologia* 1 : 139–60.

HENDERSON, J. 1924. Mollusca of Colorado, Utah, Montana, Idaho and Wyoming. *University of Colorado Studies* 13 : 65–223.

HENDERSON, J. 1936. Mollusca of Colorado, Utah, Montana, Idaho and Wyoming – Supplement. *University of Colorado Studies* 23 : 81–145.

HOWARD, A.D. 1922. *Bulletin of the U.S. Bureau of Fisheries* 1921–1922 : 63–89.

ISOM, B.G. 1969. The Mussel Resources of the Tennessee River. *Malacologia* 7 : 397–425.

JONES, R.O. 1949. Propagation of Fresh Water Mussels. *Progressive Fish Culturist* 12 : 13–25.

LaROCQUE, A. 1953. *Catalogue of the Recent Mollusca of Canada*. Bulletin of the National Museum of Canada 129. 377 pp.

MATTESON, M.R. 1955. Studies on the Natural History of the Unionidae. *American Midland Naturalist* 53 : 126–45.

MURRAY, H.D., and LEONARD, A.B. 1962. *Handbook of Unionid Mussels in Kansas*. Lawrence, Kansas: University of Kansas. 184 pp.

MURRAY, H.D., and ROY, E.C. 1968. Checklist of Freshwater and Land Mollusks of Texas. *Sterkiana* 30 : 25–42.

OVER, W.H. 1942. *Mollusca of South Dakota*. Natural History Studies, University of South Dakota no. 5. 11 pp.

PENNAK 1978. See under Chapter II–10.

SINCLAIR, R.M., and ISOM, B.G. 1963. *Further Studies on the Introduced Asiatic Clam (Corbicula) in Tennessee*. Tennessee Department of Public Health, Stream Pollution Control Board. 76 pp.

SWINGLE, H.S. 1966. Biological Means of Increasing Productivity in Ponds. *FAO World Symposium on Warmwater Pond Fish Culture*. FAO Fisheries Reports 44(4).

VAN DER SCHALIE, H., and VAN DER SCHALIE, A. 1950. The Mussels of the Mississippi River. *American Midland Naturalist* 44(2) : 448–66.

CHAPTER II–13

Classic papers on methods of selecting species for culture:

BENNETT, G.W. 1943. Management of Small Artificial Lakes. *Illinois Natural History Survey Bulletin* 22(3) : 357–76.

SWINGLE, H.S. 1952. Farm Pond Investigations in Alabama. *Journal of Wildlife Management* 16(3): 243–49.

Summaries of life history data:

CARLANDER, K.D. 1969. *Handbook of Freshwater Fishery Biology, Vol. 1*. Ames, Iowa: The Iowa State University Press. 752 pp.

CARLANDER, K.D. 1977. *Handbook of Freshwater Fishery Biology, Vol. 2*. Ames, Iowa: The Iowa State University Press, 431 pp. (Volume 1 deals with all the North American fishes outside the order Perciformes. Volume 2 covers the family Centrarchidae. Volume 3, when available will complete the series. The amount of data in these volumes is awesome and there are thousands of literature citations. They are a very good place to begin when considering whether or not to pursue culture of a species.)

Descriptive books on freshwater fishes of North America:

EDDY, S., and UNDERHILL, J.C. 1978. *How to Know the Freshwater Fishes*. Dubuque, Iowa: W.C. Brown Co. 215 pp.

SCOTT, W.B., and CROSSMAN, E.J. 1973. *Freshwater Fishes of Canada*. Fisheries Research Board of Canada Bulletin 184. 966 pp.

In addition to these two books there are books on the fresh water fishes of most of the states and provinces, plus some regional books. A bibliography listing 72 such references is included in:

DEACON, J.E.; KOBETICH, G.; WILLIAMS, J.O.; and CONTRERAS, S. 1979. Fishes of North America: Endangered, Threatened or of Special Concern, 1979. *Fisheries* 4(2) : 30–44.

Publications on culture or life history of certain animals of interest:

Pacific salmon:

BARDACH, RYTHER, and McLARNEY 1972. See under Chapter I–1, in Appendix.

HINES, N.D. 1967. *Fish of Rare Breeding – Salmon and Trout of the Donaldson Strains*. Washington, D.C.: Smithsonian Institution Press. 167 pp. (Really a disguised biography of Lauren Donaldson, but also contains the basics of Pacific salmon culture.)

Whitefish:

BARDACH, RYTHER, and McLARNEY 1972. See under Chapter I–1, in Appendix.

HUET 1970. See under Chapter I–1, in Appendix.

OVCHYNNYK, M. 1963. Soviet Fish Culture. *Fishing News International* 2(3) : 279–82.

Sacramento Blackfish:

HUNER, J.V. 1982. Pond Culture of American Eel, Sturgeon and Sacramento Blackfish? *Farm Pond Harvest* 16(1) : 10–11, 16.

MURPHY, G.I. 1950. The Life History of the Greaser Blackfish (*Orthodon microlepidotus*) of Clear Lake, Lake County, California. *California Fish and Game* 36(1) : 119–33.

Crappie:

HUNER, J.V. 1982. Why Not Crappie? *Farm Pond Harvest* 16(3) : 12–13, 18, 23, 25.

Tench:

BARDACH, RYTHER, and McLARNEY 1972. See under Chapter I–1, in Appendix.

WOYNAROVICH, E. 1966. New Systems and New Fishes for Culture in Europe. *FAO World Symposium on Warmwater Pond Fish Culture.* FAO Fisheries Reports 44(4).

Eels:

ANGEL, N.B., and JONES, W.R. 1974. *Aquaculture of the American Eel (Anguilla rostrata).* North Carolina State University School of Engineering, Industrial Extension Service. 43 pp.

BARDACH, RYTHER, and McLARNEY 1972. See under Chapter I–1, in Appendix.

BROWN 1969. See under Chapter II–5, in Appendix.

BROWN 1977. See under Chapter II–5, in Appendix.

HUET 1970. See under Chapter I–1, in Appendix.

HUNER, J.V. 1982. Pond Culture of American Eel, Sturgeon and Sacramento Blackfish? *Farm Pond Harvest* 16(1) : 10–11, 16.

LING 1977. See under Chapter I–1, in Appendix.

TESCH, F.W. 1978. *The Eel.* London: Chapman and Hall. 434 pp.

USUI, A. 1974. *Eel Culture.* Translated by I. Hayashi. London: Fishing News (Books). 186 pp.

Rock bass:

REGIER 1963. See under Chapter II–3, in Appendix.

Walleye:

BARDACH, RYTHER, and McLARNEY 1972. See under Chapter I–1, in Appendix.

NIEMUTH, W.; CHURCHILL, W.; and WIRTH, T. *The Walleye, Life History, Ecology and Management.* Wisconsin Conservation Department, Publication 227. 14 pp.

Mullet:

BARDACH, RYTHER, and McLARNEY 1972. See under Chapter I–1, in Appendix.

Sturgeon:

HUNER, J.V. 1982. Pond Culture of American Eel, Sturgeon and Sacramento Blackfish? *Farm Pond Harvest* 16(1) : 10–11, 16.

Turtles:

ERNST, C.H., and BARBOUR, R.W. 1972. *Turtles of the United States.* Lexington, Kentucky: The University Press of Kentucky. 347 pp.

Important publication on fish names:

ROBINS, C.R. 1980. *A List of Common and Scientific Names of Fishes from the United States and Canada.* American Fisheries Society Special Publication No. 12. Bethesda, Maryland: American Fisheries Society. 174 pp.

CHAPTER III–1

Fish physiology and nutrition:

HALVER, J., ed. 1972. *Fish Nutrition.* New York: Academic Press. 713 pp.

HOAR, W.S., and RANDALL, D.J., eds. 1969. *Fish Physiology.* Vol. 1. New York: Academic Press. 465 pp.

NATIONAL ACADEMY OF SCIENCES. 1973. *Nutrient Requirements of Trout, Salmon and Catfish.* Washington, D.C.: National Academy of Sciences. 57 pp.

NATIONAL ACADEMY OF SCIENCES. 1977. *Nutrient Requirements of Warmwater Fishes.* Washington, D.C.: National Academy of Sciences. 78 pp.

RUSSELL-HUNTER, W.D. 1970. *Aquatic Productivity.* New York: MacMillan. 306 pp.

Information of feeding habits of fishes in nature:

CARLANDER 1969. See under Chapter II–13, in Appendix.

CARLANDER 1977. See under Chapter II–13, in Appendix.

CHAPTER III–2

References dealing with aquatic fertilization:

BALL, R.C. 1952. The Biological Effects of Fertilizer on Warmwater Lake. *Michigan State College Agricultural Experiment Station Technical Bulletin* 16(3) : 266–69.

BARDACH, RYTHER, and McLARNEY 1972. See under Chapter I–1, in Appendix.

Boyd, C. 1979. *Water Quality in Warmwater Fish Ponds*. Alabama Agricultural Experiment Station, University of Alabama Press. 366 pp.

Boyd, C., and Snow, J. 1975. *Fertilizing Farm Fish Ponds*. Auburn University Agricultural Experiment Station Leaflet 88. 6 pp.

Buck, D.H.; Baur, R.J.; and Rose, C.R. 1978. Utilization of Swine Manure in a Polyculture of Asian and North American Fishes. *Transactions of the American Fisheries Society* 107 : 216–22.

Davis, H.S., and Wiebe, A.H. 1931. Experiments on The Culture of Black Bass and other Pondfish. Appendix 9 to the *Report of the U.S. Commissioner of Fisheries* 1930(1931). 29 pp.

Gooch, B.C. 1967. Appraisal of North American Fish Culture Fertilization Studies. *World Symposium on Warmwater Pond Fish Culture*. FAO Fisheries Reports 44(4) : 13–26.

Hepher, B. 1967. Ten Years Research in Fish Pond Fertilization in Israel. *Bamidgeh* 14(2) : 2–38.

Hickling 1971. See under Chapter I–1.

Huet 1970. See under Chapter I–1.

Maciolek, J.A. 1954. *Artificial Fertilization of Lakes and Ponds, a Review of the Literature*. U.S. Fish and Wildlife Service, Special Scientific Report Fisheries 113. 41 pp.

Meehean, O.L., and Marzulli, F. 1945. The Relationship Between Production of Fish and the Carbon and Nitrogen Contents of Fertilized Fish Ponds. *Transactions of the American Fisheries Society* 73(1943) : 262–73.

Neess, J.C. 1949. Development and Status of Pond Fertilization in Central Europe. *Transactions of the American Fisheries Society* 76 : 335–58.

Olsen, C. 1970. On Biological Nitrogen Fixation in Nature, Particularly by Blue-green Algae. *Compte Rendu des Travaux due Laboratoire de Carlsberg* 37(12) : 269–83.

Schaperclaus 1933. See under Chapter I–1, in Appendix.

Schroeder, G. 1974. Use of Cowshed Manure in Fish Ponds. *Bamidgeh* 26 : 84–96.

Schroeder, G. 1978. Agricultural Wastes in Fish Farming. *Commercial Fish Farmer* 4(6) : 33–34.

Stickney, R.R., and Hesby, J.H. 1978. Tilapia Culture in Ponds Receiving Swine Waste. In *Culture of Exotic Fishes Symposium Proceedings*, edited by R.O. Smitherman, W.L. Shelton and J.H. Grover, pp. 90–101. Bethesda, Maryland: Fish Culture Section, American Fisheries Society.

Swingle, H.S. 1947. *Experiments on Pond Fertilization*. Alabama Polytechnic Institute Agricultural Experiment Station Bulletin No. 264. 34 pp.

Wahby, S.O. 1974. Fertilizing Fish Ponds. I—Chemi-stry of the Waters. *Aquaculture* 3(3) : 245–59.

Wheaton 1977. See under Chapter I–1, in Appendix.

Wiebe, A.H. 1930. Investigations on Plankton Production in Fish Ponds. *U.S. Bureau of Fisheries Bulletin* 46 : 137–76.

Woynarovich, E. 1967. New Systems and New Fishes for Culture in Europe. *FAO World Symposium on Warmwater Pond Fish Culture*. FAO Fisheries Reports 44(4).

Wrobel, S. 1966. The Effects of Nitrogenous-Phosphatic Fertilization on the Chemical Composition of Water, Primary Production of Phytoplankton and Fish Yield in Ponds. In *Selected Articles from Acta Hydro-biologica*. Translated (from Polish) by H. Massey-Kornobis. Springfield, Virginia: Clearing-house For Federal Scientific and Technical Information. pp. 1–46.

General references on fertilizers, including compost, primarily for terrestrial use:

Frieden, E. 1972. The Chemical Elements of Life. *Scientific American* 227(1) : 52–59.

Pratt, C.J. 1965. Chemical Fertilizer. *Readings from Scientific American: Plant Agriculture*, edited by J. Janick, R.W. Schery, F.W. Woods, and V.W. Ruttan, pp. 152–62.

Rodale, J.I., ed. 1960. *The Complete Book of Composting*. Emmaus, Pennsylvania: Rodale Farms. 1007 pp.

CHAPTER III–3.

Frost, S.W. 1957. The Pennsylvania Insect Light Trap. *Journal of Economic Entomology* 47(1) : 81–86.

Heidinger, R.C. 1971. Feeding Fish With Light. *The American Fish Farmer*, December, 1971 : 12–13.

Heidinger, R.C. 1971. Use of Ultraviolet Light to Increase the Availability of Aerial Insects to Caged Bluegill Sunfish. *Progressive Fish Culturist* 33(4) : 187–92.

Johnson, V. 1976. The Will-o'-the Wisp Bug Lite Fish Feeder. *Farm Pond Harvest* 10(1) : 8, 22.

Merkowsky, A.J.; Handcock, A.J.; Newton, S.H. 1977. Attraction of Aerial Insects as a Fish Food Supplement. *Arkansas Academy of Science Proceedings*. Vol. XXXI : 75–76.

Newton, S.H., and Merkowsky, A.J. 1977. Attracting Insects as Supplemental Food for Channel Catfish: A Two-Season Evaluation. *Arkansas Farm Research* XXVI(3) : 9.

The following papers, while not dealing with trapping insects as feed, suggest the appropriateness of insect protein, as compared to other animal proteins, in fish diets:

1972. Unsaturated Fatty Acids in Trout Diets. *Nutri-*

tion Review 30(6) : 144–47.

CASTELL, J.D.; SINNHUBER, R.O.; WALES, J.H.; and LEE, D.J. 1972. Essential Fatty Acids in the Diet of Rainbow Trout (*S. gairdneri*): Growth, Feed Conversion and Some Gross Deficiency Symptoms. *Journal of Nutrition* 102(1) : 77–86.

CHAPTER III–4

HALVER 1972. See under Chapter III–1, in Appendix.

HUET 1970. See under Chapter I–1, in Appendix.

Sedgwick 1976. See under Chapter II–6, in Appendix.

CHAPTER III–5

There are three books containing information on many organisms and their culture and use as feed. They are referred to repeatedly in the list that follows. Further useful information may be found in the magazines devoted to the aquarium hobby. Also be sure to check the references in this appendix for Chapter III–7, dealing with the use and nutritive value of specific cultured foods.

General books on cultured foods:

IVLEVA, I.V. 1969. *Mass Cultivation of Invertebrates: Biology and Methods.* Academy of Sciences of the U.S.S.R., All-Union Hydrobiological Society. Published for the U.S. National Marine Fisheries Service by the Israel Program for Scientific Translations. 148 pp.

MASTERS, C.O. 1975. *Encyclopedia of Live Foods.* Neptune City, New Jersey: TFH Publications. 336 pp.

NEEDHAM, J.G., ed. 1959. *Culture Methods for Invertebrate Animals.* New York: Dover Publications. 590 pp.

Aquarium magazines: See under Chapter IX–4, in Appendix.

Nematodes:

IVLEVA 1969.

MASTERS 1975.

NEEDHAM 1959.

Tubifex worms:

MASTERS 1975.

NEEDHAM 1959.

Enchytraeid worms:

IVLEVA 1969.

MASTERS 1975.

NEEDHAM 1959.

Earthworms:

GADDIE, R.E., and DOUGLAS, D.E. 1975. *Earthworms for Ecology and Profit. Vol. 1: Scientific Earth-*

worm Farming. Ontario, California: Bookworm Publishing Co. 176 pp.

EDWARDS, C.A., and LOFTY, J.R. 1972. *Biology of Earthworms.* Ontario, California: Bookworm Publishing Co. 283 pp.

Amphipods:

MASTERS 1975.

Fairy Shrimp:

IVLEVA 1969.

Brine Shrimp:

IVLEVA 1969.

MASTERS 1975.

SORGELOOS, P. 1973. High Density Culturing of the Brine Shrimp, *Artemia salina* L. *Aquaculture* 1(4) : 385–91.

Daphnia:

DEWITT, J.W., and CANDLAND, W. 1971. The Water Flea. *The American Fish Farmer:* 8–11.

EMBODY, G.C., and SADLER, W.O. 1934. Propagating Daphnia and Other Forage Organisms Intensively in Small Ponds. *Transactions of the American Fisheries Society* 64 : 205.

MASTERS 1975.

NEEDHAM 1959.

Cyclops:

MASTERS 1975.

NEEDHAM 1959.

Cypris:

MASTERS 1975.

NEEDHAM 1959.

Cockroaches:

MASTERS 1975.

NEEDHAM 1959.

Grasshoppers:

MASTERS 1975.

NEEDHAM 1959.

Crickets:

SWINGLE, H.S. 1961. *Raising Crickets for Bait.* Leaflet 22, Department of Publications, Agricultural Experiment Station, Auburn University, Auburn, Alabama. 4 pp.

MASTERS 1975.

NEEDHAM 1959.

Mosquitoes:

MASTERS 1975.

Midges:

IVLEVA 1969.

MCLARNEY, W.O. 1974. An Improved Method for Culture of Midge Larvae for Use as Fish Food. *The Journal of the New Alchemists* (2) : 118–19.

MCLARNEY, W.O.; HENDERSON, S.; and SHERMAN, M.M. 1974. A New Method for Culturing *Chironomus tentans* Fabricius Larvae Using Burlap Substrate in Fertilized Pools. *Aquaculture* (4) : 267–76.

YASHOUV, A. 1970. Propagation of Chironomid Larva as Food for Fish Fry. *Bamidgeh* 22(4) : 101–5.

Fly maggots:

GRESSARD, B. 1971. The Trail Lake Fish Feeder. *Farm Pond Harvest* 5(3) : 14.

MASTERS 1975.

NACE, G.W. 1968. See under Chapter II–9.

Mealworms:

MASTERS 1975.

NEEDHAM 1959.

"Minnows":

JACOBSON, S. 1974. Livebearing Fishes: An Inexpensive and Easy Way to Culture Supplementary Food for Tilapia in the Backyard Fish Farm. *The Journal of the New Alchemists* (4) : 106–7.

MARTIN, M.J. 1977. Culture Methods for Raising Forage Fishes. *The Proceedings of the 1976 Fish Farming Conference and Annual Convention of the Catfish Farmers of Texas.* College Station, Texas: Texas A & M University Press.

NEWTON, S.H.; MERKOWSKY, A.J.; HANDCOCK, A.J.; and MEISCH, M.W. 1977. Mosquitofish, *Gambusia affinis* (Baird and Girard) Production in Extensive Polyculture Systems. *Arkansas Academy of Sciences Proceedings.* Vol. XXXI : 77–78.

Aquatic Plants:

EDWARDS, P. 1980. *Food Potential of Aquatic Macrophytes.* Manila, Phillipines. ICLARM. 51 pp.

Comfrey:

HILLS, L. 1976. *Comfrey: Fodder, Food and Remedy.* New York: Universe Books. 253 pp.

Grains:

LOGSDON, G. 1977. *Small-Scale Grain Raising.* Emmaus, Pennsylvania: Rodale Press. 305 pp.

Amaranth:

COLE, J.N. 1979. *Amaranth: From the Past for the Future.* Emmaus, Pennsylvania: Rodale Press. 311 pp.

CHAPTER III–6

For information on specific commercial prepared feeds, I suggest you contact the various feed companies. Specifics, usually in rather technical language, on the nutrient requirements of aquaculture animals, and information on the way in which commercial feeds are formulated and processed can be found in:

HALVER 1972. See under Chapter III–1, in Appendix.

NATIONAL ACADEMY OF SCIENCE 1973. See under Chapter III–1, in Appendix.

NATIONAL ACADEMY OF SCIENCE 1977. See under Chapter III–1, in Appendix.

PRICE, K.S., JR.; SHAW, W.N.; and DANBERG, K.S., eds. 1976. *Proceedings of the First International Conference on Aquaculture Nutrition.* Newark, Delaware: College of Marine Studies, University of Delaware. 323 pp.

Technical journal dealing with prepared feeds:

Feedstuffs. Miller Publishing Co., 2501 Wayzata Blvd., Box 67, Minneapolis, Minnesota 53440.

Publications dealing with specific aspects of prepared feeds:

HASTINGS, W.H.; MEYERS, S.P.; and BUTLER, D.P. 1971. A Commercial Process for Water-Stable Fish Feeds. *Feedstuffs* 43(47) : 38.

HUBLOU, W.F. 1963. Oregon Pellets. *Progressive Fish Culturist* 23 : 175–80. (Development of the moist pellet.)

LEWIS, W.M.; WEHR, L; and KOEHL, D. 1973. A Preliminary Evaluation of a Fish Diet Based on Roasted Soybeans and Fresh Fish. *Proceedings of the 27th Annual Conference of the Southeastern Association of Game and Fish Commissioners:* 460–64. (Suggestions for a prepared diet which could be manufactured by the individual farmer.)

CHAPTER III–7

It is difficult to draw a line between references which should be cited for this chapter and the previous one. Therefore, please also see Chapter III–6, in Appendix. The information presented in Tables III–7–1 and III–7–2 was drawn from many sources. The following publications accounted for more than a few citations:

HALVER 1972. See under Chapter III–1, in Appendix.

IVLEVA 1969. See under Chapter III–5, in Appendix.

LEUNG, W.T.W., and FLORES, M. 1961. *Tabla de Composición de Alimentos para Uso en America Latina.* Guatemala City: INCAP-ICNND. 132 pp.

LING, S.W. 1966. Feeds and Feeding of Warmwater Pond Fishes in Asia and the Far East. *FAO World Symposium on Warmwater Pond Fish Culture.* FAO Fisheries Reports 44(4).

MANN, H. 1961. Fish Cultivation in Europe. In *Fish as Food, Vol. 1; Production, Biochemistry and Microbiology.* Edited by G. Borgstrom, pp. 77–102. New York: Academic Press.

MILLER, D.F. 1958. Composition of Cereal Grains and Forages. National Academy of Sciences—National Research Council Publication No. 585. Washington: National Academy of Sciences. 663 pp.

TAYLOR, R.L. 1975. *Butterflies in my Stomach.* Santa Barbara, California: Woodbridge Press. 224 pp.

U.S. FISH AND WILDLIFE SERVICE. 1982. Trout Feeds and Feeding. (Excerpt from "Manual of Fish Culture") National Fisheries Center, Kearneysville, West Virginia.

A number of references pertinent to particular feeding methods and strategies may be found in the section of this appendix covering Part II. The following publications cover specific aspects of feeding. I also recommend you keep up with Tom Lovell's regular column in *Aquaculture Magazine* (formerly *Commercial Fish Farmer;* see Chapter I–1 in Appendix for address).

SWINGLE, H.S. 1968. Estimation of the Standing Crop and Rates of Feeding Fish in Ponds. *Proceedings of the World Symposium on Warmwater Pond Fish Culture.* FAO Fisheries Reports 44(4): 416–23.

TAL, S., and HEPHER, B. 1966. Economic Aspects of Fish Feeding in the Near East. *Proceedings of the World Symposium on Warmwater Pond Fish Culture.* FAO Fisheries Reports 44(4).

CHAPTER IV–1

For references on classical Chinese polyculture or North American farm pond culture, see first the references listed for Chapters I–1 and II–1 in Appendix, respectively.

Theoretical papers on polyculture:

GRYGIEREK, E. 1973. The Influence of Phytophagous Fish on Pond Zooplankton. *Aquaculture* 2(1973): 197–208.

STICKNEY, R.R. 1977. The Polyculture Alternative in Aquatic Food Production. In *Drugs and Food from the Sea—Myth or Reality?* Edited by P.N. Kaul and C.J. Sindermann, pp. 385–92. Norman, Oklahoma: University of Oklahoma Press.

TANG, Y.A. 1970. Evaluation of Balance between Fishes and Available Fish Foods in Multispecies Fish Culture Ponds in Taiwan. *Transactions of the American Fisheries Society* 99(4): 708–18.

YASHOUV, A. 1966. Mixed Fish Culture—an Ecological Approach to Increase Pond Productivity. *Proceedings of the World Symposium on Warmwater Pond Fish Culture.* FAO Fisheries Reports 44(4): 258–71.

YASHOUV, A. 1969. Mixed Fish Culture in Ponds and the Role of Tilapia in it. *Bamidgeh* 21(3): 75–82.

Papers on North American polycultures:

BUCK, BAUR, and ROSE 1978. See under Chapter III–2, in Appendix.

HUNER 1976. See under Chapter II–10, in Appendix.

KILGEN, R.A. 1974. Food Habits of the White Amur, Largemouth Bass, Bluegill and Redear Sunfish Receiving Supplemental Feed. *Proceedings of the Southeastern Association of Game and Fish Commissioners* 27: 620–24.

PERRY, W.G., JR., AND AVAULT, J.W. 1976. Polyculture Studies with Channel Catfish and Buffalo. *Proceedings of the Annual Conference of South-eastern Game and Fish Commissioners:* 91–101.

PRETTO-MALCA, R. 1976. Polyculture Systems with Channel Catfish as the Principal Species. Ph.D. thesis, Auburn University, Auburn Alabama. 202 pp.

PRETTO, R., and SMITHERMAN, R.O. 1976. Polyculture Systems with Channel Catfish as the Principal Species. Proceedings of the National Fish Culture Workshop, Fish Culture Section, American Fisheries Society, January 13–15, 1976, Springfield, Missouri.

SWINGLE, H.W. 1966. Biological Means of Increasing Productivity in Ponds. *FAO World Symposium on Warmwater Pond Fish Culture.* FAO Fisheries Report 44(4): 243–57.

CHAPTER IV–2

The integration of aquaculture with agriculture and terrestrial ecosystems is characteristic of much of Oriental aquaculture, and the reader should study that, including references given for Chapters I–1, I–2, and IV–1. References dealing with specific aspects of this integration follow:

Livestock as a source of fertilizer:

BUCK, BAUR, and ROSE 1978. See under Chapter III–2, in Appendix.

SCHROEDER 1974. See under Chapter III–2, in Appendix.

SCHROEDER, G.L. 1978. Agricultural Wastes in Fish

Farming. *Commercial Fish Farmer* 4(6) : 33–34.

STICKNEY and HESBY 1978. See under Chapter III–2, in Appendix.

WOYNAROVICH 1967. See under Chapter III–2, in Appendix.

Nitrogen fixation by aquatic plants:

GALSTON, A.W. 1975. The Water Fern — Rice Connection. *Natural History Magazine* 84(10) : 10–11.

NATIONAL ACADEMY OF SCIENCES. 1976. *Making Aquatic Weeds Useful: Some Perspectives for Developing Countries.* Washington, D.C.: National Academy of Sciences. 175 pp.

OLSEN 1970. See under Chapter III–2, in Appendix.

Irrigation with aquaculture water:

ERVIN, S. 1977. Fertile Fish Pond Water Irrigation Trials. *The Journal of the New Alchemists* 4 : 59–60.

McLARNEY, W.O. 1974. New Alchemy Agricultural Research Report No. 2. Irrigation of Garden Vegetables with Fertile Fish Pond Water. *The Journal of the New Alchemists* 2 : 73–76.

McLARNEY, W.O. 1976. Further Experiments in the Irrigation of Garden Vegetables with Fertile Fish Pond Water. *The Journal of the New Alchemists* 3 : 53.

Removal of excess nutrients by aquatic plants:

EDWARDS 1980. See under Chapter III–5, in Appendix.

NATIONAL ACADEMY OF SCIENCES 1976. See under "Nitrogen fixation by aquatic plants," above.

Aquatic plants as human food (including rice-fish culture):

BHANTUMNAVIN, K., and McGARRY, M.G. 1971. *Wolffia arrhiza* as Possible Source of Inexpensive Protein. *Nature* 232(5311) : 495.

BARDACH, RYTHER, and McLARNEY 1972. See under Chapter I–1, in Appendix.

BOYD, C. 1968. Freshwater Plants: A Potential Source of Protein. *Economic Botany* 22 : 359–68.

CLAUSEN, P.W. 1919. A Possible New Source of Food Supply. *Scientific Monthly*, August 1919.

COCHE, A.G. 1967. Fish Culture in Rice Fields: A World-wide Synthesis. *Hydrobiologia* 30(1) : 1–44.

DORE, W.G. 1969. *Wild-rice.* Canada Department of Agriculture Publication 1393. 84 pp.

EDWARDS 1980. See under Chapter III–5, in Appendix.

FINFROCK, D.C., and MILLER, M.D. 1959. *Wild Rice.* Leaflet No. 116, Agricultural Extension Service, University of California, Davis, California.

HERKLOTS, G.A.C. 1979. *Vegetables in South-East Asia.* New York: Hafner Press. 525 pp.

HODGE, W.H. 1956. Chinese Water Chestnut or Matai — a Paddy Crop of China. *Economic Botany* 10(1) : 4–65.

HUET 1970. See under Chapter I–1, in Appendix.

INTERNATIONAL RICE RESEARCH INSTITUTE. 1975. *International Rice Research Institute Reporter* 3. (Special issue dealing exclusively with floating rice.)

LOGSDON 1977. See under Chapter III–5, in Appendix.

LOGSDON 1978. See under Chapter I–1, In Appendix.

LOYACANO, H.A., JR., and GROSVENOR, R.B. 1974. Effects of Chinese Waterchestnut in Floating Rafts on Production of Channel Catfish in Plastic Pools. *Proceedings of the Southeastern Association of Game and Fish Commissioners* 27 : 471–73.

MEYER, SNEED, and ESCHMEYER 1973. See under Chapter II–3, in Appendix.

MOYLE, J.B., and KRUEGER, P. 1969. *Wild Rice in Minnesota.* State of Minnesota Department of Conservation, Division of Game and Fish, Informational Leaflet No. 5. 6 pp.

NATIONAL ACADEMY OF SCIENCES. 1975. *Underexploited Tropical Plants with Promising Economic Value.* Washington, D.C.: National Academy of Science. 188 pp.

NATIONAL ACADEMY OF SCIENCES. 1976. See under "Nitrogen fixation by Aquatic Plants," above.

WILLIAM, A.K., and LEEPER, G.F. 1976. *Chinese Waterchestnut Culture.* Information sheet available from Horticultural Crops Laboratory, Richard B. Russell Agriculture Research Center, Athens, Georgia. 4 pp.

ZAIGER, D. 1965. *Growing Healthy Watercress.* Agricultural Extension Leaflet, no. 9. Division of Agriculture, Department of Resources and Development, Trust Territory of the Pacific Islands, Saipan, Mariana Islands. 2 pp.

Aquatic plants as feed for aquatic animals:

LITTLE, E.C.S. 1968. *Handbook of Utilization of Aquatic Plants.* Rome: FAO.

See references for Chapters II–8, II–10, III–5, and VIII–1, in Appendix.

Hydroponics:

BAUM, C.M. 1981. Gardening in Fertile Waters. *New Alchemy Quarterly* No. 5, pp. 3–8.

DEKORNE, J.B. 1975. *The Survival Greenhouse: An Eco-System Approach to Home Food Production.* El Rito, New Mexico: The Walden Foundation. 165 pp.

DOUGLAS, J.S. 1973. *The Beginner's Guide to Hydroponics.* London: Pelham Books. 156 pp.

SNEED, K. 1975. Fish Farming and Hydroponics.

Aquaculture and the Fish Farmer 2(1) : 11, 18–20.

Aquaculture in solar greenhouses:

DEKORNE 1975. See under "Hydroponics," above.

HEAD, W., and SPLANE, J. 1979. *Fish Farming in Your Solar Greenhouse.* Eugene, Oregon: Amity Foundation. 43 pp.

TODD, J.H. 1977. Tomorrow is our Permanent Address. *The Journal of the New Alchemists* 4 : 85–106.

Aquaculture in the context of an entire farm:

HO, R. 1961. Mixed Farming and Multiple Cropping in Malaya. *Proceedings of the Symposium on Land Use and Mineral Deposits in Hong Kong, Southern China and Southeast Asia.* Paper No. 11 : 88–104.

LEMARE, D.W. 1952. Pig-rearing, Fish-farming and Vegetable Growing. *Malayan Agriculture Journal* 35 : 156–66.

MARTIN, M. 1979. You Asked for It. *Commercial Fish Farmer* 5(5) : 14–17. (Rotation of fish ponds with terrestrial crops.)

TODD, J. 1979. Dreaming in My Own Backyard. *Commercial Fish Farmer* 5(4) : 23–27.

CHAPTER IV–3

General treatments of cage culture:

MCLARNEY, W.O. 1983. Fish Farm With Cages! *The Mother Earth News* 81 : 38–41 and 102–3.

MCLARNEY, W.O., and PARKIN, J. 1983. *The New Alchemy Back Yard Fish Farm Book: Growing Fish in Floating Cages.* Andover, Massachusetts: Brick House Publishing Co. 77 pp.

NEFF, G.N., and BARRETT, P.C. 1976. *Profitable Cage Culture.* Homestead, Florida: Inqua Corporation. 30 pp.

Cage culture bibliography:

KOSCH, A.G. 1978. Culture of Fish in Cages. FAO Fisheries Circular 704. FAO, Rome. 43 pp.

Other publications dealing with cage culture:

ARMBRESTER, W., JR. 1972. The Growth of Caged *Tilapia aurea* (Steindachner) in Fertile Farm Ponds. *Proceedings of the Southeastern Association of Game and Fish Commissioners* 25 : 446–51.

BARDACH, RYTHER, and MCLARNEY 1972. See under Chapter I–1, in Appendix.

BROWN 1977. See under Chapter I–1, in Appendix.

BROWN and CHESNESS 1974. See under Chapter II–6, in Appendix.

BUCK, D.H.; BAUR, R.J.; and ROSE, C.R. 1970. *An Experiment in the Wintertime Culture of Trout in Cages Floated in a Southern Illinois Pond.* Illinois Natural History Survey. 6 pp.

COLLINS, R.A. 1971. Cage Culture of Catfish in Reservoir Lakes. *Proceedings of the Southeastern Association of Game and Fish Commissioners* 24 : 489–96.

HEIDINGER 1971. See under Chapter II–3.

KELLEY, J.P., JR. 1973. An Improved Cage Design for Use in Culturing Channel Catfish. *Progressive Fish Culturist* 35(3) : 167–69.

LIGLER, W.C. 1971. Salvaging Stunted Bluegills. *Farm Pond Harvest,* Winter, 1971: inside front cover–1, 22–23.

MCLARNEY, B., and PARKIN, J. 1979. New Alchemy's Small Scale Cage Culture Shows Promising Results. *Commercial Fish Farmer* 5(2) : 28–31.

MCLARNEY and PARKIN 1980. See under Chapter II–2.

PAGAN-FONT, F. 1970. Cage Culture of the Cichlid Fish *Tilapia aurea* (Steindachner). Ph.D. thesis, Auburn University, Auburn, Alabama. 109 pp.

PATINO, A. 1976. Cultivo Experimental de Peces en Estanques. Precis (in English) by W.O. McLarney. *The Journal of the New Alchemists* 3 : 86–90.

PENZAK, T.; GALICKA, W.; MOLINSKI, M.; KUSTO, E.; and ZALEWSKI, M. 1982. The Enrichment of a Mesotrophic Lake by Carbon, Phosphorus and Nitrogen from the Cage Aquaculture of Rainbow Trout, *Salmo Gairdneri. Journal of Applied Ecology* 1982 (19) : 371–93.

CHAPTER IV–4

The following references contain basic biological and technological information for closed systems aquaculture:

SPOTTE, S.H. 1970. *Fish and Invertebrate Culture: Water Management in Closed Systems.* New York: Wiley Interscience. 145 pp.

STICKNEY 1979. See under Chapter I–1, in Appendix.

WHEATON 1977. See under Chapter I–1, in Appendix.

There is an extensive literature on high-tech closed systems. The following list makes no attempt to be comprehensive, but consists merely of selected references to help you get the feel of such systems.

BUSS, K.; GRAFF, D.R.; and MILLER, E.R. 1970. Trout Culture in Vertical Units. *Progressive Fish Culturist* 32 : 187–91.

MEADE, T.L. 1974. The Technology of Closed System Culture of Salmonids. Marine Technology Report

30. Narragansett, Rhode Island: University of Rhode Island Sea Grant Publications. 30 pp.

Description of New Alchemy closed systems:

McLARNEY, W.O., and TODD, J.H. 1974. Walton Two: A Compleat Guide to Back yard Fish Farming. *The Journal of the New Alchemists* 2 : 79–117.

McLARNEY, W.O., and TODD, J.H. 1979. Aquaculture on a Reduced Scale. *Commercial Fish Farmer* 3(6) : 10, 12–17.

ZWEIG, R.D. 1977a. The Saga of the Solar-Algae Ponds. *The Journal of the New Alchemists* 4 : 63–68.

ZWEIG, R.D. 1977b. Three experiments with Semi-Enclosed Fish Culture Systems. *The Journal of the New Alchemists* 4 : 6–76.

ZWEIG, R.D. 1979a. Investigations of Semi-Enclosed Aquatic Ecosystems. *The Journal of the New Alchemists* 5 : 93–104.

ZWEIG, R.D. 1979b. The Birth and Maturity of an Aquatic Ecosystem. The *Journal of the New Alchemists* 5 : 105–9.

ZWEIG, R.; WOLFE, J.; and ENGSTROM, D. 1980. Solar Aquaculture. *The Journal of The New Alchemists* 6 : 93–107.

References on other small scale closed systems:

HEAD and SPLANE 1979. See under Chapter IV–2, in Appendix.

LOGSDON 1978. See under Chapter I–1, in Appendix.

SERFLING, S., and MENDOLA, D. 1978. Aquaculture Using Controlled Environment Ecosystems. *Commercial Fish Farmer* 4(6) : 15–18.

WELSH, G., ed. 1977. *Essays on Food and Energy.* Catonsville, Maryland: Foundation for Self-sufficiency. 184 pp.

WELSH, G., and NEILS, K. 1978. Fish Farming on the Home Front. *Commercial Fish Farmer* 4(6) : 19–22.

CHAPTER V–1

Look for references on breeding of particular species in the sections of this appendix dealing with Part II. The basic source for information on reproduction of fishes is:

BREDER, C.M., and ROSEN, D.E. 1966. *Modes of Reproduction in Fishes.* Garden City, N.Y.: Natural History Press. 941 pp.

Information on induced spawning by pituitary injection can be found in publications listed in:

ATZ, J.W., and PICKFORD, G.E. 1964. *The Pituitary Gland and its Relation to the Reproduction of Fishes in Nature and in Captivity: An Annotated*

Bibliography for the years 1956–1963. FAO Fisheries Biology Technical Paper No. 37.

See also:

CLEMENS, H.P., and SNEED, K.E. 1962. Bioassay and Use of Pituitary Materials to Spawn Warm-water Fishes. U.S. Fish and Wildlife Service Research Report 61. 30 pp.

WUNDER, W. 1972. Hypophysization in Pisciculture. Fisheries and Marine Service Translation Series 3174. 8 pp.

CHAPTER V–2

For references on sunfish, buffalofish and tilapia hybrids see under Chapters II–2, II–4, and II–8, in Appendix respectively.

Three key publications on selective breeding:

HINES 1967. See under Chapter II–12, in Appendix.

MOAV, R., and WOHLFARTH, G.W. 1966. Genetic Improvement of Yield in Carp. *FAO World Symposium on Warmwater Pond Fish Culture.* FAO Fisheries Reports 44(4).

NOSHO, T.Y., and HERSHBERGER, W.K., eds. 1976. Salmonid Genetics: Status and Role in Mariculture. University of Washington Sea Grant Report WSG WO 76-2. University of Washington, Seattle. 42 pp.

CHAPTER VI–1

BENNETT, G.W. 1970. *Management of Lakes and Ponds.* New York: Van Nostrand Reinhold Co. 283 pp.

BOUSSU, M.F. 1967. New "Live Car" Improves Catfish Harvesting and Handling. *Commercial Fisheries Review*, December, 1967: 33–35.

BROWN 1969. See under Chapter I–1, in Appendix.

COON, K.L.; LARSEN, A.; and ELLIS, J.E. 1968. Mechanized Haul Seine for Use in Farm Ponds. *Fishery Industrial Research* 4(2) : 91–108.

FAO. 1977. *Fishing with Electricity.* London: Fishing News (Books). 304 pp.

GREENLAND, D.C. 1974. Recent Developments in Harvesting, Grading, Loading and Hauling Pond Raised Catfish. *Transactions of the American Society of Agricultural Engineers* 17(1) : 59–62.

GREENLAND, D.C., and GILL, R.L. 1974. Trapping Pond-Raised Channel Catfish. *Progressive Fish Culturist* 36(2) : 72–76.

HUET 1970. See under Chapter I–1, in Appendix.

LAGLER 1956. See under Chapter I–2, in Appendix.

McClane, A.J. 1965. *McClane's Standard Fishing Encyclopedia and International Angling Guide.* New York: Holt, Rinehart and Winston. 1057 pp. (The literature on hook and line fishing is voluminous. This is the single most complete and authoritative source.)

Pachner, L.C. 1973. Tons of Carp—Feed 'em, Trap 'em, Eat 'em. *Farm Pond Harvest,* Winter, 1973 : 4–6.

Posey, L., and Schafer, H. 1964. Evaluation of Slat Traps as Commercial Fishing Gear in Louisiana. *Proceedings of the 18th Annual Conference of the Southeastern Association of Game and Fish Commissioners:* 517–22.

Stramel, G. 1970. Trapping Catfish Can be Successful. *The American Fish Farmer* 1(3) : 17–18.

Starrett, W.C., and Barnickol, P.G. 1965. Efficiency and Selectivity of Commercial Fishing Devices Used on the Mississippi River. *Bulletin of the Illinois Natural History Survey* 26(4) : 325–66.

Tarrant, R.W. 1974. Recapturing Channel Catfish in Cages from which They were Released. *Proceedings of the 1974 Fish Farming Conference and Annual Convention; Catfish Farmers of Texas.* Texas Agricultural Extension Service, Texas A&M University: 33–39.

CHAPTER VI–2

See references in Appendix I.

CHAPTER VI–3

Bell, G.R. 1964. A Guide to the Properties, Characteristics and Uses of Some General Anesthetics for Fish. *Fisheries Research Board of Canada Bulletin* 148. 4 pp.

Boussu 1967. See under Chapter VI–1, in Appendix.

Greenland 1974. See under Chapter IV–1, in Appendix.

Hudson, S. 1977. From the Pond Bank: Special Precautions Needed in Hauling Fish. *The Commercial Fish Farmer* 3(5) : 31–33.

Johnson, S.A. 1979. Transport of Live Fish. *Aquaculture Magazine* 5(6) : 20–24.

McCraren, J.P., and Millard, J.L. In press. *Transportation of Warmwater Fishes.* San Marcos, Texas: Fish Cultural Development Center, U.S. Fish and Wildlife Service.

McFarland, W.N. 1960. The Use of Anesthetics for the Handling and Transport of Fishes, *California Fish and Game* 46(4) : 407–32.

Randall, D.J., and W.S. Hoar. 1971. Special Techniques. In *Fish Physiology.* Vol. VI. Edited by W.S. Hoar and D.J. Randall, pp. 511–28. New York: Academic Press.

Schaperclaus 1933. See under Chapter I–1, in Appendix.

CHAPTER VI–4

General:

Bardach, Ryther, and McLarney 1972. See under Chapter I–1, in Appendix.

Brown 1977. See under Chapter I–1, in Appendix.

Fisheries Branch Library, FAO. 1977. *A Selected Bibliography on the Economic Aspects of Aquaculture, 1969 to 1977.* Rome: FAO. 14 pp.

Hudson, S. 1979. From the Pond Bank: Fish Farmers Advised to Set Own Prices. *The Commercial Fish Farmer* 5(2) : 32–34.

National Academy of Sciences 1978. See under Chapter I–1, in Appendix.

Tal, S., and Hepher, B. 1966. Economic Aspects of Fish Feeding in the Near East. *FAO World Symposium on Warmwater Pond Fish Culture.* FAO Fisheries Reports 44(4).

Todd, J.; Nugent, C.; McLarney, B. 1978. Caught in a Paradigm Shift: Economics at New Alchemy. *New Alchemy Newsletter,* Fall-Winter 1978: 2–7.

Vondruska, J. 1976. Aquacultural Economics Bibliography. NOAA Technical Report NMFS SSRF–703 Washington, D.C.: National Marine Fisheries Service. 123 pp.

Channel catfish:

1976. Catfish Farming: How to Make it a Profitable Operation. *Commercial Fish Farmer* 2(3) : 9–10, 12–15.

Brown 1977. See under Chapter I–1, in Appendix.

Lee 1973. See under Chapter II–3, in Appendix.

Mack 1971. See under Chapter II–3, in Appendix.

Trout:

1972. Bob Erkins Talks About Trout Marketing. *Fish Farming Industries* 3(2, 3, and 5) : 20–24, 31–32, and 34.

Araji, A.A. 1972. An Economic Analysis of the Idaho Rainbow Trout Industry. University of Idaho, College of Forestry, Wildlife and Range Science, Department of Agricultural Economics, AE Series No. 118. 9 pp.

MacCrimmon, Stewart, and Brett 1974. See under Chapter I–1, in Appendix.

Scott, C.A., and Fessler, F.R. 1970. There's Profit

in Trout Production. *Fish Farming Industries* 1(3) : 18, 21–22.

Crayfish:

AVAULT 1975. See under Chapter II–10, in Appendix.

Fresh water shrimp:

SHANG and FUJIMURA 1977. See under Chapter II–11, in Appendix.

Fee fishing ponds:

GRIZZELL 1972. See under Chapter IX–1, in Appendix.

Bait minnows:

HUDSON 1974. See under Chapter IX–2, in Appendix.

Prices of fishery products:

NATIONAL MARINE FISHERIES SERVICE. 1978. *Fishery Statistics of the United States, 1975*. Washington: U.S. Department of Commerce. 418 pp. (Revised annually.)

CHAPTER VII–1

1979. Syphoning Spillways for Longer Pond Life. *Agricultural Research*, USDA, July 1979 : 3–5.

AVAULT, J.W., JR. 1978. You Can Cut Costs on Levee Construction. *Commercial Fish Farmer* 4(5) : 42.

BARD, DE KIMPE, LAZARD, LeMASSON, and LESSENT 1976. See Chapter II–8, in Appendix. (Low cost, labor-intensive methods of pond construction.)

BARDACH, RYTHER, and McLARNEY 1972. See under Chapter I–1, in Appendix

CHAKROFF, M. 1976. See under Chapter I–1, in Appendix. (Low, cost, labor-intensive methods of pond construction.)

EDMINSTER 1947. See under Chapter II–1, in Appendix.

FOSTER, T.H., and WALDROP, J.E. 1972. *Cost-size Relationships in the Production of Pond-raised Catfish for Food*. Mississippi State University Agricultural and Forestry Experiment Station Bulletin 792. 69 pp.

HUET 1970. See under Chapter I–1, in Appendix.

LEE 1973. See under Chapter II–3, in Appendix.

LOGSDON 1978. See under Chapter I–1, in Appendix.

MACK 1971. See under Chapter II–3, in Appendix.

McLARNEY, W.O., and HUNTER, R.J. 1976. A New Low-Cost Method of Sealing Fish Pond Bottoms. *The Journal of the New Alchemists* 3 : 85.

MARGOLIN, M. 1975. *The Earth Manual*. Boston: Houghton Mifflin. 208 pp.

MITCHELL and USRY 1967. See under Chapter II–3, in Appendix.

RENFRO, G., JR. 1959. *Sealing Leaking Ponds and Reservoirs*. U.S. Soil Conservation Service, SCS-TP-150. 6 pp.

SCHAPERCLAUS 1933. See under Chapter I–1, in Appendix.

SOIL CONSERVATION SERVICE. 1971. *Ponds for Water Supply and Recreation*. Agriculture Handbook 387, U.S. Soil Conservation Service, Washington, D.C. 55 pp.

STICKNEY 1977. See under Chapter I–1, in Appendix.

WHEATON 1977. See under Chapter I–1, in Appendix.

CHAPTER VII–2

STICKNEY 1979. See under Chapter I–1, in Appendix.

WHEATON 1977. See under Chapter I–1, in Appendix.

CHAPTER VIII–1

The literature on aquatic weed control is enormous. Here I have listed only general reviews of the subject or publications dealing with specific weeds or control methods, plus a few books which may be helpful in identifying aquatic plants.

Books and manuals on aquatic plants:

FASSETT, N.C. 1960. *A Manual of Aquatic Plants*. Madison, Wisconsin: University of Wisconsin Press, 405 pp.

JACKSON, D.F., ed. 1964. *Algae and Man*. New York: Plenum Press. 434 pp.

KLUSSMAN, W.G., and LOWMAN, F.G. 1975. *Common Aquatic Plants*. College Station, Texas: Texas Agricultural Extension Service. 15 pp.

SMITH, G.M. 1950. *Fresh-water Algae of the United States*. New York: McGraw-Hill. 719 pp.

STODOLA, J. 1967. *Encyclopedia of Water Plants*. Neptune City, New Jersey: TFH Publications. 368 pp.

WELDON, L.W.; BLACKBURN, R.D.; and HARRISON, D.S. 1969. Common Aquatic Weeds. *U.S. Department of Agriculture Handbook 352*. 43 pp.

A variety of papers on aquatic weed control may be found in the following:

Aquatic Botany 3 (1977). (Special issue devoted to biological control of aquatic weeds.)

Hyacinth Control Journal.

Publications devoted to aquatic weed control in general:

BLACKBURN, R.D. 1968. Weed Control in Fish Ponds in the United States. *Proceedings of the World Symposium on Warmwater Pond Fish Culture*. FAO Fisheries Reports 44(4) : 1–17.

HUET 1970. See under Chapter I–1, in Appendix.

LAWRENCE, K.J.M. 1968. Aquatic Weed Control in Fish Ponds. *Proceedings of the World Symposium on Warmwater Pond Fish Culture.* FAO Fisheries Reports 44(4) : 76–91.

LOGSDON 1978. See under Chapter I–1, in Appendix.

NATIONAL ACADEMY OF SCIENCES 1976. See under Chapter IV–2, in Appendix.

Papers dealing with specific weeds and/or control methods:

APPLIED BIOCHEMISTS 1976. *How to Identify and Control Water Weeds and Algae.* Mequon, Wisconsin: Applied Biochemists. 64 pp.

BOYD, C.E.; PRATHER, E.E.; and PARKS, R.W.; 1975. Sudden Mortality of a Massive Phytoplankton Bloom. *Weed Science* 23 : 61–67.

DEAN, J.L. 1969. Biology of the Crayfish *Orconectes causeyi* and its Use for Control of Aquatic Weeds in Trout Lakes. *Technical Papers of the Bureau of Sport Fisheries and Wildlife* 24. 15 pp.

GLEASON, F.K. 1977. Toxic Blue-green Algae: It can be More than a Nuisance. *The Journal of Freshwater* 1(3) : 12–15.

HICKLING 1967. See under Chapter II–8, in Appendix.

KILGEN and SMITHERMAN 1971. See under Chapter II–8, in Appendix.

OPUSZYNSKI, K. 1972. Use of Phytophagous Fish to Control Aquatic Plants. *Aquaculture* 1(1) : 61–74.

SEAQUIST, B. 1977. Aquatic Farms Clean up Waste Water. *Popular Science* 211(3) : 88–89.

SILLS, J.B. 1970. A Review of Herbivorous Fish for Weed Control. *The Progressive Fish Culturist* 32(3) : 158–61.

TUCKER, C.S., and BOYD, C.E. 1978. Consequence of Periodic Applications of Copper Sulfate and Simazine for Phytoplankton Control in Catfish Ponds. *Transactions of the American Fisheries Society* 107(2) : 316–20.

VAN DEUSEN, R.D. 1973. Ornamental Allies: Swans Control Plants. *American Fish Farmer* 4(2) : 4–6.

YOUNT, J.L., and CROSSMAN, JR., R.A. 1970. Eutrophication Control by Plant Harvesting. *Journal of Water Pollution Control,* May 1970, Part 2 : R173–R183.

CHAPTER VIII–2

The chapter emphasizes prevention of disease. For that very reason I have included a fairly ample list of references on disease control. The following list includes only general works; in them one can find fur-

ther references dealing with specific diseases or parasites. Be aware that "fish doctoring" is as prone to fads as human medicine; both old and more recent references are included for this reason. Works best suited for the lay reader are marked with an asterisk.

*AMLACHER, E. 1970. *Textbook of Fish Diseases.* Translated by D.A. Conroy and R.L. Herman. Neptune City, New Jersey: TFH Publications. 302 pp.

AVAULT, J.W., JR. 1981. "Prevention" of Fish Diseases: Some Basics Reviewed, Part One. *Aquaculture Magazine,* July-August, 1981 : 40–41.

AVAULT, J.W., JR. 1981. "Prevention" of Fish Diseases: Some Basics Reviewed, Part Two. *Aquaculture Magazine,* September-October, 1981 : 40.

BUSCH, R.A. 1978. Protective Vaccines for Mass Immunization of Trout. *Salmonid* 1(6): 10–14, 32.

*DAVIS 1953. See under Chapter II–6, in Appendix.

Fish Health News. Supervisor of Documents, U.S. Government Printing Office, Washington, D.C. 20402. (A periodical).

HALVER 1970. See under Chapter III–1, in Appendix.

*HUET 1970. See under chapter I–1, in Appendix.

JOHNSON, S.K. 1977. *Crawfish and Freshwater Shrimp Diseases.* College Station, Texas: Center for Marine Resources, Texas A & M University. 19 pp.

*KLONTZ, G.W. 1973. *Syllabus of Fish Health Management.* College Station, Texas: Center for Marine Resources, Texas A & M University. 165 pp.

MAWDESLEY-THOMAS, L. E. 1972. Diseases of Fish. *Proceedings of a Symposium organized jointly by the Fisheries Society of the British Isles and the Zoological Society of London.* London: Academic Press. 380 pp.

*McGREGOR, E.A. 1963. Publications on Fish Parasites and Diseases. 330 B.C.–A.D. 1923 Special Scientific Reports—Fisheries, 474. Washington, D.C.: U.S. Fish and Wildlife Service. 84 pp.

*MEYER, F.P., and HOFFMAN, G.L. 1976. Parasites and Diseases of Warmwater Fishes. U.S. Fish and Wildlife Service Resource Publication 127 : 20 pp.

REICHENBACH-KLINKE, H.H. 1973. *Fish Pathology.* Neptune City, New Jersey: TFH Publications. 512 pp.

REICHENBACH-KLINKE, H., and ELKAN, E. 1972. *The Principal Diseases of Lower Vertebrates. Book 1. Diseases of Fishes.* Neptune City, New Jersey: TFH Publications. 205 pp.

RIBELIN, W.E., and MIGAKI, G., eds. 1975. *The Pathology of Fishes.* Madison, Wisconsin: University of Wisconsin Press. 1004 pp.

*ROBERTS and SHEPHERD 1974. See under Chapter II–6, in Appendix.

SARIG, S. 1971. *Diseases of Fishes, Book III. Diseases of Warmwater Fishes.* Neptune City, New Jersey: TFH Publications. 127 pp.

SCHAPERCLAUS, W. 1954. *Fischkrankheiten.* Berlin: Akademie-Verlag. 708 pp. (In German.)

SNIESZKO, S.F., ed. 1970. *A Symposium on Diseases of Fishes and Shellfishes.* American Fisheries Society Special Publication No. 5. Washington, D.C.: American Fisheries Society. 526 pp.

*STICKNEY 1977. See under Chapter I–5, in Appendix.

*U.S. BUREAU OF SPORT FISHERIES AND WILDLIFE. *Fish Disease Leaflets.* (A series of leaflets organized by disease organism.)

U.S. FISH AND WILDLIFE SERVICE. 1982. Current List of Fish Disease Leaflets. National Fisheries Center, Kearneysville, West Virginia.

VAN DUIJN C., JR., 1967. *Diseases of Fishes.* London: Illiffe Books. 309 pp.

WEDEMEYER, G.A.; MEYER, F.P.; and SMITH, L. 1976. *Diseases of Fishes Book V: Environmental Stress and Fish Diseases.* Neptune City, New Jersey: TFH Publications. 192 pp.

CHAPTER VIII–3

Most of the material in this chapter does not seem to require further referencing. A few references on the use of fish poisons and mosquitofish follow.

Fish poisons:

HOOPER, A.D., and CRANCE, J. 1960. Use of Rotenone in Restoring Balance to Overcrowded Fish Populations in Alabama Lakes. *Transactions of the American Fisheries Society* 69(4) : 351–57.

MEYER, F.A. 1966. Chemical Control of Undesirable Fishes. In *Inland Fisheries Management,* edited by A. Calhoun. Sacramento, California: California Department of Fish and Game.

Mosquitofish:

DAVEY, R.B.; MEISCH, M.V.; GRAY, D.L.; MARTIN, J.M.; SNEED, K.E.; and WILLIAMS, F.J. 1974. Various Fish Species and Biological Controls for the Dark Rice Field Mosquito in Arkansas Rice Fields. *Environmental Entomology* 3(5) : 823–26.

NAVY ENVIRONMENTAL AND PREVENTIVE MEDICINE UNIT No. 2. 1972. *Guide to the Use of the Mosquito Fish, Gambusia affinis, for Mosquito Control.* Norfolk, Virginia: Navy Environmental and Preventive Medicine Unit No. 2.

CHAPTER VIII–4

The following is a very incomplete listing, with a few

of the key publications on general pollution issues, plus some references of particular interest to aquaculturists.

General:

ODUM, W.E. 1973. The Potential of Pollutants to Adversely Affect Aquaculture. *University of Miami, Gulf and Caribbean Fisheries Institute Proceedings* 25 : 163–74.

SCIENTISTS' INSTITUTE FOR PUBLIC INFORMATION. *Water Pollution, A Scientists' Institute for Public Information Handbook.* New York: Scientists' Institute for Public Information.

Acid rain:

EHRLICH, A., and EHRLICH, P. 1983. Poison from Above. *Mother Earth News,* January/February, 150–51.

JONES, B. 1979. Acid Rain: Trout Fishing's Greatest Threat? *Trout* 20(4): 10–19.

MACROBERT, A. 1977. Look What They've Done to the Rain. *Mother Jones,* December, 65–67.

Heavy metals:

HUTCHINSON, T.H., ed. 1975. *Proceedings of the Symposium of the International Conference on Heavy Metals in the Environment.* Three volumes. Toronto: University of Toronto Press.

Nuclear:

GOFMAN, J.W. and TAMPLIN, A.R. 1979. *Poisoned Power: The Case Against Nuclear Power Plants Before and After Three Mile Island.* Emmaus, Pennsylvania: Rodale Press. 353 pp.

GYORGY, A. 1979. *No Nukes: Everyone's Guide to Nuclear Power.* Boston: South End Press. 478 pp.

Pesticides:

1974. Fish Kill: "Spinning Death" Believed Caused by Agri Chemicals, Takes Heavy Toll in Arkansas, Louisiana Minnow Crops. *Commercial Fish Farmer* 1(1) : 25–28.

MCLARNEY, W.O. 1970. Pesticides and Aquaculture. *American Fish Farmer* 1(10) : 6–7, 22–23.

PCB's:

ORGANIZATION FOR ECONOMIC COOPERATION AND DEVELOPMENT, ENVIRONMENTAL DIRECTORATE. 1973. *Polychlorinated Biphenyls, Environmental Aspect.* Paris: OECD. 44 pp.

CHAPTER VIII–5

Information on legal matters is better obtained by consulting appropriate authorities than by reading. Other good sources are the various trade associations (see Appendix IV–4) and publications (*Aquaculture*

Magazine and *U.S. Trout News*). Overviews of the legal situation, especially as regards commercial aquaculture can be found in the following two references:

MacCRIMMON, STEWART, and BRETT 1974. See under Chapter I–1, in Appendix.

NATIONAL ACADEMY OF SCIENCES 1978. See under Chapter I–1, in Appendix.

CHAPTER IX–1

GRIZZELL, R.A., JR. 1972. All You Want to Know about the Fee Fishing Business "but Were Afraid to Ask." *Fish Farming Industries* 3(3) : 14–17.

ORAVETZ, C.A. Pay Lake Operators Directory. *Catfish Farmer and World Aquaculture News* 6(2) : 24–27.

WATT, C.E. 1974. Location and Design of a Commercial Fish-out Operation. *Proceedings of the 1974 Fish Farming Conference and Annual Convention of the Catfish Farmers of Texas.* College Station, Texas: Texas Agricultural Extension Service. 77–79.

CHAPTER IX–2

ALTMAN, R.W., and IRWIN, W.H. No date. *Minnow Farming in the Southwest.* Oklahoma City, Oklahoma: Oklahoma Department of Wildlife Conservation. 35 pp.

COYKENDALL, R.L. 1977. Aquacultural studies of mosquitofish, *Gambusia affinis*, in Earthen Impoundments: Stocking Rate Optimization for Yield, Protection of Overwintering Fish Stocks. *Proceedings of the California Mosquito and Vector Control Assocation* 45 : 80–82.

DOBIE, J.R.; MEEHEAN, O.L.; SNIESZKO, S.F.; and WASHBURN, G.N. 1956. *Raising Bait Fishes.* U.S. Fish and Wildlife Circular 35. 124 pp.

FORNEY 1968. See under Chapter II–10, in Appendix.

GIUDICE, J.J.; GRAY, D.L.; and MARTIN, J.M. No Date. *Manual for Baitfish Culture in the South.* Washington D.C.: U.S. Fish and Wildlife Service and University of Arkansas Cooperative Extension Service. 50 pp.

HUDSON, S. 1974. Minnow Farming, an American Enterprise, Then–Now–and the Future. *Catfish Farmer and World Aquaculture News* 6(1) : 31–32, 37–38.

HUNER, J.V. 1979. Mosquitofish. *Farm Pond Harvest* 13(3) : 11–12.

JOHNSON, S.K., and DAVIS, J.T. 1978. Raising Minnows. Texas Agricultural Extension Service Publication MP–783. College Station, Texas: Texas A & M University. 15 pp.

MARTIN 1977. See under Chapter III–5.

NEWTON, S.H., and MEISCH, M.V. 1977. Mosquitofish Production Systems in Arkansas. *Arkansas Farm Research Journal* 23(4) : 4.

NEWTON et al. 1977. See under Chapter III–5, in Appendix.

PRATHER, E.E.; FIELDING, J.R.; JOHNSON, M.C.; and SWINGLE, H.S. 1953. *Production of Bait Minnows in the Southwest.* Auburn, Alabama: Alabama Agricultural Experiment Station, Auburn University. 71 pp.

STICKNEY 1979. See under Chapter I–1, in Appendix.

TEXAS A & M UNIVERSITY 1968. *Proceedings of the Commercial Bait Fish Conference.* College Station, Texas: Texas A & M University.

WHITE, J.T. 1970. Minnows by the Million. *American Fish Farmer* 1(9) : 8–11, 27.

CHAPTER IX–3

Bibliographic citations for culture methods for growing goldfish, crayfish and frogs as laboratory animals may be found, respectively, in the portions of this Appendix covering Chapters IX–4, II–10, and II–9.

CHAPTER IX–4

Publications dealing with commercial culture of ornamental fishes:

1972. Ozark's Goldfish Breed in Artificial Nests. *American Fish Farmer* 3(2): 5 pp.

BOOZER, D. 1973. Tropical Fish Farming. *American Fish Farmer* 4(8) : 4–5.

CLEMENS, H.P. 1965. *A Goldfish Bibliography.* Aquarium Publishing Co. (Sponsored by Ozark Fisheries, Stoutland, Missouri). 41 pp.

MARTIN, M. 1983. You Asked For It: Goldfish Farming Parts I and II. *Aquaculture Magazine* 9(3) : 38–40 and 9(4) : 38–40.

MERYMAN, C.D. 1978. Farm Production of Introduced Ornamental Fishes in Florida. *Symposium on Culture of Exotic Species.* Fish Culture Section, American Fisheries Society.

MEYER, SNEED, and ESCHMEYER 1973. See under Chapter III–3, in Appendix.

NEILS, K.E. 1979. The Technical and Commercial Aspects of the Goldfish Industry. *Aquaculture Magazine* 6(1) : 22–26.

STICKNEY 1979. See under Chapter I–1, in Appendix.

Periodicals dealing with ornamental fishes:

American Currents, (official publication of the North American Native Fishes Association, dealing with native ornamental fishes), Marine Education Center, 1650 East Beach Blvd., Biloxi, Mississippi 39530.

Freshwater and Marine Aquarium, P.O. Box 187, Sierra Madre, California 91024.

Tropical Fish Hobbyist, TFH Publications, Neptune City, New Jersey.

Books on ornamental fishes:

AXELROD, H.R., and VORDERWINKLER, W. *Encyclopedia of Tropical Fishes*. Neptune City, New Jersey: TFH Publications. 631 pp. (Loose-leaf form, updated continually.)

BAUGH, T. 1983. *A Net Full of Natives: Some North American Fishes*. Sierra Madre, California: RCM Publications. 84 pp.

INNES, W.T. 1956. *Exotic Aquarium Fishes*. Neptune City, New Jersey: TFH Publications. 446 pp.

CHAPTER IX–5

BARDACH, RYTHER, and McLARNEY 1972. See under Chapter I–1, in Appendix.

BROWN 1977. See under Chapter II–3, in Appendix.

DAVIS 1953. See under Chapter II–6, in Appendix.

HUET 1970. See under Chapter I–1, in Appendix.

LEITRITZ and LEWIS 1976. See under Chapter II–6, in Appendix.

MEADE 1971. See under Chapter II–6, in Appendix.

SEDGWICK 1976. See under Chapter II–6, in Appendix.

WESTERS and PRATT 1977. See under Chapter II–6, in Appendix.

WILLOUGHBY, H. 1971. A New Look at Hatcheries. In *Bureau of Sport Fisheries and Wildlife, Sport Fishing U.S.A.*: 359–70.

CHAPTER X–1

Aquaculture and limnology texts covering all physical aspects of the aquatic environment:

FREY 1963. See under Chapter I–2, in Appendix.

RUTTNER 1953. See under Chapter I–2, in Appendix.

STICKNEY 1979. See under Chapter I–1, in Appendix.

WETZEL 1975. See under Chapter I–2, in Appendix.

WHEATON 1977. See under Chapter I–1, in Appendix.

Temperature:

BRETT, J.R. Some Principles in the Thermal Requirements of Fishes. *Quarterly Review of Biology* (1956) 31(2) : 75–87.

BRETT, J.R. 1959. Thermal Requirements of Fish— Three Decades of Study, 1940–1970. In *Publication W60-3*, 160 : 17, edited by C.M. Tarzwell. U.S.

Public Health Service, Division of Water Supply and Pollution Control.

BRETT, J.R.; SHELBOURN, J.E.; and SHOOP, C.T. 1969. Growth Rate and Body Composition of Fingerling Sockeye Salmon, *Oncorhynchus nerka*, in Relation to Temperature and Ration Size. *Journal of the Fisheries Research Board of Canada* 15(4) : 587–605.

CARLANDER 1969. See under Chapter II–13, in Appendix.

CARLANDER 1977. See under Chapter II–13, in Appendix.

Light:

SPIELER, R.E. 1977. Diel and Seasonal Changes in Response to Stimuli: a Plague and a Promise for Mariculture. *Proceedings of the Eighth Annual Meeting of the World Mariculture Society:* 865–82.

Solar aquaculture:

DEKORNE 1975. See under Chapter IV–2, in Appendix.

HEAD and SPLANE 1979. See under Chapter IV–2, in Appendix.

McLARNEY and TODD 1974. See under Chapter IV–4, in Appendix.

TODD 1977. See under Chapter IV–2, in Appendix.

Turbidity:

BUCK, H.D. 1956. *Effects of Turbidity on Fish and Fishing*. Oklahoma Fisheries Research Laboratory Report 56. 62 pp.

CAIRNS, J., JR. 1967. Suspended Solids Standards for the Production of Aquatic Organisms. *Purdue University Engineering Bulletin* 129 : 16–27.

WALLEN, I.E. 1951. The Direct Effect of Turbidity on Fishes. *Bulletin of the Oklahoma Agricultural and Mechanical College* 48 : 1–27.

CHAPTER X–2

Fish physiology as related to water chemistry:

BROWN, M.E. ed. 1957. *The Physiology of Fishes* Vol. II. New York: Academic Press. 447 pp.

HOAR, W.S., and RANDALL, D.J., eds. 1971. *Fish Physiology*. Vol 6. New York: Academic Press. 559 pp.

General references on water chemistry, including limnology texts:

FREY 1963. See under Chapter I–2, in Appendix.

RUTTNER 1953. See under Chapter I–2, in Appendix.

STICKNEY 1979. See under Chapter I–1 in Appendix.

STUMM, W., and MORGAN, J. 1970. *Aquatic Chemistry*. New York: John Wiley and Sons. 583 pp.

WHEATON 1977. See under Chapter I–1, in Appendix.

Water chemistry analysis:

AMERICAN PUBLIC HEALTH ASSOCIATION. 1976. *Standard Methods for the Examination of Water and Wastewater*. Washington, D.C.: American Public Health Association. (Usually referred to simply as "Standard Methods.")

RAINWATER, G.H., and THATCHER, L.L. 1960. Methods for the Collection and Analysis of Water Samples. U.S. Geological Survey Water-Supply Paper 1454. 301 pp.

WEAST, R.C., ed. 1975–76. *CRC Handbook of Physics and Chemistry*. Cleveland Ohio: CRC press. (Not a water analysis text, but extremely handy in the lab.)

Water chemistry in closed recirculating systems:

MEADE 1974. See under Chapter IV–4, in Appendix.

SPOTTE 1970. See under Chapter IV–4, in Appendix.

Water chemistry in algal systems:

ENGSTROM, D. In preparation. (This as yet untitled work will summarize what has been learned at New Alchemy about the chemistry of solar-algae ponds and the like.)

ZWEIG 1979a. See under Chapter IV–4, in Appendix.

ZWEIG 1979b. See under Chapter IV–4, in Appendix.

Aeration:

BOYD, C.E. 1982. New Aeration Tests may Provide Better Basis for Comparison. *Aquaculture Magazine* 8(4): 28–31.

CAIN, B. 1979. At Lake George, New York: Wind-powered Bubbler Keeps Dock Ice-free. *Popular Science*, December, 1979: 15–16.

GHOSH, S.R., and MOHANTY, A.N. 1981. Observations on the Effect of Aeration on Mineralization of Organic Nitrogen in Fish Pond Soil. *Bamidgeh* 33(2): 50–56.

RAPPAPORT, R.; SARIG, S.; and MAREK, M. 1976. Results of Tests of Various Aeration Systems on the Oxygen Regime in Genosar Experimental Ponds and Growth of Fish Therein. *Bamidgeh* 28(3): 35–49.

ROMAIRE, R.P., and BOYD, C.E. 1979. Management of Dissolved Oxygen in Pond Fish Culture. *Farm Pond Harvest* 13(3): 8–10, 20–21, 25.

SEALE, J. 1978. How a Windmill Technology for Aquaculture May Evolve. *The Commercial Fish Farmer* 4(6): 11–14.

TODD, J.H. 1978. Windmills: An Appropriate Technology for Aquaculture. *The Commercial Fish Farmer* 4(6): 8–10.

The nitrogen cycle and treatment of nitrogenous wastes:

BAIRD, R.; BOTTOMLEY, J.; and TAITZ, H. 1979. Ammonia Toxicity and pH Control of Fish Toxicity of Bioassays of Treated Wastewaters. *Water Research* 13(2): 181–84.

DELWICHE, C.C. 1970. The Nitrogen Cycle. *Scientific American* 223(3): 137–46.

MEADE 1974. See under Chapter IV–4, in Appendix.

SHARMA, B., and AHLERT, R.C. 1977. Nitrification and Nitrogen Removal. *Water Research* 11: 897–925.

SPOTTE 1970. See under Chapter IV–4, in Appendix.

WHEATON 1977. See under Chapter I–1, in Appendix.

Hydrogen sulfide:

MATHIS, W.P.; BARDY, L.E.; and GILBREATH, W.J. 1962. Preliminary Report on the Use of Potassium Permanganate to Alleviate Acute Oxygen Shortage and Counteract Hydrogen Sulfide in Fish Ponds. *Proceedings of the Southeastern Association of Game and Fish Commissioners* 16: 357–59.

Fertilization and water chemistry:

WAHBY 1974. See under Chapter III–2, in Appendix.

Appendix IV–2 Where To Write

Please don't write to me or anyone else at New Alchemy unless you are certain this is the only place the answers you seek can be found. New Alchemy is truly swamped; we are only about twenty people, but as a result of a single television appearance we received about 1500 letters, and that is not an isolated instance. If you do write New Alchemy, please consider

joining as an Associate Member, Contributing Member or Patron of the Institute. (For address see p. 550.) Our ability to serve the public is largely supported by the membership program.

As for the names on this short list, these noble souls have volunteered to deal with written inquiries, but please be as considerate of them as of us. Do the best literature search you

can before taking pen in hand. Then think
about local information sources. Local offices
of the U.S. Soil Conservation Service, the Fish
and Game or Conservation Departments of the
states and provinces and some university extension services may be able to help. If all else
fails, you can write to the following:

Aquaculture Magazine,
P.O. Box 2329,
Asheville, North Carolina 28802

Fish Farming Experimental Station,
U.S. Fish and Wildlife Service,
P.O. Box 860
Stuttgart, Arkansas 72160

John Guidice, Fishery Biologist
U.S Fish and Wildlife Service
P.O. Box 4389
Jackson, Mississippi 39216

Al Lopinot,
Farm Pond Harvest,
P.O. Box AA,
Momence, Illinois 60954

National Marine Fisheries Service,
Number One Union National Plaza,
Suite 1160,
Little Rock, Arkansas 72201

The U.S. Bureau of Sport Fisheries and Wildlife "Second Report to the Fish Farmers" listed
the following additional offices which offer
advisory services:

Regional Offices (For general advice on warmwater fish farming):

Pacific Region: 730 Northeast Pacific Street,
Portland, Oregon 97208.

Denver Region: Federal Center, Building 07,
Denver, Colorado 80225.

Southwest Region: Federal Building, 500
Gold Avenue, Albuquerque, New Mexico
87103.

North Central Region: Federal Building,
Fort Snelling, Twin Cities, Minnesota 55111.

Southeast Region: 17 Executive Park Drive,
N.E., Atlanta, Georgia 30320.

Sport Fishery Research Laboratories
(For technical problems on warmwater fish
farming):

Fish Farming Experimental Station,
Post Office Box 860,
Stuttgart, Arkansas 72160.

Southeastern Fish Cultural Laboratory,
Marion, Alabama 36756.

Fishery Services Offices (For general advice on
warmwater fish farming):

Leader, Cooperative Fishery Unit—
Auburn University, Auburn, Alabama,
36830.
University of Arizona, Tucson, Arizona
87521.
Humboldt State College, Arcata, California
95521.
University of Georgia, Athens, Georgia
30001.
Louisiana State University, Baton Route,
Louisiana 70803.
University of Missouri, Columbia, Missouri
65201.
North Carolina State University, Raleigh,
North Carolina 27607.
Oklahoma State University, Stillwater,
Oklahoma 74074.
Virginia Polytechnic Institute, Blacksburg,
Virginia 24061.

Fishery Biologist—
Central States Fishery Station,
Post Office Box 18,
Princeton, Indiana 47570.

Fort Niobrara National Wildlife Refuge,
Hidden Timber Star Route,
Valentine, Nebraska 69201.

BSFW Biologist,
Soil Labratory,
Bureau of Indian Affairs,
Gallup, New Mexico 87301.

Hatchery Biologist Offices
(For disease and parasite problems):

Coleman National Fish Hatchery, Route 1,
Box 2105, Anderson, California 96007.

National Fish Hatchery, Post Office Box 158, Pisgah Forest, North Carolina 28768.

National Fish Hatchery, Springville, Utah 84663.

National Fish Hatchery, Post Office Box 252, Genoa, Wisconsin 54632.

Hatchery Biologist, Box 292, Stuttgart, Arkansas 72160.

The following offices of the National Marine Fisheries Services offer advice on fish farming:

Regional Office:

Federal Building,
144 First Avenue South, St.
Petersburg, Florida 33701.

Marketing Office:

Post Office and Courts Building,
600 West Capitol Avenue,
Little Rock, Arkansas 72201

In Canada, information may be obtained from the following federal and provincial agencies:

NEWFOUNDLAND
Department of Fisheries and Oceans,
P.O. Box 5667
St. John's, Newfoundland
A1C 5X1

MARITIME PROVINCES
Department of Fisheries and Oceans,
P.O. Box 550,
Halifax, Nova Scotia
B3J 2S7

NOVA SCOTIA
Provincial Department of Fisheries,
P.O. Box 700,
Pictou, Nova Scotia

NEW BRUNSWICK
Provincial Department of Fisheries,
Fredericton, New Brunswick

PRINCE EDWARD ISLAND
Provincial Department of Fisheries,
P.O. Box 2000,
Charlottetown, Prince Edward Island
C1A 7N8

QUEBEC
Department of Fisheries and Oceans
P.O. Box 15,500
901 Cap Diamant
Quebec, P.Q.
G1K 7X7

Provincial Ministry of Recreation, Fish and Game,
Place de la Capitale,
150 St-Cyrille Blvd.,
Quebec, P.Q.

Provincial Ministry of Agriculture,
Aquaculture Division,
200 Chemin St-Foie,
Quebec, P.Q.
G1R 4X6

ONTARIO
Provincial Ministry of Natural Resources,
Fisheries Branch,
Parliament Buildings,
Toronto, Ontario
M7A 1W3

PRAIRIE PROVINCES
Department of Fisheries and Oceans,
Freshwater Institute,
501 University Crescent,
Winnipeg, Manitoba
R3T 2N6

MANITOBA
Provincial Department of Renewable Resources and
Transportation Services,
1495 St. James Street,
Winnipeg, Manitoba
R3G 0W9

SASKATCHEWAN
Provincial Department of Tourism and Renewable
Resources,
Fisheries and Wildlife Branch,
Room 202, Provincial Office Building,
Prince Albert, Saskatchewan
S6V 1B5

ALBERTA
Provincial Alberta Department of Energy and
Natural Resources,
10363–108 Street,
Edmonton, Alberta
T5J 2L8

BRITISH COLUMBIA
Department of Fisheries and Oceans,
Pacific Biological Station,
P.O. Box 100,
Nanaimo, British Columbia
V9R 5K6

Provincial Department of Recreation and Travel Industry,
Fish and Wildlife Branch,
1019 Wharf Street,
Victoria, British Columbia
V8W 2Y9

For recent information in print on aquaculture, particularly technical publications, you can also contact the National Aquaculture Information Service (a computerized literature service) (see under Chapter I–1).

Appendix IV–3   Sources of Fish, Tools, and Supplies.

This service is largely provided by the *Aquaculture Magazine* (formerly *Commercial Fish Farmer's*) "Buyer's Guide." (See reference list for Chapter I–1 in Appendix for address.) Rather than duplicate that service, I have included here only certain categories not covered by the Buyer's Guide, plus a few outstanding suppliers I have had personal dealings with.

AIRSTONES:

Carolina Biological Supply Co.,
Burlington, North Carolina 27215

Environmental Management and Design, Inc.,
132 North 7th St.,
Ann Arbor, Michigan 48103

ANESTHETICS:

Crescent Research Chemicals, Inc.,
5301 N. 37th Pl.,
Paradise Valley, Arizona 85253

DYE FOR SHADING WATER:

Aquashade, Inc.,
Box 117,
Dobbs Ferry, New York 10522

EARTHWORMS:

Check advertisements in such magazines as *Organic Gardening and Farming* or *Outdoor Life*. Select suppliers located near you and which guarantee live delivery.

FEEDERS:

(Most types of fish feeder are listed in the *Aquaculture Magazine* Buyers' Guide. However, two feeders mentioned in the text are not included.)

Farm Pond Harvest,
P. O. Box AA,
Momence, Illinois 60954
(Floating feeder for sunfishes, for use with dry feed.)

Hedlunds of Medford,
Box 305T,
Medford, Wisconsin 54451
(The best of the U-V bug light fish feeders.)

FIBERGLASS FOR MAKING SOLAR-ALGAE PONDS:

The Kalwall Corp.,
P.O. Box 237,
Manchester, New Hampshire 03105

FILTER MEDIA:

National Filter Media Corp.,
1717 Dixwell Avenue,
Hamden, Connecticut 06514

FISH:

(The *Aquaculture Magazine* Buyers' Guide lists many sources of stock for the aquaculturist. However, an even more comprehensive listing for a particular state may be obtained by writing the address below.)

U.S. Department of the Interior,
Fish and Wildlife Service,
Division of Fish Hatcheries,
Washington, D.C. 20240

METERS AND TEST KITS FOR WATER CHEMISTRY:

Advanced Marketing, Inc.,
P.O. Box 21036,
Cleveland, Ohio 44121
(Inexpensive pH meter.)

Bausch and Lomb,
820 Linden Ave.,
Rochester, New York 14625
(Portable and lab colorimeters.)

Cole-Parmer Instrument Co.,
7425 North Oak Park Ave.,
Chicago, Illinois 60648
(All types of meters, including some low priced ones.)

Corning Scientific Instruments.
Medfield, Massachusetts 02052
(pH and D.O. meters and electrodes.)

Fisher Scientific Co.,
711 Forbes Ave.,
Pittsburgh, Pennsylvania 15219
(Carries most brands of all types of test equipment.)

Hach Chemical Corp.,
P.O. Box 389
Loveland, Colorado 80537
(Low to medium priced kits and meters.)

Hydrolab Corp.,
P.O. Box 9406,
Austin, Texas 78766
(Excellent expensive meters.)

LaMotte Chemical Products Co.,
Box 329,
Chestertown, Maryland 21620
(Colorimeters.)

Markson Science, Inc.,
577 Oak St.,
Box 767,
Del Mar, California 92014
(Inexpensive colorimeter for nitrogen.)

Orion Research,
380 Putnam Ave.,
Cambridge, Massachusetts 02139
(pH and D.O. probes and meters.)

VWR Scientific Inc.,
P.O. Box 232,
Boston, Massachusetts 02101
     or
P.O. Box 3200,
San Francisco, California 94119
(Most brands of all types of test equipment.)

NETS:
In addition to the suppliers listed in the *Aquaculture Magazine* Buyers' Guide, I have personally had good dealings with these:

Memphis Net and Twine Co.,
P.O. Box 8331
Memphis, Tennessee 38108

The Netcraft Co.,
2800 Tremainsville Road,
Toledo, Ohio 43613

OTABS:
Pemble Laboratories,
River Falls, Wisconsin 54022

SEALANTS FOR PONDS:
International Enzymes, Inc.,
1706 Industrial Rd.,
Las Vegas, Nevada 89102

SECCHI DISCS:
Forestry Suppliers,
Box 8397,
Jackson, Mississippi 39204

InterOcean Systems,
3510 Kurtz St.,
San Diego, California 92110

Kahl Scientific Instruments,
Box 1166,
El Cajon, California 92022

Lamotte Chemical Products,
Box 329
Chestertown, Maryland 21620

Plas-Labs,
917 E. Chilson St.,
Lansing, Michigan 48906

TWR Scientific, Inc.,
P.O. Box 232,
Boston, Massachusetts 02101
     or
P.O. Box 3200,
San Francisco, California 94119

Wildlife Supply,
301 Cass St.,
Saginaw, Michigan 48602

SILKWORM EGGS:
Marguerite Shimmin,
2470 Queensbury Rd.,
Pasadena, California 91105

TIES FOR CAGE CONSTRUCTION:
Panduit Corp.,
17303 South Ridgeland Ave.,
Tinley Park, Illinois 60477

VITAMIN PREMIX:
Nutritional Biochemicals,
26201 Miles Rd.,
Cleveland, Ohio

WEED HARVESTERS:
Aquamarine Corporation,
Waukesha, Wisconsin 53186

WINDMILLS AND WIND PUMPS FOR AQUACULTURE:
Aermotor,
Industrial Park—Box 1364,
Conway, Arkansas 72032
(Water pumping windmills.)

Bowjon,
2829 Burton Ave.,
Burbank, California 91504
(Water pumping windmill, also usable for aeration.)

Brace Research Institute,
Macdonald College of McGill University,
Ste. Anne de Bellevue,
Quebec, Canada H0A 1C0
(Plans for a Savonius rotor windmill, usable for water pumping or aeration.)

Dempster Industries, Inc.,
P.O. Box 848,
Beatrice, Nebraska 68310
(Water pumping windmills.)

Heller-Aller Co.,
Perry and Oakwood Streets,
Napoleon, Ohio 43545
(Water pumping windmills.)

Lake Aid Inc.,
Dept. FPH, R.R. #2,
Bismark, North Dakota 58501
(De-icing windmill, usable for aeration.)

The Mishler Windmill Co.,
Beverly, Kansas 67423
(Water pumping windmills.)

New Alchemy Institute,
237 Hatchville Road,
East Falmouth, Massachusetts 02536
(Plans for an inexpensive water pumping windmill.)

Sparco,
c/o Enertech,
P.O. Box 420,
Norwich, Vermont 05055
(Water-pumping windmills.)

Wadler Manufacturing Co., Inc.,
Rt. 2, Box 76,
Galena, Kansas 66739
(Savonius rotor for de-icing or aeration.)

Appendix IV–4   National and State Aquaculture Organizations and Associations

This list by no means includes all the groups devoted to some aspect of aquaculture, but has been narrowed down to those few most likely to be of interest to the practical fish farmer.

UNITED STATES

Catfish Farmers of America,
P.O. Box 1609,
Jackson, Mississippi 39205

Fish Culture Section,
   American Fisheries Society,
Box 127,
Linesville, Pennsylvania 16424

United States Trout Farmers
   Association,
P.O. Box 171,
Lake Ozark, Missouri 65049

ALABAMA

Alabama Catfish Farmers,
P.O. Box 11000,
Montgomery, Alabama 36198

ARKANSAS

Catfish Farmers of Arkansas
Rt. 2, Box 122,
Lake Village, Arkansas 71653

CALIFORNIA

California Aquaculture
   Association,
P.O. Box 110,
Monterey, California 93942

FLORIDA

Florida Tropical Fish Farms
   Association,
P.O. Box 1519,
Winter Haven, Florida 33880

HAWAII

Hawaii Prawn Producers
   Association,
335 Merchant Street, Room 359,
Honolulu, Hawaii 96813

KANSAS

Kansas Commercial Fish
   Growers Assoc.,
Box 237,
Pretty Prairie, Kansas 67570

LOUISIANA

Louisiana Crawfish Farmers
   Association,
University Station, Knapp Hall,
Baton Rouge, Louisiana 70803

MAINE

Maine Aquaculture Association,
P.O. Box 535,
Damariscotta, Maine 04543

MINNESOTA

Central Minnesota Fish Farmers
   Association,
Box 39,
Waverly, Maine 55390

MISSISSIPPI
Catfish Farmers of Mississippi
P.O. Box 1609,
Jackson, Mississippi 39205

MISSOURI
Missouri Fish Farmers
    Association
Columbia, Missouri 65201

RHODE ISLAND
Rhode Island Aquaculture
    Association
P.O. Box H,
Kingston, Rhode Island 02881

TEXAS
Fish Farmers of Texas,
P.O. Box 2948,
College Station, Texas 77841

WISCONSIN
Wisconsin Trout Growers,
Box 115,
Lewis, Wisconsin 54851

Appendix IV–5   Schools Offering Aquaculture Training.

Getting a university education with emphasis on aquaculture is no easy matter in the United States or Canada. Although the importance, or I would say the urgency, of the subject suggests that we should be turning out aquaculture managers, administrators, educators, researchers, technicians, and aquatic farmers with strong academic backgrounds in and out of aquaculture, the fact is that we are not.

The largest and best known university aquaculture program, and certainly the one with the broadest scope, is that at Auburn University. A real danger in many programs is that the student may be indoctrinated in "aquabusiness" and high technology approaches. And most academic aquaculture programs are handicapped by lack of aquatic facilities where students can observe and participate in the growing of aquatic crops. The fact is that aquaculture "laboratory facilities" require too much space for most campuses. The list which follows does not imply any endorsement, but contains all schools known to me which offer programs, courses or outstanding faculty members with an interest in aquaculture. The best tactic in choosing a school is simply to write for the school's catalog. Also consider the geographical

location; best is one that is similar to the one in which the student expects to work.

For many students, particularly undergraduates, I do not necessarily recommend an academic emphasis on aquaculture at all. It is often better to acquire a good background in biological sciences, with emphasis on aquatic biology and ecology, and pick up practical experience by employment in aquaculture. Among other academic disciplines of particular value to the prospective aquaculturist are agriculture, chemistry, economics, nutrition, and physiology.

The New Alchemy Institute offers a very few one month to one year internships in aquaculture, with emphasis on management of semi-closed systems. In the future they may also offer short courses, and one-hour workshops in aquaculture are gven every Saturday through the "Farm Saturday" visitors' program. For further information write to Denise Backus, The New Alchemy Institute, 237 Hatchville Road, East Falmouth, Massachusetts 02536.

The U.S. Fish and Wildlife Service publishes a list of college level training programs in aquaculture. Contact the National Fisheries Center, Kearneysville, West Virginia 25430.

Institutes and universities offering courses or programs in aquaculture and mariculture:

ALABAMA
Auburn University
Auburn, Alabama 36830

ALASKA
University of Alaska
Anchorage, Alaska 99504

University of Alaska
College, Alaska 99701

ARIZONA
University of Arizona
Tucson, Arizona 85721

ARKANSAS
Arkansas State College
Conway, Arkansas 72032

Arkansas State University
State University, Arkansas 72467

University of Arkansas at Pine
Bluff
Pine Bluff, Arkansas 71601

University of Arkansas -
Monticello
Monticello, Arkansas 71655

CALIFORNIA
University of California–
Bodega Marine Laboratory
Bodega Bay, California 94923

University of California
Davis, California 95616

California Polytechnic State
University
San Luis Obispo, California
93401

California State University
Chico, California 95929

Humboldt State University
Arcata, California 95521

San Diego State University
San Diego, California 92182

COLORADO
Colorado State University
Fort Collins, Colorado 80523

DELAWARE
University of Delaware
Lewes, Delaware 19958

University of Delaware
Newark, Delaware 19711

FLORIDA
Florida Institute of Technology
Jensen Beach, Florida 33457

Florida State University
Tallahasse, Florida 32306

University of Miami
Miami, Florida 33149

GEORGIA
University of Georgia
Athens, Georgia 30602

University of Georgia
Tifton, Georgia 31793

HAWAII
College of Tropical Agriculture
and Human Resources
University of Hawaii
Honolulu, Hawaii 96813

University of Hawaii at Manoa
Kaneohe, Hawaii 96744

Honolulu Community College
Honolulu, Hawaii 96817

Windward Community College
Kaneohe, Hawaii 96744

ILLINOIS
Southern Illinois University
Carbondale, Illinois 62905

KANSAS
Kansas State University
Manhattan, Kansas 66506

KENTUCKY
Murray State University
Murray, Kentucky 42071

University of Louisville
Louisville, Kentucky 40292

LOUISIANA
University of Southwestern
Louisiana
Lafayette, Louisiana 70504

Louisiana State University
Baton Rouge, Louisiana 70803

Southern University
Baton Rouge, Louisiana 70813

MARYLAND
University of Maryland
College Park, Maryland 20742

MASSACHUSETTS
Woods Hole Oceanographic
Institution
Woods Hole, Massachusetts
02543

MICHIGAN
University of Michigan
Ann Arbor, Michigan 48109

Michigan State University
East Lansing, Michigan 48814

MISSISSIPPI
Gulf Coast Research Laboratory
Ocean Springs, Mississippi 39564

Mississippi State University
Mississippi State, Mississippi
39762

MISSOURI
University of Missouri
Columbia, Missouri 65211

NEW HAMPSHIRE
University of New Hampshire
Durham, New Hampshire 03824

NEW YORK
City University of New York
New York, New York 10031

Cornell University
Ithaca, New York 14853

Long Island University/C.W. Post
College
Greenvale, New York 11548

NORTH CAROLINA
University of North Carolina
Chapel Hill, North Carolina
27514

University of North Carolina
Wilmington, North Carolina
28401

East Carolina University
Greenville, North Carolina 27834

Haywood Technical College
Clyde, North Carolina 28721

North Carolina State University
Raleigh, North Carolina 27650

OHIO
Ohio State University
Columbus, Ohio 43210

OKLAHOMA
University of Oklahoma
Norman, Oklahoma 73019

OREGON
Oregon State University
Corvallis, Oregon 97331

PENNSYLVANIA
Mansfield State College
Mansfield, Pennyslvania 16933

Pennsylvania State University
University Park, Pennsylvania
   16802

RHODE ISLAND
University of Rhode Island
Kingston, Rhode Island 02881

SOUTH CAROLINA
South Carolina Sea Grant
   Consortium
Charleston, South Carolina 29401

Clemson University
Clemson, South Carolina 29631

TENNESSEE
University of Tennessee
Knoxville, Tennessee 37901

Memphis State University
Memphis, Tennessee 38152

TEXAS
Texas A&M University
College Station, Texas 77843

VERMONT
Goddard College
Plainfield, Vermont 05667

VIRGINIA
Old Dominion University
   Institute of Oceanography
Norfolk, Virginia 23508

Virginia Institute of Marine
   Science
Gloucester Point, Virginia 23062

Virginia Institute of Marine
   Science
Wachapreague, Virginia 23404

Virginia Tech
Blacksburg, Virginia 24061

WASHINGTON
University of Washington
Seattle, Washington 98195

Grays Harbor College
Aberdeen, Washington 98520

Peninsula College,
Port Angeles, Washington 98362

WISCONSIN
University of Wisconsin -
   Madison
Madison, Wisconsin 53706

DISTRICT OF COLUMBIA
American University
Washington, D.C. 20016

PUERTO RICO
University of Puerto Rico,
   Mayaguez
Mayaguez, Puerto Rico 00708

VIRGIN ISLANDS
College of the Virgin Islands
St. Thomas, U.S.V.I. 00801

CANADA
University of Guelph
Guelph, Ontario N1G 2W2

Appendix IV-6   Diagnostic Services Available

U.S.A.

ALABAMA
O.L. Green
Farm Fresh Catfish Co.
P.O. Box 188
Greensboro 36744

Southeastern Cooperative
   Fish Disease Laboratory
Auburn University
Department of Fisheries and
   Allied Aquacultures
Auburn 36830
(205) 826-4786

ALASKA
Rober S. Grischkowsky, Ph.D.
Fish Pathologist III
333 Raspberry Road
Anchorage 99502
(907) 344-0541

ARIZONA
Ronald D. Major
(Limited diagnostic services
   available)
Hatchery Biologist-Area
P.O. Box 39
Pinetop 85935
(602) 336-1933

ARKANSAS
John K. Beadles, Ph.D.
Professor of Biology
Chairman, Division of Biological
   Sciences,
Arkansas State University
Jonesboro
State University 72467
(501) 972-3082

David A. Becker, PH.D
(Parasitology only)
Dept. of Zoology, SE632
University of Arkansas

Fayetteville 72701
(501) 575-3251

Jimmy E. Camper, Hatchery
   Biologist
(Limited diagnostic services
   available)
Greers Ferry National Fish
   Hatchery
Rt. 3, Box 71
Heber Springs 72543
(501) 362-6038

Rex Flagg
Fish Health Services
P.O. Box 674
Pocahontas, 72455
(501) 892-8357

Dr. Roy Grizzell
Rt. 1, Box 496
Monticello 71655
(501) 367-8163

Scott Henderson, Fishery
  Biologist
Arkansas Game and Fish
  Commission
P.O. Box 178
Lonoke 72086
(501) 676-7963

Andrew J. Mitchell
U.S. Fish & Wildlife Service
Fish Farming Experimental
  Station
P.O. Box 860
Stuttgart 72160
(501) 673-8761

James H. Stevenson, Ph.D., Dean
College of Science
Arkansas State University
Jonesboro
State University 72467
(501) 972-3079

CALIFORNIA

George C. Blasiola, Jr.
Aquatic Biologist/Parasitologist
California Academy of Sciences
Steinhart Aquarium, Golden Gate
  Park
San Francisco 94118
(415) 221-5100, ext. 242 or 245

Renee Rosemark
University of California
Bodega Marine Laboratory
Bodega Bay 94923
(707) 875-3662

Robert LeBleu, President
Monterey Bay Research Institute
Box 343
Soquel 95073
(408) 476-9497

Robert Toth
Fish Disease Laboratory
California Deprtment of Fish &
  Game
407 West Line Street
Bishop 93514
(714) 872-2791

Mel Wills
Fish Disease Unit
California Department of Fish &
  Game
Mojave River Hatchery
P.O. Box 938
Victorville 92392
(714) 245-9981

Harold Wolf
California Department of Fish &
  Game
Fish Disease Laboratory
2111 Nimbus Road
Ranch Cordova 95670
(916) 355-0809

COLORADO

Dennis Anderson
Paul Janeke, Rex Flagg
(Limited diagnostic services
  available)
Fish Disease Control Center
U.S. Fish & Wildlife Service
P.O. Box 917
Fort Morgan 80701
(303) 867-9474

Lawrence J. Griess
Fisheries Research Center
P.O. Box 2287
Fort Collins 80522
(303) 484-2836

George Post, Ph.D.
Colorado State University
Diagnostic Laboratory
Fort Collins 80523
(303) 491-1101

Wildlife Vaccines, Inc.
11475 West 48th Ave.
Wheat Ridge 80033
(303) 422-8323

DELAWARE

James E. Harvey, Fisheries
  Biologist
850 E. Schuylkill Road, EC2
Pottstown, PA 19464
(215) 327-2631

FLORIDA

Fisher Doctor Clinical Center,
  Inc.
(Exclusive of viral work)
P.O. Box 765
Brandon 33511
(821) 626-1805

Lanny R. Udey, Ph.D.
(Exclusive of viral work)
Department of Microbiology
  (R138)
University of Miami School of
  Medicine
Fish and Shellfish Pathology Lab
P.O. Box 016960

Miami 33101
(305) 547-6563

Ralph Van Blarcom, F.D.D.
(Exclusive of viral work)
Path-O-Tech Fish Laboratory
3924 Quixote Blvd.
Tampa 33612
(813) 971-0487

GEORGIA

Jack L. Blue, D.V.M.
North Georgia Diagnostic
  Assistance Lab
College of Veterinary Medicine
University of Georgia
Athens 30602

HAWAII

Dr. James A. Brock, D.V.M.
Aquaculture Development
  Program
Department of Land and Natural
  Resources
335 Merchant Street, Room 359
Honolulu 96813
(808) 845-9561

IDAHO

Robert A. Busch, Ph.D.
Rangen Research Hatchery
Fish Pathology Laboratory
Route 1, Hagerman 83332
(208) 837-6191 or 837-4860

G.W. Klontz, D.V.M.
Fisheries Resources, College of
  Forestry
Wildlife & Range Sciences
University of Idaho
Moscow 83843
(208) 885-6336

Otto Lynn and Mrs. Brenda Ellis
Fish Disease Diagnostics and
  Certification
Valley Trout Diagnostics
Route 3, Box 50
Buhl 83316
(208) 543-5112

Joe C. Lientz and Rick Nelson
(Limited diagnostic service
  available)
Dworshak National Fish
  Hatchery
Box 251
Ahsahka 83520
(208) 476-4591

Harold Ramsey
Idaho Dept. of Fish & Game
Fish Disease Laboratory
Hagerman State Hatchery
Hagerman 83332
(208) 837-6672

ILLINOIS
Rodney W. Horner
Fish Pthologist
RR 3, Box 398
Clearview Est.
Manito 61546
(309) 968-6114

William M. Lewis Ph.D.,
  Director, and
Roy C. Heidinger, Ph.D.,
  Asst. Director
Fisheries Research Laboratory
Southern Illinois University
Carbondale 62901
(618) 536-7761

IOWA
Dr. Ronald P. Morgan
(Referred histopathology only)
National Veterinary Services Lab
USDA, APHIS
P.O. Box 844
Ames 50010
(515) 232-0250

KANSAS
Stanton Hudson
H-D Fish Farm
Rt. 1, Box 71A
Cheney 67025
(316) 542-3686

LOUISIANA
Janice S. Hughes, Fisheries
  Biologist
Louisiana Wildlife & Fisheries
  Commission
P.O. Box 4004
Monroe 71203
(318) 343-4044

Bobby T. Walker
Fisheries Biologist Dist. 11
Louisiana Wildlife & Fisheries
  Commission
P.O. Box 4004
Monroe 71203
(318) 343-4044
(Home: Rt. 5, Box 543, Monroe
  71203)

MAINE
Pete Walker, Hatchery Biologist
Department Inland Fisheries &
  Wildlife
8 Federal Street
Augusta, Maine 04333
(207) 289-2535

Roger Dexter, Hatchery Biologist
(Limited diagnostic service
  available)
Craig Brook National Fish
  Hatchery
East Orland 04431
(207) 469-2803

David Locke, Hatchery Division
Department Inland Fisheries &
  Wildlife
284 State Street
Augusta 04333
(207) 289-3651

Stuart W. Sherburne
(Marine Fish Diseases)
Department of Marine Resources
Fisheries Research Station
West Boothbay Harbor 04575

Dr. Bruce L. Nicholson
(Exclusive of virology)
Department of Microbiology
University of Maine
Orono 04469
(207) 581-7628

MICHIGAN
Bio-aquatics International, Inc.
P.O. Box 442
Rochester 48063
(313) 652-6622

John G. Hnath
Fish Pathologist
Wolf Lake State Fish Hatchery
RR No. 1
Mattawan 49071
(616) 668-2132

Burton J. Patrick
Midwest Fish Farming
  Enterprises, Inc.
5181 Gladwin Road
Harrison 48625

MINNESOTA
Phil Economon, Senior
  Bacteriologist
Fish & Wildlife Pathology
Department of Natural

Resources
390 Centennial Office Building
St. Paul 55155
(612) 296-3043

MISSISSIPPI
Richard E. Coleman
Area Extension Wildlife &
  Fisheries Specialist
Stoneville 38776
(601) 686-9311, Ext. 269

John Giudice, Fishery Biologist
U.S. Fish & Wildlife Service
P.O. Box 4389
2531 N. West Street
Jackson 39216
(601) 354-6089

Thomas L. Wellborn, Jr., Ph.D.,
Extension Wildlife & Fisheries
P.O. Box 5405
Mississippi State 39762
(601) 325-3174

MISSOURI
Aquascience Research Group
(Primarily aquarium fish
  diseases exclusive of viral
  disease)
512 East 12th Avenue
North Kansas City 64116
(913) 621-1000

Gary W. Camenisch
Missouri Department of
  Conservation
666 Primrose Lane
Springfield 65807
(417) 883-6677

Don Livingston
(Parasites & Bacteria only)
U.S. Fish & Wildlife Service
Fish Pesticide Research
  Laboratory
Rt. 1, Columbia 65201
(314) 442-2271
(Home: 1409 Marylee Drive,
  Columbia 65201
(314) 443-5520)

Charlie Suppes
Missouri Department of
  Conservation
Blind Pony Hatchery
Route 2, Sweet Springs 65351
(816) 335-4531

MONTANA

Charlie E. Smith
(Limited diagnostic service
available)
U.S. Fish & Wildlife Service
Fish Cultural Development
Center
Rt. 2, Box 333
Bozeman 59715
(406) 587-9265

NEVADA

Robert E. Taylor, D.V.M.
(Whirling disease only)
University of Nevada-Reno
Max C. Fleischmann College of
Agriculture
Division of Veterinary Medicine
5305 Mill Street
Reno 89502
(702) 784-6135

NEW HAMPSHIRE

Jay Hendee
(Limited time available, parasitology Bacteriology and
nutritional diseases only)
State of New Hampshire
Fish & Game Deprtment
34 Bridge Street
Concord 03301
(603) 271-2503

NEW MEXICO

Theophilos Inslee
(Limited diagnostic service
available)
Dexter National Fish Hatchery
Dexter 88230

Jim R. Clary, Ph.D., Pathologist
Rainbow Acres Trout Farm
P.O. Box 29
McIntosh 87032
(505) 384-2944

NEW YORK

Fish Diagnostic Laboratory
Department of Avian & Aquatic
Animal Medicine
N.Y. State College of Veterinary
Medicine
Cornell University
Ithaca 14853
(607) 256-5440

George D. Ruggieri, Ph.D.,
New York Aquarium and Osborn

Laboratories of Marine Sciences
Boardwalk & West 8th Street
Brooklyn 1124
(212) 266-8500

John Schachte, Jr., Ph.D.
Associate Fish Pathologist
New York State Dept. of
Environmental Conservation
Rome Fish Pathology Laboratory
8314 Fish Hatchery Road
Rome 13440
(315) 337-0910

NORTH CAROLINA

Charles P. Carlson
(Limited diagnostic service
available)
Fish Hatchery Biologist
P.O. Box 158
Pisgah Forest 28768
(704) 877-3122

Edward J. Noga, D.V.M.
(Limited diagnostic service
available)
Aquatic Medicine, Dept. of CASS
North Carolina State University
4700 Hillsborough Street
Raleigh 27606
(919) 829-4230

OHIO

Francis H. Bezdek, Biologist
6354 Low Road
Lisbon 44432
(216) 277-3242

Clay Edwards
Ohio Cooperative Fish Unit
1735 Neil Avenue
Columbus 43210

Roland L. Seymour, Ph.D.
(External fungi only)
Department of Botany
The Ohio State University
1735 Neil Avenue
Columbus 43210
(614) 422-0564

OKLAHOMA

Mr. Lonnie Cook
Medicine Park State Fish
Hatchery
Star Route B, Box 66-E
Lawton, 73501
(405) 529-2795

Mr. Bobby Dorr
Durant State Fish Hatchery
Route 1, Box 188
Caddo 74729
(405) 924-4085

Mr. Jack Harper
Southeast Region
Route 1, Box 188
Caddo 74729
(405) 924-4087

Mr. Thomas Kingsley
Byron State Fish Hatchery
Route 1, Box 67
Byron 73723
(405) 474-2663

Mr. Leon Moore
Holdenville State Fish Hatchery
Route 3, Box 45
Holdenville 74848
(405) 379-5408

OREGON

J.D. Conrad, Chief Pathologist
(Bacteriology and Parasitology
only)
Oregon Department of Fish &
Wildlife
17330 S. E. Evelyn Street
Clackamas 97015
(503) 657-2014

J.L. Fryer, Ph.D., Professor
Department of Microbiology
Nash Hall, Oregon State
University
Corvallis 97331
(503) 754-4441

J.E. Sanders, R.A. Holt, Associate
Fish Pathologists and W.J.
Groberg, Asst. Fish Pathologist
Oregon Department of Fish &
Wildlife
Department of Microbiology
Nash Hall, Oregon State
University
Corvallis 97331
(503) 754-4441

Jim Winton, Robert Olson,
Cathy Lannan
(Costs available upon request)
Oregon State University
Marine Science Center
Newport 973765
(503) 867-3011

RHODE ISLAND
R.E. Wolke, D.V.M., Ph.D.
Marine Pathology Laboratory
University of Rhode Island
Kingston 02881
(401) 364-7057

SOUTH DAKOTA
Richard C. Ford
Fisheries Management Specialist
3305 West South Street
Rapid City 57701
(605) 394-2391

TENNESSEE
W.A. Dillon
(Limited diagnostic services
    available)
Department of Biology
University of Tennessee
Martin 38238

TEXAS
Ray A. Bendele, Ph.D.
(Histopathology, Bacteriology,
    Toxicology, and Transmission
    Electron Microscopy)
Veterinary Pathologist
Texas Veterinary Medical
    Diagnostic Laboratory
Drawer 3040
College Station 77840
(713) 845-3414

David G. Huffman, Ph.D.
(Parasites only)
Aquatic Station, SWTSU
San Marcos 78666
(512) 245-2284

Ken Johnson, Ph.D.
Extension Fish Disease Specialist
Department of Wildlife &
    Fisheries
Room 202, Nagle Hall
Texas A&M University
College Station 77843
(713) 845-7471

Don H. Lewis, Ph.D.
(Bacteriology)
Department of Veterinary
    Microbiology
College of Veterinary Medicine
Texas A&M University
College Station 77843
(713) 845-5941
Home: (713) 693-6517

Stewart McConnell, Ph.D.
(Virology only)
Virologist, Department of
    Microbiology
School of Veterinary Medicine
Texas A&M University
College Station 77840
(713) 845-5941

Thomas G. Meade, Ph.D.
(Parasitology only)
Department of Life Sciences
Sam Houston State University
Huntsville 77341
(713) 295-6211, ext. 1551

UTAH
Ron Golde
Fish Experiment Station
Rt. 1, Box 254
Logan 84321

Richard Heckmann, Ph.D.
153 WIDB, Zoology Dept.
Brigham Young University
Provo 84602
(801) 374-1211, ext. 2495

VERMONT
Rick Nelson
(Limited diagnostic service
    available)
White River National Fish
    Hatchery
Rt. 2, Box 107
Bethel 05032

VIRGINIA
Frank O. Perkins
(Diseases of cultured molluscs
    only)
Virginia Institute of Marine
    Science
P.O. Box 162
Gloucester Point 23062
(804) 642-4083

WASHINGTON
Lee W. Harrell, D.V.M.
(1. Bacterial disease and
    nutritional pathology of
    salmonids; 2. Vaccines and
    immunology of salmonids)
NMFS Aquaculture Experiment
    Station
Manchester 98353
(206) 842-5434

Marsha Landolt, Ph.D.
College of Fisheries WH-10
University of Washington
Seattle 98195
(206) 543-4290

Steve Leek, Eric Pelton
(Limited diagnstic service
    available)
Fish Hatchery Biology Lab
U.S. Fish & Wildlife Service
Box 17
Cook 98605
(509) 538-2755

John J. Majnarich, Ph.D.
    Steven G. Newman, Ph.D., &
    Robert Tedrow
(Exclusive of viral work)
BioMed Research Laboratory
1115 East Pike Street
Seattle 98122
(206) 324-0380

Gilbert B. Pauley, Ph.D.
(Fish and Shellfish)
Associate Professor, College of
    Fisheries
University of Washington
Seattle 98195
(206) 543-6475

Steven Roberts
Washington Dept. of Game
1421 Anne Ave.
E. Wenatchee 98801
(509) 884-0970

Tavolek, Inc.
Donald F. Amend, Ph.D.; Keith
    Johnson Ph.D.
(Assistance with major problems.
    Routine diagnostics not
    accepted. Contact before
    submitting samples.)
2779-152 Avenue, NE
Redmond 98052
(206) 883-2150

Warner Taylor, Ray Brunson
(Limited diagnostic service
    available)
Olympia National Fish Hatchery
2625 Parkmont Lane, Bldg. A.
Olympia 98502

WEST VIRGINIA
National Fish Health Laboratory
(Exclusive of routine service—

will handle only referrals from
field diagnosticians for special
diagnostic confirmations and
undefined or unusual diseases
requiring research input.)
Rt. 3, Box 40E (Leetown)
Kearneysville 25430
(304) 725-8461

WISCONSIN
J.W. Warren, H.M. Jackson
(Limited diagnostic service
    available)
Fish Disease Control Center
U.S. Fish & Wildlife Service
P.O. Box 1595
La Crosse 54601
(608) 783-6451

John Held, Ph.D.
(Warmwater only)
Biology Department, University
    of Wisconsin at LaCrosse
LaCrosse 54601
(608) 784-6050

Fred P. Meyer, Ph.D., and
    Roland B. Henry
(Limited avilability)
U.S. Fish & Wildlife Service
Fish Control Laboratory
P.O. Box 862
LaCrosse 54601
(608) 784-9666

WYOMING
Douglas L. Mitchum
Wyoming Game & Fish
    Laboratory
University of Wyoming
P.O. Box 3312
Laramie 82071
(307) 745-5865

CANADA

BRITISH COLUMBIA
Hilda Lei Ching, Ph.D.
Envirocon Ltd.
c/o B.C. Research
3650 Wesbrook Mall
Vancouver V6S 2L2
(604) 228-1614

Mr. Gary Hoskins, Dr. G.R. Bell,
    and Dr. J.P.T. Evelyn
Fish Health Program
    (Pacific Region)
Pacific Biological Station
Department of Fisheries &
    Oceans
Nanaimo V9R 5K6
(604) 758-5200

NOVA SCOTIA
John Cornick
Fish Health Unit Maritimes
Dept. of Fisheries and Oceans
Resource Branch

P.O. Box 550
Halifax B3J 2S7
(902) 426-8381

ONTARIO
R.A. Sonstegard, Ph.D.
Dept. of Biology-Pathology
McMaster University
Hamilton
(416) 525-9140, ext. 4378

Fish Pathology Laboratory
Department of Pathology
Ontario Veterinary College
University of Guelph
Guelph, N1G 2W1
(519) 824-4120, ext. 2681

QUEBEC
Dr. Jean-Louis Frechette
Department of Pathology &
    Microbiology
Faculty of Veterinary Medicine
University of Montreal
P.O. Box 5000
St. Hyacinthe J2S 7C6
(514) 773-8521, ext. 304

Appendix IV–7   Soil Conservation Service Biologists

NATIONAL BIOLOGIST
Carl H. Thomas
Soil Conservation Service
P.O. Box 2890
Washington, D.C. 20013

**NATIONAL TECHNICAL CENTER
BIOLOGISTS**

NORTHEAST
Robert W. Franzen
160 E 7th Street
Chester, Pennsylvania 19013
215/499-3919

MIDWEST
Rex Hamilton
100 Centennial Mall North
Federal Building
U.S. Courthouse, Room 393
Lincoln, Nebraska 68508

SOUTH
E. Ray Smith, Jr.
P.O. Box 6567
Fort Worth, Texas 76115

WEST
L. Dean Marriage
511 NW Broadway

Federal Building
Portland, Oregon 97209

FIELD BIOLOGISTS

ALABAMA
Hugh D. Kelly
P.O. Box 311
Auburn, Alabama 36830
205/821-8070

Robert E. Waters
P.O. Box 311
Auburn, Alabama 36830
205/821-8070

ALASKA
Devony Lehner-Welch
2221 E Northern Lights Blvd.
Suite 129, Professional Bldg.
Anchorage, Alaska 99504
907/276-4246

ARIZONA
David W. Seery
2717 N Fourth
Suite 140
Flagstaff, Arizona 86001

Donald W. Welch
3241 Romero Road
Tucson, Arizona 85705

John C. York
230 North First Avenue
3008 Federal Bldg.
Phoenix, Arizona 85025
602/261-6711

ARKANSAS
Paul M. Brady
Federal Office Bldg.
Room 2405
700 West Capitol Avenue
Little Rock, Arkansas 72201

Donald R. Linder
Federal Office Bldg.
Room 2405
700 West Capitol Avenue
Little Rock, Arkansas 72201

Robert Price
Federal Office Bldg.
Room 2405
700 West Capitol Avenue
Little Rock, Arkansas 72201

CALIFORNIA
Randall L. Gray
2828 Chiles Road
Davis, California 95616
916/449-2874

Larry H. Norris
621 J Street, Suite 2
Sacramento, California 95814

David W. Patterson
1350 N Main Street, Suite 1
Red Bluff, California 96080
916/527-2667 or 2668

Ronald F. Schultze
2828 Chiles Road
Davis, California 95616
916/449-2853

Glenn I. Wilcox
344 Salinas Street, Suite 203
Salinas, California 93901

CARIBBEAN AREA (See FLORIDA)

COLORADO
Eddie W. Mustard
Diamond Hill, Building A
Third Floor
2490 West 26th Avenue,
Room 309
Denver, Colorado 80211
303/837-5651

Edward L. Nielson, Jr.
Crossroads Plaza
Suite 204
2784 Crossroads Blvd.
Grand Junction, Colorado 80502

CONNECTICUT
Vacant (contact SRC)
Mansfield Professional Park
Storrs, Connecticut 06268
203/429-9361

DELAWARE
Bruce A. Kirschner
(NJ, DE, and MD Tri-state Water
Resources Planning Staff)
Treadway Towers, Suite 210
9 East Loockerman Street
Dover, Delaware 19901

FLORIDA and CARIBBEAN AREA
John F. Vance, Jr.
401 SE 1st Avenue
Gainesville, Florida 32601
904/377-0958

GEORGIA
Louis A. Justice
Federal Building
355 East Hancock Avenue
Athens, Georgia 30613

Jesse Mercer, Jr.
Federal Building
355 East Hancock Avenue
Athens, Georgia 30613

IDAHO
Mike W. Anderson
304 North 8th Street
Room 345
Boise, Idaho 83702
208/334-1611

ILLINOIS
Stephen J. Brady
Springer Federal Building
301 North Randolph Street
Champaign, Illinois 61820
217/398-5277

INDIANA
William F. Beard
Corporate Square-West,
Suite 2000
5610 Crawfordsville Road
Indianapolis, Indiana 46224

Jame D. McCall
Corporate Square-West,
Suite 2000
5610 Crawfordsville Road
Indianapolis, Indiana 46224
317/269-6462

IOWA
Dave W. Beck
210 Walnut Street
Room 693, Federal Building
Des Moines, Iowa 50309

Bill D. Welker
210 Walnut Street
Room 693, Federal Building
Des Moines, Iowa 50309
515/284-4260

KANSAS
Richard L. Hager
760 South Broadway
Salina, Kansas 67401
913/823-4548 or 4549

Robert J. Higgins
760 South Broadway
Salina, Kansas 67401
913/823-4548 or 4549

KENTUCKY
William H. Casey
333 Waller Avenue
Lexington, Kentucky 40504
502/781-3090

LOUISIANA
Billy R. Craft
3737 Government Street
Alexandria, Louisiana 71301
318/473-7803

Martin D. Floyd
555 Goodhope Drive
Norco, Louisiana 70079
504/764-2228

Richard W. Simmering
3737 Government Street
Alexandria, Louisiana 71301
318/473-7180

MAINE
Robert J. Wengrzynek, Jr.
USDA Office Building
University of Maine
Orono, Maine 04473
207/866-2132

MARYLAND
Anne M. Lynn
4321 Hartwick Road
College Park, Maryland 20740
301/344-4135

MASSACHUSETTS
Vacant (contact SRC)
451 West Street
Amherst, Massachusetts 01002
413/256-0441

MICHIGAN
Charles M. Smith
Room 101
1405 South Harrison Road
East Lansing, Michigan 48823
517/337-6850

MINNESOTA
Frank J. Fink
200 Federal Bldg.
 & U.S. Courthouse
316 North Robert Street
St. Paul, Minnesota 55101

George L. Pollard
200 Federal Bldg.
 & U.S. Courthouse
316 North Robert Street
St. Paul, Minnesota 55101
612/725-7670

MISSISSIPPI
Ramon L. Callahan
Suite 1321, Federal Building
100 W Capitol Street
Jackson, Mississippi 39269
601/960-5199

Michael J. Hinton
306 Federal Building
200 East Washington Street
P.O. Box 1160
Greenwood, Mississippi 38930
601/453-2762

Charles E. Hollis
Suite 1321, Federal Building
100 W Capitol Street
Jackson, Mississippi 39269

Harvey G. Huffstatler
P.O. Box 1789
Tupelo, Mississippi 38802
601/844-2341

David R. Thomas
W.M. Colmer Federal Bldg.
Room 323
Main & New Orleans Streets
Hattieburg, Mississippi 39401
601/544-4511 or 4512

MISSOURI
Edward A. Gaskins
555 Vandiver Drive
Columbia, Missouri 65201
314/442-2271

John P. Graham
555 Vandiver Drive
Columbia, Missouri 65201

MONTANA
Ronald F. Batchelor
Federal Building
Room 443
10 East Babcock Street
Bozeman, Montana 59715
406/587-5271

NEBRASKA
Robert O. Koerner
Room 345, Federal Bldg. & U.S.
 Courthouse
Lincoln, Nebraska 68508
402/471-5303

NEW HAMPSHIRE
Allan P. Ammann
(New England Water Resources
 Planning Staff)
Federal Building
Durham, New Hampshire 03824

Judy L. Tumosa
Federal Building
Durham, New Hampshire 03824
603/868-7581

NEW JERSEY
David L. Smart
1370 Hamilton Street
Somerset, New Jersey 08873
201/246-1205

NEW MEXICO
William J. Slone
200 East Griggs
P.O. Box D-202
Las Cruces, New Mexico 88001
505/523-8159

Edwin A. Swenson, Jr.
517 Gole Avenue, SW
Room 3301
Albuquerque, New Mexico 87102
505/766-3277

NEW YORK
Robert E. Myers
James M. Hanley Federal
 Building
100 South Clinton Street
Syracuse, New York 13260
315/423-5519

NORTH CAROLINA
John P. Edwards
Federal Building, Fifth Floor
310 New Bern Avenue
Raleigh, North Carolina 27601
919/755-4375

NORTH DAKOTA
Gary A. Kilian
Box 1458
Bismarck, North Dakota 58502

Erling B. Podoll
Box 1458
Bismarck, North Dakota 58502
701/255-4011, ext 425

OHIO
Gene P. Barickman
Van Wert Complex
147 E Main Street
Van Wert, Ohio 45891

Sally L. Griffith
Federal Building, Room 522
200 North High Street
Columbus, Ohio 53215
614/469-6980

Jerome S. Myszka
Federal Building, Room 522
200 North High Street
Columbus, Ohio 53215

Clifford T. Turik
Rush Creek Project Office
160 Carter Street
Bremen, Ohio 43107

OKLAHOMA
Steve Elsener
2700 South 4th Street
Chickasha, Oklahoma 73018
405/224-2421 or 2442

Billy M. Teels
USDA Building
Stillwater, Oklahoma 74074
405/624-4437

Stephen R. Tully
1820 North Sioux Street
Claremore, Oklahoma 74017
918/341-7373 or 7374

OREGON
Clyde A. Scott
Federal Building
1220 SW Third Avenue
Portland, Oregon 97204
503/221-2991

PENNSYLVANIA
Vacant
Federal Bldg. & U.S. Courthouse
P.O. Box 985
Federal Square Station
Harrisburg, Pennsylvania 17108
717/782-2268

SOUTH CAROLINA
James E. Lewis, Jr.
Room 950, Strom Thurmond
Federal Building
1835 Assembly St.
Columbia, South Carolina 29201
803/765-5681

William J. Melven
Room 950, Strom Thurmond
Federal Building
1835 Assembly St.
Columbia, South Carolina 29201
803/765-5681

SOUTH DAKOTA
Connie M. Vicuna
Federal Building
200 Fourth Street, SW
Huron, South Dakota 57350

TENNESSEE
Gerald L. Montgomery
675 Estes Kefauver Federal Building,
U.S. Courthouse
801 Broadway
Nashville, Tennessee 37207

Stephen A. Sewell
675 Estes Kefauver Federal Building,

U.S. Courthouse
801 Broadway
Nashville, Tennessee 37207

TEXAS
Gary Bates
101 S Main Street
Temple, Texas 76501–7682

James Henson
101 S Main Street
Temple, Texas 76501–7682

Willard Richter
33 E Twohig, Room 108
San Angelo, Texas 76903
915/658-6269

Edward M. Schwille
Terrell Area Office
808B Westmore Avenue
Terrell, Texas 75160
214/563-6431 or 6432

Frank Sprague
101 S Main Street
Temple, Texas 76501–7682

Jerry M. Turrentine
Uvalde Area Office
1022 Garner Field Road
Uvalde, Texas 78801
512/287-7444

Gary L. Valentine
101 S Main Street
Temple, Texas 76501–7682

Michael E. Zeman
855 Interstate 10, South
Room 135
Beaumont, Texas 77701
713/839-2675

UTAH
Donald J. Goins
Nile Chapman Building
West Highway 40
P.O. Box 295
Roosevelt, Utah 84066
801/722-4621 or 4622

Paul A. Obert
Nile Chapman Building
West Highway 40
P.O. Box 295
Roosevelt, Utah 84066
801/722-4621 or 4622

Robert Sennett
4012 Federal Building
125 South State Street
Salt Lake City, Utah 84138
801/524-5054

William W. Wood, III
Nile Chapman Building
West Highway 40
P.O. Box 295
Roosevelt, Utah 84066
801/722-4621 or 4622

VERMONT
Vacant (contact SRC)
One Burlington Square
Suite 205
Burlington, Vermont 05401
802/951-6795

VIRGINIA
Gregory H. Moser
(Watershed Planning Staff)
Federal Building
400 North 8th Street
Richmond, Virginia 23240

Lawrence H. Robinson
Federal Building
400 North 8th Street
Richmond, Virginia 23240
804/771-2461

WASHINGTON
Ivan L. Lines, Jr.
360 U.S. Courthouse
West 920 Riverside Avenue
Spokane, Washington 92201
509/456-3772

WEST VIRGINIA
Gary A. Gwinn
75 High Street
Morgantown, West Virginia 26505
304/599-7151

Lynn M. Shutts
75 High Street
Morgantown, West Virginia 26505

WISCONSIN
Thomas P. Thrall
4601 Hammersley Road
Madison, Wisconsin 53711
608/264-5582

WYOMING
Richard C. Rintamaki
Federal Building
100 East B Street, Room 3124
Casper, Wyoming 82602
307/265-5550

INDEX